Engineering Hydrology

About the Authors

Sharad K. Jain, Ph.D., is currently the Director of the National Institute of Hydrology, Roorkee, India. He has published more than 250 technical papers and more than 30 book chapters, and has served as author or editor of 9 books. He is a member of editorial boards of several journals and of many committees related to water management. Dr. Jain has made significant contributions to the water sector in India through R&D, providing solutions to real-life problems, and capacity building.

Vijay P. Singh, Ph.D., D.Sc., is Distinguished Professor, Regents Professor, and Caroline & William N. Lehrer Distinguished Chair in Water Engineering in the Department of Biological and Agricultural Engineering at Texas A&M University. He has published nearly 1100 journal articles and 115 book chapters, and has served as author or co-author of 28 books and editor of over 70 books, including *Handbook of Applied Hydrology*, Second Edition, and *Encyclopedia of Snow, Ice and Glaciers*.

Engineering Hydrology

An Introduction to Processes, Analysis, and Modeling

Sharad K. Jain, Ph.D.
Vijay P. Singh, Ph.D., D.Sc.

McGraw Hill Education

New York Chicago San Francisco
Athens London Madrid
Mexico City Milan New Delhi
Singapore Sydney Toronto

Library of Congress Control Number: 2018966284

Engineering Hydrology: An Introduction to Processes, Analysis, and Modeling

1 2 3 4 5 6 7 8 9 QVS 24 23 22 21 20 19

ISBN 978-1-259-64197-8
MHID 1-259-64197-X

Sponsoring Editor
Lauren Poplawski

Editorial Supervisor
Stephen M. Smith

Production Supervisor
Lynn M. Messina

Acquisitions Coordinator
Elizabeth M. Houde

Project Manager
Touseen Qadri, MPS Limited

Copy Editor
Susan McClung

Proofreader
Kirti Dogra, MPS Limited

Art Director, Cover
Jeff Weeks

Composition
MPS Limited

To the loving memory of my father;
and to my mother, wife Shikha, son Sanchit,
daughter Surabhi, family, and well-wishers.

SKJ

To wife Anita, son Vinay, daughter Arti,
daughter-in-law Sonali, son-in-law Vamsi,
and grandchildren Ronin, Kayden, and Davin.

VPS

Contents

Part 4 Analysis and Modeling

Preface

O Water, because of your presence, the Atmosphere is so refreshing,
and imparts us with vigour and strength.
We revere you who gladdens us by your pure essence.
The Rig Veda

Hydrology is a science whose roots go back many thousands of years. Ancient texts from as far back as about 2500 BC and remnants of structures built around that time and thereafter are testimony to this fact. However, the growth in hydrological sciences was low-key during the Middle Ages. Fortunately, it picked up from the sixteenth century onward. The last two centuries have seen numerous exciting developments, driven by the needs of society, technological developments in the allied fields, and the inquisitiveness of hydrologists. We believe that these three factors will continue to provide the thrust to drive the growth in hydrological sciences in the future.

An increasing world population and rising demand have put water at the center stage of decision-making these days. Knowledge of hydrology is fundamental to finding technically sound, practical, and workable solutions to water-related problems. This engineering hydrology textbook offers a thorough overview of all important aspects of hydrology, combining detailed coverage of scientific principles with real-world applications and technologies. The book follows a logical course that builds on fundamental concepts and developments with modern hydrologic techniques.

The subject matter of the book is covered in five parts. The first part begins with an introductory chapter on hydrology, which is followed by chapters covering two fundamental concepts in hydrology: the hydrologic cycle and water balance.

Part 2 describes the major components of the hydrologic cycle, which are covered in six chapters on precipitation, evapotranspiration, infiltration, surface water, groundwater, and water quality. Water quality, of course, is not a component of the hydrologic cycle, but this topic has been included here since the quality of water determines its possible uses. In view of the numerous instances of degradation of the quality of water, its importance is likely to further rise in the future.

Hydrologists deal with many types of data, both spatial and temporal, that are collected through various types of sensors and devices. Thus measurements and analysis of data have a central role in hydrology. The third part of this book covers measurements. It begins with a chapter on hydrologic data observation networks, followed by a chapter on streamflow measurement. Remote sensing by now has established itself as a powerful technique for observation of spatial data that are analyzed and stored by the use of GIS. These are covered in the third chapter of Part 3. Measurement of several hydrometeorologic variables is covered in relevant chapters in Part 2.

After measurement, the data needs to be analyzed and modeled. This is the focus of the fourth part of the book, which covers rainfall-runoff modeling, the unit hydrograph, flow routing, and pollutant modeling. The concept of environmental flows is used to keep rivers in a healthy condition, and the methods required to compute environmental flows are covered in the last chapter in Part 4.

Hydrologic design is covered in the final part of this book, which has five chapters. This part discusses the tools that a hydrologist commonly uses. Two chapters cover statistical analysis and regression and correlation, which are commonly employed in hydrologic analysis and design. The next chapter describes hydrologic modeling in greater detail and is followed by a chapter on computation of design storm and design flood. These days hydrologists and all those who deal with water are concerned about the likely impacts of climate change on the water sector. The final chapter of the book covers this topic.

This book has been designed with the needs of students taking their first course in hydrology in mind. It also covers many topics that are part of the syllabus of graduate programs in universities across the globe. In each chapter we briefly describe recent advances that will be of interest to research scholars who want to pursue higher studies in hydrology. The book will also be of interest to engineering hydrology practitioners. Basic hydrologic processes are explained in the book along with current techniques for modeling and analyzing data. Example problems in various chapters help build important concepts.

We hope that this book will be useful to readers and will quench the thirst of students. At the same time, we hope that the book will amplify their curiosity and motivate them to further explore this exciting field.

Sharad K. Jain, Ph.D.
Vijay P. Singh, Ph.D., D.Sc.

Acknowledgments

A book is a compilation of information and knowledge about a topic that the authors have amassed over a long period of their professional careers, organized in a logical sequence. This book is no exception.

During the course of writing this book, we have consulted many books, technical papers, reports, and other material. Discussions with colleagues and students provided valuable insights. It is not possible to name all these people but we would like to acknowledge the inputs and advice from all of them. We are thankful to colleagues at the National Institute of Hydrology and Texas A&M University who helped in many different ways. Brijesh Kumar is thanked for help with editing.

Thanks are conveyed to McGraw-Hill Education, which enabled us to consolidate and share our ideas about hydrology, and particularly to Lauren Poplawski, for her keen interest in this book. We thank the team at MPS Limited for efficiently and promptly carrying out copy editing and formatting, and addressing many requests for changes.

Our families have very graciously coped with our absences during after-office hours and on weekends and holidays while we were occupied with the book.

PART 1

Preliminaries

CHAPTER 1
Introduction

Since the early days when hydrology was considered as an appendage of hydraulics, hydrology has come a long way; it has developed as a full-fledged geophysical science. Defining hydrology and reflecting on its scope, this chapter briefly traces the history of hydrology and reflects on what the future may hold. It also discusses the objective of this book and its organization.

1.1 What Is Hydrology?

Hydrology is the science of water. The word *hydrology* comes from the Greek words *hydro* (water) and *logos* (study). According to the glossary from the United Nations Educational, Scientific, and Cultural Organization (UNESCO; http://webworld.unesco.org/water/ihp), hydrology is the science that deals with (1) the waters above and below the land surfaces of the Earth, their occurrence, circulation, and distribution (in both time and space), their biological, chemical, and physical properties, and their reaction to their environment, including their relationship with living beings; and (2) the processes governing the depletion and replenishment of the water resources of the land areas of the Earth, and the various phases of the hydrological cycle. Hydrology is an earth science that is closely related to other natural sciences.

Hydrology deals with the space, time, and frequency characteristics of the occurrence, movement, distribution, and storage of water quantity and quality in all phases of water above and below the Earth's surface.

1.2 A Brief History of Hydrology

Many civilizations grew in river valleys around 3,000–5,000 years ago, including the Indus Valley civilization, the Egyptian civilization in the Nile Valley, the Chinese civilization that grew on the banks of the Huang-Ho River, and the Tigris and Euphrates civilizations in Mesopotamia. These civilizations developed near rivers because the water demands were readily met by the rivers. Clearly, humans have been observing water-related phenomena for a very long time. These observations, accompanied by humans' desire to use the river resources to benefit themselves, gradually evolved into the science of hydrology. Hydraulic engineers from these civilizations constructed canals, levees, dams, and wells to get water that is a necessity for life, as well as to avoid disaster (Hubbart, 2008).

Agriculture in Egypt was highly dependent on the regular inundation of the fields adjoining the Nile River. Therefore, the timing and extent of inundation were very important for the Egyptians, who had started measuring the water levels of the Nile River around 3000–3500 BC (Biswas, 1970; Hubbart, 2008). The device used to measure

Nile water levels was known as the *Nilometer*. Three types of Nilometers were in use; among these, the Roda Nilometer was the most notable. Biswas (1970) noted that the Egyptian King Menes dammed the Nile River near his capital Memphis and diverted water to a newly dug channel for societal use. During the middle kingdom (2160–1788 BC), artificial lakes (some of these were quite big) were used in Egypt to store and regulate the high flows of the Nile.

The idea of a global hydrologic cycle dates back at least 3,000 years. Early Greek philosophers, including Thales, Anaxagoras, Herodotus, Hippocrates, Plato, and Aristotle, discussed the processes involved (Hubbart, 2018). Some of their ideas about the hydrologic cycle were reasonable, but some, such as the mechanisms by which water from the sea came back to the rivers, were not correct. However, the Romans constructed aqueducts to transport water from the Tiber River, and those aqueducts are still in use today, which implies that their basic understanding of fluid mechanics, hydraulics, and hydrology was correct.

Hydrological concepts were highly developed in ancient India. The Vedic texts (more than 3,000 years old), various Puranas, Vrhat Samhita (550 AD), Meghmala (900), and other literary works contain extensive material on water. Knowledge of water is scattered in Sanskrit verses, which take the form of hymns and prayers addressed to various deities and stars, many of which were natural forces. For example, the Rig Veda verse I.32.8 explains that all the water that goes to the sky with wind by the heat of the Sun's rays gets converted to clouds, and then again after the penetration by the Sun's rays, it rains and get stored into rivers, ponds, oceans, and other bodies of water. Vrhat Samhita 54.1–2 says that the water veins beneath the Earth are like veins in the human body, some higher and some lower. The water falling from the sky assumes various colors and tastes due to differences in the nature of the Earth. More details about this can be found in the book *Hydrology in Ancient India* (NIH, 1990).

Plato studied the origins of rivers and springs. He believed that there was a huge reservoir beneath the Earth's surface and numerous interconnected spaces and passages, and the water entering the Earth filled these spaces. For his part, Aristotle wrote a treatise on meteorology entitled *Meteorologica,* which covered a range of hydrologic topics, including rain, dew, snow, and hail, as well as the hydrologic cycle. This book also dealt with astronomy, chemistry, geology, and physics.

The Indians and Greeks were the first to begin hydrometeorological observations. Kautilya (fourth century BC), a minister under King Chandragupta, began to take measurements of rain, and the data were used to compute taxes on land and for the planting of crops. The concept of a hydrologic cycle was also developed around that time. Systematic measurements of rainfall were started in Palestine. Mishnah, a Jewish religious book, has discussions indicating that rainfall measurements were carried out in Palestine from the second century BC to about the second century AD (Biswas, 1970).

Rain gauges were used as early as 1247 AD in China and 1441 AD in Korea. Snow gauges were also developed in China around this time. Devices to measure flow velocity were developed in Europe in the eighteenth century.

Leonardo da Vinci, the famous Italian scientist and philosopher (1452–1519), wrote a number of notes on the hydrologic cycle, and many of his statements are valid even today. He also developed a good understanding of the principles of open channel flow and used floats to measure streamflow velocity and discharge. Roman scientists are credited for many discoveries during the seventeenth century, including the continuity principle.

1.2.1 Hydrology in Ancient Times—The Case of India

The Indus Valley civilization, which flourished in the Indus River basin and other parts of North-West India about 5,000 years ago, had sophisticated urban water supply and sewage systems. Ruins of the civilization found at places such as Mohenjodaro and Harappa provide evidence that this civilization was well acquainted with hygiene and sanitation, including covered drains running beneath the streets of the cities. All this was not feasible without a sound understanding of the principles of fluid mechanics, hydraulics, and hydrology.

Like other sciences, the science of water was well developed in ancient India. A review of ancient Indian literature shows that it contains a gold mine of knowledge about hydrology. The hydrological concepts that have been discovered in the last three centuries were known and well documented in ancient Indian literature, published as early as 3000 BC. Important concepts of modern hydrology are scattered throughout various verses of Vedas, Puranas, Meghmala, Mayurchitraka, Vrhat Samhita, and other ancient Indian works. These were written in Sanskrit.

In the Vedic age, Indians developed the concept that water gets divided into minute particles due to the effect of sun rays and wind. The Vedic texts, which are more than 3,000 years old, contain valuable references on water; the details of the hydrologic cycle, the most important concept on which the modern science of hydrology is founded, are scattered in Rig Veda in verses that take the form of hymns and prayers addressed to deities such as Indra (firmament), Agni (fire), and Maruts (wind). Verses RV.I, 27.6 and 32.8 describe the formation of cloud by evaporation due to the Sun and wind:

> Vibhaktasi chitrabhano sindhoroorma upak aa
> Sagho dashushay chharasi.
>
> RV. I, 27.6

> Nadam na bhinnamuya shayanam mano ruhana atim yantyapah
> Yashchidwatro mahina paryatishthattasamhih patsutah shirbbhoova.
>
> RV.I, 32.8

These verses say that all the water that goes to the sky with wind by the heat of the Sun gets converted to clouds, and then again after the penetration by the Sun's rays, it rains and get stored into rivers, ponds, oceans, and other bodies of water.

Two verses (RV.V, 54.2 and 55.5) explain that the cloud-bearing winds are the cause of rainfall:

> Pra vo marootaststavisha udnyavo vayovridho ashwayujah parijayah
> San vighuta dadhati vashati tritah swarntyapoivana parijayah.
>
> RV.V, 54.2

> Udeeryatha marootah samudrato youam vrishtim varshyatha pureeshidam
> Na vo dastra up dasyanti dhanayvah shubam yatamanu ratha avratsat.
>
> RV.V, 55.5

This says: "O cloud-bearing winds, your troops are rich in water, they are strengtheners of life, and are your strong bonds; they shed water and augment food, and are harnessed with waves that wander far and spread everywhere. Combined with lightning, the triple-group (of wind, cloud, and lightning) roars aloud, and the waters fall upon the Earth." This explains that the cloud-bearing winds uplift water from the oceans and the clouds charged with water shower down the rain.

Similarly, hymns in Vayu Purana (51.14–16) state the following:

> Aadityapateetam suryaganeha somam sankramatay jalam
> Nadeebhirvayuyuktabhirlokadhanam pravartatay.
>
> Vayu, 51.14
>
> Yatsomatstravatay surya tadbhayshvavatishthatay
> Megha vayunighatain visrajant jalam bhuvi.
>
> Vayu 51.15
>
> Evamutikshapyatay chaiva patatay cham punarjalam
> Na nashmu udkasyasti tadev parivartatay.
>
> Vayu 51.16

That is, the water evaporated by the Sun ascends to the atmosphere through the capillarity of air, and it gets cooled and condensed there. After clouds form, it rains by the force of air. Thus, water is not lost in all these processes, but rather gets converted continuously from one form to other.

Devices to measure rainfall were developed during the time of Kautilya, which worked on the same principle as that of modern hydrology, except that weight measures were adopted instead of modern linear measurements of rainfall. A well-organized water-pricing system was prevalent during the time of Kautilya. Numerous references exist in Vedic literature, Arthashastra, Puranic sources, Vrhat Samhita, Mayuracitraka, Meghmala, Jain, Buddhist, and other ancient Indian literature, all of which enumerate the status of hydrology and water resources in ancient India.

In arid regions, an all-out effort was made to conserve every drop of rainwater. Jhalaras (Fig. 1-1) are typically rectangular stepwells that have tiered steps on three or four sides, which were built centuries ago to collect the subterranean seepage of an upstream reservoir or a lake and to supply water for religious rites, royal ceremonies, and community use.

Biswas (1970) has rightly remarked that the growth of modern science in Europe would hardly have been possible without the background of pioneering contributions from India, China, and Arabic countries, well up to the twelfth century AD.

1.2.2 Development of Hydrology, from 1850 Onward

The term *hydrology* arrived at its current meaning around 1750 (Hubbart, 2008). Maidment (1993) opined that the year 1850 might be regarded as marking the beginning of the development of today's methods in hydrologic practice. In 1922, the Section of Scientific Hydrology was added by the International Union of Geodesy and Geophysics, recognizing thereby that hydrology had a scientific status. The period from the beginning of 1900 to 1950 saw great advancements in the hydrological sciences.

In 1851, Mulvaney first described the concept of the time of concentration and developed the rational method for runoff computation. Through experiments on flow

Figure 1.1 A jhalara in the Rajasthan state of India. Source: https://en.wikipedia.org/wiki/Chand_Baori.

in sands in 1856, Henry Darcy established the basic law describing the flow through porous media. In 1871, Saint-Venant derived the equations of one-dimensional (1D) surface water flow. Many advances in the hydrological sciences were made throughout the 1800s. Several European contributions during the seventeenth and early eighteenth centuries to hydrology are notable, including velocity measurement by Pitot tube, Bernoulli's equation, and Chezy's equation. Poiseuille, DuPuit, DuBoys, Stokes, Manning, Reynolds, and others made substantial contributions to fluid mechanics, hydraulics, and sediment transport during this period (Rouse and Ince, 1957/1980; Hubbart, 2008; Hager, 2014).

In 1891, Robert Manning, an Irish engineer, developed the famous equation to compute river-flow velocity and discharge that is named after him (Hager, 2014). Green and Ampt (1911) produced an infiltration model that is still popular today; and Streeter and Phelps (1925) developed the dissolved oxygen sag curve for rivers. The widely used Muskingum routing method was developed in 1938 by McCarthy. Horton (1933) developed a theory of infiltration to estimate rainfall excess and improve hydrograph separation techniques. The Theis groundwater equation was developed in 1935, which laid the foundation of quantitative groundwater hydrology. Sherman (1932) developed the concept of the unit hydrograph, which gave a new direction to hydrological modeling.

Several more ideas followed in quick succession. Notable among these were the Clark (1945) unit hydrograph model and Horton's (1945) law of streams. Penman (1948) developed the combination method for the computation of evaporation from water bodies. Thornthwaite (1948) and Penman (1948) made important contributions to models of evapotranspiration. Horton's seminal work on geomorphology and erosional processes in 1948, in which he proposed empirical laws that were later known as *Horton's laws of stream numbers and stream area*, laid the foundation of the theory of stream ordering.

In 1956, the Soil Conservation Service (SCS), now called the Natural Resources Conservation Service (NRCS) of the U.S. Department of Agriculture, developed the SCS-Curve Number method for computing the amount of storm runoff, taking abstractions into account. Lighthill and Whitham (1955) developed the kinematic wave theory, which laid the foundation of physically based modeling of hydrological processes. Nash (1957) developed the linear cascade model for Unit Hydrograph (UH), and Dooge (1959) proposed the General Unit Hydrograph theory. Nash proposed a method in 1959 to estimate the parameters of UH by using the theory of moments. The 1960s saw the development of a large number of computer hydrologic models, such as the series of computer models by the Hydrologic Engineering Center (HEC), in Davis, California.

Rodrigues-Iturbe and Valdes developed the geomorphic unit hydrograph concept in 1979. In the Soviet Union, scientists completed a mammoth exercise of world water balance in 1978. Another noteworthy development was the linear theory of hydrologic systems, by Dooge (1973). Dunne (1970) advanced the concept of saturated overland flow. Some prominent hydrological models currently in use will be described later in this book.

Regarding the field of hydrologic statistics, Gumbel (1941) developed a distribution of flood flows. After extensive work on the Nile River, Hurst (1951) discovered long-term persistence in river flows, which led to what is now called the *Hurst phenomena*. At the same time, Chow (1951) developed the frequency factor method. This was followed by the Thomas–Fiering (1962) model for streamflow synthesis, the Mandelbrot and Wallis (1969) concept of fractional Gaussian noise, and the disaggregation model of Valencia and Schaake (1973). The analysis of plotting positions by Cunnane (1978) and the development of the probability-weighted moments by Landwehr, Matalas, and Wallis (1979) were important contributions to the area of hydrologic statistics. The concept of probability-weighted moments was advanced by Hosking (1990). Many physically based, semidistributed catchment models, such as SWAT and Mike-SHE, were developed from the 1990s onward, and the profession saw a large-scale use of commercial hydrological software.

1.2.3 Future Directions

While much has been accomplished in the field of hydrology, there is still a great deal about the hydrologic processes that are unknown or uncertain. With the advent of the twenty-first century, many new challenges have emerged. Future climate change and the realization that its impacts will be felt mainly through water have unleashed an unprecedented interest in hydrologic research. Hopefully, we will have a better understanding of the upscaling of hydrologic processes from plot to large basin, the feedback between atmosphere and land, regional evapotranspiration and the hydrological cycle, the interaction of hydrological processes and human systems, and ecohydrology. "Burges" (2011) mentions a long-term grand challenge for the world community as "the coupled modeling of the ocean-atmosphere-landform hydrologic cycle for the purpose of long–lead time hydrologic prediction." A state-of-the-art compendium of the hydrologic sciences and technology recently came out as the *Handbook of Applied Hydrology* (Singh, 2016).

In recent times, hydrologists have shown a keen interest in embracing technological advances in other fields for the benefit of this science. Advances in data-driven techniques, such as artificial neural networks (ANN), fuzzy theory, and big data, have found numerous applications. The acquisition and use of spatial data have undergone a complete change with the advent of robust sensors for hydrometeorological data, remote sensing, geographic information systems (GISs), Internet of Things (IOT), and web-based computing. Many existing mathematical models, such as the aforementioned series of models by the HEC and MODFLOW by the U.S. Geological Survey (USGS), were refurbished to take advantage of advancements in technology. As a result, hydrologists these days have access to versatile models in the public domain as well as commercial offerings.

1.3 Why This Book—and Why Now?

With the growing scarcity of water and increasing changes in the magnitudes, intensities, variabilities, and uncertainties of inputs and factors governing the occurrence, distribution, movement, storage, and management of water and its quality, there is a renewed interest in hydrology. Hydrology provides the foundation for analyzing water systems and making appropriate plans and decisions. For those involved in the planning and management of water resources, flood forecasting and management, town planning, hydropower generation, and environmental management and restoration, knowledge of hydrologic principles is essential. Knowledge of hydrology is also considered vital for a range of other disciplines. For instance, without an understanding of hydrologic principles, agricultural engineers cannot manage crop production systems efficiently; hydropower engineers require knowledge of hydrology in planning, design, and operation of their projects; and disaster managers need to apply hydrologic principles to deal with water-related disasters.

A number of authors have written excellent textbooks on hydrology from various perspectives. However, as expected, the contents of these books vary in both scope and detail. Authors have covered the topics of hydrologic data observations, hydrologic processes, hydrologic analysis, and modeling techniques in varying levels of detail. Viewing hydrology in a broad sense, this book covers the basic hydrologic concepts, data analysis, modeling, design, climate change, and environment-related topics from the perspective of engineering applications. Included in this treatment are topics related not only to surface water hydrology but also to unsaturated zone hydrology, groundwater hydrology, and water-quality hydrology. Hydrology is much more than a tool for engineering design—it is also equally important for environmental and watershed management and also for societal development, for that matter. This is the view that this book is attempting to reflect.

1.3.1 Who Is the Intended Readership of This Book?

This book is an introductory text targeted toward the students in their junior or senior undergraduate year. However, due to the emphasis on fundamentals and emerging concepts, the book should also be useful for graduate students, professionals, and anyone else who would like to understand the basic concepts of hydrological science and engineering.

1.3.2 Organization of This Book

Keeping the objective of the book in mind, the chapters have been organized into five parts. Part 1 aims to introduce the subject matter of the book and consists of three chapters. In this chapter, Chapter 1, hydrology has been defined and a brief history given. Chapter 2 describes the hydrologic cycle, along with quantitative details. Water balance is an important part of any hydrologic analysis. It can be carried out for a small microwatershed, for a medium-to-large river basin, and even for the Earth as a whole. This forms the subject matter of Chapter 3.

Part 2 provides a detailed discussion on the components of the hydrologic cycle. All the main components, such as precipitation, evapotranspiration, infiltration, surface water, and subsurface water, have been covered in detail. After reading this part, the reader will have a good understanding of all these processes. Water quality, although not a component of the hydrologic cycle, is extremely important from the point of view of both the use and the management of water, and the fundamentals of this topic are covered in this book.

To manage any resource, it is essential to measure it accurately. Part 3 of the book covers measurements of hydrometeorological variables (WMO, 2008). It begins with a chapter on techniques to analyze networks from the adequacy perspective. Guidelines to fix locations of additional stations are described. Streamflow is one of the most important components of the hydrologic cycle, and techniques for its measurement and data analysis have been discussed in a separate chapter. Remote sensing and GIS are emerging tools for capturing data about hydrologic and allied variables. GIS is used these days for the storage and processing of spatial data. These two techniques have been briefly described in this part.

At this stage, with adequate coverage of the basics accomplished, the book turns to engineering aspects and solutions of practical problems. Part 4 deals with the analysis and modeling of catchments. It begins with a chapter on rainfall runoff modeling, followed by unit hydrograph analysis, which is a widely used technique for rainfall runoff modeling and design flood estimation. River and reservoir routing forms an important component of hydrologic analysis and design. These are backbones of flood forecasting and project formulation. Thus, the techniques of river and reservoir routing have been covered in detail. The next chapter deals with pollutant modeling, which is necessary for water quality management and river rejuvenation. Worldwide, many rivers are being degraded because a large quantity of water is being withdrawn for meeting societal needs. Environmental flows are provided in rivers so that the minimum needs of river ecosystem are met. These flows, described by the various techniques for the assessment of environmental flows and issues involved in their implementation, are covered in this book.

Part 5 of this book, on hydrologic design, contains two chapters that describe data-based statistical techniques used for hydrologic analysis. They cover probability distributions, their parameter estimation, regression and correlation analysis, hypothesis testing, and trend analysis. A chapter on hydrologic modeling describes the various types of models and the calibration and validation of these models. A water resources development project should be able to negotiate the incoming design flood safely. This requires that storm data are carefully analyzed and design floods are calculated by following the national standards. Underlying techniques for these have been discussed in the chapter on design storms and design floods. Finally, the impact due to the change in climate is likely to place additional stress on hydroinfrastructure and water resources management. The topic of climate change has been included, which describes its causes, the techniques to predict its impacts on the water sector, and how we can adapt to it and mitigate the effects.

The book covers the topic of engineering hydrology, with sufficient weightage on basic aspects, with explanation using figures and diagrams and examples. In each chapter, recent advances in each topic have been covered briefly so that interested readers can pursue these topics in greater detail.

1.4 Summary

Water is central to the environment and is considered to be the driver of nature. There is a large mismatch in water availability in terms of space and time. Extreme hydrometeorological events trigger disasters and quality of water is a concern. Hence, management of water resources is necessary so that water of right quantity and quality is provided at the right time. Society is to be protected from disasters and we need to protect/rejuvenate environment. All this requires a sound understanding of hydrologic principles. Various chapters of this book have been written to that end.

1.5 Questions

1.1 A large number of definitions of hydrology have been given by different organizations and books. Find five of these definitions and identify their differences.

1.2 What key points emerge from these definitions of hydrology?

1.3 In your opinion, which historical period was most important from the point of view of the development of hydrological sciences?

1.4 Study the Egyptian civilization in the Nile Valley from the perspective of water management. What can we learn from this?

1.5 Study the Chinese civilization that grew on the banks of the Huang-Ho River and the Tigris and Euphrates civilizations in Mesopotamia from the water management point of view. What lessons can be learned from looking at these peoples?

1.6 Study the Indian civilization that developed along the banks of the Ganges River and the Indus River from the water management point of view. What lessons can be learned from this?

1.7 List the main reasons why ancient civilizations mostly grew near rivers.

1.6 References

Biswas, A. K. *History of Hydrology.* Amsterdam: North Holland Publishing, 1970.

Burges, S. J. "Invited Perspective: Why I Am an Optimist," *Water Resource Research,* 47:W00H11, doi:10.1029/2010WR009984, 2011.

Clark, C. O. Storage and the unit hydrograph. Transactions, ASCE, Vol. 110, Paper No. 2261, 1419–1446, 1945.

Chow, V. T. "A General Formula for Hydrologic Frequency Analysis," *Transactions of the American Geophysical Union,* 32:231–237, 1951.

Cunnane, C. "Unbiased Plotting Positions—A Review," *Journal of Hydrology,* 37(3/4):205–222, 1978.

Dooge, J. C. "A general theory of the unit hydrograph," *Journal of Geophysical Research,* 64(2):241–256, 1959.

Dooge, J. C. I. *Linear Theory of Hydrologic Systems.* Washington, DC: Agricultural Research Service, U.S. Department of Agriculture, 1973.

Dunne, T. *Runoff Prediction in Humid Areas.* ARS 41-160. Washington, DC: Agricultural Research Service, U.S. Department of Agriculture.

Green, W. H., and C. A. Ampt. "Studies on Soil Physics: 1. Flow of Water and Air through Soils," *Journal of Agricultural. Science,* 4:1–24, 1911.

Gumbel, E. J. "The Return Period of Food Flows," *Annals of Mathematical Statistics,* 12:163–190, 1941.

Hager, W. *Hydraulicians in Europe 1800–2002.* 2 vols. Barcelona: International Association for Hydraulic Engineering and Research (IAHR), 2014.

Horton, R. E. "The Role of Infiltration in the Hydrologic Cycle," *Transactions of the American Geophysics Union,* 145:446–460, 1933.

Horton, R. E. "Erosional Development of Streams and Their Drainage Basins: Hydrophysical Approach to Quantitative Geomorphology," *Bulletin of the Geological Society of America,* 56:275–370, 1945.

Hosking, J. R. M. L-moments: Analysis and estimation of distributions using linear combinations of order statistics, *J. Roy. Statistical Soc.*, Series B, 52:105–124, 1990.

Hubbart, J. *History of Hydrology,* 2008, http://www.eoearth.org/view/article/153525 Oct. 24, 2008.

Hubbart, J. A. "Considering the Future of Water Resources: A Call for Investigations that Include the Cultural Anthropology of Water," *Global Journal of Archaeology & Anthropology*, 3(2), GJAA.MS.ID.555607, 2018.

Hurst, H. E. "Long-Term Storage Capacities of Reservoirs," *Transactions of ASCE,* 116:776–808, 1951.

Lighthill, M. J., and G. B. Whitham. "On Kinematic Waves: Flood Movement in Long Rivers," *Proceedings of the Royal Society, London,* 22(A):281–316, 1955.

Landwehr, J., N. C. Matalas, and J. R. Wallis, "Probability Weighted Moments Compared with Some Traditional Techniques in Estimating Gumbel Parameters and Quantiles," *Water Resources Research,* 15(5):1055–1064, 1979.

Mandelbrot, B. B., and J. R. Wallis. "Computer Experiments with Fractional Gaussian Noises, Part 1: Averages and Variances," *Water Resources Research,* 5(1): 228–241, 1969.

Maidment, D. (Ed). *Handbook of Hydrology.* New York: McGraw-Hill, 1993.

Nash, J. E. "The Form of the Instantaneous Unit Hydrograph," *Hydrological Science Bulletin,* 3:114–121, 1957.

National Institute of Hydrology (NIH). *Hydrology in Ancient India.* Roorkee, India: National Institute of Hydrology (NIH), 1990.

Penman, H. L."Natural Evaporation from Open Water, Bare Soil and Grass," *Proceedings of the Royal Society of London, Series A,* 193:120–145, 1948.

Rouse, H., and S. Ince. *History of Hydraulics.* Iowa City, IA: Iowa Institute of Hydraulic Research, University of Iowa, 1957/1980.

Sherman, L. K. "Stream Flow from Rainfall by the Unit Graph Method," *Engineering News Record,* 108:501–505, 1932.

Singh, V. P. (Ed.). *Handbook of Applied Hydrology.* 2nd ed. New York: McGraw-Hill, 2016.

Streeter, H. W., and E. E. Phelps. A study of the pollution and natural purification of the Ohio River. Bulletin No. 146, U.S. Public Health Service, Government Printing Office, Washington, DC, 1925.

Thomas, H. A., and M. B. Fiering. "Mathematical synthesis of stream flow sequences for the analysis of river basins by simulation." In *Design of Water Resources Systems*, edited by A. Masss et al. (Ed). Cambridge, MA: Harvard University Press:459–493 1962.

Thornthwaite, C. W. "An Approach Toward a Rational Classification of Climate," *Geography Review*, 38:55–94, 1948.

Valencia, D. R., and J. L. Schaake Jr. "Disaggregation Processes in Stochastic Hydrology," *Water Resources Research*, 9(3):580–585, 1973.

World Meteorological Organization (WMO). *Guide to Hydrological Practices, Vol. I: Hydrology—From Measurement to Hydrological Information*. WMO-No. 168. Geneva, Switzerland: World Meteorological Organization (WMO), 2008.

CHAPTER 2

The Hydrologic Cycle

The hydrologic cycle (HC) constitutes the basis of hydrology and also defines its scope, as well as its relation to other sciences, geophysical and otherwise. It is also closely connected to carbon, nitrogen, and other biogeochemical cycles. This chapter presents a short discussion of the hydrologic cycle and its components and its relation to other cycles.

2.1 Introduction

A number of natural cycles in which large quantities of matter continuously move are operating on the Earth, including the hydrologic cycle, the carbon cycle, the nitrogen cycle, and several biogeochemical cycles. The *hydrologic cycle (HC)*, or the water cycle, is the cycle of circulation of water, involving changes in the physical state of water between the liquid, solid, and gas phases. This cycle is the continuous exchange of water among atmosphere, land (involving surface and subsurface waters), oceans and seas, and organisms. HC forms the fundamental concept in hydrology. The National Research Council (NRC, 1982) has defined the hydrologic cycle as "the pathway of water as it moves in its various phases to the atmosphere, to the Earth, over and through the land, to the ocean, and back to the atmosphere." Being a cycle, it does not have any beginning or end, and water is present in all three states (solid, liquid, and gas). A pictorial view of the HC is given in Fig. 2.1.

Hydrological sciences are mainly concerned with the land portion of the HC; interactions with the oceans and atmosphere are also studied. NRC (1991) characterized the HC as the integrating process for the fluxes of water, energy, and the chemical elements. Incidentally, there are four *spheres*: the hydrosphere (water), the pedosphere (soils), the lithosphere (rocks), and the atmosphere (air).

The HC can be subdivided into three major systems: the oceans and seas, constituting the hydrosphere, the major reservoir and source of water; the atmosphere, functioning as the carrier and deliverer of water; and the land, constituting the pedosphere and lithosphere as the users of water. On a global basis, the HC is a closed system, but on a local basis it is an open system. The presence of water vapor in the atmosphere is one of the factors that make the Earth livable for humans. The presence of water vapor in the atmosphere helps keep the temperature in the range of about −40°C to 55°C.

The water evaporating from the land and the oceans enters the atmosphere. From the atmosphere, water falls on the land as precipitation. The oceans receive water from the atmosphere by precipitation, and from the land through rivers and

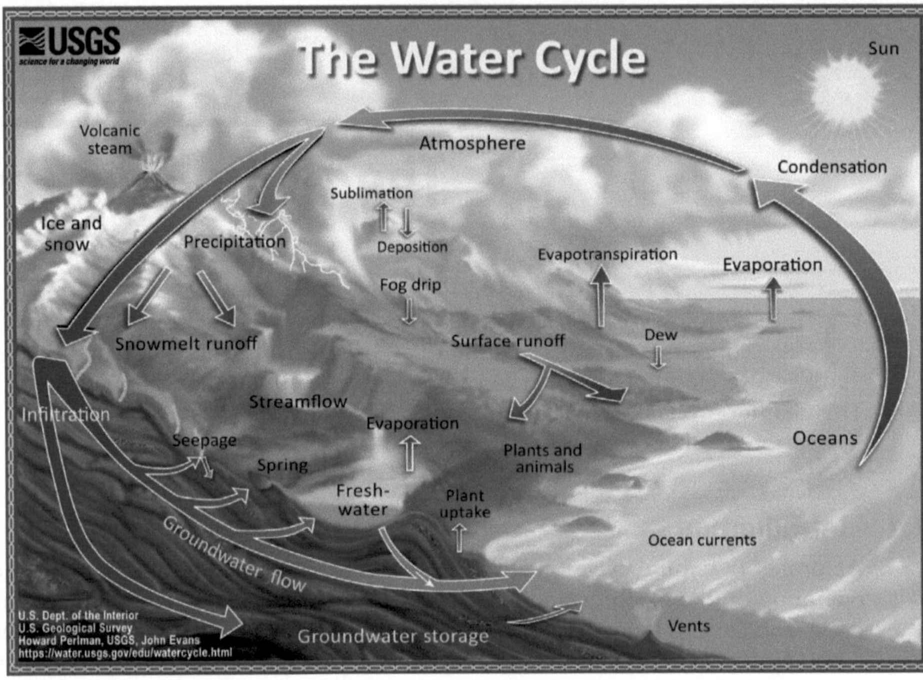

FIGURE 2.1 A pictorial representation of the HC. (Source: http://water.usgs.gov)

groundwater flow. The water leaves land through multiple means—evapotranspiration (ET), streamflow, and groundwater flow whereas it leaves oceans through evaporation. Precipitation, evaporation, and infiltration mostly occur in the vertical plane, while overland flow, streamflow, and groundwater flow dominantly occur in the horizontal plane.

In the HC, each phase usually has (1) transport of water, (2) temporary storage, and (3) change of state. For example, in the atmospheric phase, there are vapor flow and vapor storage and condensation (phase change from vapor to the liquid or solid state). Moreover, water is present in the atmosphere in the form of vapor, while it mostly takes the form of saline liquid in the oceans. From the point of view of use, the land phase of the HC is the most important. This may explain the scope of hydrology being primarily confined to the land phase.

One way to visualize the HC is a perpetual distillation and pumping system. In this endless water circulation, the glaciers and snowpacks are replenished, and the quantity of river water is refilled by good-quality water. Shiklomanov (1999) called the exchange of water among the oceans, land, and the atmosphere "turnover." This turnover affects the global patterns of the movement of ocean waters and gases in the atmosphere, thereby greatly influencing climate.

Because water is a very good solvent, chemistry is an integral part of the HC. Usually, rain and snow are pure forms of water, although these may also be mixed with pollutants present in the atmosphere. During the movement on the Earth, many chemical compounds are mixed with water, and consequently, the water quality degrades.

Reservoir	Volume (10^6 km^3)	Percent of Total
Oceans	1,370	97.25
Ice caps and glaciers	29.0	2.05
Groundwater	9.5	0.68
Lakes	0.125	0.01
Soil moisture	0.066	0.005
Atmosphere	0.013	0.001
Streams and rivers	0.0017	0.0001
Biosphere	0.0006	0.00004

TABLE 2.1 Water On and Below the Earth's Surface

The HC is, thus, the integrating process for the fluxes of water, energy, and chemical elements (NRC, 1991).

In view of the complexities and coverage, the study of the complete HC is truly interdisciplinary. For instance, meteorologists study the atmosphere, hydrologists study the land surface, soil scientists study the pedosphere, geologists and hydrologists study the lithosphere, and oceanographers study the oceans. In addition, a host of other professionals study HC, including agricultural engineers, energy managers, disaster managers, environmentalists, public health officers, and inland navigation managers. Of course, it is hard to demarcate the various domains in such an interdisciplinary subject, and often the boundaries are fuzzy. The domain of hydrologists mainly consists of the land phase and the lithosphere phase, which by themselves are vast.

The HC can be visualized as composed of storages and processes that move water among these storages. Oceans are the largest reservoirs on Earth, holding about 97.5% of the planet's water (Table 2.1). Of the remaining 2.5% freshwater, about 70% is stored in ice at the poles and in Greenland (WWAP 2009). The distribution of land and ocean surfaces on the Earth is highly uneven. In the Southern Hemisphere, four times more surface is covered by ocean than by land. The ratio between land and ocean coverage is almost equal in the Northern Hemisphere. About 21% of Earth's freshwater is stored in groundwater below the land surface. Rivers and lakes hold less than 1% of the freshwater on Earth.

2.2 Components of the Hydrologic Cycle

The Sun's energy, in the form of light and heat, evaporates water from the oceans, rivers, lakes, soil, and vegetation. Therefore, the Sun is the prime mover of the HC. The six main processes in the HC are precipitation, interception, evaporation, infiltration, runoff, and subsurface flow. These processes may occur simultaneously during a precipitation event, and for some time after the event is over.

Water is evaporated from open water bodies such as rivers, ponds, and oceans by using solar energy. Transpiration of water from vegetation adds a substantial quantity of moisture to the atmosphere. Gravity moves water on the Earth's surface from higher elevations to lower elevations.

We now very briefly discuss the components of the HC. A detailed treatment of the major components will be given in subsequent chapters of this book.

2.2.1 Atmospheric Component

Precipitation is the most important atmospheric component of the HC.

Precipitation

In this context, *precipitation* can be defined as any water deposit (liquid or solid form) that generally falls from clouds. Precipitation is classified in different types that include rain, freezing rain, snow, ice pellets, snow pellets, fog, dew, mist, and hail. The distribution of precipitation on the Earth's surface is generally controlled by the mechanisms that lift moist air masses to cause saturation. Precipitation also depends on the amount of water vapor held in the air, which depends on air temperature.

Water vapor condenses to form clouds, which result in precipitation when conditions are suitable. Precipitation is received on the land in the form of rain, snow, hail, frost, and dew, among others. The pattern and magnitude of precipitation depend on climatic factors such as cloud cover, temperature, radiation, pressure, humidity, and wind speed. Since these factors vary with space and time, precipitation is a function of time and space. In most places, rainfall is the predominant form of precipitation. In many places, rainfall is synonymous with precipitation and is primarily responsible for streamflow generation or floods in natural rivers.

A part of precipitation falling on the surface infiltrates in the soil and some part flows on the land. Evaporation from surface water (e.g., lakes, streams, and oceans) return moisture to the atmosphere, while plants return water to the atmosphere by transpiration. Warm air rising from the Earth's surface carry these water vapors to the atmosphere. When these vapors reach the cooler zone of the atmosphere, the moisture condenses on the fine particles present in the air. This is called *condensation*. Because ocean water covers around 70% of the Earth's surface, most precipitation falls on the oceans. The precipitation that falls on land takes one of several paths through the HC.

Precipitation at a place depends on large-scale as well as local processes. Hence, a large variability in precipitation is observed over a range of spatial and temporal scales. Among the components of the HC, precipitation is one of the most difficult to model. In addition to the amount, the frequency, intensity, form (liquid or solid), and number of events are important in water management.

Each year, approximately 505,000 km^3 of water falls as precipitation. Of this, about 78% (approximately 398,000 km^3) falls over the oceans. The precipitation falling as rain is about 107,000 km^3, and as snow, it is about 1,000 km^3.

Chapter 4 discusses precipitation in greater detail.

2.2.2 Surface Components

In this section, we will discuss the surface components of the HC.

Interception

The precipitation that does not fall on the bare soil is caught by vegetation or other surface covers. This trapped amount is known as *interception loss*. A part of this water may be evaporated back to the atmosphere. The remainder of the precipitation eventually reaches the ground, but with some delay after temporarily staying on the surface cover. The amount of water stored on the wetted surface of the land cover is the interception storage. Interception is more important for low-intensity rainstorms.

The distribution of precipitation falling on the ground surface can be modified by the presence of vegetation that intercepts some of the falling rain. *Canopy interception* is the portion of precipitation that falls on the vegetation canopies, is stored upon them, and is evaporated. The amount of water intercepted by vegetation depends on the structure of branches and leaf density. Some of the intercepted water does not reach the ground surface and is evaporated back from the vegetation surface to the atmosphere.

Evaporation and Transpiration

Evaporation is the transfer of water from liquid to vapor and back to the atmosphere. It depends on the temperature of the evaporating surface and the ambient air, as well as the vapor pressure deficit—the difference in vapor pressure between the water surface and the atmosphere. Evaporation from a vegetated surface also depends on soil moisture.

Transpiration is the loss of water from the stomatal openings in the plant leaves. *Stomata* are pores on the undersurface of a leaf that open in presence of sunshine. When the stomata open, water vapors diffuse from cells of leaves into the atmosphere.

When computing water loss from a vegetated surface, the processes of transpiration and evaporation from the soil surface and water bodies are usually considered together, and this is known as *evapotranspiration (ET)*. Thus, ET is the total loss of water by evaporation from a land surface and transpiration from its vegetation. The amount of ET varies according to the type of vegetation, its ability to transpire, and the availability of water in the soil. ET is difficult to measure, partly because this process is not visible.

Oceans contribute about 86% of global evaporation that lowers oceanic temperature through evaporative cooling. Chapter 5 provides a detailed discussion of ET.

Infiltration

A portion of the precipitation reaching the Earth's surface seeps into the ground through a process called *infiltration,* which is the movement of water from precipitation into the soil. Infiltration is the transition from surface water to groundwater. The infiltration rate depends upon soil or rock permeability, as well as other factors. The infiltrated water may finally join *groundwater*.

The *infiltration rate* is the rate at which the water enters the ground. Infiltration rate varies with space and time due to a number of factors. After a spell of rain, infiltration can saturate the soil column. With time, a portion of this water may drain due to the forces exerted by gravity; the water is known as *gravitational water.* The term *field capacity,* broadly speaking, denotes the maximum amount of water that can be stored in the soil against gravitational forces. When soil becomes dry, the permanent wilting point is reached, which is the lower limit of water available for use by plant roots.

The amount of infiltration depends on the land slope, the amount and type of vegetation, soil type, and the amount of moisture present in the soil. The more are the openings (cracks, pores, joints) in the soil, the more is the infiltration.

Infiltrated water is useful for plant growth. When soil moisture content becomes so low that plants are unable to extract water through their roots, irrigation needs to be applied

Water that does not soak into the soil moves across the surface as runoff, eventually flowing into streams and rivers and may ultimately join the ocean. Precipitation that falls as snow in glacial regions is accumulated there. Snow and glaciers slowly melt and the meltwater joins a river.

Chapter 6 provides a detailed discussion of infiltration.

Groundwater

Groundwater is the water contained in the zone below the ground that is permanently saturated. The top of groundwater is known as the *water table.* Groundwater flows toward lower gradients because of gravity. In the groundwater zone, pressures are greater than or equal to atmospheric pressure.

A geologic stratum that has porosity to store water and hydraulic conductivity to store and transmit significant quantities of water is called an *aquifer*. Aquifers store water in the underground reservoirs and also act as pathways to pass water. Some aquifers are more efficient as pathways (e.g., cavernous limestones), and some are more effective as storage reservoirs (e.g., sandstones). Most aquifers perform both functions, but to different degrees.

Depending on the presence or absence of the water table, aquifers may be classified as unconfined or confined. For an unconfined aquifer, the water table is the upper surface of saturation zone. The water table in an unconfined aquifer is in contact with the atmosphere through pores in the unsaturated soil.

To appreciate the significance of groundwater, it may be interesting to compare it quantitatively with other components of the HC. Of all the global freshwater resources, which include ice caps and glaciers, groundwater constitutes about two-thirds. If ice caps and glaciers are excluded and only utilizable freshwater resources are considered, then other components are negligibly small compared with groundwater. However, not all groundwater is easily accessible because nearly half of it is located too far below the ground. Even if only active groundwater regimes are considered, groundwater is about 95% of total freshwater resources.

Nearly 97% of the world's potable water comes from groundwater, which is also an important (and sometimes the only) source of irrigation. The study of groundwater requires hydrologic and geologic investigations and modeling over areas that may range from a few hundred hectares to thousands of square kilometers, and may even span transboundary aquifers.

Chapter 8 provides a detailed discussion of subsurface water.

Overland and Channel Flow

Precipitation that reaches the surface of the Earth but does not infiltrate flows on the surface and is called *runoff*. Runoff can also come from melting of snow and ice. Not all precipitated water returns to the sea as runoff; much of it evaporates before reaching the ocean or an aquifer.

The term *runoff* also refers to the surface flow of water to areas of lower elevation. All runoff ultimately flows from the land to oceans. The Earth's continents experience runoff because of the imbalance between precipitation and evaporation.

The flow of water through a river channel is commonly called *streamflow* or *stream discharge.* Based on the path taken by water, it may be divided into surface flow, interflow, and base flow.

Overland flow is important in urban watersheds and paved areas, such as highways, airport runways, and parking lots; its quantity is not high in grasslands and forestlands.

Base Flow

Base flow is the runoff that reaches the stream or river by passing first through the underlying aquifer, rather than by flowing on the land surface (ASCE, 1996). Thus, base flow is that portion of streamflow that is gradually withdrawn from groundwater storage or other delayed sources. Synonyms for *base flow* are *groundwater flow, seepage flow, low flow,* and *fair weather flow.*

Base flow contribution to streamflow depends on the geologic nature of the water-table aquifer. The lateral movement of groundwater is slower than the vertical movement because the hydraulic gradient is smaller for lateral movement. The supply from groundwater to the channel will continue as long as the necessary gradient is present. If there is no additional percolation through infiltration to the aquifer, the hydraulic gradient falls as water moves to the stream from higher elevations. Consequently, lesser water will travel to the stream with time. This process is called *base flow recession.*

In humid climate and a shallow water table that is hydraulically connected to the stream, base flow may be present around the year. Base flow is absent in semiarid climates and areas of deep groundwater. Sources such as lakes, marshes, snowpack, and channel bank may also supply water for base flow. At a given time, more than one source may be supporting base flow.

The various components of the water cycle (precipitation, runoff, groundwater, etc.) greatly differ in their spatial and temporal variability; susceptibility to pollution and biochemical qualities; predictability; resilience; vulnerability to stresses (overexploitation, land use, and climate change); and the capacity to provide services in a sustainable manner. Due to these differences, various components respond to the external stimuli in widely different ways. The uneven distribution of water resources over time and space and their modification through human use and abuse are sources of water crises in many parts of the world (WWAP 2009).

2.2.3 Importance of the Components of the Hydrologic Cycle

The relative importance of the various components of HC depends on time and space scales. If we consider HC components during a storm at short time and space scales, dominant components are precipitation, infiltration, and overland flow; evaporation and interception have only small effects. Over annual scales for a large arid region, ET and streamflow become dominant.

HC is a closed system strictly at global scale, not at local scales or over short time periods. Complexities in modeling also depend upon the time and space scales.

2.3 Scales for the Study of the Hydrologic Cycle

Depending on the purpose of the research, the HC is studied on many spatial scales. Horton (1931) recognized the diversity of scales when he stated: "Any natural, exposed surface may be considered as a unit area on which the HC operates. This includes, for example, an isolated tree, even a single leaf or twig of a growing plant, the roof of a building, the drainage basin of a river system or any of its tributaries, an undrained glacial depression, a swamp, a glacier, a polar ice-cap, a group of sand dunes, a desert playa, a lake, an ocean, or the Earth as a whole" (quoted in NRC, 1991).

Regarding space, two scales are readily distinct: the global scale and the catchment scale. These topics are discussed next.

2.3.1 Global Scale

From a global perspective, the HC can be considered to be comprised of three major systems: the oceans, the atmosphere, and the landsphere. Precipitation, runoff, and evaporation are the main processes that move water among these systems, as illustrated in Fig. 2.2. Assessments at large scale are required to understand the global fluxes and global circulation patterns.

Fig. 2.3 shows the schematic of the HC on the global scale in the Earth system, and Fig. 2.4 shows the microscale view of the cycle in the land system.

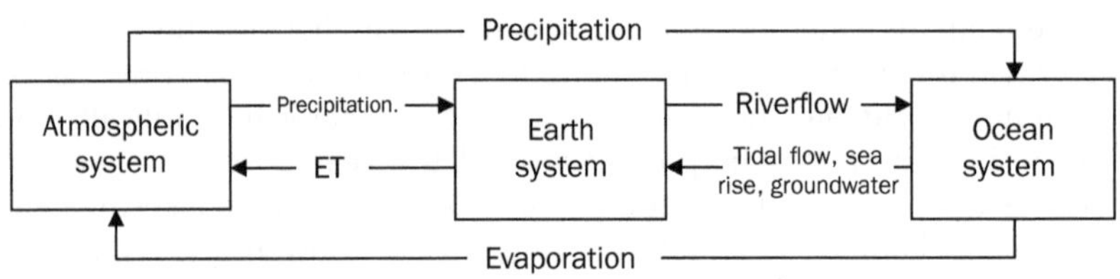

FIGURE 2.2 A global schematic of the HC. (Source: Singh (1992).)

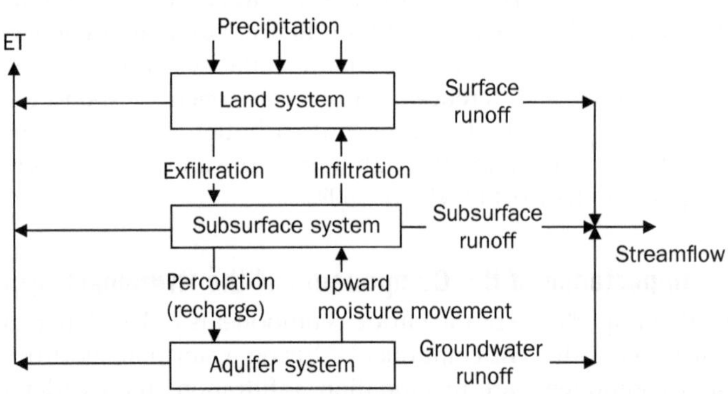

FIGURE 2.3 A schematic of the HC of the Earth system. (Source: Singh (1992).)

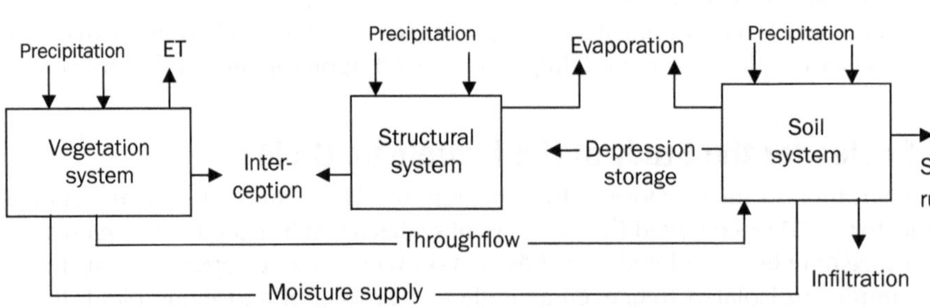

FIGURE 2.4 A detailed schematic of the HC in the land system. (Source: Singh (1992).)

2.3.2 Catchment Scale

Here, the size of a catchment scale can vary from a few square kilometers to thousands of square kilometers and the study period can vary from a few hours to one year or more. A catchment system can be further subdivided into the land surface subsystem, the subsurface subsystem, and the aquifer subsystem.

2.3.3 Time Scales in the Hydrologic Cycle

There is a large variation in time taken by water to move through various subsystems of the HC. Speed of movement of water in a river is much higher compared to that in subsurface zone. The time-step size for a study will depend on this factor. If the time-step size of computations for a fast-moving river is large, important variations may be missed and hence, an important factor is how detailed results are needed. Depending upon the kind of detailing required, the time-step should be small enough so that all the important variations in key components of HC are computed in requisite detail. However, a too small time-step means that more computations will have to be carried out. In addition, the purpose of the study and the availability of data are also important. If forecasting of floods in a small river basin is the objective of a study, time-step size would typically be an hour or even shorter. To compute water availability in a large basin, daily time-step would normally suffice. Further, if the flow data are measured once a day, not much additional insight would be obtained by adopting a time-step of, say, 3 hours. The estimated periods for the renewal of water resources in water bodies on the Earth are given in Table 2.2.

The dominant processes for the land subsystem are precipitation, ET, infiltration, and surface runoff. The range of spatial and temporal dimensions of many processes related to the HC is shown in Fig. 2.5.

Hydrologists frequently use the concept of *water year,* which typically begins with the so-called wet season. This concept is used to minimize the effects of interannual changes in storage-related quantities (e.g., soil moisture) in water balance. In India, the water year commonly runs from June 1 to May 31 of the next calendar year; in the United States, the water year begins on October 1.

Water of Hydrosphere	Period of Renewal
World ocean	2,500–4000 years
Groundwater	Weeks to thousands years
Polar ice	9,700 years
Mountain glaciers	1,600 years
Ground ice of the permafrost zone	10,000 years
Lakes	1 to 10 years
Bogs	5 years
Soil moisture	2 weeks to several year
Channel network	A few weeks
Atmospheric moisture	8 days
Biological water	Several hours to one week

TABLE 2.2 Periods of Water Resource Renewal on the Earth (Source: Shiklomanov and Rodda (2003) and others.)

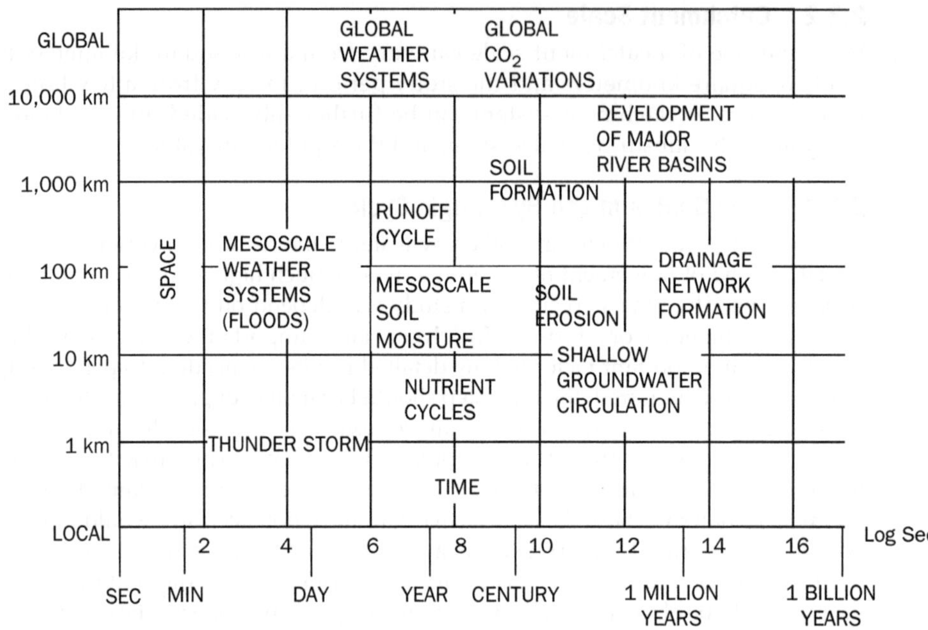

FIGURE. 2.5 An illustration of the range of process scales in hydrology. (Source: NRC (1991).)

2.4 Carbon and Nitrogen Cycles and Their Relation with the Hydrologic Cycle

Several natural cycles are affected by the HC and also affect the HC in a feedback system. Two important cycles, the carbon cycle and the nitrogen cycle, are briefly described next.

2.4.1 Carbon Cycle

All living things are made of carbon, which is also part of many nonliving things, such as ocean, air, and land. In nature, carbon is found in coal, fossil fuels, and wood. Carbon exists in the atmosphere in two main forms: carbon dioxide (CO_2) and methane (CH_4). These gases absorb and retain heat in the atmosphere and are partially responsible for the greenhouse effect. Carbon cycle describes the movement of carbon in various forms between the biosphere, atmosphere, oceans, and geosphere (Fig. 2.6). It is one of the biogeochemical cycles that includes sinks and stores of carbon and processes by which they exchange carbon. The carbon cycle is complex, and each constituent process has an impact on other processes.

Many processes govern the exchange of carbon between the atmosphere and vegetation. *Photosynthesis* is the process by which plants use CO_2 and sunlight to make their food; this is also known as *primary production*. Plants and animals that are buried below the Earth's surface may turn into fossil fuels (e.g., coal and oil) over millions of years. When humans burn fossil fuels, most of the carbon enters the atmosphere as CO_2.

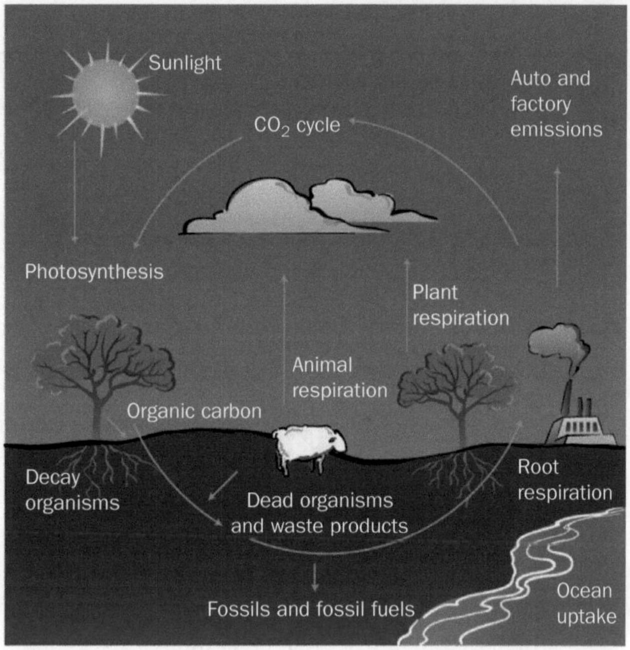

FIGURE 2.6 The carbon cycle. (Source: https://eo.ucar.edu, accessed on March 16, 2018.)

Living beings also release CO_2 to the atmosphere during respiration. Large exchanges of CO_2 also take place between the oceans and the atmosphere. The dissolved CO_2 in the oceans is used by marine biota in photosynthesis. Analysis of ice cores shows that there is now more CO_2 in the atmosphere than it was in the last 420,000 years. Over the last 50 years, the concentration of CO_2 in the atmosphere has been rising monotonically (Fig. 2.7). Because carbon uptake in the terrestrial biosphere depends on biotic factors, it follows a diurnal and seasonal cycle (see Fig. 2.7). This cyclic variation is the strongest in the Northern Hemisphere.

The carbon cycle is a closed system; the amount of carbon in the world is fixed. It is neither created nor destroyed, but its distribution may change with time. Carbon leaves the terrestrial biosphere in different ways and time scales. The combustion or respiration of organic carbon releases CO_2 rapidly into the atmosphere. It can also be exported into the oceans through rivers. Carbon stored in soil can remain for thousands of years before it is released. Because forests absorb CO_2 from the atmosphere, changing land use is changing CO_2 concentration in the atmosphere. Another important process in the carbon cycle is the burning of fossil fuels (i.e., coal, petroleum products, and natural gas).

Oceans contain the greatest quantity of actively cycled carbon in the world. The lithosphere contains the largest amount of carbon, buried in sedimentary rocks in the form of carbonate minerals and organic compounds such as oil, natural gas, and coal (Carlson et al., 2001). Carbon enters the ocean mainly through the dissolution of atmospheric CO_2, which is converted into carbonate. Absorption of CO_2 by oceans is one of the most important forms of carbon sequestering, which helps limit the rise of CO_2 in the atmosphere.

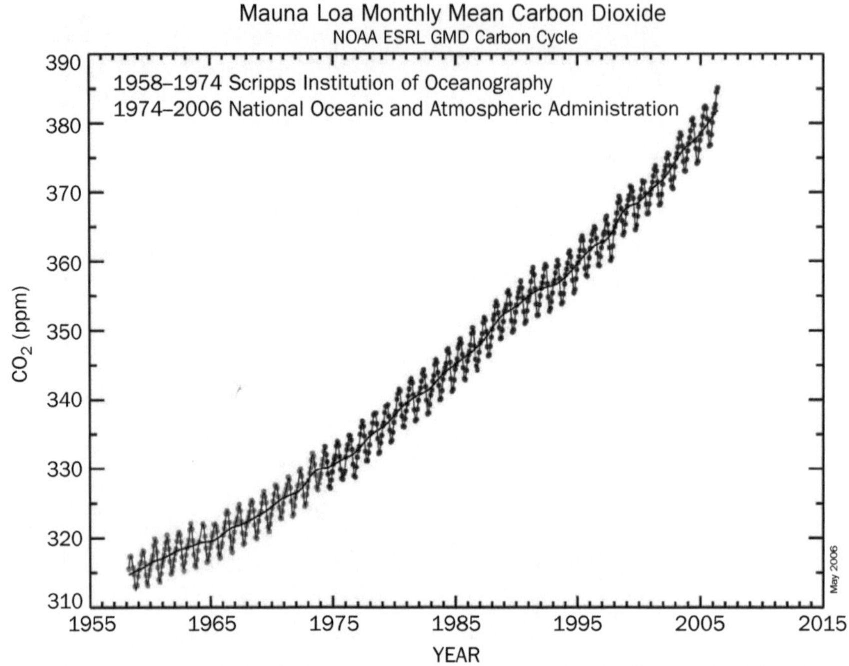

FIGURE 2.7 Concentration of atmospheric CO_2, measured at the Mauna Loa Observatory in Hawaii. (Source: https://celebrating200years.noaa.gov/datasets/mauna/image3_full.jpg.)

Human Influence on the Carbon Cycle

Human activities, particularly since the industrial era, have significantly increased the amount of CO_2 in the atmosphere by burning fossil fuels. This is the largest and most direct human influence on the carbon cycle. Over the past several centuries, human-caused land use and land cover changes (LULCs) have also led to changes in the carbon cycle. Deforestation removes forests (which hold large amounts of carbon) and replaces them, generally with agricultural or urban areas that store comparatively small amounts of carbon. So the net effect is that more carbon remains in the atmosphere. In addition, increased amounts of CO_2 in the atmosphere can also lead to higher gross primary production. Carbon increases photosynthesis rates by allowing plants to use water more efficiently. Humans also indirectly influence the carbon cycle by changing the terrestrial and oceanic biosphere. Changes in carbon cycle have impacts on the HC.

Scientists have compiled data from a variety of sources, ice cores, and observations to create a long series of atmospheric CO_2 concentration over the last 400,000 years. It is found that the recent rise in CO_2 is unlike anything seen during that time, both in terms of the rate of increase and the levels to which it is rising. Before the recent rise in concentration, CO_2 fluctuated by about 80 ppm in tandem with the Ice Ages (which had a regular occurrence). In recent years, this pattern has been breached.

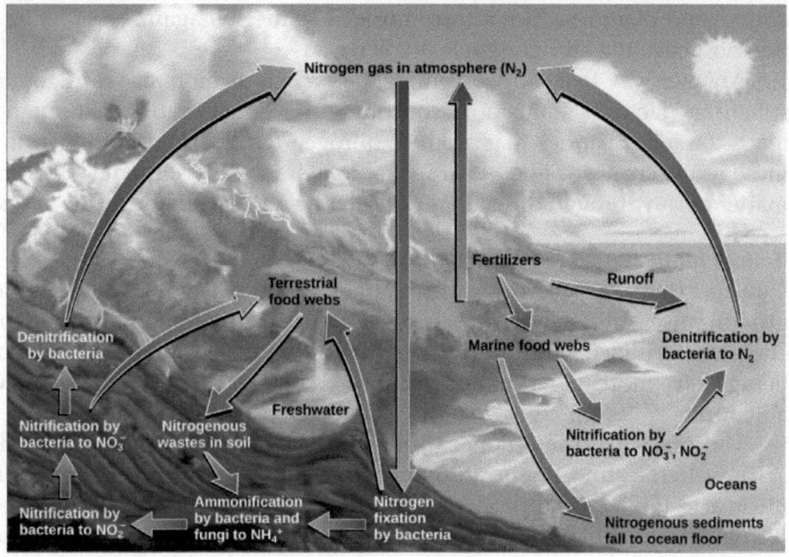

Figure 2.8 The nitrogen cycle. (Source: http://cnx.org and USGS. Accessed on June 13, 2017.)

2.4.2 Nitrogen Cycle

Nitrogen is a colorless, odorless, and nontoxic gas that is necessary for all known life forms on Earth. Nitrogen gas (N_2) is the largest constituent of the Earth's atmosphere (close to 78%), but this form is relatively nonreactive and unusable by plants. Much of the nitrogen in plants is used in chlorophyll molecules, which are essential for photosynthesis and plant growth. Chemical processing or natural fixation (through processes such as bacterial conversion) is necessary to convert gaseous nitrogen into compounds such as nitrate or ammonia that can be used by plants.

Like the carbon cycle, the nitrogen cycle is an important part of the Earth's ecosystem. It consists of biological and physical processes by which nitrogen moves between its chemical forms. Important processes in the nitrogen cycle include fixation, ammonification, nitrification, and denitrification (Fig. 2.8). Atmospheric nitrogen is of limited biological use, leading to a scarcity of usable nitrogen in many ecosystems. The nitrogen cycle is of interest to hydrologists because nitrogen availability can affect the rate of key ecosystem processes, including primary production and decomposition. Human activities, such as making fertilizers and burning fossil fuels, have significantly altered the amount of fixed nitrogen in the Earth's ecosystems (Bernhard, 2010).

Nitrogen from the air enters the nitrogen cycle through several unique microorganisms that convert N_2 gas to inorganic forms that are usable by plants. Nitrogen also can enter the cycle from manure, decaying plant materials, and the application of commercial nitrogen fertilizers. It can be lost from the cycle to the atmosphere, removed by harvesting crops, or lost to surface water or groundwater. The nitrogen cycle (Fig. 2.8) shows nitrogen changing from organic matter in the soil to bacteria, plants, and back to the organic matter. Plant and animal wastes decompose, adding nitrogen to the soil. Plants use the nitrogen in the soil to grow. People and animals eat plants. Subsequently, animal and plant residues return nitrogen to the soil again, completing the cycle. For the most part, the nitrogen cycle is soil based.

Another way that nitrogen enters the cycle is as inorganic nitrogen from the atmosphere and factories. Rainstorms contribute atmospheric nitrogen through raindrops that reach the soil. Factories that produce nitrogen fertilizers add nitrogen to the soil when these fertilizers are applied to crops. Nitrogen in the air becomes part of the biological matter mostly through the actions of bacteria and algae in the nitrogen fixation process.

Human Influence on the Nitrogen Cycle

Due to extensive cultivation of legumes (particularly soy, alfalfa, and clover), the growing use of the Haber-Bosch process (to create ammonia) in the manufacture of chemical fertilizers, and pollution emitted by vehicles and industrial plants, human beings have more than doubled the annual transfer of nitrogen into biologically available forms.

The presence of ammonia (NH_3) in the atmosphere has tripled due to human activity. In the atmosphere, ammonia clings to water droplets, eventually resulting in nitric acid (HNO_3) that produces acid rain. Ecosystem processes can be enhanced by nitrogen fertilization, but anthropogenic input can also result in nitrogen saturation, which weakens productivity and can damage the health of plants, animals, fish, and humans. On-site sewage facilities, such as septic tanks, release large amounts of nitrogen into the environment. Human alterations to the global nitrogen cycle are most intense in places where vehicle emissions and industrial agriculture are the highest (Gupta, 2012).

2.4.3 Linkages Among Hydrologic, Carbon, and Nitrogen Cycles

Hydrologic, carbon, and nitrogen cycles are linked mainly through the biosphere. In particular, the HC is a biogeochemical cycle, which means that the HC affects other biogeochemical cycles, such as the carbon cycle and the nitrogen cycle. In turn, the HC also is affected by these other cycles in a feedback system. For example, eroded sediment and phosphorus are carried by surface runoff to rivers and water bodies. Phosphorus is applied to agricultural lands through fertilizers. Its quantity over and above that consumed by plants is carried by surface runoff to water bodies and causes eutrophication. Similarly, nitrogen is also transported to water bodies. The transport of eroded sediment and dissolved salts is responsible for the salinity of oceans and seas. For example, the water in the Gulf of Mexico is facing hypoxia (a condition in which there is inadequate oxygen supply) as a result of nitrates from fertilizer being transported there by rivers, including the Mississippi.

2.5 Influence of Human Activity and Land-Use Changes on the Hydrologic Cycle

Watersheds are subjected to many types of changes, both major and minor, due to various causes. Some of these are natural changes, and others are due to human activity. Watershed changes affect virtually all elements of the HC. There is a need to understand and quantify how human-induced changes alter the HC so that timely actions can be taken to minimize damages caused by adverse impacts as well as harness the benefits due to positive changes. The classical and the modern viewpoints of the role of humans in the HC are shown in Figs. 2.9 and 2.10.

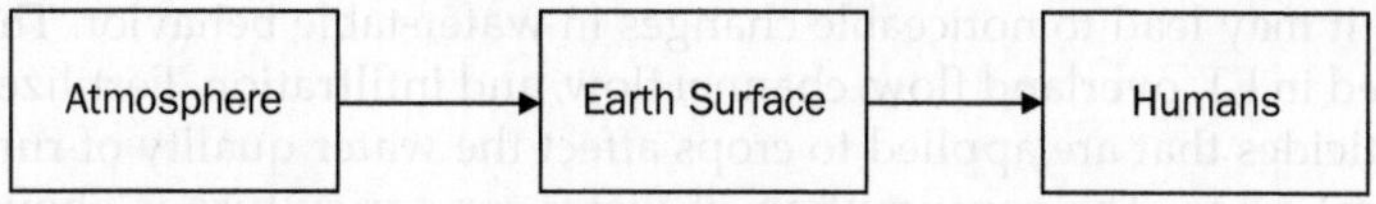

FIGURE 2.9 The classical viewpoint on the place of humans in the HC.

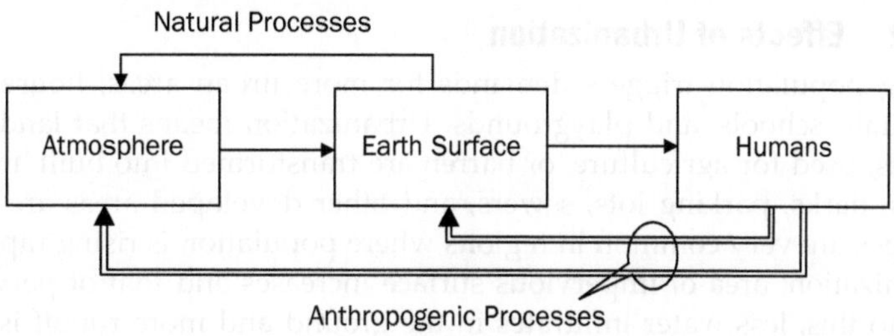

FIGURE 2.10 The modern viewpoint on the role of humans in the HC. (Source: NRC (1982).)

A number of factors impact the hydrology of a river basin. Variability in the various components of HC arises due to natural causes as well as anthropogenic reasons. At most places, natural hydrologic regimes have undergone substantial changes due to large scale water withdrawals and diversions to meet the societal needs as well as due to changes in land use. Human intervention, such as, dams, diversions, interbasin transfers, have significantly impacted river flows and sediment transport in many river reaches. Interventions can change the partition of water among infiltration and surface runoff; peaks of river flow hydrographs may increase (e.g., due to urbanization) or may fall (e.g., due to afforestation). Due to climate change, water and other cycles are being impacted. It will be important to understand and isolate the impacts of individual causes so that appropriate mitigation measures can be taken.

Watershed changes can be distinguished as point changes or nonpoint changes. Structural changes, such as dam construction, channel improvement, and detention storage, are examples of point changes. These may change relative magnitudes of components of HC such as evaporation, seepage, residence/travel time, etc. Afforestation, agriculture, mining, and urbanization are nonpoint changes that affect catchment response.

A qualitative discussion of the hydrologic consequences due to watershed changes follows next.

2.5.1 Effects of Agricultural Changes

The changes discussed here imply that a land area that was earlier forested or barren is now being cultivated. As a result, the vegetal cover changes, soil crusting and infiltration characteristics change, and artificial bunds may be placed, causing changes in water retention and infiltration. Cultivators may withdraw water from surface sources (i.e., rivers or canals) and subsurface sources and apply to the cultivated areas. The effect of these changes and intervention in the hydrologic regime is pronounced and may be multiplicative. When water is withdrawn from the groundwater zone, it may lead to noticeable changes in water-table behavior. The impacts are also noticed in ET, overland flow, channel flow, and infiltration. Fertilizers, pesticides, and insecticides that are applied to crops affect the water quality of runoff from the agricultural areas. The consumption of water for agriculture is about 70% of the total global consumption, and in some countries it is as high as 90% (e.g., Pakistan), about 80% (e.g., India), and about 50% (e.g., the United States). This use has implications for almost the entire HC.

2.5.2 Effects of Urbanization

Rising population triggers demands for more urban areas, houses, roads, markets, hospitals, schools, and playgrounds. Urbanization means that lands that were under forests, used for agriculture, or barren are transformed into built areas where houses, roads, parks, parking lots, sewers, and other developed areas are constructed. Such changes are very common in regions where population is rising rapidly. As a result of urbanization, area of impervious surface increases and that of pervious surface falls. Due to this, less water infiltrates in the ground and more runoff is produced. Urban development usually increases the volume and peak of direct runoff for a given rainfall event. The time of travel of water is reduced, resulting in a lower lag time and a shorter time of concentration. In brief, the hydrologic effects of urbanization are (1) increased water withdrawals from surface and subsurface sources, (2) increased peak flow and diminishing base flow of streams, (3) reduced infiltration, and (4) increased pollution of rivers and aquifers, endangering the ecology. Urban authorities in many places are now using pervious pavement, which helps increase infiltration. In some cases, paved places are being modified into grasslands to reduce runoff and surface albedo.

Built areas are also heat islands, and changes in the local microclimate are noticeable as a result. The temperature in urban areas can be about 2–4°C higher than nearby rural areas, which can affect precipitation.

2.5.3 Effects of Forest Activities

Forest-related activities involve the planting as well as cutting of trees, soil erosion control and water management, and the use of chemical fertilizers and sprays. The immediate effect of these activities is change in vegetal cover. When a forested area is deforested and forest litter is removed, the interception of precipitation is reduced considerably. Litter removal changes the infiltration capacity of soil and has a pronounced effect on the raindrop impact and the resulting soil erosion. ET generally decreases with the removal of vegetation. These changes amount to increased production of direct runoff, reduced surface roughness, and decreased recharge to groundwater. The hydrograph of direct runoff rises more quickly and the time-to-peak reduces. However, when additional trees are planted in an area, the effect is the reverse, although the impact takes place gradually as trees grow. In general, infiltration is more in forests. During low-intensity rains, lesser flow is produced from forested areas compared to other land uses. Thus, forests can have a moderating impact on flood generation.

2.5.4 Effects of Structural Changes

Typical structural changes in hydrosystems include dams, weirs, levees, diversions, and channel improvement works. Regardless of the intended functions, such changes affect the hydrology of the stream on which these features are built. In general, the peak of outflow from a reservoir is less, and the flow may be more uniform than the prestorage condition. The volume of flow downstream may be considerably less in the after-project scenario if the reservoir water is diverted elsewhere.

Channel improvements have an opposite effect on the hydrograph. Decreasing channel roughness, for example, increases flow velocity and discharge for the same channel cross section. The removal of vegetation, lining of the channel, and proper maintenance can greatly reduce roughness. When the river channels are straightened and bends are removed or smoothened, the flow velocity increases and travel time reduces. Depending upon the bed material, infiltration through the bed and banks also modifies flow characteristics. Construction of highways, railways, and airports can affect the hydrologic processes significantly at the local scale.

2.5.5 Changes in Water Quality

Even if an adequate quantity of water exists in a place, its use may be limited because of poor quality. The quality of water is significantly deteriorating in many places due to human, industrial, and agricultural activities. Changes in the quality of water in natural bodies affect aquatic life, environment, and water use.

2.6 Impact of Climate Change on the Hydrologic Cycle

The increased emission of greenhouse gases due to anthropogenic activity is believed to be the cause of the gradual warming of Earth's lower troposphere. Analysis of measured hydrometeorological data, proxy indicators, and paleoclimatological records suggests that the climate of the Earth may be undergoing significant changes. The major changes that could affect climate are changes in vegetation (changes in ET, soil moisture, albedo, and radiation balance), increased use of water for day-to-day needs, burning of fossil fuels, and industrial activities. These changes may be taking place due to both natural and human causes, but in the recent times, human actions are responsible for most of the changes in HC. Climatic change will have important impacts on the HC, and vice versa.

Atmospheric and oceanic circulations are two of the major factors that determine the distribution of climatic zones over the Earth. Changes in these cycles or circulations can cause major climatic shifts. An increase in the temperature of the atmosphere would influence wind patterns; changes in precipitation pattern, timing, and distribution; melting of polar ice caps; and recession of glaciers. Because the HC is a thermally driven system, rises in atmospheric temperatures will affect many of its components. The change manifests itself in changed patterns of spatial and temporal variability in the components of the HC. Additionally, the acceleration of the HC (higher temperatures mean more evaporation, and thus more precipitation) may result in more severe weather conditions. Note that some scientists do not agree with this statement. Some scientists believe that the increased frequency and severity of El Niño events in recent decades have occurred due to the acceleration of the HC induced by global warming. It will also cause higher ET.

If average global temperatures continue to increase as they have in recent decades, water that is currently trapped as ice in the polar ice sheets will melt, causing a rise in sea levels. Water also expands as it gets warmer, further exacerbating the sea level rise. Many heavily populated coastal areas, like New Orleans and Bangladesh, will be inundated by a mere 1-m rise in sea levels.

Global warming is likely to lead to higher ET; changes in precipitation pattern, timing, and distribution; and increased melting and recession of glaciers. Most climate scientists agree that climate warming will intensify, accelerating or enhancing the global HC. Enhancement could be caused by increasing rates of ET, precipitation, and streamflow. There are likely to be associated changes in the atmospheric water content, soil moisture, ocean salinity, and glacier ice content.

The mode of precipitation is as important as the magnitude in determining the hydrologic impacts, and precipitation variability at multidecadal scales can mask long-term trends. Widespread increases in heavy precipitation events have been observed in some places where total precipitation has decreased. In addition, more precipitation now falls as rain rather than snow in northern regions. The results of the reported studies suggest that over large areas of Asia and North America, actual ET is increasing on average, even though pan evaporation is decreasing.

2.7 Summary

All studies dealing with water are concerned with some part(s) of the hydrologic cycle. The availability and quality of water at many places is linked with the growth of human civilization. As a result of human activities, the features/characteristics of many components of the hydrologic cycle are undergoing changes. It is necessary to predict the magnitude and extent of these changes so that the required mitigation measures to minimize harmful impacts can be initiated well in time.

2.8 Questions

2.1 What percentage of freshwater on the Earth is contained in rivers, streams, and lakes?

i. More than 50% ii. Nearly 100%
iii. Less than 1% iv. Between 10% and 20%

2.2 From which source does the energy to move the HC come?

i. Movement of the Earth ii. Solar energy
iii. Both movement of the Earth and the Sun iv. Fossil fuels

2.3 From where does transpiration of water take place?

i. Lakes and reservoirs ii. All types of plants (vegetation)
iii. Plants and oceans iv. All plants, oceans, and lakes

2.4 Explain the significance of HC in water management.

2.5 What are the major components of HC? Which of these are critical to arid and humid areas?

2.6 Besides HC, several other cycles operate in nature. Write a short note on what these cycles are and how they interact with the HC.

2.7 What will the impact of global warming be on carbon and nitrogen cycles?

2.8 Give five main human actions that affect the HC, and identify the components affected by each of these.

2.9 A barren area is converted to forest by planting trees. How the HC will be affected in the area?

2.10 The significance of different components of the HC changes based on the time scale. For a given watershed (the space scale is constant), consider time scales of 1 hour, 1 day, 1 week, 1 month, 1 season, and 1 year. Which component will be more significant at which scale?

2.11 The significance of different components of the HC changes with space scale. For a given time scale, consider space scales such as a small hillslope, a small watershed, a medium-size watershed, and a large watershed. Which component will be more significant at the given scale?

2.12 Can hydrology be defined as the study of the HC?

2.13 Hydrology is sometimes called the science of water. Is that true?

2.14 In hydrology, water quantity has historically received far more emphasis than water quality. Is this focus justified? If yes, then why?

2.9 Activities

2.1 Visit some water bodies in and around you, note down the components of HC, and try to ascertain their magnitude.

2.2 Take a survey to find the primary sources of water in your area. Find out about the quality of water. Has it undergone significant change with time? If yes, why?

2.3 Has human activity affected the HC in your area? If yes, in what ways, and has the impact been good or bad?

2.10 References

ASCE. *Handbook of Hydrology*. ASCE Manual and Reports on Engineering. Practice No. 28. Reston, VA: ASCE, 1996.

Bernhard, A. The nitrogen cycle: processes, players, and human impact. *Nature Education Knowledge*, 3(10):25, 2010.

Carlson, C. A., N. R. Bates, D. A. Hansell, and D. K. Steinberg. *Carbon Cycle, in Encyclopedia of Ocean Sciences.* 2nd ed (pp. 477–486). New York: Academic Press, 2001.

Committee on "Opportunities in the Hydrologic Sciences" of the National Research Council (NRC). *Opportunities in the Hydrologic Sciences.* Washington, DC: National Academy Press, 1991.

Creative Commons. Home page. http://cnx.org. Accessed on June 13, 2017.

Dessler, A. E. *Introduction to Modern Climate Change.* New York: Cambridge University Press, 2012.

Dunn, T., and L. B. Leopold. *Water in Environmental Planning.* San Francisco: W. H. Freeman, 1978.

Gupta, R. C. Energy and Environmental Management in Metallurgical Industries. PHI Learning Pvt. Ltd., New Delhi, 2012.

Horton, R. E. The field, scope, and status of the science of hydrology. in Trans. AGU, Reports and Papers, Hydrology, National Research Council, Washington, D.C, pp. 189–202, 1931.

National Resource Council (NRC). *Scientific Basis of Water Resource Management.* Washington, DC: National Academy Press, 1982.

National Resource Council (NRC). *Global Change in Geosphere-Biosphere.* Washington, DC: National Academy Press, 1986.

Seckler, D., U. Amarasinghe, D. Molden, R. de Silva, and R. Barker. *World Water Demand and Supply, 1990 to 2025: Scenarios and Issues.* Research Report 19. Colombo, Sri Lanka: International Water Management Institute, 1998.

Shiklomanov, I. A. *World Water Resources: Modern Assessment and Outlook for the 21st Century. Prepared in the Framework of IHP, UNESCO.* St. Petersburg: State Hydrology Institute, 1999.

Shiklomanov, I. A. and John C. Rodda. World Water Resources at the Beginning of the Twenty-First Century, Cambridge University Press, Cambridge, UK, 2003.

Singh, V. P. (Ed.). *Handbook of Applied Hydrology.* 2nd ed. New York: McGraw-Hill, 2016.

United Nations Educational, Scientific, and Cultural Organization (UNESCO). *World Water Balance and Water Resources of the Earth.* Paris: United Nations Educational, Scientific, and Cultural Organization (UNESCO), 1978.

University Corporation for Atmospheric Research (UCAR). Home page. https://eo.ucar.edu, accessed on March 16, 2018.

WWAP (2009). World Water Development Report 3. Water in a Changing World. World Water Assessment Programme. The United Nations.

CHAPTER 3
Water Balance

Water balance is one of the fundamental concepts in hydrology. Discussing the basic concepts of water balance, this chapter presents scales in hydrology and water balance equations for various space-time scales. It shows how the formulation of the balance equation changes with scale, for the significance of components varies with the scale. At large time scales, evapotranspiration (ET) becomes one of the dominant components, but it is quite small at small time scales, such as a minute or an hour, particularly in humid conditions. Similarly, with small watersheds, rainfall can be considered as quite uniform, but not on large watersheds. The chapter concludes with a discussion of global water balance.

3.1 Water Balance—The Mathematical Representation of the Hydrologic Cycle

Every hydrologic system follows the law of conservation of mass, which is expressed as water balance, or more popularly known as the *continuity equation.* The hydrologic system under study can be a catchment, a reservoir or lake, a glacier, a channel or its segment, or an aquifer. In terms of space, the hydrologic system may cover a small watershed in one or more countries, a region, a continent, or the entire Earth. Depending on the spatial scale, the size of the time step can be taken hourly, daily, weekly, monthly, or yearly.

The quantities of water going through the components of the hydrologic cycle can be evaluated by the so-called hydrologic equation, which is a simple, spatially lumped continuity or water budget equation, expressed as

$$I - Q = \Delta S \tag{3.1}$$

where I is the inflow of water to a given area during any given time period, Q is the outflow of water from the area during the selected time period, and ΔS is the change in storage of water in the given area during the time period. If I and Q vary continuously with time, then Eq. (3.1) can be written as

$$dS(t)/dt = I(t) - Q(t) \tag{3.2}$$

Integration of Eq. (3.1) yields

$$\int dS(t) = \int [I(t) - Q(t)]\, dt$$

or

$$S(t) - S(0) = \int_0^t I(t)dt - \int_0^t Q(t)dt = V_I(t) - V_0(t) \tag{3.3}$$

where $S(0)$ is the initial storage at time $t = 0$, $S(t)$ is the storage at time t, and $V_0(t)$ and $V_I(t)$ are the volumes of outflow and inflow at time t. Each of the terms of this lumped equation is the result of a number of other terms. These can be subdivided or even eliminated from the equation, depending upon the temporal and spatial scale of the study (Jain and Singh, 2006).

The continuity equation is one of the governing equations for almost all hydrologic problems. For a watershed, Eq. (3.1) may be written as

$$P + Q_{SI} + Q_{GI} - E - Q_{SO} - Q_{GO} - S - \varepsilon = 0 \tag{3.4}$$

where P is the precipitation over the watershed, Q_{SI} is the surface inflow, Q_{GI} is the groundwater inflow, E is the evaporation from the watershed, Q_{SO} is the surface water outflow, Q_{GO} is the groundwater outflow, ΔS is the change in the storage of water in the watershed, and ε is the discrepancy term. For large watersheds, Q_{GI} and Q_{GO} are usually negligible. The discrepancy term is included in Eq. (3.4) because the sum of all the other terms may not be zero due to measurement errors, simplifying assumptions, or both. A small value of the discrepancy term, however, does not necessarily mean that all other terms have been correctly measured or estimated.

Depending on a specific problem, the terms of Eq. (3.4) may be further subdivided. For example, when applying the hydrologic equation for short time intervals, the change in the total water storage (ΔS) may be subdivided into several parts: changes of moisture storage in the soil (ΔM), in aquifers (ΔG), in lakes and reservoirs (ΔL), in river channels (ΔS_C), in glaciers (ΔS_G), and in snow cover (ΔS_S). Thus, ΔS can be expressed as

$$\Delta S = \Delta M + \Delta G + \Delta L + \Delta S_C + \Delta S_G + \Delta S_S \tag{3.5}$$

The water balance equation may be applied for any time interval; the computation of mean annual water balance for a basin is the simplest, since it is possible to disregard changes in water storages in the basin (ΔS), which are difficult to measure and compute. In general, the shorter the time interval, the more stringent are the requirements for measurement or computation of the components and the more subdivided are the terms of Eq. (3.1). This results in a complex equation that is difficult to close with an acceptable level of error (Jain and Singh, 2006).

The hydrologic equation may be applied to areas of any size, but the complexity of computation greatly depends on the extent of the area under study. The smaller the area, the more complicated is its water balance because it is difficult to estimate some components of the equation. Finally, the components of the hydrologic equation may be expressed in terms of any consistent unit, such as the mean depth of water (mm), or the volume of water (m^3), or in the form of flow rates (m^3/s or mm/s).

3.2 Forms of the Water Balance Equation

Many forms of the water balance equation are possible by subdividing, consolidating, or eliminating some of the terms, depending upon the purpose of computation. Water balance techniques are also a means for solving important theoretical and practical hydrological problems. Water balance enables a quantitative evaluation of water resources and changes in them under the influence of human activity and climate change.

The relative and absolute magnitude of the water balance of lakes, river basins, and groundwater basins forms a basis for the hydrological substantiation of projects for the rational use, control, and redistribution of water resources in time and space (interbasin water transfer, streamflow control, etc.). Knowledge about water balance assists in the prediction of consequences of artificial changes in the regime of streams, lakes, and groundwater basins.

An understanding of the water balance is also important for the study of the hydrological cycle. With water balance data, it is possible to compare individual sources of water in a basin over periods of time. Further, analysis of the individual water balance components and the interaction of these components in the balance equation makes it possible to identify deficiencies in the distribution of observational stations and discover systematic measurement errors.

It is also highlighted that different users of water are interested in different water balance components. For example, farmers are interested in soil moisture, disaster relief agencies are interested in flood data, and hydropower utilities are interested in river flow and reservoir storage. Current information on the water balance of river and lake basins for short-time intervals (season, month, week, and day) is used for the operational management of reservoirs and for the compilation of hydrological forecasts for flood management (Sokolov and Chapman, 1974).

Finally, water balance provides an indirect evaluation of an unknown water balance component from the difference between the known components (e.g., long-term evaporation from a river basin may be computed by the difference between precipitation and runoff). In general, the use of water balance implies the measurement of both storage and fluxes (rates of flow) of water. Sokolov and Chapman (1974) note that by judicious selection of the volume and period of time for which the balance will be applied, some measurements can be eliminated.

The water balance equation [Eq. (3.1)] may be simplified or made more complex, depending upon several factors: the available data, the purpose of computation, the type and size of the water body (river basin or artificially separated administrative district, lake or reservoir, etc.), and hydrographic and hydrologic features. The water-balance time interval and the phase of the hydrological regime (flood, low flow) for which the water balance is computed are also important considerations.

In large river basins, Q_{GI} and Q_{GO} are small compared with other terms and can be ignored. In other words, subsurface water exchange with neighboring basins is assumed to be zero. There is no surface water inflow into a river basin with a distinct watershed divide (assuming no human-induced exchange of water with other basins), and therefore, Q_{SI} is not included in the water balance equation of a river basin. Thus, for a large river basin, Eq. (3.1) can be simplified as

$$P - E - Q_{SO} - \Delta S - \varepsilon = 0 \tag{3.6}$$

The water balance may be computed for any time interval. Sokolov and Chapman (1974) classified water balance in two categories: mean water balances and balances for specific periods (such as a year, season, month, or number of days), sometimes called *current* or *operational water balances*. Computations of mean water balances are typically performed for a calendar year or a hydrological year. Of course, such computations may be performed for any other time interval, a season or a month. Note that a month is not a natural time unit; in different calendars, a month is interpreted differently.

The computation of mean annual water balance is the simplest because it is possible to disregard changes in water storage in the basin (ΔS), which are difficult to measure and compute. Over a long period, positive and negative water storage variations for individual years tend to balance out, and their net impact over a long period may be assumed to be negligible.

The computations of water balance for short time intervals are somewhat difficult. The shorter the time interval, the more precise are the requirements for measurement or computation of water-balance components and the more subdivided should be the values of S and other elements. This results in a complex water balance equation that is difficult to close with an acceptable level of error. The term ΔS must also be considered in the computation of the mean water balance for seasons or months.

A river basin is the only natural area for which large-scale water balance computations can be simplified. The accuracy of computation generally increases with the increase in the river basin area. In general, the smaller the basin area, the more complicated is its water balance. This is because it is difficult to estimate secondary components of the balance, such as groundwater exchange with adjacent basins; water storage in lakes, reservoirs, swamps, and glaciers; the dynamics of water balance of forests; and irrigated and drained lands.

In contrast to a river basin, the water balance computations for lakes, reservoirs, and glaciers become more complex as the areas of these bodies increase. This happens due to several reasons. In a large lake, there may be a number of inflowing channels which can't be measured individually. It is difficult to correctly measure/estimate change in water storage of a large lake. Same is the case with a large glacier whose measurements pose additional difficulties due to remoteness and accessibility. Estimation of precipitation input to a large water body with high accuracy is also difficult.

3.3 Closure of the Water Balance Equation

Here the term closure implies correct estimation of individual components of the applicable water balance equation so that the right-hand size is zero or close to it. To properly close the water balance equation, individual water balance components should be measured/estimated independently as feasible. However, due to limitations of measurement methods and practical reasons, the equation may not always balance. The discrepancy of water balance (ε) is given as a residual term of the water balance equation. This term includes errors in the determination of the components considered and the values of components not taken into account by the particular form of the equation being used. A low value of ε may indicate only that component parts of water balance equation tend to approximately balance out, or that all the components have been included and correctly estimated.

If a component of water balance equation can't be measured or estimated, its value can be estimated by using the water balance equation if values of all the other components are available and the right-hand side is assumed as zero. This, however, is not a preferred solution since the value so estimated will contain an error which is the net error in estimation of all the remaining components. At times, this magnitude of this net error may be even bigger than the magnitude of the component that has been estimated by following this approach.

3.4 Global Water Balance

Table 3.1 shows quantitative values of water present in various forms on the Earth. Volumewise, glaciers and permanent snow cover have more water than groundwater, and atmosphere has six times more water than river flows.

	Distribution Area (10^3 km^2)	Volume (10^3 km^3)	Layer (m)	Percentage of Global Reserves	
				Of total water	Of freshwater
World ocean	361,300	1,338,000	3,700	96.5	–
Groundwater	134,800	23,400	174	1.7	–
Freshwater		10,530	78	0.76	30.1
Soil moisture		16.5	0.2	0.001	0.05
Glaciers and permanent snow cover	16,227	24,064	1,463	1.74	68.7
Antarctic	13,980	21,600	1,546	1.56	61.7
Greenland	1,802	2,340	1,298	0.17	6.68
Arctic islands	226	83.5	369	0.006	0.24
Mountainous regions	224	40.6	181	0.003	0.12
Ground ice/ permafrost	21,000	300	14	0.022	0.86
Water reserves in lakes	2,058.7	176.4	85.7	0.013	–
Fresh	1,236.4	91	73.6	0.007	0.26
Saline	822.3	85.4	103.8	0.006	–
Swamp water	2,682.6	1.47	4.28	0.0008	0.03
River flows	148,800	2.12	0.014	0.0002	0.006
Biological water	510,000	1.12	0.002	0.0001	0.003
Atmospheric water	510,000	12.9	0.025	0.001	0.04
Total water reserves	510,000	1,385,984	2.718	100	N/A
Total freshwater reserves	148,800	35,029	235	2.53	100

TABLE 3.1 Water Reserves in Various Phases of the Hydrologic Cycle (Source: UNESCO (1978))

Table 3.1 shows that although the Earth is known as "the blue planet," the usable freshwater resources are finite and limited. Approximately 2.5% of the world's water resources makes up the total water supply. Depending on the source of water, the residence time, as shown in Table 3.1, varies widely. River flow has about 2 weeks of residence time, but groundwater can take thousands of years to recharge naturally. Likewise, the residence time in glaciers can also vary from 10 to thousands of years.

An approximate distribution of freshwater on the Earth is shown in Table 3.2. Interestingly, only limited freshwater is available as groundwater and surface water. Although small in comparison with freshwater stored in ice caps and glaciers, groundwater in the upper 800 m of depth is almost 35 times greater than the freshwater stored in lakes, rivers, and inland seas, which is only about 0.32% of world's freshwater. If this

Source	Volume ($\times 10^6$ km^3)	Percentage
Polar ice caps and glaciers	29.0	77.33
Groundwater at depths up to 800 m	4.15	11.07
Groundwater at depths below 800 m	4.15	11.07
Lakes, rivers, and streams	0.12	0.32
Soil moisture and seepage	0.067	0.18
Atmospheric water vapor	0.013	0.03
Total	37.50	100

TABLE 3.2 Approximate Distribution of Freshwater on the Earth

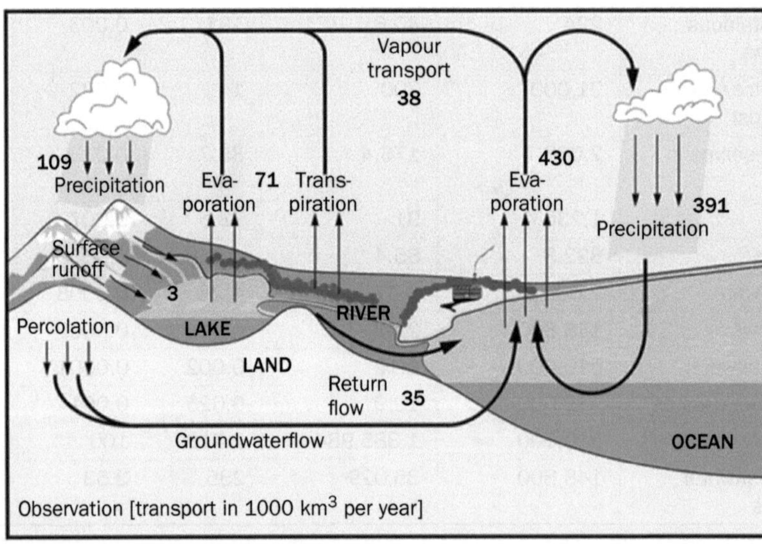

FIGURE 3.1 The global water cycle following. Annual values are in units of 10^3 km^3 year-1. (Source: http://iopscience.iop.org/article/10.1088/1748-9326/5/2/025202/meta)

table is compared to Table 2.1 and 3.1, some difference in estimates will be noted. This highlights the uncertainties in such global estimates from different sources which arise due to use of different methods, data, assumptions, etc.

The storages and fluxes of the hydrologic cycle are shown in Fig. 3.1. According to the estimates (Oki and Kanae, 2006), the annual average volume of precipitation on the land surface is about 111×10^3 km^3. Out of this amount, about 65.5×10^3 km^3 is returned to the atmosphere as ET and the runoff from land to oceans is 45.5×10^3 km^3. Regarding the water balance of oceans, the volume of precipitation over them is about 391×10^3 km^3, 45.5×10^3 km^3 of water is received as runoff from the land, and 436.5×10^3 km^3 is lost

Continent	Precipitation (mm)	Precipitation (km³)	Evaporation (mm)	Evaporation (km³)	Runoff (mm)	Runoff (km³)
Europe	790	8,290	507	5,320	283	2,970
Asia	740	32,200	416	18,100	324	14,100
Africa	740	22,300	587	17,700	153	4,600
North America	756	18,300	418	10,100	339	8,180
South America	1,600	28,400	910	16,200	685	12,200
Australia and Oceania	791	7,080	511	4,570	280	2,510
Antarctica	165	2,310	0	0	165	2,310
Land as a whole	800	119,000	485	72,000	315	47,000
Areas of external runoff	924	110,000	529	63,000	395	47,000[a]
Areas of internal runoff	300	9,000	300	9,000	34	1,000[b]

[a] Including underground water not drained by rivers.
[b] Lost in the region through evaporation.

TABLE 3.3 Water Balance at the Continental Scale (Source: Shiklomanov (1993))

as evaporation. If we consider the water balance of atmosphere, 436.5×10^3 km³ of water is received as evaporation from oceans and 65.5×10^3 km³ from land. Out of the 502×10^3 km³ of water received every year, 391×10^3 km³ is the precipitation over oceans and 111×10^3 km³ is over land.

3.4.1 Continental Water Balance

Table 3.3 presents mean water-balance components (i.e., precipitation, evaporation, and runoff) for seven continental regions and the land as a whole. The water balance of continents gives an estimate of renewable resources available for meeting the needs of the society. As expected, water is not uniformly distributed across the continents.

Asia receives the largest volume of precipitation, followed by South America and Africa. Antarctica receives the least volume of precipitation. Evaporation in Antarctica is nil, while on all the other continents, it makes up more than 50% of the total volume of precipitation. Africa loses about 80% of the precipitation due to evaporation. The total evaporation from the land as a whole is equal to about 60% of the total precipitation.

South America is the wettest continent, Asia the second wettest, and Oceania is the driest region. Asia receives the highest volume of precipitation and also produces the highest runoff, followed by South America. It may also be noted that large continents, such as Africa, have immense variability, which is hidden in the average values.

3.4.2 River Basin Water Balance

Table 3.4 shows the average annual flows of large rivers of the world. Clearly, the world's largest river is the Amazon, with nearly one-fifth of the world's total river flow.

Management of river basin water resources depends on the components of water balance. For example, consider the Krishna River basin in peninsular India, which

No	Continent	River	Length (km)	Drainage area (km²)	Average discharge (m³/s)	Outflow
1	S. America	Amazon	6,400	7,050,000	209,000	Atlantic Ocean
2	Africa	Congo	4,371	4,014,500	41,200	Atlantic Ocean
3	S. America	Orinoco	2,140	880,000	40,000	Atlantic Ocean
4	Asia	Yangtze (Chang Jiang)	6,418	1,808,500	30,166	East China Sea
5	S. America	Negro	2,230	691,000	28,400	Amazon River
6	S. America	Río de la Plata	290	3,100,000	22,000	Atlantic Ocean
7	Asia	Brahmaputra	3,848	712,035	19,800	Till Padma River
8	Asia	Yenisei	3,438	2,580,000	19,600	Kara Sea
9	S. America	Paraná	4,880	2,582,672	17,290	Río de la Plata, Atlantic Ocean
10	Asia	Lena	4,472	2,490,000	16,871	Laptev Sea
11	N. America	Saint Lawrence	3,058	1,030,000	16,800	Gulf of Saint Lawrence
12	N. America	Mississippi	3,730	3,202,230	16,792	Gulf of Mexico
13	S. America	Marañón	1,737	358,000	16,708	Amazon River
14	Asia	Ganges	2,510	907,000	16,648	Till Padma River
15	Asia	Mekong	4,023	811,000	16,000	South China Sea
16	S. America	Tocantins	2,699	920,000	13,600	Atlantic Ocean
17	Asia	Irrawaddy	2,170	411,000	13,000	Andaman Sea
18	Asia	Ob'-Irtysh	5,410	2,990,000	12,475	Gulf of Ob
19	Asia	Amur	4,352	1,855,000	11,400	Sea of Okhotsk
20	N. America	Mackenzie–Peace–Finlay	4,241	1,790,000	10,300	Beaufort Sea
21	Asia	Pearl–Xi Jiang	2,400	453,700	9,500	South China Sea
22	Asia	Volga	3,692	1,380,000	8,060	Caspian Sea
23	Europe	Danube	2,860	817,000	7,130	Black Sea
24	Africa	Zambezi	2,574	1,331,000	7,070	Indian Ocean
25	Asia	Indus	3,200	1,165,000	6,600	Arabian Sea

Table 3.4 Length, Drainage Area, and Average Annual Discharge of Selected Large Rivers in the World (Source: https://en.wikipedia.org/wiki/List_of_rivers_by_discharge)

covers an area of 259,000 km^2. The average annual rainfall for the river basin is 926 mm. The Krishna Water Dispute Tribunal assessed the 75% dependable yield of the basin as 61,500 million cubic meters (MCM) and the average yield as 73,000 MCM. On this basis, allocation of water was decided among warring factions.

3.4.3 The Impact of Natural and Human-Induced Changes on Water Balance

As the state of a river basin changes due to natural and human-induced causes, the magnitude of various water balance terms also changes. Due to natural causes, the magnitude, pattern, and form of precipitation (solid or liquid) may change, and this will affect evaporation, river flow, and groundwater recharge.

With climate change looming on the horizon, the water balance components may experience significant changes. Different regions will be affected differently. Wet areas may become wetter and dry areas drier, or wet areas may become less wet and dry areas les dry. Seasonality of precipitation may change as well. Human actions are also changing land use and land cover, which also will affect the hydrologic response of catchments. Many studies, such as Schmied et al. (2016), have investigated variations of global and continental water-balance components as affected by climate change and human intervention.

3.5 Summary

Water is central to the environment and is considered to be the driver of nature. All studies dealing with water are concerned with some part or another of the hydrologic cycle. As a result of many human activities, the relative magnitude of many components of the hydrologic cycle are undergoing changes. It is necessary to predict the magnitude and extent of these changes so that the required ameliorative measures can be initiated well in time.

The hydrologic cycle in a true sense applies on a global scale because it is a closed system only on that scale. At all other scales (e.g., river basin or continental scale), it is not a closed system. The relative magnitude and importance of water balance terms depend on the time scales in question and the period. During intense storms, ET may be very small or negligible; during dry season, infiltration is typically zero because there is no precipitation.

3.6 Questions

3.1 For the country in which you are currently residing, find out the main components of water balance: average precipitation, average river flow, and average ET.

3.2 Find out the river basin in which you are living and the components mentioned in Question 3.1 for that river basin. Compare these values for the continent in which you are living, as given in Table 3.3.

3.3 If in a catchment with area of 20,000 km^2, the dominant land use changes from forest to agriculture over a period of 10 years, which water-balance component will change the most, and why?

3.4 Compare Tables 2.1 and 3.2 and note the differences. See that when we try to obtain estimates of a quantity from different sources, there are differences. Can you obtain better estimates of these quantities?

3.5 For a small, rural watershed in a semiarid climate, tabulate the relative significance of rainfall, surface runoff, interflow, baseflow, groundwater flow, infiltration, and ET at hourly, daily, monthly, and yearly time scales. Do the same for a similar-size watershed in a humid climate.

3.6 Consider a large watershed and do the same exercise as in Question 3.5.

3.7 Consider a small urban watershed and repeat the same exercise as in Question 3.5.

3.8 Consider a large urban watershed and repeat the same exercise as in Question 3.5.

3.9 When doing the water balance of a watershed, what key assumptions one makes with regard to space-time variability of different hydrologic-cycle components?

3.10 What are the main sources of error when computing the water balance of a watershed?

3.11 Consider a small watershed with an area of 1,000 hectares. This watershed received rainfall that occurred at an average rate of 2 cm/h for 2 hours. If all this water were collected in a reservoir whose size is 100 m by 100 m, then what would be the depth of water in this reservoir?

3.12 A large rainstorm producing rainfall of 200 mm/h for 10 hours occurs in an area, and about 25% of this storm becomes runoff. What would be the size of the reservoir in ha-m that would capture the entire runoff?

3.13 Consider rainfall, interception, depression storage, infiltration, ET, and runoff. Which components are most affected and which are not affected by high urbanization? Which components are most affected and which are not affected by forestation?

3.14 Which of the components listed in Question 3.13 is most important on an interstate highway?

3.15 Consider a watershed changing from rural to part urban, and part rural to part urban, and part forested to fully urban. Discuss the relative significance of the components of the hydrologic cycle from one state of the watershed to the other.

3.7 References

Oki, T., and S. Kanae. Global hydrological cycles and world water resources. *Science,* 313(5790): 1068–1072, doi:10.1126/science.1128845, 2006.

Jain, S. K., and V. P. Singh. Hydrologic Cycle. Van Nostrand's Scientific Encyclopaedia, https://doi.org/10.1002/0471743984.vse8624, 2006.

Schmied, H., L. Müller, S. Adam, G. Eisner, M. Fink, H. Flörke, et al. Variations of global and continental water balance components as impacted by climate forcing uncertainty and human water use. *Hydrology and Earth System Sciences Discussions,* doi:10.5194/hess-2015-527, 2016.

Shiklomanov, I. A. *World Water Resources: Modern Assessment and Outlook for the 21st Century. Prepared in the Framework of IHP, UNESCO.* St. Petersburg: State Hydrology Institute, 1999.

Shiklomanov, I. World water resources. In P. Gleick (Ed.), *Water in Crisis.* New York: Oxford University Press, 1993.

Sokolov, A. A., and T. G. Chapman. *Methods for Water Balance Computation. Studies and Research in Hydrology #17.* Paris: The UNESCO Press, 1974.

PART 2

Hydrologic Components

CHAPTER 4

Precipitation and Interception

Precipitation is one of the most important components of the hydrologic cycle, and that is the case at all space-time scales, unlike some other components of the hydrologic cycle. It also constitutes the input for hydrologic models. This chapter deals with different aspects of precipitation, including cloud formation and rainfall generation; measurement, validation, and processing of precipitation; areal averaging; spatial interpolation; filling-in of missing values; and disaggregation. Methods have been explained with illustrative examples to enable the reader to perform analyses.

4.1 Precipitation

Water vapors that form clouds come from evaporation from surface water bodies, from irrigated areas, and evapotranspiration (ET) from vegetation. The term *precipitation* refers to water released from clouds in the form of rain, freezing rain, sleet, snow, or hail. It is the primary mechanism by which atmospheric water is delivered to the Earth. Most precipitation in the vast areas of the world falls as rain.

Precipitation is a highly variable phenomenon; it does not fall in the same amount over a large region, or even in a city. Thunderstorms may deliver several centimeters of rain in one locality, while another area a few kilometers away may be entirely dry. At Cherrapunji, India, it rained 2,300 cm in 1861. In contrast to this excessive rainfall, some parts of Africa and Chile received no rainfall from October 1903 to January 1918 (more than 14 years)! In terms of time, rain falls fairly evenly during the year in some places. In others, it falls only during specific seasons. For example, in Asian countries, most annual rainfall occurs during the monsoon season. The frequency of rainfall varies from place to place, as well as from one amount to another. Similarly, the duration of rainfall, the number of days that rain falls (or rain events), and the number of days between rainfalls vary from place to place and year to year.

4.2 Cloud Formation

Condensation is the process by which drops of liquid water are formed from water vapor. It is necessary for the formation of clouds and precipitation. In the atmosphere, condensation occurs as a parcel of rising air expands and cools until some water vapor molecules join together. The heat released in this process is the heat that was absorbed when water was evaporated; this is known as the *latent heat* of condensation. This transfer of heat from the Earth's surface through evaporation to higher elevations in the atmosphere keeps the Earth's surface cooler than it would be if there were no water on the Earth.

When water evaporates from the Earth's surface, it is carried into the atmosphere as invisible vapor. Atmosphere always contains water vapor, and the maximum amount of water that it can contain depends on the temperature. The saturation level of an air parcel depends on its temperature; the temperature at which a given air parcel becomes saturated and begins to condense is known as the *dew point.* Air cools either through expansion or by coming into contact with a cool object such as a cold landmass or an ice-covered expanse.

Because vapor-laden air cools as it rises, after a certain elevation, the air is unable to hold water vapor. As a volume of unsaturated air cools, its relative humidity increases. If sufficiently cooled, the relative humidity becomes 100% and the corresponding temperature equals the dew point. If cooled further, the excess water vapor starts to deposit onto surfaces and objects in the atmosphere or on condensation nuclei (small particles or aerosols or ice particles), leading to the formation of droplets.

4.2.1 Classification of Precipitation

Precipitation is classified in many ways; the most common in hydrology is based on the factors that cause the lifting of moist air masses. Several mechanisms that cause air parcels to rise have been identified. In the orographic lift, the air is pushed against rising landforms, such as mountains. As an air mass rises, it cools down adiabatically (in an adiabatic process, there is no transfer of heat or matter between air mass and its surroundings), which may raise the relative humidity of air parcel to 100% and create clouds. The resulting precipitation is called *orographic precipitation. Convergent lift* takes place when air streams flow toward each other, which happens commonly on islands or peninsulas. Convection lift takes place with a parcel that is warmer and less dense than the surrounding air. In many places, daytime heating in summer season can produce convection clouds. Convective precipitation has high variability and is difficult to forecast. A *front* is formed when two air masses meet. A *cold front* is formed when a cold air mass moves over a warm one. A *warm front* is the opposite: a warm air mass advances onto a cold one. In frontal lifting, an air mass along a front is lifted over the other. Cyclonic precipitation is caused from lifting of air that converges in a low-pressure area.

Once saturated air is ready for condensation, particulate matter or particles in the air, such as dust or salt, is needed for raindrops to form on. These particles are called *condensation nuclei* because moisture condenses around them in layers. Cloud droplets are very small (say about 1/100 mm in diameter) and very light, and thus they float in the air. As the droplets collide, some of them grow and start to fall. In the process, they join more droplets and grow further. If the droplets become bigger than about 1/10 mm in diameter, they fall on the ground as rain. The larger drops fall at a faster velocity. On their downward path, they encounter several smaller raindrops. Some of these smaller drops would be captured most probably by bigger drops to form larger drops. Of course, this is a very simplified description of a highly complex rain formation process.

4.2.2 Cloud Types

Clouds are typically classified by their structure. Luke Howard, in 1803, introduced Latin names, such as *cirrus, cumulus, nimbus,* and *stratus,* for different types of clouds.

Based on appearance, clouds are basically divided into two classes: stratiform (single-layer) clouds and cumuliform (accumulated) clouds. Stratiform clouds are formed through the large-scale rising of entire layers of air, while cumulus clouds are primarily formed through the action of small-scale, vertical mixing processes.

Cumulus clouds appear like puffs of cotton; they are not very tall and are indicators of fair weather. The bottom of such clouds is frequently fairly close to the ground. If cumulus clouds grow tall, they can turn into thunderstorms. Stratus clouds appear as flat sheets of clouds. Their presence indicates an overcast day or steady rain. Sometimes stratus clouds may stay in one place for several days. Cirrus clouds are high, feathery clouds. When they are scattered in a clear blue sky, they are indicators of fair weather. *Nimbus* is a word associated with clouds; when this word is added to *cloud,* it typically means that precipitation is falling from the cloud. Figure 4.1 shows a system of classification of clouds.

The method of vertical lifting and the stability of the atmosphere also determine the type of cloud that may form. Cumulus clouds tend to form in unstable atmospheres. Layered clouds form in more stable environments in which large parcels of air are lifted slowly.

The formation of cloud may be localized. Typically, on a hot day in summer, air rising over a moist area may lead to the formation of cumulus clouds, which darken to rain clouds as more moisture condenses. Artificial rainmakers typically induce cumulus clouds to produce rain by seeding them with condensation nuclei such as silver iodide or dry ice, but the chances of successful cloud seeding vary. Rain-bearing clouds usually have large vertical growth; their large height takes them to higher levels with subzero temperatures, where ice particles are commonly present.

Cumulus and cumulonimbus clouds are also associated with thundershowers. These are shower clouds in which electrical discharges can be seen as lightning or heard

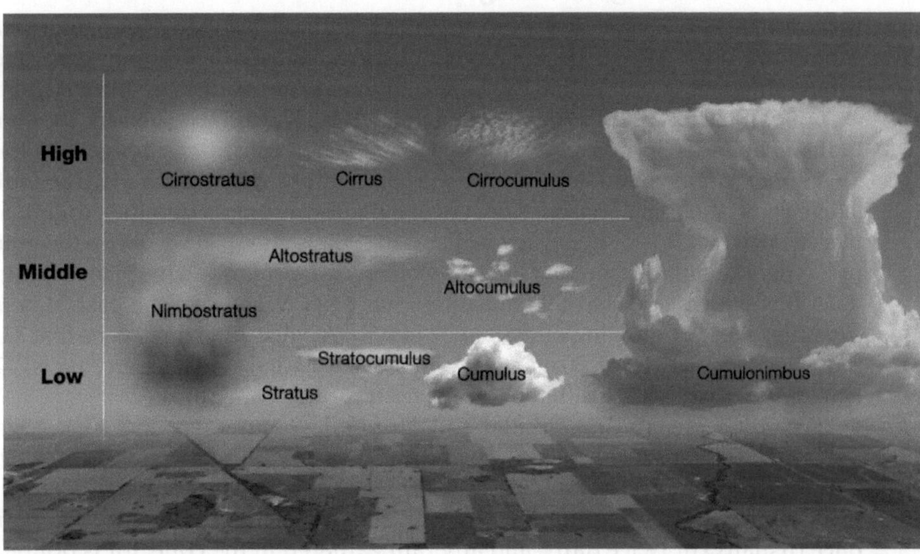

FIGURE 4.1 A system of classification of clouds. (Source: https://en.wikipedia.org/wiki/List_of_cloud_types#/media/File:Cloud_types.jpg)

as thunder. Thunderstorms develop when large amounts of cloud water, in the form of liquid or solid, are carried upward to heights where the temperature is frequently less than –20°C. Thunderstorms represent regions of strong convective motion in the atmosphere.

4.3 Observation of Precipitation

Precipitation is the most important meteorological input in hydrological studies, in which four features are of interest: amount, duration (intensity), frequency, and spatial distribution. The total amount of precipitation that reaches the ground in a given period is expressed as the depth at which it would cover a horizontal projection of the Earth's surface. If any part of the precipitation is snow or ice, its depth when melted is recorded. The unit of precipitation is length, and daily amounts should be read to the nearest 0.1 mm. Thus, rainfall of 1.51 cm over an area means that if all the rainfall in the given time is retained, it will form a 1.51-cm-thick water layer over that area. Weekly, fortnightly, and monthly amounts, however, should be read to the nearest 1 mm at least. Precipitation and other meteorological variables are commonly measured at fixed time(s) every day. For more information about observation and processing of precipitation data, refer to Jain and Singh (2003) and Jain (2011).

4.3.1 Rainfall Measurement

Rainfall is measured by a gauge that consists of a funnel to delineate the collection area and leads rainwater to a storage jar. Basically, there are three types of rain gauges: (1) standard or ordinary rain gauges (ORGs) are manually read, commonly once a day; (2) self-recording rain gauges (SRRGs) record rainfall depth in the form of a continuous plot; and (3) automatic rain gauges with data loggers.

4.3.2 Nonrecording Rain Gauges

Because the size, shape, and exposure affect the rain trapped by a gauge, standard gauges are used so that observations are correctly used in computations and compared. Different countries have adopted different rain gauges as their standard. For example, IS 5225: Meteorology—Raingauge, non recording, issued by the Bureau of Indian Standards specifies the requirements for non recording rain gauges for India. The Symon's rain gauge, for example, has been adopted as the standard in India. This gauge is installed on a masonry or concrete platform sunken into the ground, and it is placed such that it is perfectly leveled and the rim is about 30 cm above the ground. In India, the rainfall measured at 8:30 a.m. on any particular date is entered against that date. It is understood that this much rain has been received in the past 24 hours. For example, a record showing rainfall of 22 mm for August 12, 2015, means that 22 mm of rain has fallen between 8:30 a.m. on August 11, 2015 and 8:30 a.m. on August 12, 2015. Different countries have different norms for recording rainfall.

An ORG consists of a circular collector and a funnel (see Fig. 4.2). A standard Symon's rain gauge consists of a collector funnel having a rainfall collection area of either 200 cm^2 or 100 cm^2. The funnel leads to a base unit, partly embedded in the ground and containing a plastic or glass collector bottle. The gauge is read once (usually) or twice daily, and any rain gathered in the collector is poured into a graduated measuring glass cylinder to determine the rainfall depth in millimeters.

Because rainfall received is measured and noted manually, an observer may make errors. He or she may wrongly read the measuring cylinder; may wrongly note the

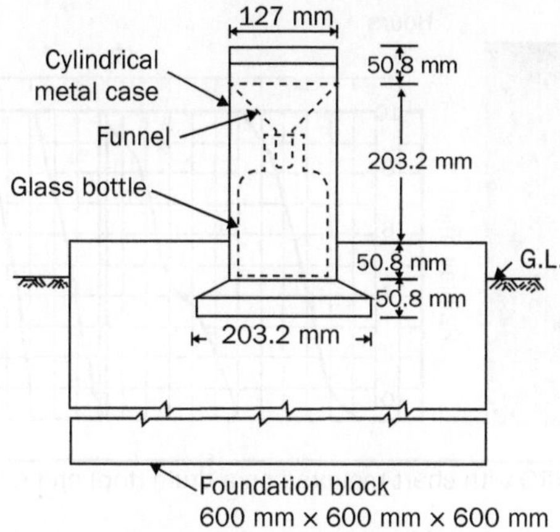

FIGURE 4.2 A typical nonrecording rain gauge installation.

amount; may read the gauge at the wrong time; may write the amount for the wrong date; or may use the wrong measuring glass (e.g., a 200-cm^2 glass for a 100-cm^2 gauge). Errors also may arise due to a faulty instrument: for instance, a damaged gauge rim will change the collection area; water will not reach the collection bottle if the funnel is blocked; or a collector bottle may be broken or leaking. Errors from most of these causes are difficult to detect unless the magnitude of error is large. Errors in the observations at a station are easy to detect if there is a concurrent record from a recording rain gauge at the same station or a nearby station.

4.3.3 Self-Recording Rain Gauge (SRRG)

The natural siphon rain gauge is commonly used to measure rainfall variability. It consists of a circular collector funnel and rainfall recording mechanism. The funnel leads to a chamber where a float is located. This float rises when rainwater enters the chamber. Attached to the float is a pen that records rainfall (in the form of rise of the float) on a chart mounted on a drum. A siphon chamber is attached to the float chamber. After 10 mm of rain has fallen, a siphon action is initiated. After the water is drained, the float returns to the original position.

The drum and the chart move with the help of a mechanical clock to complete one rotation in 1 day (24 hours). The horizontal axis of the chart is marked for hours, and the vertical axis represents the depth of rainfall. The chart is changed every day at the predetermined time, and hourly rainfall values are read and recorded on data sheets. On days with no rain, the pen traces a horizontal line on the chart. During a rainstorm, the pen produces a sloping line; the higher the rainfall intensity, the steeper the slope of the line will be. Rainwater drained by the siphon action may be collected in a chamber and measured; this can be a check for the total rainfall. Figure 4.3 shows an SRRG assembly.

Measurement errors can arise due to instrumental fault or observer error. If the float is wrongly adjusted, the siphon action may be initiated at a rainfall depth different than 10 mm. In very intense rainfall, the curves showing rainfalls may overlap and it might be difficult to distinguish them. Further, the clock may malfunction or may

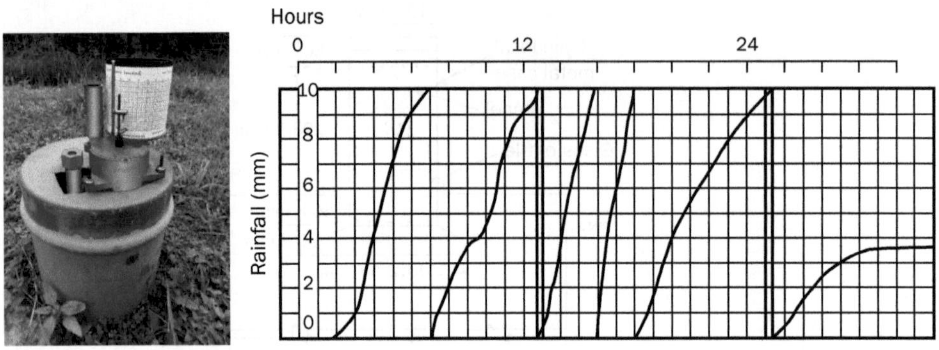

FIGURE 4.3 An SRRG with chart mounted on a drum (top) and rainfall recorded by the instrument on a chart.

stop and the timings will be incorrect. If the float sticks to the walls of the chamber, rainfall will be recorded wrongly. Further, the observer may read the data from the chart wrongly. Commonly, a nonrecording rain gauge is also installed near the recording gauge, and the daily rainfall tally recorded by the two is compared. Traditionally, if there is a discrepancy, ORG data are considered more reliable.

4.3.4 Tipping-Bucket Rain Gauge

A tipping-bucket rain gauge consists of a circular collector funnel that directs rain into a pair of tipping buckets that sit on a knife edge (Fig. 4.4). After rainwater has filled one side of the bucket with a small amount of rain (say, 0.025 cm), the assembly tips. An electrical pulse is generated on each tilt and is recorded. After the tilt, rainwater begins to fill the other side of the bucket, and the pattern continues. A data logger records the occurrence of each tilt to provide data about rainfall over time; this information may be downloaded later for use. Because the entire operation is automatic, the chance of errors is small.

Measurement errors may occur when the funnel is partially blocked such that water enters the buckets at a rate that is different from that of rainfall. If the buckets are damaged or out of balance, they may tip before or after nonstandard rainfall, or the tipping may be recorded incorrectly. A faulty switch may fail or incorrectly register tips.

These days, SRRGs with data loggers and a communication channel to transmit data to a central computer are common. Data loggers store data in digital form, either as rainfall at a fixed interval or as timing for each event of rainfall of fixed depth.

4.3.5 Precipitation Measurement by Weather Radar

Although rain gauges have been and continue to be in use for a very long time, they suffer from a limitation: a rain gauge gives the point value of rainfall, but in almost

Figure 4.4 A tipping-bucket recording rain gauge (top view).

all hydrologic applications, spatially averaged values are required. Weather radars have been developed to overcome this limitation. Radars are being employed these days to observe precipitation. Using data consisting of radar pulses reflected by raindrops, certain types of radar can provide spatial estimates of rainfall rates over areas of diameter of 40–200 km (WMO, 2008). The effective radar range depends on the type of radar, power output, and receiver configuration. Doppler radar is radar that uses the Doppler effect to measure precipitation. The receiver system of these radars is sensitive to the changes in frequency induced by moving raindrops when these changes are small.

The word *radar* stands for "radio detection and ranging." A radar transmits microwaves which, upon hitting an object, are returned back to the radar. Radars specifically designed to measure precipitation are commonly used these days to get areal estimation of precipitation. The intensity of the energy returned (or echo) at the radar depends on the number and size of raindrops present in the atmosphere. If many large drops are present, reflectivity is high, but if the drops are smaller, it will be lower. The time taken by the echo to reach the radar indicates the distance or location of the rainfall field with respect to the radar.

Radar is calibrated using the data of rain gauges and a relationship between the radar reflectivity Z (mm^6/m^3 or the sixth power of the diameter of raindrops in mm/m^3 of the atmosphere) and rainfall intensity R (mm/h), known as the *Z-R relationship*, is established. A commonly used empirical equation is

$$Z = 200\, R^{1.6} \tag{4.1}$$

The areal coverage of a radar depends on hardware; typical ranges are a few hundreds of kilometers. A weather radar gives precipitation estimates over wide areas and the movement of storms can also be determined. Radars can easily measure precipitation over areas that are remote and difficult to access. It has been found that improved precipitation estimates are obtained by combined use of rain gauges and radar over and above what any single of them could provide. On the negative side, radars are expensive and require skilled persons to operate. The measurements by using weather radar are also affected by echoes from the ground. But the benefits of radars far outweigh the costs.

By the use of radars, spatial variability of rainfall can be displayed on a monitor in real time. Due to these benefits, weather radars are finding increasing application in precipitation measurement and flood forecasting.

These days, use of satellites to estimate precipitation over large areas is also increasing. This topic is further discussed in a later section.

4.4 Processing of Rainfall Data

Raw precipitation data may have gaps and inconsistent values. Before the data are stored in a database or used in further analysis, it is necessary to carry out preliminary checks, scrutiny, and validation. Rainfall data are processed with two major objectives: to examine the data for errors and remove them, and to organize the data in a form that is helpful for subsequent analyses. Rainfall data processing consists of a series of procedures whose steps are briefly described here.

The rainfall data may have the following errors:

- Entries on the wrong day—shifted entries
- Entries made as accumulations
- Missed entries
- Rainfall measurement missed on days of low rainfall

4.4.1 Internal Consistency Check

Observed data should be first checked for reasonableness based on experience and statistics of the station/region. Some of the statistical indices used to check rainfall data include normal rainfall, highest observed rainfall, and rainfall corresponding to a 25-, 50-, or 100-year return period.

Example 4.1

The rainfall value reported from a station for a particular day is 308.6 mm, and the statistics of the reporting station are:

i. Normal monthly rainfall of the corresponding month	:	350.0 mm
ii. Highest observed 1-day rainfall	:	285.3 mm
iii. Standard deviation (σ) of maximum 1-day rainfall	:	51.0 mm
iv. Value of 1-day maximum rainfall of 100-year return period	:	300.0 mm
v. Value of 1-day probable maximum precipitation (PMP)	:	370.8 mm

Should this high rainfall value be accepted or rejected?

Solution

Since the reported rainfall for the day is more than the normal monthly rainfall of the corresponding month, it is doubtful. This value is more than the highest observed 1-day rainfall. The reported daily value is compared with the 1-day PMP value. Because it is less than the PMP (370.8 mm), it is not advisable to reject this data. This data should be flagged and further examined by other tests.

The internal consistency or self-consistency checks use statistical information of the station and current data. The procedure will be illustrated with the help of an example in a subsequent section.

4.4.2 Scrutiny of Precipitation Data by Multiple Time Series Analysis

Plots of time series data of multiple stations on a single graph are helpful and can be drawn for data of any time interval. Comparison of daily rainfall values at two stations is shown in Fig. 4.5. This type of variation appears to be normal, given the spatial structure of rainfall. In addition, monthly and annual total rainfall values at nearby stations are frequently compared to detect the presence of systematic or very large errors in the observed data.

To scrutinize rainfall data, the data of various rain gauge stations are tabulated in columns in a table or spreadsheet. A careful examination of the table helps to reveal anomalies that may be present in the data and that may be difficult to see in multiple time series graphs.

Magnitude of many systematic errors is so small that it becomes difficult to detect them. If such data are accumulated for a long duration of time, the error may be easier to detect. Therefore, to check the data, time series cumulative data, say, monthly totals or annual totals are prepared. These are then compared with the corresponding historical values, say, the maximum monthly rainfall for the month, to detect possible errors. It is noted that in this strategy, it may not be possible to detect the exact value which is erroneous.

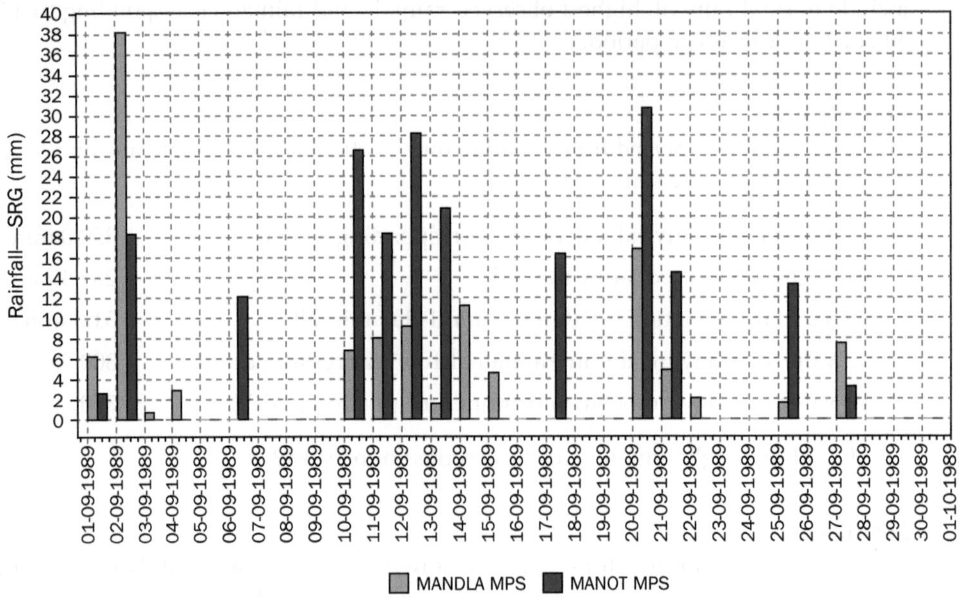

FIGURE 4.5 Comparison of daily rainfall at multiple stations.

4.4.3 Scrutiny by Statistical Tests

Two statistical tests, test for means and test for variance, are commonly performed.

Test for Means (t-Test) The most common parametric test to check whether two samples are from the same population is the student's t-test. The main assumptions of this test are that (1) the observations are independent, (2) the observations are drawn from normally distributed populations, and (3) these populations have the same variance. Hence, this test is conducted to check whether mean values of two samples under consideration are significantly different from each other or not and also to find if the two series can be assumed to belong to the same population. To conduct this test, the t statistic of the samples is determined by

$$t = \left|\overline{X_1} - \overline{X_2}\right| \Big/ S\sqrt{\frac{1}{n_1} + \frac{1}{n_2}} \tag{4.2}$$

where $\overline{X_1}$ and $\overline{X_2}$ are the arithmetic mean values of the two samples of size n_1 and n_2, respectively; and S is the unknown population standard deviation estimated from the sample variances s_1 and s_2 as

$$S = \frac{(n_1 - 1)s_1 + (n_2 - 1)s_2}{n_1 + n_2 - 2} \tag{4.3}$$

If the value of statistic t is less than the tabulated value of Student's distribution at some chosen significance level α and $(n_1 + n_2 - 1)$ degrees of freedom, then the hypothesis that "the means of both the samples are not significantly different" may be accepted at the chosen significance level.

Example 4.2

The average values of annual rainfall of two adjacent basins are 926 and 1,044 cm, and their standard deviations are 141 and 165 cm, respectively. These values have been computed by using data series of 36 and 39 years, respectively. By using the t-test, find out if the two data series belong to the same population at the 5% significance level.

Solution

We first compute S [Eq. (4.3)]:

$$S = \frac{(36-1)\times 141 + (39-1)\times 165}{36+39-2} = \frac{4935+6270}{73} = 153.49.$$

$$\text{Now, } t = \left|\overline{X}_1 - \overline{X}_2\right| \Big/ S\sqrt{\frac{1}{n_1}+\frac{1}{n_2}} = \frac{|926-1044|}{153.49\times(1/36+1/39)} = \frac{118}{8.196} = 14.39.$$

From standard t-tables, the value of t at the 5% significance level and 74 degrees of freedom is 1.66. Because the value obtained is much higher, we can say that the data series do not belong to the same population.

Test for Variances (*F*-Test) The *F*-test is commonly used for testing whether the variances of two samples are significantly different. According to this test, the *F* statistic of the samples is determined as:

$$F = s_1^2/s_2^2 \tag{4.4}$$

If the computed F is less than the tabulated value of F distribution at a chosen significance level α, and $n_1 - 1$ and $n_2 - 1$ degrees of freedom, then the hypothesis that "the variances of both the samples are not significantly different" may be accepted at the chosen significance level.

Statistical tests are discussed in Chap. 19 in more detail.

4.4.4 Correction and Completion of Rainfall Data

Primary and secondary validation tests reveal most errors and missing data in a series. Erroneous and missing values can be identified. Such incorrect and missing values are commonly replaced and filled by values that are estimated by using observations at the same or neighboring stations. The procedure adopted for correction and completion of data depends on the type of error observed/indicated and the availability of suitable data. Some typical cases are discussed next.

Accumulated Rainfall Data If ORG data are observed after several days, the recorded total represents the accumulated rainfall over a period of several days. A practical solution is to distribute the accumulated totals among the various days in proportion of the rainfall measured at the nearest neighboring station over the same period.

Long-Term Shift in Rainfall Data Consistency of a rain gauge station data is examined to determine if the factors governing rainfall at a station have significantly changed during the period of record. If there is a significant change, inconsistency would appear in the rainfall data of that station. This inconsistency would be noticed from the time when the significant change was initiated. Some of the common causes for inconsistency are:

i. Rain gauge instrument was shifted to a new nearby location.
ii. The surroundings of the station have undergone a marked change (e.g., a tall building was constructed near the station) so that the exposure conditions are changed.
iii. Rain gauge develops a fault, and subsequently observational errors arise.

The double mass curve technique is used to check the consistency of a rainfall series. This is a graphical method to identify and adjust inconsistencies in a station's data by comparing the trend of data from reference stations. Double mass curve analysis technique recognizes that the data coming from the same parent population are consistent.

In a double mass curve, both axes include accumulated precipitation values (see Fig. 4.6). Usually, the accumulated seasonal or annual precipitation values of the reference stations are plotted as the abscissa (x-axis), and those of the station being tested are plotted as the ordinate (y-axis). If there is a significant change in, say, the exposure or location of the rain gauge, it leads to a change in the slope of the line. The older records are adjusted by multiplying the precipitation values by the ratio of the slope of the later period to the slope of the earlier period:

$$\text{Adjusted precipitation} = \text{Raw precipitation} \times \frac{\text{Slope of later period}}{\text{Slope of earlier period}} \tag{4.5}$$

To perform the test, long-term annual rainfall of the index station (whose data are to be tested for consistency), called station X, and that of a group of stations in the neighborhood of the index station is selected. The latter group is called base stations. Next, these data are arranged in reverse chronological order (i.e., the latest record as the first entry and the oldest record as the last entry). Now, starting from the first entry in the list, the accumulated precipitation of station X and the accumulated average precipitation of the group of base stations are calculated. Cumulative rainfall values of the index station are plotted against average cumulative values for various consecutive time periods. A change in the slope of the plot indicates a change in the precipitation regime of the index station. If a change is detected, the precipitation values at the index station prior to the change of regime are corrected by using the initial slope of the line as follows:

$$P_X(i) = m \times C_A(i) - C_X(i-1) \tag{4.6}$$

where $P_X(i)$ is the corrected rainfall at the index station X during year i; m is the initial slope of the line; $C_A(i)$ is the cumulative mean rainfall up to the year i; and $C_X(i-1)$ is the cumulative rainfall at index station X until the year $i-1$.

Example 4.3

Consider the double mass plots of Bhopal and Basoda (Figs. 4.6 and 4.7), which have been constructed using the annual rainfall data of eight other neighboring stations. Check the data about Bhopal and Basoda for consistency.

Solution

It can be observed from Fig. 4.6 that the change in slope of the mass curve for Bhopal is insignificant, whereas the same is significant in the case of Basoda (Fig. 4.7). Hence, the data for the Basoda station needs correction. The corrected series is shown in Fig. 4.7.

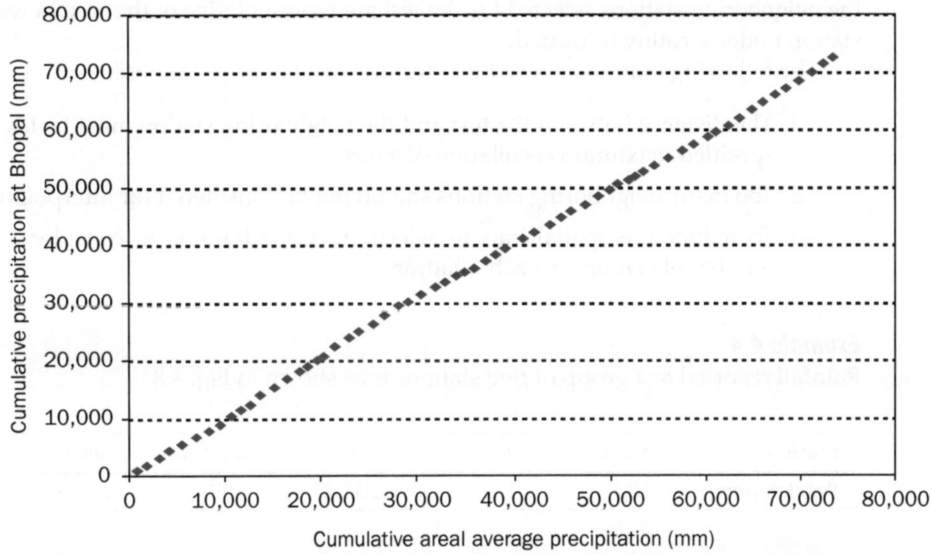

FIGURE 4.6 Double mass curve for Bhopal. The *x*-axis has the accumulated mean precipitation at a group of stations, and the *y*-axis has the accumulated precipitation at Bhopal.

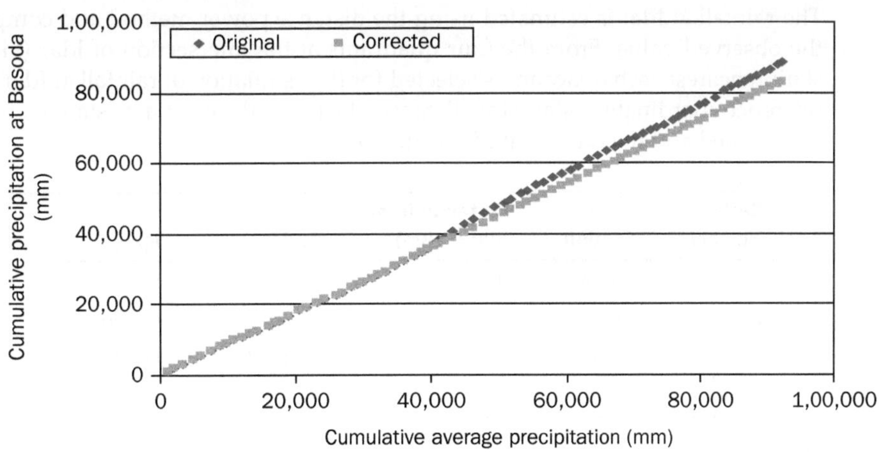

FIGURE 4.7 Double mass curve for Basoda. The *x*-axis has the accumulated mean precipitation at group of stations and on the *y*-axis, accumulated precipitation at Basoda has been plotted. Both original and corrected data have been plotted.

4.4.5 Spatial Consistency Check

Spatial pattern observed in the rainfall data forms the basis to investigate rainfall data for spatial consistency. For this purpose, an estimate of rainfall at the station under investigation is obtained by computing weighted average of rainfalls observed at nearby stations. If the difference between the observed values at the station and the estimated values for the same duration exceeds certain limits, the observed values become doubtful and are flagged for further investigations. It needs to be ensured that the neighboring stations selected for the test are representative of the area in which the station under scrutiny is located.

The following criteria are used to select the neighboring stations:

i. The distance between the test and the neighboring station must be less than a specified maximum correlation distance.
ii. Too many neighboring stations should not be considered for interpolation.
iii. To reduce the spatial bias in selection, it is advisable to consider an equal number of stations in each quadrant.

Example 4.4

Rainfall reported at a group of five stations is as shown in Fig. 4.8.

Station	Kundla	dar	Virpur	Chandop	Sabli
Rainfall (mm)	132.1	12.1	103.3	125.7	149.8

During the quality control process, the data at Idar are identified as doubtful. Check this information for spatial consistency.

Solution

The rainfall at Idar is estimated using the distance power method and compared with the observed value. From the four quadrants at the intersection of Idar (Fig. 4.8), the station nearest each quadrant is selected for the estimation of rainfall at Idar. Using the reference coordinate system, the distance of each of the estimator stations from Idar is determined and the rainfall at Idar estimated:

Serial Number	**Station**	**Distance from Idar D_i (km)**	**$1/D_i^2$**	**R_i/D_i^2**
1.	Kundla	42	5.67×10^{-4}	0.075
2.	Virpur	39	6.57×10^{-4}	0.068
3.	Sabli	75	1.78×10^{-4}	0.027
Total			14.02×10^{-4}	0.170

Rainfall at Idar $= [\Sigma(R_i/D_i^2)] / [\Sigma(1/D_i^2)] = 0.17/14.02\times10^{-4} = 121.25$ mm.

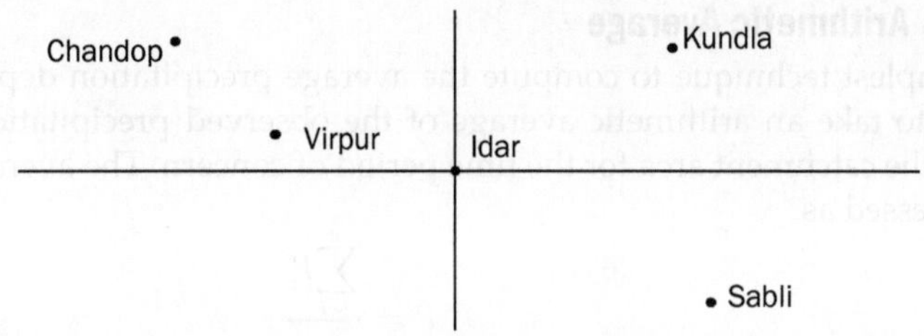

FIGURE 4.8 Locations of stations for spatial consistency check.

Because the observed value is very much different from the estimated value, it is rejected and replaced by the estimated value. Note that there is a possibility that the decimal point was wrongly placed while recording the data at Idar.

4.5 Spatial Averaging of Rainfall Data

Precipitation observations at the rain and snow gauges are point measurements. However, in the hydrological analysis and design, normally mean areal precipitation data are used. Precipitation process exhibits large spatial variation although the values recorded at shorter spacing typically have high correlation. Many methods of computing areal rainfall from point measurements have been developed.

Let the precipitation data be available at n stations, spread out over an area, and P_i be the observed depth of precipitation at the ith station. Using linear interpolation, areal average precipitation over an area can be computed by

$$P^* = \sum_{i=1}^{n} P_i W_i \tag{4.7}$$

where W_i is the weight of the ith station. The spatial averaging techniques differ in the way these weights are computed. Weights of an optimal interpolation technique are decided such that the variance of error in the estimation is the minimum.

The most commonly used methods to compute spatial average precipitation are as follows:

i. Arithmetic average
ii. Normal ratio method
iii. Distance power method
iv. Thiessen polygon method
v. Isohyetal method
vi. Kriging

A particular method is selected depending upon the available data, use of results and precision required, skills of the staff, and the availability of time and other resources. Some of the commonly used methods are described next.

4.5.1 Arithmetic Average

The simplest technique to compute the average precipitation depth over a catchment area is to take an arithmetic average of the observed precipitation depths at gauges within the catchment area for the time period of concern. The average precipitation can be expressed as

$$P = \frac{\sum_{i=1}^{n} P_i}{n} \tag{4.8}$$

where P is the average catchment precipitation from the data of n stations and P_i is the precipitation at station i. If the gauges are relatively uniformly distributed over the catchment that is not in the hilly region and the precipitation values do not have a wide variation, this technique yields good results.

4.5.2 Thiessen Polygon Method

This method is based on the concept of proximal mapping. The value of precipitation at a point is likely to be closer to the value at the neighboring location compared to that at a far location. In the Thiessen polygon method, weights are assigned to neighboring stations based on catchment area that is closer to that station than any other station. To determine this area, first lines joining the stations are drawn and then perpendicular bisectors to these are drawn to form polygons around each station (Fig. 4.9). It is assumed that the precipitation over the entire polygon is same as that observed at the station lying inside the polygon. Next, the area of each polygon is measured. Weight of each station is calculated as the area of the polygon surrounding the station divided by the total catchment area. The weighted average precipitation is computed

Thiessen polygon plot for Betwa up to Basoda (SEP-1995)

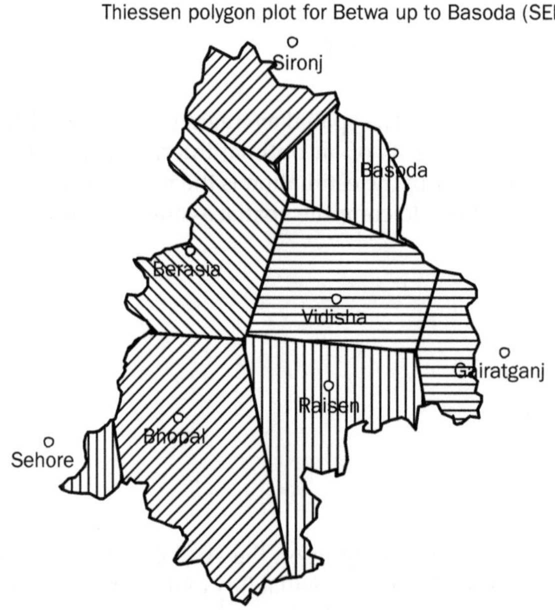

Sl.	Station Name	Weight
1	Basoda	0.0849
2	Berasia	0.1678
3	Bhopal	0.2573
4	Gairatganj	0.0578
5	Raisen	0.1685
6	Sehore	0.0220
7	Sironj	0.0804
8	Vidisha	0.1611

Avg. rainfall: 170.801 (estimated using a combination of 8 raingauges)

Figure 4.9 Thiessen polygon method for computing mean areal rainfall.

by multiplying the precipitation measured at each station by its weight and summing as follows:

$$P = \sum_{i=1}^{n} P_i W_i \tag{4.9}$$

in which $W_i = A_i/A$, where A_i is the area represented by the station i and A is the total catchment area. Clearly, the weights will sum to unity.

An advantage of this method is that the data of stations outside the catchment also may be used if this will help in capturing the variation of rainfall in the catchment. The method works well with nonuniform spacing of stations. Once determined, the weights of the stations remain the same, and it is easier to computerize the method. A major drawback of this method is the assumption that precipitation between two stations varies linearly and the method does not make allowance for variation due to orography. In this method, the precipitation depth changes abruptly at the boundary of polygons. Also, whenever new stations are added to or removed from the network, a new set of polygons must be drawn. The method fails to give any idea as to the accuracy of the results. If a few observations are missing, it may be more convenient to estimate the missing data than to construct a new set of polygons. Software has been developed to compute Thiessen weights and average rainfall by using this method (see Panigrahy, Jain, Kumar, and Bhunya, 2009).

Example 4.5

For a catchment, the rainfall data from six stations for the month of July, along with weights, are as given in Table 4.1. Find the weighted average rainfall for the catchment by using the Thiessen polygon method and compare it with arithmetic mean rainfall.

Solution

Using the observed rainfall and station weight, weighted rainfall at each station is computed. Summation of these values gives the weighted average rainfall for the catchment. The computations are shown in Table 4.1.

Arithmetic mean of rainfalls = (262.0 + 521.0 + 177.0 + 338.0 + 158.0 + 401.6)/6
= 1,857.6/6 = 309.6 mm

Serial Number	Station Name	Station Weight	Rainfall (mm)	Weighted Rainfall (mm)
1.	Sohela	0.06	262.0	15.7
2.	Bijepur	0.12	521.0	62.5
3.	Padampur	0.42	177.0	74.3
4.	Paikmal	0.28	338.0	94.6
5.	Binka	0.04	158.0	6.3
6.	Bolangir	0.08	401.6	32.1
Weighted average catchment rainfall				285.5

TABLE 4.1 Estimation of the Mean Areal Rainfall by the Thiessen Polygon Method

The average rainfall computed by the Thiessen method is lower than the arithmetic mean because the station having the highest weight has reported very low rainfall. The Thiessen estimate is more realistic because a fairly large part of the catchment falls in low rainfall zone.

4.5.3 Isohyetal Method

In the isohyetal method, rainfall values are plotted at their respective stations on a suitable base map and contours of equal rainfall, called isohyets, are drawn. These isohyets are drawn after taking into account the spacing of stations, the quality, and variability of data. In regions of little or no physiographic influence, drawing of isohyets is relatively a simple matter of interpolation. But in places of pronounced orography, the analyst should take into consideration the orographic effects, storm orientation, etc. to adjust or interpolate between station values.

Computer software are available these days to draw isohyetal maps. As an example, the isohyetal map for an area is shown in Fig. 4.10. The total depth of precipitation is computed by measuring the area between successive isohyets, multiplying this area by the average rainfall of the two ishohyets, and totaling. The average depth of precipitation is obtained by dividing this sum by the total area. The average depth of precipitation (P_i) over this area is obtained by

$$P = \frac{\sum_{i=1}^{n} P_i \ A_i}{\sum_{i=1}^{n} A_i} \tag{4.10}$$

where A_i is the area between successive isohyets and P_i is the average rainfall between the two isohyets.

Example 4.6

Using the point rainfall data for a catchment, isohyetal lines are drawn as shown in Fig. 4.10. The area enclosed by each isohyet is calculated as given in Table 4.2. Compute the average rainfall for the catchment.

Solution

For each isohyet, the average rainfall is computed (the maximum observed rainfall is 521 cm and the minimum 158 cm). This amount, multiplied by the area enclosed by that isohyet, gives the volume of rainfall for that isohyet. Next, the volumes of rainfall for the different isohyets are summed and the sum is divided by the area of the catchment to yield the average catchment rainfall. The computations are shown in Table 4.2.

Average catchment rainfall = 1,304,415/5,600 = 232.9 mm

4.5.4 Kriging

Kriging is a geostatistical interpolator that can be used for the areal estimation of rainfall. It is based on the principle that the value at an unknown location should be the weighted average of the known values of its neighbors weighted by the distance of the neighbor from the unknown location. Rain gauges should be located in zones of similar climatic characteristics. In Kriging, a function, named a *variogram* or *semivariogram,* plays an important role—it determines when the difference between values becomes significant as the time between measurements becomes longer. The Kriging estimate

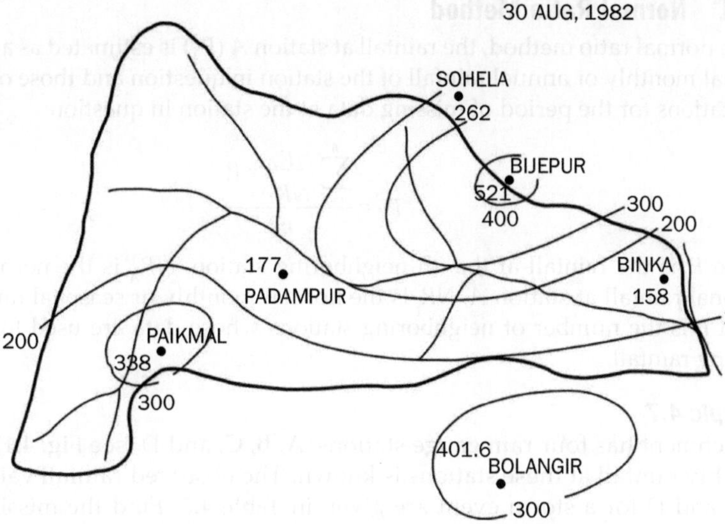

FIGURE 4.10 The isohyetal method for computing mean areal rainfall.

Isohyetal Range (mm)	Average Value (mm)	Area (km²)	Volume (mm-km²)
521.0–500.0	510.5	70	35,735
500.0–300.0	400.0	530	212,000
338.0–300.0	319.0	100	31,900
200.0–300.0	250.0	2,080	520,000
158.0 – 200.0	179.0	2,820	504,780
		5,600	1,304,415

TABLE 4.2 Estimation of Mean Areal Catchment Rainfall Using the Isohyetal Method

is a linear combination of the known values in which the weights are a function of the assumed variogram and are selected so that the estimator is unbiased, with minimum variance for prediction errors. Different models of variogram have been used. Thus, Kriging consists of a group of methods. For more details, reference is made to Bras and Rodriguez-Iturbe (1985).

4.6 Estimation of Missing Data

While using precipitation data, one often comes across situations in which some data are missing, which may arise due to several causes, including instrument malfunction, an absent observer, and loss of records. Missing rainfall data can be filled by using various techniques. Due to the spatial structure of precipitation data, some interpolation technique that uses the data of nearby stations is commonly adopted. Generally, rainfall for the missing period is estimated either by using the normal ratio method or the inverse distance power method. One can also use the entropy method for estimating missing values (Singh and Harmancioglu, 1997).

4.6.1 Normal Ratio Method

In the normal ratio method, the rainfall at station A (P_A) is estimated as a function of the normal monthly or annual rainfall of the station in question and those of the neighboring stations for the period of missing data at the station in question:

$$P_A = \frac{\sum_{i=1}^{n} \frac{NR_A}{NR_i} \times P_i}{n} \tag{4.11}$$

where P_i is the rainfall at the ith neighboring station, NR_A is the normal monthly or seasonal rainfall at station A, NR_i is the normal monthly or seasonal rainfall at station i, and n is the number of neighboring stations whose data are used for estimation of missing rainfall.

Example 4.7

A catchment has four rain gauge stations, A, B, C, and D (see Fig. 4.11). The normal monthly rainfall at these stations is known. The observed rainfall values at stations B, C, and D for a storm event are given in Table 4.3. Find the missing rainfall for station A.

Solution

The ratio of normal rainfall at station A to normal rainfall at station i or NR_A/NR_i is calculated based on the values given in Table 4.3.

The estimated rainfall at station A is

$$P_A = \frac{1.14 * 98.9 + 1.02 * 120.5 + 0.92 * 110.0}{3} = 112.3 \ mm$$

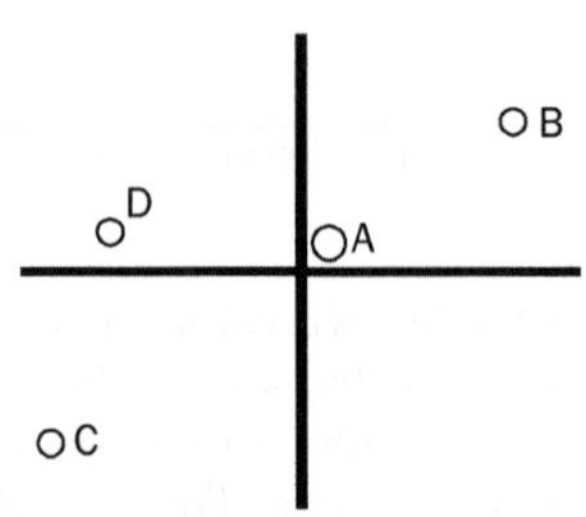

Figure 4.11 The location of stations A, B, C, and D.

Station	A	B	C	D
Normal Rainfall (mm)	331.3	290.8	325.9	360.5
NR_A/NR_i	1	331.3/290.8 = 1.14	331.3/325.9 = 1.02	331.3/360.5 = 0.92
Event Rainfall (mm)	?	98.9	120.5	110.0

Table 4.3 Normal Rainfall Values for Example 4.7

4.6.2 Inverse Distance Power Method

Here, the weighted average of measured rainfall data at a group of nearby stations is used to estimate the missing rainfall at a station. The weight of a station is computed as some power of the reciprocal of the distance of the estimator stations from the station where rainfall is to be estimated. This power can be unity. Let D_i be the distance of the estimator station from the estimated station. If the weights are an inverse square of distance, the estimated rainfall at station A is

$$P_A = \sum_{i=1}^{n} P_i / D_i^2 \Big/ \sum_{i=1}^{n} 1/D_i^2 \tag{4.12}$$

Note that the weights go on reducing with increasing distance. A major shortcoming of this method is that the orographic features and spatial distribution of the variables are not considered.

The procedure for estimating the rainfall data by this technique is illustrated through an example. If A, B, C, and D are the locations of stations discussed in the example of the normal ratio method, the distance of each estimator station (B, C, and D) from station A, whose data is to be estimated, is computed with the help of the coordinates using the formula

$$D_i^2 = [\,(x - x_i)^2 + (y - y_i)^2] \tag{4.13}$$

where x and y are the coordinates of the station whose data is estimated, and x_i and y_i are the coordinates of stations whose data are used in the estimation.

Example 4.8

Using the data from Example 4.5, estimate the rainfall at station A using the inverse distance power method.

Solution

Because the coordinates of the stations are known, their distances from station A can be calculated. The weight $1/D_i^2$ is then computed for each station, and the rainfall at station A is estimated as in Table 4.4.

$$\text{Rainfall at station A} = \frac{570.7 \times 10^{-3}}{5.01 \times 10^{-3}} = 113.9 \text{ mm}$$

4.6.3 Relation Between Point and Areal Rainfall

A point-rainfall measurement does not represent the mean areal rainfall correctly. Eagleson (1970) has pointed out that the difference between the mean rainfall over an area about the

Station	Distance from Station A	$1/D_i^2$	Rainfall P_i (mm)	Weighted Rainfall $P_i \times (1/D_i^2)$ (mm)
B	28.0	1.29×10^{-3}	98.9	125.6×10^{-3}
C	17.7	3.19×10^{-3}	120.5	384.6×10^{-3}
D	42.5	0.55×10^{-3}	110.0	60.5×10^{-3}
Total		5.01×10^{-3}		570.7×10^{-3}

Table 4.4 Computation of rainfall for Example 4.8.

storm center and the point rainfall measured at the storm center increases with a decrease in the total rainfall depth, decreases with increasing duration, is larger for convective and orographic precipitation than for cyclonic storms, and increases with increasing area.

Point-rainfall measurements have been (and continue to be) utilized in deriving rainfall depth-frequency curves. The U.S. National Weather Service developed a set of area-depth curves for various storm durations for watersheds up to 400 mi², which can be used with frequency curves to transform the point-rainfall depth to areal-average depth. Leclerc and Schaake (1972) have expressed the relationship algebraically as

$$K = \frac{P_A}{P} = 1 - \exp(-1.1D^{1/4}) + \exp(-1.1D^{1/4} - 0.003863A) \tag{4.14}$$

where D is the rainfall duration in hours and A is the area in square kilometer. Many organizations, such as the U.S. Soil Conservation Service and the U.S. Bureau of Reclamation, have developed procedures to convert given point PMP rainfall to areal rainfall. Many equations have been derived for this purpose, but most are area specific, and their goal is to express K as a function of area (and possibly rainfall duration as well). Horton (1924) expressed storm areal mean depth P_A as

$$P_A = P \exp(-aA^b) \tag{4.15}$$

where P is the maximum point rainfall depth at the center of the storm, A is the watershed area, and a and b are empirical constants. Using the least squares method, Dhar and Bhattacharya (1977) found the values of a and b for the North Indian plains as follows:

Duration	Value of *a* (km²)	Value of *b* (km²)
1 day	0.0016	0.6614
2 days	0.0018	0.6306
3 days	0.003	0.5691

It is emphasized that the areal precipitation estimates derived on the basis of point-rainfall observations from a network of sparsely and unevenly distributed precipitation gauges could have large uncertainty. Radar-sensed echo intensity reflected by precipitation could provide useful areal estimates of precipitation over areas with few or no precipitation gauges. For this purpose, a dense network of precipitation gauges can be set up on a temporary basis. Data from these stations could be related to the intensities of echo reflected by falling rain and sensed by radar to develop relationships similar to Eq. (4.1) for the area of interest.

4.7 Disaggregation of Rainfall Data

Many hydrological applications, such as flood forecasting, require rainfall data of shorter duration, whereas the network of rain gauges providing short-duration data is usually sparse in comparison to that of daily rain gauges. Hence, it is often necessary to disaggregate (or divide) daily rainfall data into shorter time intervals (usually 1 hour). The observed daily rainfall is distributed in hourly values to follow the same pattern as the observed hourly rainfall. For this purpose, daily data is used along with short-interval rainfall data at nearby stations. Mass curve method is commonly employed in such computations.

A *mass curve* is a graph of accumulated rainfall at a station versus time. Mass curves plot of rainfall data of a recording station can be prepared by accumulating rainfall values for the duration of storm under analysis (or any other interval) and plotting them against time. To decide which recording rain gauge is representative of which of the nonrecording rain gauges for distributing daily rainfall values, mass curves of the data of recording rain-gauge stations are compared with those of the nonrecording stations.

Assume that the daily rainfall observed at 0800 hours is to be distributed into hourly values. First, the hourly rainfall from 0900 hours on a day to 0800 hours on the next day is accumulated to obtain total rainfall for 24 hours. Next, the rainfall during each hour is divided by the sum to obtain hourly rainfall as a fraction of total rainfall for 24 hours (0800 hours to 0800 hours). The sum of all these fractions will be one. Now, the daily rainfall is multiplied by the fraction for a particular hour to obtain rainfall for that hour. The procedure is explained with the help of an example.

Example 4.9

Daily rainfall data (expressed in millimeters) of four stations for the period August 28 to August 30, 1973 is given in Table 4.5.

Hourly rainfall data of two SRRG stations (Mandla and Jabalpur) are given in Table 4.6 for August 27–29, 1973. Disaggregate the daily rainfall data into hourly values.

Solution

The hourly rainfall data at each of the two SRRG stations is accumulated and plotted on a graph to prepare the mass curve of hourly rainfall as shown in Fig. 4.12. The daily rainfall data at each of the four stations is also cumulated and plotted on a graph in Fig. 4.13. The mass curves of daily rainfall are compared with those of hourly rainfall to determine which of the daily rain-gauge stations are representative of which of the SRRG stations. Based on the match of curves, it is decided that the daily rainfall stations at Pendra Road and Dindori are represented by Mandla, while the variation of rainfall at Jabalpur matches with the rainfall at Niwas and Ghansore stations.

To convert the daily rainfall to hourly rainfall, the hourly rainfall from 0900 hours on a day to 0800 hours on the next day is accumulated, and the rainfall during each hour is expressed as a ratio of the total rainfall during 24 hours (0900 hours to 0800 hours). These ratios are used to distribute the daily rainfall at the rain-gauge stations which are represented by the particular SRRG. The daily rainfall distributed for 1 day (i.e., corresponding to August 28–29, 1973) is given in Table 4.7.

Station Name	August 28, 1973	August 29, 1973	August 30, 1973
Pendra Road	65.3	80.8	1.1
Dindori	23.2	58.4	1.0
Ghansore	42.0	98.0	20.6
Niwas	30.4	203.0	33.4

Table 4.5 Daily Rainfall Data for Stations in Example 4.9

Hours	Mandla			Jabalpur		
	August 27, 1973	August 28, 1973	August 29, 1973	August 27, 1973	August 28, 1973	August 29, 1973
1	0.0	0.0	0.0	0.0	0.0	23.5
2	0.0	0.0	0.0	0.0	0.0	30.5
3	0.0	0.0	0.0	0.0	0.3	10.2
4	0.0	0.0	0.0	0.0	0.5	30.8
5	0.0	0.0	0.0	0.0	0.1	16.9
6	0.0	0.0	0.0	0.0	1.0	1.9
7	3.0	1.7	0.0	0.0	24.4	0.6
8	0.0	9.0	0.0	0.0	0.2	3.1
9	0.0	2.3	0.5	0.0	0.0	5.3
10	0.9	0.0	0.4	0.0	0.0	3.8
11	0.0	1.5	0.1	0.0	23.8	0.8
12	0.0	5.5	0.0	0.0	1.3	1.5
13	0.0	0.0	0.0	0.0	0.2	0.0
14	0.0	0.0	0.0	0.0	0.0	0.3
15	0.0	0.0	0.0	0.0	0.1	0.9
16	0.0	2.5	0.3	0.0	0.6	1.5
17	0.2	1.5	0.4	0.0	29.0	0.4
18	2.1	0.8	0.3	0.0	11.0	0.5
19	0.7	3.4	0.0	25.7	7.1	0.1
20	0.3	0.6	0.0	0.0	0.4	0.5
21	0.1	2.5	0.0	0.1	2.7	0.7
22	0.1	0.0	0.0	1.0	6.8	0.0
23	4.7	0.0	0.0	6.0	20.6	0.0
24	0.0	0.0	0.0	1.3	0.0	0.0

Table 4.6 Hourly Rainfall Data of Mandla and Jabalpur on August 27–29, 1973

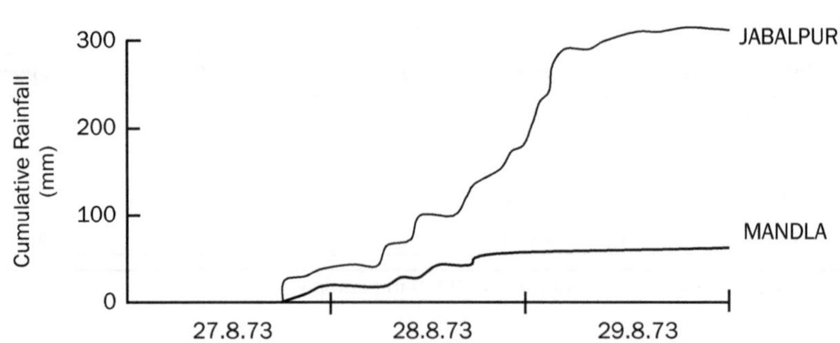

Figure 4.12 Mass curve of hourly rainfall for Mandla and Jabalpur.

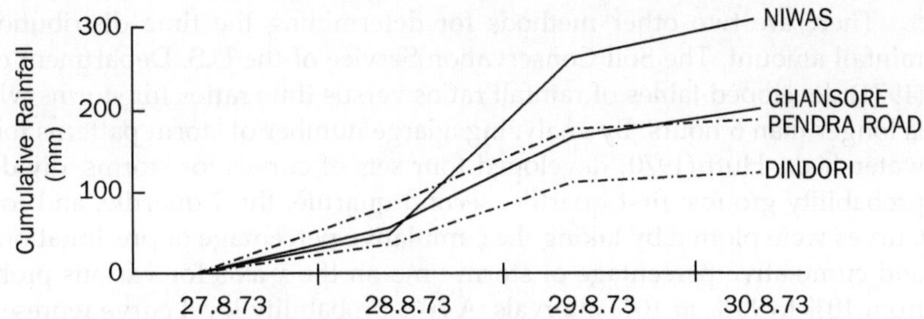

FIGURE 4.13 Mass curve of daily rainfall at various stations.

Hour	Pendra Road		Dindori		Niwas		Ghansore	
	August 28, 1973	August 29, 1973	August 28, 1973	August 29, 1973	August 28, 1973	August 29, 1973	August 28, 1973	August 29, 1973
1		0		0		21.6		10.4
2		0		0		28.0		13.5
3		0		0		9.4		4.5
4		0		0		28.3		13.7
5		0		0		15.5		7.5
6		0		0		1.7		0.8
7		0		0		0.6		0.3
8		0		0		2.8		1.4
9	9.0		6.5		0.0		0.0	
10	0.0		0.0		0.0		0.0	
11	5.9		4.3		21.9		10.5	
12	21.6		15.6		1.2		0.6	
13	0.0		0.0		0.2		0.1	
14	0.0		0.0		0.0		0.0	
15	0.0		0.0		0.1		0.0	
16	9.8		7.1		0.6		0.3	
17	5.9		4.3		26.6		12.9	
18	3.1		2.3		10.1		4.9	
19	13.3		9.6		6.5		3.1	
20	2.4		1.7		0.4		0.2	
21	9.8		7.1		2.5		1.2	
22	0.0		0.0		6.2		3.0	
23	0.0		0.0		18.9		9.1	
24	0.0		0.0		0.0		0.0	
Sum	80.8		58.4		203.0		98.0	

TABLE 4.7 Daily Rainfall Distributed into Hourly Rainfall on August 28–29, 1973

There are two other methods for determining the time distribution of a given rainfall amount. The Soil Conservation Service of the U.S. Department of Agriculture (1971) developed tables of rainfall ratios versus time ratios for storms whose duration is longer than 6 hours. By analyzing a large number of storm patterns for Midwestern watersheds, Huff (1970) developed four sets of curves for storms, divided into equal probability groups: first quartile, second quartile, third quartile, and fourth quartile. Curves were plotted by taking the cumulative percentage of precipitation on the y-axis and cumulative percentage of storm time on the x-axis for various probability levels from 10% to 90%, at 10% intervals. A 70% probability level curve represents the distribution occurring in 30% of the storms or less, and a 30% probability level curve represents the distribution occurring in 70% of the storms or less.

4.8 Snow Data

Three features related to snow are important in hydrological studies: snow depth, snow cover area, and snow water equivalent. These data are used to estimate the water content of the snowpack and are of immense use in the planning of irrigation and hydropower projects, reservoir regulation, and disaster management.

4.8.1 Observation of Snow Data

In snowfall measurements, only freshly fallen snow is measured. To ensure this, if there is no snow on the ground, a small area is cleared or the existing snow surface is covered with wood, tarpaulin, etc. so that old and new snow can be segregated, and the depth from top surface to this layer is measured. In case there is old snow on the ground, it would be wrong to determine the depth of new snow as the difference between total depth and the depth of old snow since snow keeps on compacting.

A *snow pillow* is used to measure the water equivalent of snow. It is a flat, circular container that is filled with nonfreezing liquid. The weight of snow on the pillow can be measured by hydrostatic techniques or pressure transducer. The depth of snow may also be measured in a standard container of uniform cross section after leveling the snow (without compressing it). Snow pillows of various dimensions and materials are used to measure the weight of accumulating snow. Common pillows are flat circular containers of 3.7 m in diameter and made of rubberized material filled with a nonfreezing liquid. The pillow is installed on the ground, flush with the ground, or buried under a thin soil layer. To prevent damage to the equipment and to preserve the snow cover in its natural condition, snow pillow site should be fenced.

In regions where snowfalls are heavy, snow stakes are commonly installed at different accessible, representative locations and are observed to determine the depth of fallen snow. Snow stakes should be installed at those locations where they can be viewed and read easily from a distance. It needs to be ensured that the site is representative of the area and is free from human disturbance. Snow stakes should be painted white to minimize the increase in the melting of surrounding snow. The stakes are clearly graduated in meters and centimeters, and in inaccessible areas, crossbars are fitted so that the snow depth can be read from a distance.

Snow Water Equivalent (SWE) is the amount of water contained in the snowpack. It is the depth of water that would result if the complete snowpack is melted where it has fallen. SWE can be determined by melting the snow collected in a gauge and measuring the melted water in the same way as rainfall. It is a good practice to take several representative samples. Alternatively, cylindrical samples of fresh snow are collected with a suitable snow sampler and either weighed or melted. Weighing-type recording gauges are used to determine the water content of snowfall. Isotope techniques also are used to estimate snow water equivalents. However, the instruments used are expensive, and skilled personnel are needed to collect and analyze samples and interpret results.

In catchments that receive considerable snowfall, snow surveys are made on permanent snow courses which are permanently marked lines. The snow courses should be representative of the areas of snowfall. These should be located in accessible areas where snow falls to the ground without being intercepted by vegetation, and the site should be protected from strong winds. During surveys, snow samples are taken, and depths are measured at various places at a number of points along the course.

Snowfall data are prescribed in the same way as the rainfall data.

4.8.2 Measurement of Snow by Remote Sensing

Snow-related data such as snow covered area, SWE, etc. can be measured by remote sensing. Data from various bands of the electromagnetic (EM) spectrum, such as gamma rays, visible and near-IR (NIR), thermal infrared (IR), and microwaves, are used for this purpose. Among these, the microwave band is best suited for snow-related applications, followed by visible and NIR bands. Many algorithms and models have been developed to determine snow-covered area, snow depth, and SWE from satellite data. Remote-sensing data has many advantages: repeated coverage of the area, ability to get information from remote and inaccessible areas, and cost-effectiveness.

The microwave region of the EM spectrum includes wavelengths in the range of 1 mm to 1 m. Microwave remote sensing has an additional advantage. Depending on the wavelength, it can see through clouds, haze, light snow, and smoke. Active microwave sensors used in remote sensing supply their own source of energy or illumination. Passive sensors, on the other hand, are those that sense the naturally available microwave energy within their field of view. The potential of both active and passive microwave regions is similar. An advantage of active microwave over passive microwave is higher resolution. However, the analysis of active microwave data is more complex than of passive microwave data.

4.9 Regional/Global Meteorological Data Sets

Many attempts have been made to develop regional/global meteorological data sets by using instrumented and data acquired from space-based platforms. The Asian Precipitation–Highly Resolved Observational Data Integration Towards the Evaluation of Water Resources (APHRODITE) project has provided a high-resolution, daily gridded precipitation data set covering nearly the whole Eurasian continent. The APHRODITE project (http://www.chikyu.ac.jp/precip/english/) has developed state-of-the-art daily precipitation data sets with high-resolution grids for Asia. The data sets are created primarily with data obtained from a rain-gauge observation network. The project closed in 2007 and the data are available from 1951 to 2011. In addition, the project has developed a high-resolution, gridded daily mean temperature product for Asia.

Tropical Rainfall Measuring Mission (TRMM) was a project to collect global precipitation data from space. The overall objective of TRMM was to use active and passive sensors to describe tropical rainfall quantitatively. TRMM data were used to obtain multiyear sets of tropical and subtropical rainfall observations; data are available from 1998 onward. The Tropical Rainfall Measuring Mission (TRMM) Multisatellite Precipitation Analysis (TMPA) combines precipitation estimates from multiple satellites. TMPA merges rainfall estimates from all microwave sensors (e.g., TMI, SSMI, AMSU) and calibrated by TRMM precipitation radar. It is available at fine scales (0.25 degrees × 0.25 degrees and 3-hourly intervals). The data set covers the latitude band 500 N-S for the period from 1998 to 2015.

With the success of TRMM (which ended in 2015), the Global Precipitation Measurement (GPM) program was developed to provide not only data continuity, but also an improvement on the TRMM data. More details can be seen at https://earthdata.nasa.gov/trmm-to-gpm. It is an international network of satellites building upon the success of the TRMM, which collects rain, snow, and other precipitation data worldwide every 3 hours. The Level 3 Integrated Multi-Satellite Retrievals for GPM (IMERG) gridded product, developed by the National Aeronautics and Space Administration (NASA), has a resolution of 0.1 degree and is updated every 30 minutes from 600 N–600 S. IMERG will combine precipitation measurements from various international satellites.

The Global Historical Climatology Network (GHCN) (https://www.ncdc.noaa.gov/) is an integrated database of climate summaries from land-surface stations across the globe. GHCN collects data from more than 20 sources, and the time period of some data series is more than 150 years.

4.10 Rainstorm Analysis

While designing a dam, it is necessary that the outlet capacity is adequate to safely pass a high flood of certain magnitude. This critical flood is known as the *design flood* for the structure. The type and magnitude of design flood depends upon the type and size of structure. For this purpose, the hydro-projects are classified among the following categories:

i. Large or medium dams
ii. Medium structures, such as barrages, road and railway bridges
iii. Small or minor structures, such as cross-drainage works, minor irrigation tanks, and minor road bridges

If long-term runoff data are not available, rainfall data that are generally available for a longer period are used to estimate the *design storm*. This design storm, together with a suitable rainfall-runoff model, is used to obtain a design flood. The rainstorm analysis is the first step in the design storm estimation procedure. The design storm (rainfall) is a magnitude of rainfall and its distribution that is developed for the design of specific types of structures. It has three components—namely, the rainfall amount, the areal distribution of rainfall, and the time distribution of rainfall. This topic is discussed in greater detail in Chapter 21.

4.11 Interception

Where precipitation does not directly fall on bare soil, it is caught by vegetation or other objects on the Earth's surface. This is known as *interception*, the amount of precipitation caught in this process is called *interception loss*, and the amount of water stored on the surface of the cover is called *interception storage*. Its magnitude depends on the shape, density, and properties of surfaces of leaves and stems, as well as the vertical layering of canopies. Part of intercepted water is evaporated back into the atmosphere without reaching the Earth's surface. The remainder of the water may eventually reach the soil with some delay after temporary storage on the surface cover. As interception involves two processes, precipitation and surface cover, it depends on the characteristics of these two.

Falling rain moves through the vegetation canopy in two ways: (1) rain that falls through spaces between the leaves or by dripping from leaves and branches, which is known as *throughfall*; and (2) rain that reaches the ground by flowing along stems and trunks of the trees, known as *stemflow*. Water reaching the ground by the first process has to pass through any litter layer that exists before entering the soil. Generally, stemflow is much smaller than throughfall.

Precipitation characteristics that affect interception are intensity, storm duration, total volume of precipitation, and evaporation conditions during the storm. Comparatively a significant fraction of precipitation is intercepted during low-intensity rainstorms. During large and intense rainstorms, the amount of water intercepted compared to the total precipitation is insignificant. Thus, interception does not have much impact during major floods. Note that canopy interception can be important in controlling soil from erosion by the impact of raindrops.

The major vegetation-related factors that affect interception include vegetation type, density, storage capacity of leaves, age, and shape of canopy. The season of the year affects interception, as many trees shed leaves in autumn. Trees whose branches slope outward and upward direct the falling water toward the trunk, and thus are likely to have more stemflow than trees with horizontal branches.

Although interception is commonly much smaller than precipitation, it can be significant in some cases. Carlyle-Moses (2017) reported that canopy interception accounted for 25% of rain that fell at a lowland tropical rainforest in Australia over a 3-year period. Annual and season-long canopy interception loss can exceed 50% of rainfall under certain conditions.

4.12 Summary

Undoubtedly, precipitation remains the most important and frequently used hydrometeorological variable in studies dealing with water resources. Precipitation is also the variable whose longest measuring data series are available at a large number of stations. This chapter gives detailed coverage of the acquisition and processing of precipitation data and interception.

4.13 Questions

4.1 Write four errors that can take place while entering rainfall data into records.

4.2 Briefly describe the types of precipitation.

4.3 Describe how clouds are formed.

4.4 With the help of diagrams, describe an ORG and a SRRG.

4.5 Briefly explain how rainfall can be measured by radars and by the use of satellites.

4.6 For a 1-hour rainstorm, depths of rainfall are given as follows:

Time (min)	Rainfall depth (mm)
10	5
20	25
30	40
40	20
50	10
60	4

Construct the rainfall hyetograph, with rainfall intensity on the ordinate.

4.7 Suppose that the rainstorm in question 6 fell on a watershed with an area of 100 ha. What would be the volume of water (in cubic meters) generated by this rainstorm?

4.8 What is the peak rainfall intensity (in mm/h) of the rainstorm in question 6?

4.9 When rainfall reported at a group of six stations in a basin was being examined for spatial consistency, doubts arose about the value reported at Majgawn. The distance between Majgawn and the other five stations is given in the following table:

Station	Rampur	Hansraj	Rehli	Majgawn	Borina	Jabera
Rainfall (mm)	112.1	99.5	103.3	220.7	89.7	92.1
Distance from Majgawn (km)	61	43	51	0	25	36

Check the values at Majgawn for spatial consistency.

4.10 For a catchment, the rainfall data at six stations for the month of September, along with their Thiessen weights, are as given in the following table:

Station	Rampur	Hansraj	Rehli	Majgawn	Borina	Jabera
Rainfall (mm)	112.1	99.5	103.3	220.7	89.7	92.1
Distance from Majgawn (km)	0.21	0.05	0.33	0.18	0.14	0.09

Find the weighted average rainfall for the catchment by using the arithmetic average and Thiessen polygon methods.

4.11 A catchment has six rain-gauge stations: Rampur, Hansraj, Rehli, Majgawn, Borina, and Jabera. The normal monthly rainfall at these stations and the observed rainfall at these stations for a storm event are given in the following table.

Station	Rampur	Hansraj	Rehli	Majgawn	Borina	Jabera
Normal Rainfall (mm)	112.1	99.5	103.3	120.0	89.7	92.1
Event Rainfall (mm)	61	43	51	?	52	65

Find the missing rainfall at Majgawn using the normal ratio method.

4.12 Find the missing value at Majgawn after removing Rampur (with 112.1 mm normal rainfall) in question 11. Compare value obtained here with that obtained in question 11. Now do the same by also removing Jabera (with 92.1 mm normal rainfall). Compare three of the values and comment on the effect of removing one and two stations on the computed missing values.

4.13 Using the data from question 11, estimate rainfall at Majgawn station by using the reciprocal distance power method.

4.14 References

Bras, R. L., and I. Rodriguez-Iturbe. *Random Functions and Hydrology.* Reading, MA: Addison-Wesley, 1985.

Carlyle-Moses, D. E. Rainfall interception, detention, and depression storage. Chapter 43 in *Handbook of Applied Hydrology,* 2nd ed. (V. P. Singh, editor-in-chief), pp. 43.1–43.4. New York: McGraw-Hill, 2017.

Dhar, O. N., and B. K. Bhattacharya. Relationship between central rainfall and its areal extent for severemost rainstorms of north Indian Plain. *Irrigation and Power,* 28:2245–2250, 1977.

Eagleson, P. S. *Dynamic Hydrology.* New York: McGraw-Hill, 1970.

Huff, F. A. Time distribution characteristics of rainfall rates. *Water Resources Research,* 6:447–454, 1970.

Horton, R. E. Estimates on determining the mean precipitation of a drainage basin. *Journal of the New England Water Works Association,* 38:1–47, 1924.

Jain, S.K., and Singh, V.P. *Water Resources Systems Planning and Management.* Amsterdam: Elsevier, 2003.

Jain, S.K. "Hydrological Measurements and Analysis of Data". Web-based course under National Program on Technology Enhanced Learning (NPTEL). http://nptel.ac.in/courses/105107129/, 2011.

Leclerc, G., and J. C. Schaake. *Derivation of Hydrologic Frequency Curves.* Report 142. Cambridge, MA: Massachusetts Institute of Technology, 1972.

Panigrahy, N., S. K. Jain, V. Kumar, and P. K. Bhunya. Algorithms for computerized estimation of Thiessen weights. *Journal of Computing in Civil Engineering,* ASCE, 23(4), 239–247, 2009.

Singh, V. P., and N. B. Harmancioglu. Estimation of missing values with the use of entropy. In Integrated Approach to Environmental Data Management Systems (N. B. Harmancioglu, M. C. Alpaslan, S. D. Ozkul, and V. P. Singh, eds., pp. 267–274). NATO ASI Series, 2. Environment-Vol. 31, Dordrecht, Netherlands: Kluwer Academic Publishers, 1997.

CHAPTER 5

Evapotranspiration

Evaporation and transpiration, which transfer water from land and vegetation to the atmosphere, are key components of the hydrologic cycle, especially at large space-time scales. These components are fundamental to agricultural irrigation, forest management, land use, land cover design, drought management, and water resources planning. This chapter discusses these components and their measurements and estimation.

5.1 Introduction

Large amounts of water are evaporated from ponds and lakes, particularly in arid climates. Crops and trees transpire large portions of water that is applied through irrigation or is received through rainfall. Evaporation and transpiration together constitute what is called *evapotranspiration (ET)*. The importance of ET depends upon the time scale and the type of hydrologic event. For an intense, short-storm event, abstractions due to evaporation and transpiration are usually small and often are neglected in hydrologic balance computations. The bulk of evaporation and transpiration from a catchment takes place during the time between runoff events, which is usually long. Hence, ET abstractions are more important between storm events. ET from large areas in semiarid and temperate climates may be about two-thirds of the annual precipitation, and it may be up to 90% in arid climates. ET links the atmospheric, hydrologic, and agricultural sciences.

At the catchment scale, ET represents the largest water flux after precipitation, but it is very challenging to measure ET at this scale due to the heterogeneity of the topography. In practice, ET is often calculated as the residual from precipitation and other water fluxes (e.g., runoff, change in soil water storage) and is subject to large errors at a short temporal scale and in watersheds with poorly defined surface and subsurface watershed boundaries.

5.1.1 Definitions

Evaporation is the process by which liquid water from water bodies (rivers, ponds, lakes) or soil is converted to vapor. It is commonly expressed in mm/h.

Potential evaporation is the quantity of water evaporated per unit time per unit area of an idealized extensive free water surface under existing atmospheric conditions (Shuttleworth, 1992). The time scale for measurement or estimation can be an hour, a day, a month, or a year. ET is usually assumed to be uniform over the area under consideration and is expressed in depth (millimeters, centimeters, or inches) or rate (mm/h, cm/month, in./day).

Reference evapotranspiration is the ET rate from a reference surface, not short of water. It is also called the *reference crop ET* and is denoted as ET_0 (UN Food and Agriculture Organization; www.fao.org). The reference surface is a hypothetical grass reference crop with specific characteristics. It is commonly expressed in mm/day or cm/month.

Aridity index (AI), as defined by the United Nations Environmental Programme (UNEP), is the annual precipitation (mm) divided by the annual potential ET (mm).

5.1.2 Evaporation Process

Because the conversion of liquid water into water vapor requires energy, the rate of evaporation depends on the availability of energy, as well as how easily water vapor can diffuse into the air. Thus, evaporation involves the transfer of energy, and hence the molecules that have energy to break all bonds with their neighbors and move into the air will evaporate. The heat required (in calories, or cal) to convert 1 g of water at 1 atmosphere (atm) pressure to vapor is the latent heat of vaporization. Its value is 539 cal/g at 1 atm pressure at 100°C, which means that 539 cal of heat is required to evaporate 1 g of water. The heat energy required to raise the temperature of water to its boiling point is less than that required to convert the liquid water to vapor. The water vapor is lighter than other atmospheric gases, and hence it tends to rise above air.

When warmed, water molecules become more active and move from the liquid water to the atmosphere. As more molecules move into the atmosphere, the density of water molecules near the water surface increases. This leads to collision between molecules emerging from the water with those already in the air, and it causes some water molecules to return to the water. Thus, there is a constant transfer of molecules to and from water, but the transfer from the water is dominant. Some molecules are carried away from the water body by wind. At some time, the air may become saturated and the vapor pressure of air may be equal to the vapor pressure of water. In this condition, evaporation becomes zero. Evaporation is high when the vapor pressure of air is low, which means that the gradient between the vapor pressure of air and that of water is low. The difference in the vapor pressure gradient between the water surface and the air varies based on the height above the water surface.

Sometimes water evaporates directly from the solid state (say, ice), a process called *sublimation*. However, the inflow of water vapor to the atmosphere through sublimation is much less than evaporation of liquid water from oceans, lakes, and rivers.

Heat energy, which is essential for evaporation, is derived generally from solar radiation. The relative humidity (RH) of air and the wind velocity across the water surface also influence the rate of evaporation. Near the equator, where the sea and winds are warm, the rate of evaporation is quite high. Tropical climates are typically humid, as is the case in the East Indies, the Amazon River basin, and central Africa. Near the poles, evaporation is small because the cold winds cannot hold much moisture and the weak sunshine provides little energy. Arctic climates are typically dry.

5.1.3 Physics of Evaporation

Water molecules in a water body are tightly held by intermolecular forces. In evaporation, energy is consumed to overcome these forces and move the molecules away. The energy required to move to the vapor form is called the *latent heat of vaporization* (λ). Its value is about 2.5 MJ/kg; this also depends upon temperature and is computed by

$$\lambda = 2.501 - 0.00236T \tag{5.1}$$

where T is the temperature of water in degrees Celsius.

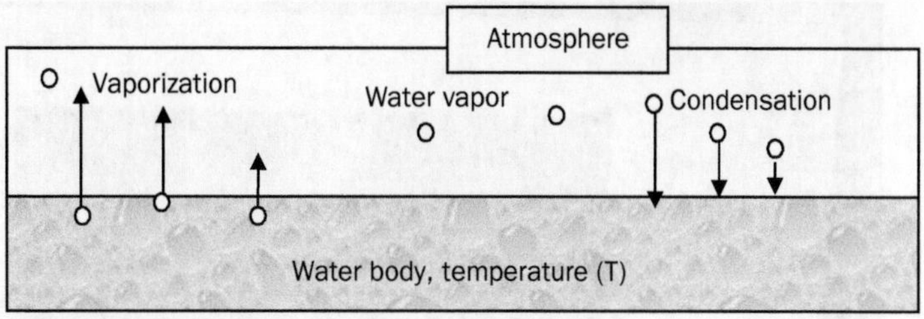

Figure 5.1 Vaporization and condensation in a body of water.

As seen in Fig. 5.1, there is a regular exchange of molecules between the water body and atmosphere for an open water body. Evaporation amounts to the difference between the vaporization rate and the condensation rate. The rate of vaporization (the rate at which water is converted into vapor) depends on water temperature and the condensation rate (the rate at which water vapor is converted to liquid water) is determined by vapor pressure. When the vapor pressure is low, vaporization takes place, and so does evaporation. If the rate of vaporization equals condensation, the air is saturated and evaporation stops.

The difference between saturated vapor pressure and vapor pressure of the air (e_a) or vapor-pressure gradient determines the rate of evaporation. As this difference increases, evaporation also increases, and vice versa. In addition, the difference in the temperature of air and that of water is important. If the temperature of water is lower than that of air, evaporation will be low because the vapor pressure of cold water is low. When air is warmer than water, evaporation continues until the vapor pressure of air equals that of water, and then no further evaporation occurs. If air is cooler than water and vapor pressure of water and air are equal, air becomes supersaturated and fog forms. Note that the vapor pressure and humidity are interdependent.

Part of the radiation energy reaching the ground warms the atmosphere near the Earth. Heat energy flux moving onward with air is called *sensible heat flux*, perhaps because it can be sensed. This term is useful in computation of ET. Usually, sensible heat flux is upward during the day and downward during the night.

5.2 Factors Affecting Evaporation

The most important factors affecting evaporation are (1) temperature, (2) solar radiation and sunshine duration, (3) RH, (4) wind speed, (5) vapor pressure difference (gradient), (6) atmospheric pressure, (7) water quality, and (8) water depth and soil type.

5.2.1 Temperature

The temperature of a place depends on its geographical location (which determines incoming solar radiation), altitude, season, and distance from the sea or ocean. Temperature is influenced by the exchange of air masses and by cloudiness (which obstructs incoming radiation) and follows diurnal and seasonal cycles. Temperature falls with altitude at approximately 0.6°C/100 m rise in elevation for moist air and 0.9°C/100 m for dry air. Hence, evaporation is less at higher altitudes. The temperature variation near large water bodies, such as seas, is less due to the moderating influence, and hence the annual and diurnal range is smaller. Generally, temperatures of nearby places are strongly correlated.

FIGURE 5.2 Thermometers installed on the Stevenson screen.

Temperature also influences the amount of dissolved gases in water and the rate of chemical and biological reactions and activities in water bodies. As the temperature increases, the kinetic energy of water molecules also increases, which allows water molecules to escape from liquid water to the air faster than otherwise. Thus, the warmer the water is, the greater is the transfer of molecules from the water to the air and evaporation. Temperature is commonly reported in degrees Celsius, but the United States and some other countries use the Fahrenheit unit of measure.

Temperature is manually measured by a set of four thermometers that are usually installed on the Stevenson screen (Fig. 5.2), which provides ventilation and shade. The screen should face north. Four types of thermometers are used: a dry bulb thermometer to measure ambient air temperature; a wet bulb thermometer to measure temperature that is attained by a volume of air if it is cooled adiabatically to saturation by evaporation of water into it (this data is used to calculate RH); maximum thermometer to measure the highest temperature reached since the last setting; and minimum thermometer to measure the lowest temperature reached since the last setting. After the temperatures from the maximum and minimum thermometers have been read, they are reset by following a standard procedure.

Wet bulb temperature is measured by using a thermometer whose bulb is wrapped by a wet cloth (called a *sock*) that is kept moist by capillary action. One end of the sock is immersed in a small cup of water, and the other covers the bulb. There must always be enough water in the cup to keep the sock wet. Wet bulb thermometers should be shielded from radiation, and should be properly ventilated.

Data from the dry bulb thermometer are the most frequent input in hydrologic studies, and ambient air temperature is best measured with a mercury-in-glass thermometer. Continuous records of temperature can be obtained by using resistance or thermocouple elements. Thermographs that produce a continuous record of temperature are used to measure and record water temperature. Typically, these comprise a mercury-in-steel sensing element immersed in the water and connected to a recorder. When the temperature of air above a water body is needed, it should be measured at 2 m above the water surface near the center of the water body (this is difficult in case of big lakes).

To measure water surface temperature, several types of thermometers, such as mercury-in-glass or mercury-in-steel (including maximum and minimum and reversing thermometers), platinum resistance or thermistor elements with electronic circuits, and meter or recorder and thermocouple thermometers, are used. In general, the precision required for water temperature is ±0.1°C. These days devices to measure and record temperature have become widely available.

If the thermometer is not properly ventilated, it will not give representative readings. An aspirated thermometer, a thermometer for which ventilation is provided by a suction fan and the surrounding body provides the requisite shade, is recommended in such conditions.

5.2.2 Processing of Temperature Data

Processing and quality control of temperature data depend on the location and site conditions, and the observed value is compared with the nearby stations. Errors may creep into the temperature data for many reasons. The observer may wrongly read the values. Reading errors frequently arise when the thermometers are old, with faint graduation. Data of the wet bulb thermometer will be erroneous if the muslin and wick of the wet bulb are not adequately saturated. There will also be an error if the muslin becomes dirty or is covered by an oily substance. These defects will result in a high value of wet bulb temperature. Errors may also arise when writing the thermometer reading.

Faults in thermometer usually result in persistent, systematic errors in data. Most of these errors can be trapped by comparing with appropriate seasonal maximum, minimum, and warning limits for a given station. Note that these limits will change with location. For obvious reasons, the maximum temperature should be more than the minimum temperature. Further, the dry bulb temperature should be greater than or (rarely) equal to the wet bulb temperature.

5.2.3 Solar Radiation

The instrument used to measure the incident solar shortwave radiation is called a *pyranometer*. These are based on multijunction thermopiles; glass domes of these devices allow only radiation in the 0.3–3-μm range to reach the pyranometer surface. A net pyrradiometer measures the difference between total (shortwave and long wave) incoming (downward) and outgoing (upward) radiation (WMO, 2008). It consists of a horizontally mounted plate with two blackened surfaces. Half the junctions of a thermophile are attached to the upper surface and the others are attached to the lower surface, so that the thermopile output is proportional to net radiation in the 0.3–100 μm band. WMO (2008) recommends that the instruments should be mounted at least 1 m above the representative vegetation cover.

5.2.4 Sunshine Duration

Sunshine duration is an indicator of radiation received at a place. The potential maximum sunshine duration is a function of latitude and longitude and season; the actual number of sunshine hours varies due to clouds, fog, and other elements. Sunshine duration is commonly measured by a sunshine recorder (Fig. 5.3), which consists of a glass sphere mounted on a section of a spherical bowl. The glass sphere focuses the Sun's rays on a paper card mounted on the bowl on which hours are indicated. When the Sun is shining, its rays burn the card. The card is changed daily after sunset. The burned traces on the card indicate the sunshine duration, and data at the required time resolution are read from the card. Note that the sunshine recorder uses the movement of the Sun to form the time base of the record. Different grooves in the bowl must be used in winter and summer, and different card types be used for each season.

Errors may arise in the sunshine duration data if the chart is wrongly placed; in addition, the observer may wrongly note the duration of sunshine from the chart. Any value of sunshine duration that is negative or more than the maximum possible value for that location should be further checked and corrected.

5.2.5 Relative Humidity

RH, typically expressed in percent, can be defined as the ratio of the actual vapor pressure of air to the saturation vapor pressure at the same pressure and temperature. Due

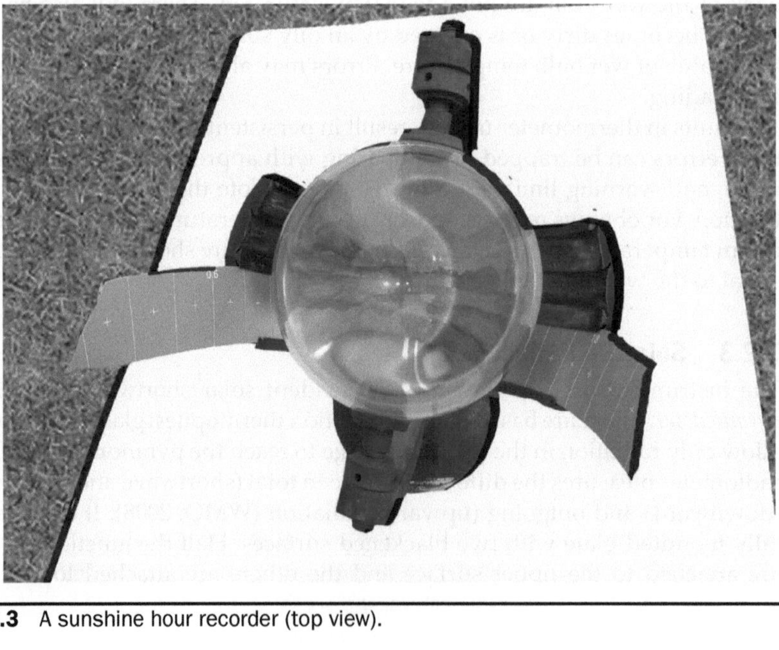

FIGURE 5.3 A sunshine hour recorder (top view).

to the availability of moisture, places close to the sea and large open water bodies have higher RH and a smaller daily variation than the other locations. RH does not vary rapidly with time.

RH is most commonly estimated by jointly measuring dry bulb and wet bulb temperatures and then calculating the dew point temperature and actual and saturated vapor pressure; it is also calculated from the wet bulb depression (difference between wet and dry bulb readings) using a set of tables. While the actual vapor pressure may vary little during the day, RH has a diurnal pattern with a minimum daily value occurring around the same time as the highest temperature (when the saturation vapor pressure is highest). RH also shows a seasonal variation.

In addition, RH may be measured continuously by means of a hygrograph, where the sensor is human/horse hair whose length changes with RH. Errors in the hygrograph may arise from bad calibration. Errors in RH may be detected by setting up upper and lower warning limits depending upon the station and season. The maximum value for RH is 100%. The graph of daily series needs to be inspected to identify and correct anomalous values.

5.2.6 Wind Speed

The wind speed at a place is controlled by the pressure gradient, which depends on temperature, topographic features, and land use/cover. Wind speed shows a wide variation with place and time. It is commonly measured by the use of a cup-type anemometer (Fig. 5.4)—cups of this device rotate due to the pressure exerted by winds. The

FIGURE 5.4 A cup-type anemometer.

rotations of cups over a time interval are measured, displayed by a counter, and/or recorded by a data logger. The number of rotations indicates the average wind speed over the chosen time interval. The usual practice is to measure the wind speed over a 3-min period, and this is considered as the instantaneous wind speed at that time. Likewise, by noting the counter reading at the beginning and end of a period, the average wind speed for that period can be determined.

The direction of wind is reported as one of the 16 points of the compass; it could be shown either as a number or as an alphabetic code. Note that wind direction is not used to compute ET—it is not a variable in formulas. Wind direction may influence evaporation at a place if the wind is blowing from a place with significantly different humidity.

Under certain conditions, a thin film of saturated vapor exists over a water surface. Left undisturbed, this film acts as an insolating buffer between the water surface and the unsaturated air above and slows the evaporation to a low rate. Usually, this condition is not seen in real life because wind tends to remove the saturated film and expose the water surface to unsaturated air. This action creates a vapor-pressure gradient that provides conditions that are favorable for evaporation. The relation between a water surface and wind drag on that surface is complex. Evaporation is believed to increase with wind velocity until the vapor-pressure gradient reaches some nearly constant value. After at this point, a further increase in wind velocity will not increase evaporation appreciably. The effect of increased wind velocity on evaporation is believed to be related to the size of the open water body. Wind removes water vapor from small bodies of water rather quickly, whereas for large bodies of water, a longer time is required before the same effect is observed.

Due to the effect of other factors, such as RH, vapor pressure, and wind speed, pan evaporation is not directly related to temperature.

5.2.7 Atmospheric Pressure

Because air is less dense at lower atmospheric pressures, the likelihood of collision of water molecules escaping in the air with air molecules reduces with atmospheric pressure. For this reason, evaporation is more when atmospheric pressure is low. Two changes take place as one moves to higher altitudes: atmospheric pressure and temperatures decrease. Consequently, the relation between altitude and evaporation is not straightforward. It has been observed that evaporation from lakes and evaporation pans increase as elevation increases.

5.2.8 Water Quality

The vapor pressure of water is reduced when solids are dissolved in water. Pure water has a higher vapor pressure than saltwater. Because there is a wide range of purity of water present in nature, it is very difficult to predict the effect of dissolved minerals in any given body of water on evaporation. The effect is probably small. A salt content of 1% slows the rate of evaporation by about 1%. Because the salt content of ocean water is a little over 3%, the seawater evaporation rate is lower than the freshwater rate by about 3%.

5.2.9 Water Depth and Soil Type

Water in shallow-water bodies gets heated more quickly than in deepwater bodies. Solar radiation penetrates shallow-water bodies and is absorbed by the soil at the bottom. This results in more heating of the water body. Therefore, evaporation from shallow-water bodies is more. Precipitation that is ponded on a drainage basin and intercepted precipitation are evaporated rapidly after rainfall, when insolation provides renewed heating. The effect of this heating is increased because dark soils and other dark objects absorb incoming radiation and convert it to heat more effectively than lighter-colored soils and objects.

This discussion on evaporation is based on the assumption that a free water surface from which evaporation can occur exists. This condition exists in the oceans, lakes, and other permanent bodies of water and perennial streams. Such a condition also nearly exists on surfaces that are wet with rain, such as soil, vegetation, and buildings; artificial surfaces, such as parking lots and roads; and animal life. Evaporation also occurs beneath the soil surface as heat energy penetrates and vaporizes the moisture contained on and between the soil grains. It is possible for evaporation to draw water from soil from a depth of >0.3 m if enough time elapses between two rainfall events.

5.3 Measurement of Evaporation

The evaporation pan, a small, shallow vessel filled with water and placed in an open area and exposed to weather, is the most widely used device to measure evaporation. The water level in the pan is measured carefully to determine how much water (measured in terms of depth) is lost by evaporation. This loss is related to the evaporation from a pond or lake.

Various agencies have stipulated the size and material of pans and the measurement procedures to bring standardization to the measurement of evaporation. Such efforts have resulted in, for example, the U.S. Weather Bureau Class A Pan, the U.S. Geological Survey (USGS) Floating Pan, and the GGI-3000 pan from Russia. Among these pans, the U.S. Weather Bureau Class A Pan is the most commonly used instrument in a large number of countries throughout the world. The World Meteorological Organization (WMO) has recommended the use of this pan, which is described next.

5.3.1 The U.S. Weather Bureau Class A Land Pan

The U.S. Weather Bureau Class A land pan is 1.22 m in diameter and 25.4 cm deep (Fig. 5.5). It is painted white on the inner base and sides and placed on a white-painted, level wooden stand above the ground. The pan is filled with water to a level at least 5.1 cm below the top and not more than 7.6 cm below the top of the pan. Many times, the pan is covered by a wire mesh to prevent birds and animals from drinking the water.

It is important to ensure that the pan is installed on a site that is level and free of obstructions. The ground cover should be as close as possible to that commonly found in the area. An obstruction (tree, building, shrub, or instrument shelter) of height h should be at least $4h$ away from the pan. The pan should never be placed on an asphalt or concrete surface, and no shadow should fall over it at any time. It should preferably be installed in a plot whose size is 15 m $\times$ 20 m. It is a good idea to fence the plot to protect the instruments and prevent animals from drinking the pan water, but the fence should not affect the wind flow over the pan. Sometimes chemical repellants are used to prevent birds and animals from drinking the water, but make sure that they do not pollute pan water.

On days without rain, when the pan evaporation is read, water is poured into the pan using a graduated cylinder to bring the level to the top of the pointer gauge. The volume of water added is recorded and represents the depth of evaporation.

Figure 5.5 The U.S. Weather Bureau class A land pan.

On the days when rain has fallen since the last observation, the rainfall may exceed the evaporation, and water may have to be removed from the pan to bring the level to the pointer level. If there is a forecast of heavy rainfall on a given day, a certain amount of water may be removed from the pan in advance to avoid overflow. This should be accounted for in computations.

The water level in the pan must be measured accurately before and after water is added. This is accomplished typically by means of a hook gauge that consists of a movable scale and vernier fitted with a hook and enclosed in a stilling well in the pan. A calibrated jar is used to add or remove water at each observation so as to maintain the water level to a prespecified depth.

The amount of evaporation (E, expressed in millimeters) between two successive observations is the difference between the water levels in the pan. It is corrected for any precipitation during the period:

$$E = P \pm \Delta d \tag{5.2}$$

where P is the depth of precipitation during the period between the two measurements and Δd is the depth of water added (+) to or removed (–) from the pan.

A good practice is to install additional instruments along with a pan. Typically an evaporation pan is a part of a weather observatory which includes a precipitation gauge, an anemometer to measure wind speed and direction, and thermometers to measure temperatures.

5.3.2 Analysis of Pan Evaporation Data

Like any other instrument and measurement, errors may arise in pan observations due to observer's mistakes and instrument errors. Leakage of water from the sides or bottom, particularly with old pans, should be promptly stopped.

Errors in the data may occur due to the following and must be corrected:

- The observer may overfill or underfill the pan on a particular day. Such values may be compensated for on the next day, but the value for the current day will be wrong.
- The maximum value of evaporation in the region is used as a first check to screen doubtful measurements and correct them.
- Among the instrument errors, leakage is the most serious and common problem, which occurs usually at the joint between the pan base and side wall. Small leaks are difficult to detect but may have a significant systematic effect on evaporation values over an extended period. Where leakage has been detected, the pan evaporation records for a period of a few months preceding the detection of leakage should be inspected carefully and should be flagged.
- Animals and birds may drink the pan water, especially if the covering wire mesh is damaged.
- Algae and dirt, if present in the water, reduce the rate of evaporation.
- Errors arise during periods of high rainfall when the depth caught by the rain gauge is different from the rain falling in the pan as a result of splashing or wind eddies.

Evaporation from the pans is affected by the size, depth, and location of the pan. To detect the error in pan evaporation data, warning and maximum limits on evaporation are fixed. Pan evaporation measurements are less accurate during periods of heavy rain, but the rate of evaporation is also less during these periods. Excessively high evaporation values should be subject to validation.

The evaporation from pans installed above the ground is influenced by the characteristics of the pan. Pans placed above the ground are subject to heat exchange through the sides and to other effects that do not occur in lakes. Leakage of water from sunken pans is difficult to detect. Additional issues with them are accumulation of floating biomass and debris on the water surface, and boundary conditions that are different from those of a large lake. Floating pans are subject to splash-in and splash-out of water and are costly to install and operate. Pans have much less heat storage than lakes and experience a different annual cycle of evaporation Masoner et al. (2008) found floating pan evaporation rates to be significantly less than land pan evaporation rates. Fairly reliable estimates of annual lake evaporation can be obtained by multiplying the annual pan evaporation by the appropriate pan coefficient.

Automatic evaporation pans are also in use these days. In these pans, the water level is kept constant by automatically releasing water into the pan from a storage tank or by removing water from the pan in the case of precipitation. Amount of water added to or removed from the pan is accounted for in computations.

5.3.3 Estimation of Lake Evaporation from Pan Measurements

Evaporation data from pans are commonly used to estimate evaporation from lakes and reservoirs and ET from an area. But size of a pan is small, and its material is different from a natural body. Hence, its heat storage characteristics and air dynamics will be quite different from a large water body, and evaporation from a pan will be higher than from such bodies.

An estimate of lake or reservoir evaporation (E_R) is obtained by multiplying the pan evaporation by a coefficient called the *pan coefficient:*

$$E_R = K_p E_{pan} \tag{5.3}$$

where K_p is the pan coefficient, E_R is the depth of evaporation from the reservoir, and E_{pan} is the pan evaporation, both measured in millimeters per day. The pan coefficient modifies E_{pan} in accordance with the physical conditions of the lake/reservoir. The value of K_p depends on climate, geographical location, season, and size and depth of the water body. It generally varies from 0.6 to 0.85 (annual scale) and from 0.3 to 1.7 (monthly scale), depending on the characteristics of the water body. Lower values are typical of dry seasons and arid climates, while higher values are for humid climates. In the absence of better estimates, a pan coefficient value of around 0.7 may be used. The pan coefficient for a particular pan can be determined by comparing with actual evaporation for a lake or with a large sunken pan to simulate a lake.

The advantages of using a pan are that (1) it is cheap, simple, and easy to install and operate, (2) the pan coefficient is stable, and (3) a less-skilled observer can handle it. Pan measurements approximate lake evaporation reasonably closely. The disadvantage is that the pan coefficient varies in time and space.

Example 5.1

Compute daily evaporation from a Class A pan if the daily rainfall and the amount of water added to bring the water level in the pan to the fixed point are as follows:

Day	1	2	3	4	5	6
Rainfall (cm)	0	0.5	0.1	0	0	0.4
Water added (cm)	1.5	1.7	0.5	1.2	0.7	1.3

Solution

The daily pan evaporation equals the amount of water required to bring its water level to the fixed point plus the water contributed by rainfall. For each day, this amount is shown as follows, and it varies from day to day, depending upon atmospheric conditions.

Day	1	2	3	4	5	6
Evaporation (cm)	1.5	2.2	0.6	1.2	0.7	1.7

Example 5.2

The surface area of a lake is 500 ha. Compute the daily evaporation for the data in Example 5.1. Assume that the pan coefficient is 0.8.

Solution

The daily lake evaporation is obtained by multiplying the daily pan evaporation by the pan coefficient:

Day	1	2	3	4	5	6
Pan evaporation (cm)	1.5	2.2	0.6	1.2	0.7	1.7
Lake evaporation (cm)	1.2	1.76	0.48	0.96	0.56	1.36

The 6-day evaporation is the sum of daily values for 6 days = 6.32 cm. The 6-day loss of water = (6.32 cm/100) × 500 × 10,000 m^2 = 316,000 m^3. It is assumed that the surface area of the lake is the same on all days.

Example 5.3

Compute the mean daily evaporation loss in hectare-meters for the month of July from a stream reach that is 100 km long and 50 m wide on average. The mean daily evaporation measured by a Class A pan for July is 0.6 cm. Assume that the pan coefficient as 0.8.

Solution

Mean daily evaporation = 0.6 cm × 0.8 = 0.48 cm

This evaporation takes place from the entire stream reach, whose area is 100 × 1000 m × 50 m = 5 × 10^6 m^2, or 500 ha. Therefore,

$$\text{Daily Evaporation} = \frac{500\,\text{ha} \times 0.48\,\text{m}}{100} = 2.4\,\text{ha} - \text{m}$$

This loss of water due to evaporation must be considered when managing the water of this stream reach.

Example 5.4

It is desired to estimate the fall in water level due to evaporation from a lake on June 24, 2015. A Class A pan is located near the lake. On this day, the rainfall amount was 0.45 cm, and the amount of water added to restore the water level to the value at the beginning of that day was 0.65 cm. Assume that the pan coefficient as 0.78.

Solution

$$\text{Class A Evaporation} = 0.44\text{ cm} + 0.65\text{ cm} = 1.1\text{ cm}$$

$$\text{Evaporation from the lake} = 1.1\text{ cm} \times 0.78 = 0.858\text{ cm}$$

Thus, on June 24, the water level declined by 0.858 cm due to evaporation.

5.4 Determination of Evaporation from Water Bodies

Evaporation from water surfaces can be determined by (1) water budget, (2) energy budget, (3) mass transfer methods, (4) combination methods, and (5) evaporation formulas. Xu and Singh (1998) evaluated the dependence of evaporation on meteorological variables at different time scales. Singh and Xu (1997) evaluated and generalized 13 mass-transfer equations for determining free-water evaporation.

5.4.1 Water Budget

The water budget equation for estimating evaporation can be written (Horton, 1943) as

$$E = I + P - O - O_s + \Delta S \tag{5.4}$$

where E is the evaporation, I is the inflow, P is the precipitation, O is the outflow, O_s is the seepage, and ΔS is the change in storage. Seepage cannot be measured or evaluated directly and accurately, and the extent to which this quantity is accurate will affect the true value of evaporation. Inflow, outflow, precipitation, and change in storage can be measured reasonably accurately. The water budget method of determining long-term evaporation can be used as a standard for comparing other methods. This method is not perfect, but it is satisfactory for practical purposes.

Example 5.5

Estimate the evaporation for a month for a lake of 500-ha surface area. The mean discharge from the lake is estimated to be 1.00 m³/s. The monthly rainfall is about 10 cm. A stream flows into the lake with an average discharge of 2.00 m³/s. The water level in the lake dropped about 5 cm during the month. The seepage losses are negligible.

Solution

$$\text{Monthly inflow} = I = 2 \times 3{,}600 \times 24 \times 30 \text{ m}^3 = 51.84 \times 10^5 \text{ m}^3$$

$$\text{Monthly outflow} = O = 1 \times 3{,}600 \times 24 \times 30 \text{ m}^3 = 25.92 \times 10^5 \text{ m}^3$$

$$\text{Monthly rainfall} = P = \frac{10}{100} \times 500 \times 10{,}000 \text{ m}^3 = 5\times10^5 \text{ m}^3$$

$$\text{Change in storage} = \Delta S = \frac{5}{100} \times 500 \times 10{,}000 \text{ m}^3 = 2.5\times10^5 \text{ m}^3$$

Apply Eq. (5.3) to estimate the monthly evaporation, as follows:

$$E = (51.84 + 5.0 - 25.92 + 2.5) \times 10^5 \text{ m}^3 = 33.42 \times 10^5 \text{ m}^3$$

$$= \frac{33.42 \times 10^5}{500 \times 10^4} \times 100 \text{ cm} = 66.84 \text{ cm}$$

5.4.2 Energy Budget

Assuming unlimited availability of water, the evaporation rate is chiefly governed by the amount of available radiant energy. The net amount of radiation available at the Earth's surface is the difference between the incoming and reflected solar (shortwave) radiation (Fig. 5.6), plus the difference between the long wave incoming and outgoing radiations:

$$R_n = S_n + L_n \tag{5.5}$$

where R_n = the net radiation, S_n = the net shortwave (solar) radiation, and L_n is the net long wave radiation, all in MJ m^{-2}/day. The net shortwave radiation is equal to the incident shortwave radiation less the part due to reflection and is given by

$$S_n = S_t(1.0 - \alpha) \tag{5.6}$$

where S_t is the incoming shortwave (0.3–3.0 μm) radiation (in MJ m^{-2}/day) and α is the reflection coefficient or albedo, which depends on the land cover and the direction of solar rays. Taller vegetation reflects less radiation than shorter. In the absence of better information, Shuttleworth (1993) suggested that albedo can be assumed to be 0.08 for open water, 0.11–0.16 for tall trees, 0.20–0.26 for short crops and grass, 0.10 (wet)–0.35 (dry) for bare soil, and 0.20 (old)–0.80 (fresh) for snow and ice. Incoming solar radiation can be estimated from the extraterrestrial radiation by

$$S_t = \left(a_s + b_s \frac{n}{N}\right) S_0 \tag{5.7}$$

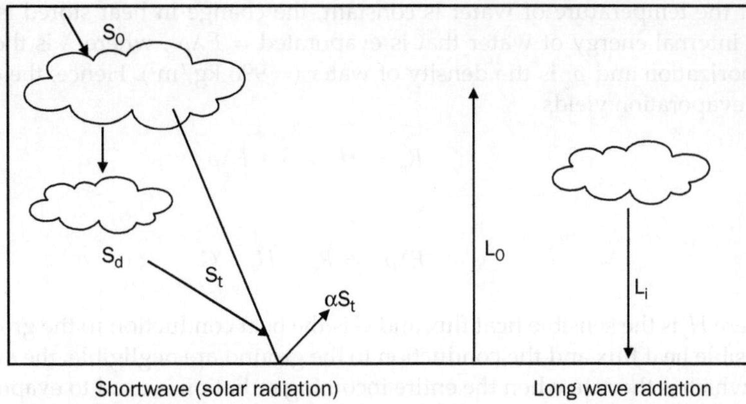

FIGURE 5.6 Radiation balance at the ground. Here, S_0 is the incoming shortwave radiation at the top of the atmosphere. After scattering and interception by the atmosphere and clouds, a part S_t comes to the Earth's surface, and part of that is reflected. L_i is the incoming long wave radiation, a part of which is reflected.

where a_s and b_s are the Angstrom coefficients, n is the number of bright sunshine hours in a day of N hours, and S_0 is the extraterrestrial radiation. Coefficients a_s and b_s can be estimated by regression analysis of data of S_0 and S_t. In the absence of locally calibrated data, one may adopt $a_s = 0.25$ and $b_s = 0.50$ (Shuttleworth, 1993).

The net outgoing long wave radiation can be estimated (Penman, 1948) as

$$L_n = L_0\left(a + b\frac{n}{N}\right) \tag{5.8}$$

in which L_0 is the net outgoing long wave radiation on a clear day and a and b are constants. Penman reported $a = 0.1$ and $b = 0.9$. The value of L_0 can be computed as

$$L_0 = \sigma T^4(\in -1) \tag{5.9}$$

where σ is the Stefan-Boltzmann constant $= 11.71 \times 10^{-8}$ cal/cm^2 °K-day or 2.0×10^{-9} mm of water/day; T is the temperature of water surface (°K); and $\in$ is the emissivity, which is defined by the ratio of long wave radiation down from the atmosphere to that upward from the Earth and can be estimated as

$$\in = a_1 + a_2(e_a)^{0.5} \tag{5.10}$$

where e_a is the mean vapor pressure of the air (millibars) at a height of 2 m, and a_1 and a_2 are constants. The value of a_1 varies from 0.43 to 0.68 and a_2 from 0.029 to 0.082. The value of emissivity varies from 0.906 to 0.986 for water surfaces. A considerable amount of radiant energy is exchanged between the terrestrial surface and atmosphere in the form of long wave (3–100 μm) radiation. The net energy received is the difference between incoming and outgoing radiation.

The energy-budget method to compute evaporation is similar to the water budget method, except that it deals with the conservation of energy. The energy available for evaporation is obtained by considering the incoming energy, the outgoing energy, and the energy stored in the water body for a given time interval. Assuming that the temperature of water is constant, the change in heat stored is the change in the internal energy of water that is evaporated $= E\lambda\rho_w$, where λ is the latent heat of vaporization and ρ_w is the density of water (= 996 kg/m³). Hence, the energy balance for evaporation yields

$$R_n = H_s + G + E\lambda\rho_w$$

or

$$E\lambda\rho_w = R_n - H_s - G \tag{5.11}$$

where H_s is the sensible heat flux and G is the heat conduction to the ground. When the sensible heat flux and the conduction to the ground are negligible, the evaporation rate E_r, which is the rate when the entire incoming radiation is used to evaporate water, can be computed as

$$E = \frac{R_n}{\lambda\rho_w} \tag{5.12}$$

Example 5.6

At a given location, the net radiation on May 1 was 180 W/m², the air temperature was 25°C, and the wind velocity at 2-m height was 2.5 m/s. Use the energy method to determine the open water evaporation rate, given $\rho_w = 996$ kg/m³.

Solution

We first compute the latent heat of vaporization by using Eq. (5.1):

$l_v = 2.501 \times 10^6 - 2{,}360T = 2.501 \times 10^6 - 2{,}360 \times 25 = 2{,}442 \times 10^3$ J/kg = 2,442 kJ/kg.

Hence,

$$E_r = \frac{180}{2{,}442{,}000 \times 996} = 7.4 \times 10^8 \text{ m/s} = 6.39 \text{ mm/day}$$

Aerodynamic Method This method computes evaporation by considering the ability of the atmosphere to transport water vapor away from the evaporating surface. Thus, evaporation is a function of the difference between the saturation vapor pressure at the surface at the ambient temperature and the vapor pressure at a height z_2 above the water surface. Applying the physical laws, the evaporation rate is given by

$$E_a = \frac{0.102u_2}{[\ln(z_2/z_0)]^2}(e_{as} - e_a) \tag{5.13}$$

Here, u_2 is the wind velocity in m/s at height z_2 (cm), and z_0 is the roughness height of water surface. If temperature (T in degrees Celsius) is known, the saturated vapor pressure e_{as} (Pa or N/m²) can be computed by

$$e_{as} = 6.11 \times \exp\left(\frac{17.27T}{237.3 + T}\right) \tag{5.14}$$

The vapor pressure e_a is given by

$$e_a = R_h \times e_{as} \tag{5.15}$$

where R_h is the RH.

Example 5.7

At a given location, the net radiation on May 1 was 180 W/m², the air temperature was 25°C, the wind velocity at 2-m height was 2.5 m/s, and RH was 70%. Use the aerodynamic method to determine the open water evaporation rate if the roughness height is 0.04 cm.

Solution

We first compute the saturated vapor pressure by Eq. (5.14):

$$e_{as} = 611 \times \exp[17.27 \times 25/(237.3 + 25)] = 3{,}169 \text{ Pa.}$$

So,

$$e_a = 0.70 \times 3169 = 2{,}218 \text{ Pa}$$

Hence,

$$E_a = \frac{0.102 \times 2.5}{[\ln(200/0.04)]^2}(3169 - 2218) = 3.34 \text{ mm/day}$$

5.4.3 Combination Method

As the name suggests, the combination method combines the energy-budget and mass-transfer methods. The most popular combination method for computing evaporation from free water surfaces was developed by Penman (1948). This method combines the physics principles and empirical concepts based on standard meteorological observations. The physical principles are the energy balance equation and the mass-transfer (or aerodynamic) equation.

The aerodynamic method can be used when the available energy is not limiting evaporation, and the energy balance method can be used when vapor transport is not a limiting factor. Normally, both of these factors are not limiting, and hence the combination method may be suitable. Here, evaporation is calculated as

$$E = \left(\frac{\Delta}{\Delta + \gamma}\right)E_r + \left(\frac{\gamma}{\Delta + \gamma}\right)E_a \tag{5.16}$$

where the first term on the right side is due to vapor transfer and the second term is due to aerodynamic mass transfer, and γ is the psychrometric constant [it is the ratio of specific heat of moist air at constant pressure to latent heat of vaporization of water (= 0.4 g/kg/K)]. Notation Δ (kPa/°C) denotes the gradient of saturated vapor pressure e_{as} [given by Eq. (5.14)], which is computed by

$$\Delta = \frac{4{,}098 e_{as}}{(237.3 + T)^2} \tag{5.17}$$

where T is the temperature in degrees Celsius.

Example 5.8

At a given location, the net radiation on May 1 was 180 W/m², the air temperature was 25°C, the wind velocity at 2-m height was 2.5 m/s, and RH was 70%. Use the combination method to determine the open water evaporation rate if the roughness height is 0.04 cm.

Solution

For the given data, the saturated vapor pressure e_{as} was calculated in the earlier example as 3,169 Pa. We now compute the gradient of e_{as}:

$$\Delta = \frac{4{,}098 e_{as}}{(237.3 + T)^2} = \frac{4{,}098 \times 3{,}169}{(237.3 + 25)^2} = 188.75 \text{ Pa/°C}$$

$$\text{Hence, } E = \left(\frac{\Delta}{\Delta + \gamma}\right)E_r + \left(\frac{\gamma}{\Delta + \gamma}\right)E_a = \left(\frac{188.75}{188.75 + 66.8}\right)6.39 + \left(\frac{66.8}{188.75 + 66.8}\right)3.34$$

$$= 5.6 \text{ mm/day.}$$

5.5 Transpiration

Transpiration is the process by which plants use water for their metabolism and growth. Plants extract water from the soil through their roots and transfer it to the atmosphere as vapor through stomata. *Stomata* are small openings in plant leaves for the exchange of moisture and gases with the atmosphere. The number of stomata on a plant leave varies with the plant species, and their function is vital to plant metabolism. Just to give an idea, the density of stomata can vary from about 8,000 to 12,500 per cm²; the number on the lower leaf surface is much more than on the upper surface. Transpiration is similar to evaporation except that water escapes to the atmosphere not from the free water surface, but through plant leaves. It is affected by plant physiological and environmental factors. The rate of movement of water through the plant is important because nutrients and minerals are carried into the plant with water.

During daylight, stomata actively transpire water vapor to the atmosphere, but they close after sunset, and then transpiration ceases. They allow carbon dioxide to enter the plant in the process of photosynthesis. Stomata respond to environmental conditions, and their opening depends on factors such as light and darkness, hot and cold weather, and other elements. Important plant physiological factors are (1) leaf structure, (2) extent and properties of protective coverings, (3) density and properties of stomata, and (4) plant health. Leaves that are exposed to direct the Sun's rays transpire more water than those shaded by other leaves. If soil moisture is adequate, the amount of moisture lost by transpiration increases with plant density. Healthy plants transpire at higher rates than do those with diseases. Only about 1% of the water taken up by the roots is retained by plant tissues; most of the moisture passes from roots to the atmosphere via leaves. Environmental factors that are most important in affecting transpiration are vapor-pressure gradient, temperature, solar radiation, wind, and available soil moisture.

The *vapor-pressure gradient* is the difference in vapor pressure between the space immediately inside the leaf and the outside air, and it is a measure of the energy required to move the water from the leaf to the air. As might be expected, transpiration is less when the vapor-pressure gradient is less, such as during a rainfall event.

However, plants transpire more when the surroundings are dry and warm, the Sun is shining, and plenty of soil moisture is available. The rate of transpiration is doubled for approximately each 10°C rise in temperature.

Solar radiation is the primary source of energy for transpiration. The solar constant is 1.94 cal/cm^2/min at the edge of the Earth's atmosphere, with about 0.30 cal/cm^2/min reaching the Earth's surface. Absorption of this energy by a leaf raises its temperature and aqueous vapor pressure. Thus, transpiration increases with increasing insolation. Different crops and different soils reflect different amounts of insolation and result in different amounts of transpiration.

Transpiration is greatly affected by soil moisture. As the soil moisture depletes, the capillary forces holding soil moisture become stronger, and it is hard for the plant roots to take moisture. This is truer when the soil moisture is near the permanent wilting point. Of course, at the permanent wilting point, transpiration stops. Wind usually increases transpiration by removing the moist air near the leaf, thereby increasing the vapor-pressure gradient. Gentle winds increase transpiration more than strong winds.

5.6 Evapotranspiration

ET and consumptive use include transpiration of water by vegetation, as well as evaporation from water surfaces, soil, snow, ice, and vegetation. For all practical purposes, the terms *consumptive water use* and *ET* are synonymous. ET converts water to a form (water vapor) that is not available to be used again. This is in contrast to uses such as hydropower generation, where water is subsequently available for use. ET is typically expressed in the unit of depth (millimeters or centimeters) for a given period.

ET is an important component of the hydrologic cycle. Its estimation is necessary in many studies, such as catchment modeling, agricultural water management, determination of water balance, and assessment of the impact of land use changes on the hydrologic response. In many watersheds, the return of moisture to the atmosphere through ET is a large proportion of input precipitation.

Despite the widespread application and use of the concept of ET, there has been considerable ambiguity in the use of various terms, such as *potential ET* and *reference crop ET*. To overcome this, the Food and Agricultural Organization (FAO) of the United Nations put out a report, commonly referred to as *FAO-56* (Allen et al., 1998). Among other things, this report introduced uniformity and standardization to the interpretation and use of various terms, such as *potential ET* and *reference crop ET*. *FAO-56* discourages the use of the former term due to ambiguities in its definition. The FAO recommended that a hypothetical surface "closely resembling an extensive surface of green grass of uniform height, actively growing, completely shading the ground, and with adequate water" (Allen et al., 1998) be adopted as the reference surface. In the FAO approach, the surface characteristics that influence ET are quantified in an unambiguous fashion (Itenfisu et al., 2003).

The ET rate from a reference surface, not short of water, is called the *reference crop ET* or *reference ET* and is denoted as ET_0 (Allen et al., 1998). The reference surface is a hypothetical grass reference crop with specific characteristics. Reference ET is expressed in the units of depth/time (e.g., mm/day). It is a climatic parameter expressing the evaporative power of the atmosphere at a given space and time. Crop and soil features are not involved in computing it. Further, crop ET under standard conditions (ET_c) refers to ET

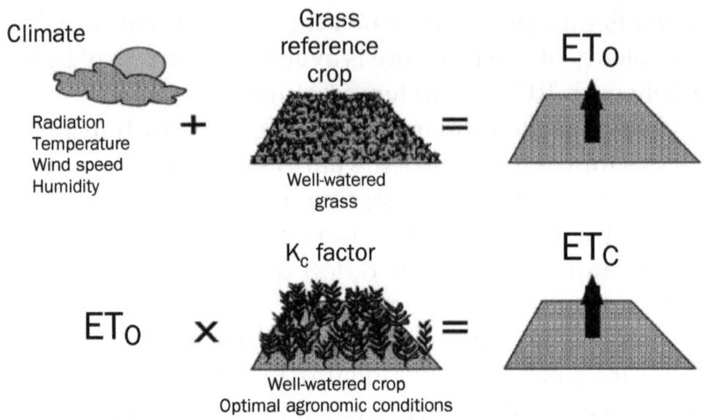

Figure 5.7 Reference crop ET or ET_0 and crop ET under standard conditions (ET_c). (Source: FAO-56.)

from excellently managed, disease-free, large, and well-watered fields that achieve full production under the given climatic conditions (see Fig. 5.7).

To estimate ET from a well-watered agricultural crop, the reference ET from a standard surface (ET_0) is first estimated. This value is multiplied by an empirical crop coefficient to obtain the ET from the crop (ET_c). The crop coefficient accounts for the difference between the standard surface and the crop.

5.6.1 Measurement of Evapotranspiration

ET is commonly measured by tanks and lysimeters and field plots. The processes are described next.

Lysimeters ET from agricultural fields can be measured directly by installing a lysimeter, a square or circular box whose area is around 8–10 m^2 and whose depth is a few meters. A lysimeter is installed in an agricultural field, filled with the same soil as in the surrounding area, and placed flush to the ground, and the same crops as grown in the surrounding area are planted. These should be placed beyond the zone of influence of individual buildings and trees and should be far enough away from the boundaries of the field to avoid the boundary effect.

Lysimeters can be classified into three groups according to their method of operation: (1) weight-based lysimeters, which use mechanical scales to determine changes in water content; (2) hydraulic-based lysimeters, which use the hydrostatic principle of weighing; and (3) volumetric-based lysimeters, where water content is kept constant and ET is estimated by the amount of water added or removed. Lysimeters are difficult and costly to install and maintain, but, if carefully installed and operated, they can give valuable and accurate data.

Field Plots Specially designed field plots are used to determine ET under field conditions. These plots are designed so that surface runoff water from the plot can be collected and measured. Deep percolation is captured by underground drain tiles. To determine ET, water input in the form of precipitation or irrigation is measured. Water losses in the form of runoff or deep percolation are subtracted from the water input, and a correction is made for soil moisture.

A variation of the field-plot method is the water-balance method, with inflow-outflow measurements from large drainage areas. This method requires measurements of the amount of water entering and leaving the area during a known time interval. These values must be adjusted based on the change in groundwater storage. Difficulty in accounting for all these variables makes this method less accurate than other methods.

5.7 Estimation of Evapotranspiration

Measurement of ET with a lysimeter is time-consuming and needs careful planning. Installation and maintenance of a lysimeter require skilled personnel, instruments, and finances. In view of the difficulties in direct measurement of ET, indirect methods are generally used to estimate ET. This requires measurements of meteorological variables that influence evaporation. Commonly, pan evaporation is multiplied by a coefficient to get the crop ET.

ET can be estimated by the water-budget or heat-budget methods; many empirical formulas have been developed that are based on meteorological data. The FAO has adopted the Penman-Monteith (PM) equation as the standard technique to compute reference ET (Allen et al., 1998).

Methods to estimate ET can be classified based on time scale [e.g., the Turc method, described later in this chapter is used for estimating annual ET, Blaney-Criddle is for monthly estimation, and the FAO-56 PM method is mainly used for estimation of daily ET (it also can be used for estimation of hourly values)]. We now describe these methods.

5.7.1 FAO-56 Penman-Monteith Method for the Estimation of ET

Numerous reference ET equations have been developed and are being used, depending upon the availability of weather data. These equations range in sophistication from empirical solar radiation– or temperature-based equations to complex methods based on physical processes such as the combination method of Penman (1948). The combination approach links evaporation dynamics with the flux of net radiation and aerodynamic transport characteristics of a natural surface. Based on observations that latent heat transfer in plant stems was influenced not only by these abiotic factors, Monteith (1965) introduced a surface conductance term that accounted for the response of leaf stomata to its hydrologic environment. This modified form of the Penman equation is widely known as the *Penman-Monteith (PM) equation.*

The PM equation is physically based because it attempts to incorporate the physiological and aerodynamic characteristics of the reference surface. While the use of the modified Penman method (Doorenbos and Pruitt, 1977) was recommended by the FAO, studies had suggested that this method overestimates ET. The FAO has recommended the use of the PM method to compute reference ET from a grass surface and has specified a grass reference ET equation (Allen et al., 1998). Studies by many scientists have shown that the reference ET computed using the PM equation yields estimates that are close to observed reference ET values.

As described in the Irrigation and Drainage Paper 56 (Allen et al., 1998), the FAO has adopted the PM equation (named here *FAO56-PM*) as the standard technique to compute reference ET. The FAO56-PM equation is stated as

$$ET_0 = \frac{0.408\Delta(R_n - G) + \gamma \dfrac{900}{T + 273} u_2(e_s - e_a)}{\Delta + \gamma(1 + 0.34u_2)} \qquad (5.18)$$

where ET_0 is the grass reference ET in mm/day, Δ is the slope of vapor pressure versus the temperature curve in kPa/°C, R_n is the net radiation at the grass surface in MJ m^{-2}/day, G is the soil heat flux density in MJ m^{-2}/day, T is the mean daily air temperature at 2-m height in °C, u_2 is the mean hourly wind speed at 2-m height in m/s, e_a is the actual hourly vapor pressure in kPa, and g is the psychrometric constant in kPa/°C.

Allen et al. (1998) have described the procedure and steps for the application of the PM equation for various time step sizes. For hourly time steps, the FAO-56 PM equation is

$$ET_0 = \frac{0.408\Delta(R_n - G) + \gamma \dfrac{37}{T_{hr} + 273} u_2(e^o(T_{hr}) - e_a)}{\Delta + \gamma(1 + 0.34u_2)} \tag{5.19}$$

where $e^0(T_{hr})$ is the saturation vapor pressure in kPa at air temperature T_{hr}, ET_0 is the grass reference ET in mm/h, R_n is the net radiation at the grass surface in MJ m^{-2}/h, and G is the soil heat flux density in MJ m^{-2}/h. Other terms are as defined earlier in this chapter.

Many scientists have studied the reliability of the PM method for estimating ET_0. Jensen, Burman, and Allen (1990) analyzed the performance of 20 different methods against lysimeter-measured ET for 11 stations located in different climatic zones around the world. The PM method was ranked as the best method for all climatic conditions. Allen et al. (1994) further stated that the FAO56-PM equation should be considered to be superior to most empirical ET_0 equations when using lysimeter-measured ET_0 data.

An application of the FAO56-PM equation requires data on solar radiation, wind speed, air temperature, vapor pressure, and humidity. However, all these input variables may not be easily available for a given location. In developing countries in particular, difficulties are often faced in collecting accurate data of all necessary climatic variables, and this can be a serious handicap in applying the FAO56-PM equation. Among the inputs needed, temperature data are routinely measured and solar radiation can be estimated with sufficient accuracy. But the other variables are mostly measured at only a few locations.

Automatic weather stations (AWSs), which are commonly used these days in developed countries to measure climatic variables, are not that common in other countries. In such circumstances, one may be forced to use the data of the nearest station, which may in fact be quite far away, and often in a completely different hydrometeorological setting. Experience shows that extrapolation of wind speed from one site to another is subject to large errors (Irmak et al., 2003). In addition, measured wind speeds over nonagricultural, arid terrain may be much faster than those over agricultural land (Burman et al., 1975). Extrapolation of the other climatic variables is equally questionable due to their unique behavior.

An example illustrating the use of the FAO56-PM method is given at the end of the chapter.

5.7.2 Bowen Ratio Energy Balance (BREB) Method

The diffusion and conduction processes that transport water vapor and sensible heat in the atmosphere are physically similar. Hence, it can be assumed that the aerodynamic resistance hindering the flow of water vapors is related to the difference in vapor pressure between two elevations and is equal to the resistance to the flow of sensible heat to the difference in temperature between the same heights. This fact was recognized by Bowen (1926) and led to the concept of the Bowen ratio.

The Bowen ratio ($\beta = H/\lambda E$) is defined as the ratio of the sensible heat H to the latent heat λE; it is directly related to the ratio of the differences in temperature and humidity measured at two different elevations:

$$\beta = \frac{H}{\lambda E} = \gamma\frac{(T_s - \mathrm{T})}{(e_{sat} - e_a)} = \gamma\frac{\Delta T}{\Delta e} \tag{5.20}$$

where γ is the psychrometric constant, T_s is the water surface temperature (°C), T is the air temperature (°C) at a reference height, and e_{sat} and e_a are the saturated and actual vapor pressures (Pa), respectively.

The energy balance for an area can be written as

$$R_n - \mathrm{G} - \lambda E - H = 0 \tag{5.21}$$

or

$$R_n - \mathrm{G} = \lambda E + H$$

Dividing all the terms by λE, we get

$$1 + \frac{H}{\lambda E} = \frac{R_n - \mathrm{G}}{\lambda E} \tag{5.22}$$

Hence, Eq. (5.22) can be written as

$$1 + \beta = \frac{R_n - \mathrm{G}}{\lambda E}$$

or

$$\lambda E = \frac{R_n - \mathrm{G}}{1 + \beta} \tag{5.23}$$

If β is known, λE can be determined from the measured values of R_n and G. Alternatively, using Eq. (5.20), it is possible to calculate the Bowen ratio from the measurement of temperature and humidity at two elevations. Also, by measuring the net radiation and soil heat flux and expressing them as water equivalent, one can use the following equation to compute evaporation:

$$E = \frac{R_n - \mathrm{G}}{1 + \gamma(\Delta T/\Delta e)} \tag{5.24}$$

This method is known as the *Bowen ratio energy balance (BREB) method.* Because diffusion of vapor and movement of sensible heat is similar, the BREB method is considered robust. A problem with this method is that the sign of the H-term changes in the morning and evening, which means that this equation is not applicable at these times.

The Bowen ratio reflects the wetness of the ground: the values of β below unity or λE exceeding H indicates ample moisture availability; the values of β over unity or H exceeding λE indicate drier surfaces. The Bowen ratio varies from 0.1 to 0.3 for humid conditions (Saxton and McGuiness, 1982), 0.4 to 0.8 temperate forests and grasslands, 2 to 6 semiarid regions with highly dry soils, and >10 for deserts.

5.7.3 Limits on Evapotranspiration

At a given place, the amount of ET is subject to three constraints: the amount of water available, the amount of energy available for ET, and the ability of the atmosphere to accept and move moisture.

5.7.4 The Budyko Curve

The Budyko curve was developed by Mikhail Budyko, a Russian climatologist who recognized that ET couples the water and energy cycles. It depicts the expected partitioning of precipitation into ET and runoff. In the graph shown in Fig. 5.8, the x-axis is the dryness index (φ = potential ET/precipitation) and the y-axis is the evaporative index (ε_B = actual ET/precipitation). A wide range of values are covered on the axes representing hydroclimatic behavior across temporal scales that are long enough to allow steady-state conditions to prevail and spatial scales large enough for the influences of climate on the land-surface/climate interactions to develop fully. Mathematically, the relation between the dryness index and the evaporative index or the Budyko curve can be expressed as (Michael and Huntington, 2017)

$$\varepsilon_B = \sqrt{\varphi \tanh\left(\frac{1}{\varphi}\right)(1 - \cosh\varphi + \sinh\varphi)} \tag{5.25}$$

A given site cannot lie above the water limit line [actual ET (AET) = precipitation] unless additional water (beyond precipitation) is being applied and cannot plot above the energy limit [AET = potential ET (PET)] unless precipitation is being lost (e.g., to groundwater). When the dryness index is <1, this is the energy-limited condition, and values >1 represent water-limited conditions. Further, actual ET cannot exceed precipitation (unless water from other sources is being added). Hence, no watershed will plot above the horizontal line with AET/P >1.0.

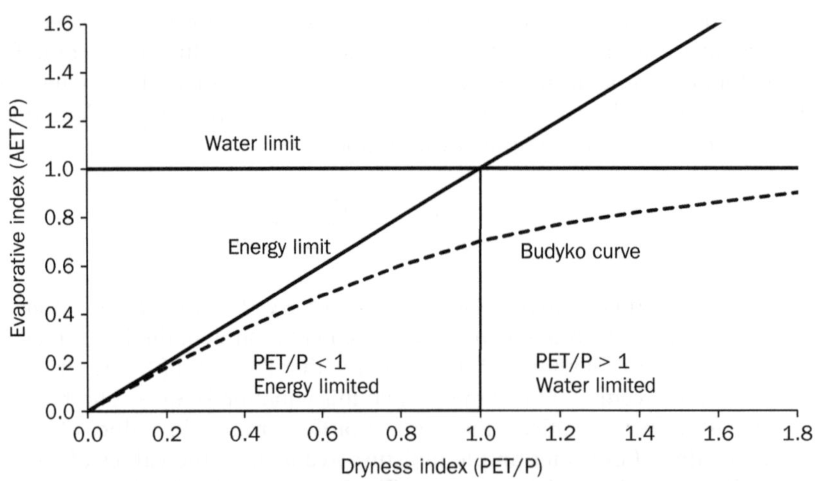

FIGURE 5.8 The Budyko curve.

As we move up on the y-axis, AET increases and runoff is less. Moving to the right on the x-axis indicates drier climate. The data from watersheds generally follow the Budyko curve. Average deviations from the Budyko curve provide insight into the inaccuracies in the water balance components. Year-to-year deviations from the Budyko curve provide insight into responsivity (resistance) and elasticity (resilience) of water yields to global change. Budyko curve has been used to separate the AET in the ET due to irrigation (blue water) and that due to rainfall (green water).

5.8 Reduced-Set Methods for Evapotranspiration Estimation

While applying FAO56-PM in practice, time and cost (instruments, etc.) involved in the acquisition and processing of the requisite meteorological data are a hindrance. All the required meteorological variables are observed at limited number of stations in the world. Further, the number of stations having reliable data for these variables is even smaller. Concerns have been expressed about the correctness of the observed meteorological data because the instruments, particularly to measure solar radiation and RH, require careful calibration and maintenance. It is common to see a drift (of as much as 10%, in pyranometers; see Samani, 2000) and hygrometers lose about 1% in accuracy per installed month (Henggeler et al., 1996). Many weather stations are often inadequately irrigated, particularly during dry periods, and thus the RH and air temperature data may not be representative. Wind data are often unavailable or lack reliability.

For these reasons, there is a need for methods to estimate ET_0 that require variables that are commonly measured and give reliable results. This has been the motivation to develop practical methods of ET_0 estimation. A number of such methods have been developed. According to the meteorological variables that play the dominant role in the method, the available methods can be grouped into three categories: (1) temperature-based, (2) radiation-based, and (3) a combination of the first two. The first two groups are also known as *reduced-set methods* because they use a small set of variables. Shahidian et al. (2012) provide a detailed discussion on this topic. For comparing the methods, lysimeter measurements or precipitation measurement missions (PMMs) are considered as the benchmark.

5.8.1 Temperature-Based Evapotranspiration Estimation Methods

Temperature is probably the easiest-to-measure and most reliable weather variable whose data are widely available. Temperature-based ET estimation methods assume that temperature is an indicator of the evaporative power of the atmosphere. While wind speed is important in arid climate, the number of sunshine hours is a more important variable in subhumid and humid climates. Xu and Singh (2001) evaluated generalized temperature-based methods for calculating evaporation.

5.8.2 The Hargreaves-Samani Method

Hargreaves (1975) found that ET can be adequately computed by using the average temperature and solar-radiation data. Hargreaves (1975) proposed the following equation to compute ET_0:

$$ET_0 = 0.0135(T + 17.78)R_s \tag{5.26}$$

where R_s is the solar radiation in terms of water evaporation (mm/day), and T is temperature in degrees Celsius. Hargreaves and Samani (1982) showed that ET can be estimated by knowing the difference between the maximum (T_{max}) and minimum (T_{min}) daily temperatures. Clear skies allow solar radiation to reach the ground easily and T_{max} is high, while long wave radiation is able to escape easily in night and T_{min} is low. On the other hand, clouds obstruct incoming solar radiation, as well as outgoing long wave radiation. Hence, T_{max} is lower and T_{min} is higher. Hence, Hargreaves and Samani (1982) proposed the following equation to estimate solar radiation using the temperature difference:

$$\frac{R_s}{R_a} = K_T(T_{max} - T_{min})^{0.5} \tag{5.27}$$

where R_a is the extraterrestrial radiation in mm/day, which can be obtained from published tables or be calculated (Allen et al., 1998). For interior regions where land mass dominates, K_T can be taken as 0.162; a value of 0.190 was suggested for coastal regions, where air is influenced by sea. R_s/R_a ranges from 0.75 on a clear day to 0.25 on a day with dense clouds. Based on Eqs. (5.23) and (5.24), Hargreaves and Samani (1985) developed an equation based on temperature, day of year, and latitude to calculate ET_o (HS equation):

$$ET_0 = 0.0135K_T(T + 17.78) \times (T_{max} - T_{min})^{0.5}R_a \tag{5.28}$$

Equation (5.24) was originally calibrated for semiarid conditions and does not explicitly account for RH. Hence, it may overestimate ET_0 in humid regions. Wind removes saturated air from the boundary layer and thus increases ET. Wind speed is a major factor affecting the performance of the HS equation. Itenfisu et al. (2003) found that the HS equation does not work well in high winds and high vapor pressure deficit situations and should be recalibrated when it is applied in areas with very high or low wind speeds. The HS equation tends to overestimate ET_0 when mean daily ET_0 is relatively low, and underestimate it when mean daily ET_0 is relatively high. This is a common issue with most of the reduced-set methods.

5.8.3 The Thornthwaite Method

Thornthwaite (1948) proposed a method of using the mean air temperature and number of hours of daylight to estimate ET_0 for short vegetation with adequate water supply. Monthly ET_0 can be estimated by the following equation:

$$ET_0 = ET_{osc} \times \frac{N}{12} \times \frac{dm}{30} \tag{5.29}$$

where N is the maximum number of sunny hours as a function of month and latitude and dm is the number of days per month. ET_{osc} is the gross ET (without corrections), which can be calculated as

$$ET_{osc} = c \times \left(\frac{10T_a}{I}\right)^a \tag{5.30}$$

where T_a is the mean daily temperature (°C), c is a coefficient which is typically 1.62, a is an exponent as a function of the annual heat index, and I is the annual heat index obtained from monthly heat indices:

$$I = \sum_{m=1}^{12}\left(\frac{T_m}{5}\right)^{1.51} \tag{5.31}$$

and

$$a = 0.492 + 1792 \times 10^{-5} I - 771 \times 10^{-7} I^2 + 675 \times 10^{-9} I^3 \quad (5.32)$$

Under dry and arid conditions, the Thornthwaite equation strongly underestimates ET_0 (Garcia et al., 2004) because it does not consider the saturation deficit of air. At high altitudes, it also underestimates the effect of radiation. Generally, the Thornthwaite method underestimates ET in humid areas.

The heat index is an integral element of Thornthwaite's classification of climates. The value of ET is modified by a factor to account for the number of daylight hours and the number of days in a month. Thornthwaite has tabulated values of this factor that correspond to various degrees and months of the year.

Assuming that each month has 30 days and each day has 12 hours of sunshine, Eq. (5.29) reduces to

$$\mathrm{ET} = 1.62\,(10T/I)^a \quad (5.33)$$

Taking its logarithm,

$$\log \mathrm{ET} = \log 1.62 + a(\log 10 + \log T - \log I) \quad (5.34)$$

clearly, ET = 1.62 when $I = 10T$ because $\log(10T) = \log I$. It has been shown by Thornthwaite that all lines obeying this equation have a common point of convergence at $T = 26.5$°C and $E = 1.35$ cm.

Example 5.9

Estimate the potential ET if mean monthly temperature is 25°C. Use $c = 1.62$ in the Thornthwaite method.

Solution

From Eq. (5.31),

$$I = \sum_{j=1}^{12}\left(\frac{25}{5}\right)^{1.51} = 12 \times 5^{1.51} = 136.341$$

From Eq. (5.32),

$$a = 0.492 + 1{,}792 \times 10^{-5} I - 771 \times 10^{-7} I^2 + 675 \times 10^{-9} I^3$$

$$a = 675 \times 10^{-9}(136.341)^3 - 771 \times 10^{-7}(136.341)^2 + 0.01792 \times 136.341 + 0.492$$

$$= 1.711 - 1.433 + 2.441 + 0.492 = 3.21$$

$$\mathrm{ET} = 1.62\left(10 \times \frac{25}{136.341}\right)^{3.21} = 11.34 \text{ cm}$$

5.8.4 The Blaney-Criddle Method

The Blaney-Criddle (1950) method computes ET based on temperature data. This method refers to mean monthly values, both for temperature and ET_0:

$$ET_o = p(0.46T_{mean} + 8) \quad (5.35)$$

where ET_o = reference crop ET (mm/day) as an average for the month, T_{mean} = mean daily temperature (°C), and p = mean daily percentage of annual daytime hours. The values of p can be read from the standard table (see Table 5.1). To compute T_{mean}, first the average of all T_{max} and T_{min} (maximum and minimum of temperatures for all the days) is determined. The average of these is T_{mean}.

$$T_{max} = \frac{\text{Sum of all } T_{max} \text{ values for the month}}{\text{Number of days in that month}}$$

$$T_{min} = \frac{\text{Sum of all } T_{min} \text{ values for the month}}{\text{Number of days in that month}}$$

$$T_{mean} = \frac{T_{max} + T_{min}}{2}$$

Example 5.10

The latitude of a place is 30°N. For the month of September, the mean T_{max} is 29.5°C and the mean T_{min} = 18.2°C. Determine the mean ET_0 for the month using the Blaney-Criddle method.

Solution

For September,

$$T_{mean} = (31.8 + 18.2)/2 = 25.0$$

For latitude 35°N and the month of September, p = 0.30 from Table 5.1.
So,

$$ET_0 = 0.30 \times (0.46 \times 25.0 + 8) = 5.85 \text{ mm/day}$$

Latitude	North	Jan	Feb	Mar	Apr	May	June	July	Aug	Sept	Oct	Nov	Dec
(°)	South	July	Aug	Sept	Oct	Nov	Dec	Jan	Feb	Mar	Apr	May	June
60		.15	.20	.26	.32	.38	.41	.40	.34	.28	.22	.17	.13
55		.17	.21	.26	.32	.36	.39	.38	.33	.28	.23	.18	.16
50		.19	.23	.27	.31	.34	.36	.35	.32	.28	.24	.20	.18
45		.20	.23	.27	.30	.34	.35	.34	.32	.28	.24	.21	.20
40		.22	.24	.27	.30	.32	.34	.33	.31	.28	.25	.22	.21
35		.23	.25	.27	.29	.31	.32	.32	.30	.28	.25	.23	.22
30		.24	.25	.27	.29	.31	.32	.31	.30	.28	.26	.24	.23
25		.24	.26	.27	.29	.30	.31	.31	.29	.28	.26	.25	.24
20		.25	.26	.27	.28	.29	.30	.30	.29	.28	.26	.25	.25
15		.26	.26	.27	.28	.29	.29	.29	.28	.28	.27	.26	.25
10		.26	.27	.27	.28	.28	.29	.29	.28	.28	.27	.26	.26
5		.27	.27	.27	.28	.28	.28	.28	.28	.28	.27	.27	.27
0		.27	.27	.27	.27	.27	.27	.27	.27	.27	.27	.27	.27

TABLE 5.1 Mean Daily Percentage (p) of Annual Daytime Hours for Different Latitudes

Note that this method is not very accurate, and the estimates need to be treated only as approximations. The underlying assumption of the Blaney-Criddle method is that the heating of air and evaporation share the heat budget in a fixed proportion. As a result, ET varies directly with the sum of the products of mean monthly air temperature and monthly percentage of daytime hours.

5.9 Radiation-Based Methods

It is known that water loss from a crop is related to the incident solar energy, and this opens the possibility to develop a simple model that relates solar radiation to ET. Over the years, many methods have been developed to estimate reference ET by using the measured net global radiation. The list includes the Priestley-Taylor method (1972), the Makkink method (1957), the Turc radiation method (1961), and the Jensen and Haise method (1963). Xu and Singh (2000) evaluated and generalized radiation-based methods for computing evaporation.

5.9.1 The Priestley-Taylor Method

The Priestley-Taylor method (Priestley and Taylor, 1972) is a simplified form of the Penman equation that needs only net radiation and temperature to calculate ET_0. This simplification is based on the observation that ET_0 is more dependent on radiation than on RH and wind (the first term in the Penman equation is much bigger than the second). The Priestly-Taylor method is the radiation-driven part of the Penman equation, multiplied by a coefficient (to be calibrated) plus an additive factor, and can be expressed as

$$ET_o = \alpha \frac{\Delta(R_n - G)}{\Delta + \gamma} + \beta \tag{5.36}$$

where α and β are calibration factors whose recommended values are 1.26 and 0, respectively. Xu and Singh (1998) calibrated this model for Switzerland and obtained values of 0.98 and 0.94 for α and β, respectively. Application of the Priestly-Taylor equation during cold weather gives unrealistic results because net radiation is often negative in winters and the method predicts dew formation, while the actual ET is positive. In general, the Priestley and Taylor method underestimates ET_0 in winters and overestimates it during the summer months.

5.9.2 The Turc Method

The Turc method was presented by Turc (1961) by using data from the humid climate of western Europe. This method uses two parameters: average daily radiation and temperature (Shahidian et al., 2012). For RH >50%, ET can be expressed as

$$ET_p = 0.31[S_n + 2.09]\left(\frac{T}{T + 15}\right) \tag{5.37}$$

and for RH <50% as

$$ET_p = 0.31[S_n + 2.09]\left(\frac{T}{T + 15}\right)\left[1 + \left(\frac{50 - RH)}{70}\right)\right] \tag{5.38}$$

where T is the air temperature (°C) and S_n is the water equivalent of net solar radiation (mm/day). The Turc method performs well in warm, humid climates. In many studies, the results of the Turc method are quite close to those from the full PM equation.

5.9.3 The Jensen and Haise Method

Assuming that ET is more closely related to net radiation than other variables such as air temperature and humidity, Jensen and Haise (1963) used measured ET and solar radiation data to derive the equation to estimate ET:

$$ET = C_t(T - T_x)R_s/\lambda \tag{5.39}$$

where ET is the reference ET in mm/day, C_T is the temperature coefficient, T is the mean temperature for a 5-day period (°C), T_x is the intercept of the temperature axis (°C), λ is the latent heat of vaporization (MJ/kg), and R_s is the solar radiation received on the Earth's surface at a horizontal surface (MJ m^{-2}/d). The temperature constant, C_T, and the variable T_x can be calculated as follows:

$$C_T = \frac{1}{C_1 + 7.3C_H} \tag{5.40}$$

where

$$C_1 = 38 - \frac{2H}{305} \quad C_H = \frac{5.0}{e_2 - e_1}$$

and

$$T_x = -2.5 - 0.14[e_2 - e_1] - \frac{h}{550} \tag{5.41}$$

where h is the altitude of the location in m; e_2 and e_1 are the saturation vapor pressure at the mean maximum temperature and the mean minimum temperature, respectively, expressed in kilo-Pascals. For the humid and rainy areas, the methods based on solar radiation are more accurate than those based only on air temperature.

Example 5.11

Compute the ET using the Jensen-Haise method if the following data are available: the mean air temperature = 25°C, latitude = 30°N, R_s = 700 ly/day, the mean maximum temperature of the warmest month = 30°C, the mean minimum temperature of the warmest month = 20°C, and the elevation = 200 m.

Solution

Estimate λ from Eq. (5.1),

$$\lambda = 2.501 - 0.00236 \times 25 = 2.442$$

$$\text{For } T = 30°\text{C}, \; e_2 = 0.6108 \times \exp\left[\frac{17.27 \times 30}{30 + 237.3}\right] = 4.24 \text{ kPa}$$

$$\text{For } T = 20°\text{C}, \; e_2 = 0.6108 \times \exp\left[\frac{17.27 \times 20}{20 + 237.3}\right] = 2.39 \text{ kPa}$$

Now,

$$C_1 = 38 - \frac{2 \times 200}{305} = 38 - 1.311 = 36.689$$

$$C_H = \frac{5.0}{e_2 - e_1} = \frac{5.0}{4.24 - 2.39} = 2.70$$

$$C_T = \frac{1}{C_1 + 7.3C_H} = \frac{1}{36.689 + 7.3 \times 2.7} = 0.0177$$

where

$$T_x = -2.5 - 0.14[e_2 - e_1] - \frac{h}{550} = -2.5 - 0.14(4.24 - 2.39) - \frac{200}{550} = -5.45$$

From Eq. (5.39),

$$\text{ET} = 0.0177 \times [25 - (-5.45)] \times 700 \text{ ly/day}$$

$$= 608.31 \text{ ly/day} = 6.45 \text{ mm/day}$$

5.9.4 Reduced-Set PM

The reduced-set PM equation, requiring only the measured maximum and minimum temperatures, uses estimates of solar radiation, RH, and wind speed. Solar radiation, R_s MJ/(m²d), can be estimated using the Angstrom formula, which relates solar radiation to extraterrestrial radiation and relative sunshine duration, or by using averages from nearby stations. For island locations, R_s can be estimated as (Allen et al., 1998)

$$R_s = 0.7R_a - b \tag{5.42}$$

where b is an empirical constant with a value of 4 MJ/(m²d). The RH can be estimated by assuming that the dew-point temperature is approximately equal to T_{min} (Allen et al., 1996, 1998), which is usually experienced at sunrise. In this case, e_a can be calculated as

$$e_a = e^o(T_{min}) = 0.611\exp\left[\frac{17.27T_{min}}{T_{min} + 237.3}\right] \tag{5.43}$$

where $e^o(T_{min})$ is the vapor pressure at the minimum temperature, expressed in mbar. Considering the impact of wind speed on the ET_0 as relatively small, except in arid and windy areas, Allen et al. (1998) recommended using average wind speed data from nearby locations or using a value of 2 m/s. The soil heat flux density, G, for monthly periods can be estimated as

$$G_i = 0.07(T_{i+1} - T_{i-1}) \tag{5.44}$$

where G_i is the soil heat flux density in month I in MJ (m – 2 d – 1); and T_{i+1} and T_{i-1} are the mean air temperatures in the previous and following months, respectively.

Allen (1995) evaluated the reduced-set PM (using only T_{max} and T_{min}) and HS. He used the mean annual monthly data from about 3,000 stations in the FAO CLIMWAT database, with the full PM serving as the benchmark and found little difference in the mean monthly ET_0 between the two methods (Shahidian et al., 2012).

5.9.5 Comments on Reduced-Set Methods

Both temperature and radiation can be used to calculate daily ET_o values with satisfactory accuracy (Shahidian et al., 2012). However, it is best to use these equations for areas having a climate similar to the one for which the original equations were developed. Most of the equations can be used with some confidence for areas with moderate conditions of humidity and wind speed. Globally, the Turc equation has produced good results for humid or semihumid areas, while the Thornthwaite equation underestimates ET_0. The Priestley-Taylor equation should not be used in the winter months in high-latitude locations, such as northern Europe. Both the Hargreaves and the reduced-set PM methods can be used effectively with only temperature measurements, although results can be improved if wind speed is taken into consideration.

The use of reduced-set equations can be very helpful in irrigation water management because the errors in applying these methods can be much smaller than those arising if data from a distant weather station are used.

CROPWAT is a software program designed for the computation of crop water and irrigation requirements based on soil, climate, and crop data. It also can be used to develop irrigation schedules for different management conditions. CROPWAT was developed by the FAO and can be downloaded for free from its website (http://www.fao.org/land-water/databases-and-software/cropwat/en/).

5.10 Emerging Methods for ET Estimation

Some researchers have used data-driven method such as artificial neural networks (ANNs) to estimate ET by the use of limited climatic data (Jain et al., 2008). Direct measurement of ET by the eddy covariance is being attempted at many places.

5.10.1 Eddy Covariance Technique

The eddy covariance method is a useful and direct method to measure and calculate turbulent fluxes within the atmospheric boundary layer. *Flux* is an amount of an entity that passes through a closed surface per unit of time. Measurements of flux are used to estimate exchange of heat, water, and gases.

Imagine air flow as a horizontal flow of many rotating eddies (Fig. 5.9); each eddy has three three-dimensional (3D) components, including vertical movement of the air. These components can be measured by a flux tower. Consider that the air flow consists of eddies of different sizes. At a given time *T1*, eddy *E1* moves air parcel *C1* upward at a speed *V1*. At another time *T2*, eddy *E2* moves air parcel *C2* downward at a speed *V2*. Each air parcel has distinct characteristics, such as temperature, humidity, and CO_2 concentration. By measuring these characteristics and the velocity of the vertical air movement, it is possible to determine the vertical upward or downward fluxes of temperature, humidity, CO_2 concentration, and other elements.

The aim behind flux measurement is to measure how many molecules move and how fast they go up and down over time. With this data, the vertical flux at this point and over this time can be calculated. An assumption of the method is that vertical flux can be presented as covariance between vertical velocity, the vertical movements, and concentration of the entity of interest. Because turbulent fluctuations evolve very

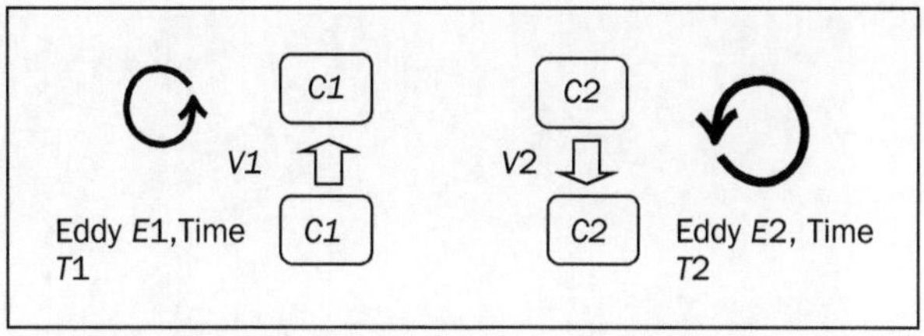

Figure 5.9 Movement of air, water, and CO_2 concentration by eddies.

quickly, and changes in concentration, density, and temperature are small, they need to be measured very quickly and accurately.

The eddy flux can be approximated by the product of mean air density and the mean covariance between deviations in instantaneous vertical wind speed and mixing ratio. Similarly, sensible heat flux is obtained by multiplying the mean air density by the covariance between deviations in instantaneous vertical wind speed and temperature. These may be converted to energy units using specific heat. Latent heat flux is computed in a similar manner by using water vapor. In the recent past, eddy flux networks such as FLUXNET have come into being, which are helping in improving and standardizing data collection and analysis. Figure 5.10 shows a flux tower that has the sensors used to measure ET. Scintillometry is another emerging technique that uses measurements of transmitted and scattered electromagnetic radiation through the atmosphere.

5.10.2 Remote Sensing for Measurement of Evaporation and Evapotranspiration

Remote-sensing observations combined with ancillary meteorological data have been used in obtaining indirect estimates of ET over a range of temporal and spatial scales (Schulz and Engman, 2000). Recently, considerable progress has been made in the remote sensing–based estimation of variables, including incoming solar radiation, surface albedo, vegetative cover, surface temperature, and surface soil moisture. These variables are estimated by measuring the electromagnetic radiation in a particular waveband reflected or emitted from the Earth's surface. Furthermore, remote sensing has a potentially important role because of its areal coverage in the spatial extrapolation process of ET.

Researchers are experimenting with satellite data (Bastiaanssen, 1998) to estimate regional actual ET. Several important parameters needed to estimate ET can now be obtained by use of remote sensing. The soil moisture may be estimated using the measurement of microwave properties of the soil (microwave emission and reflection or backscatter from soil). However, there are uncertainties in such soil moisture estimates due to factors, such as surface roughness and vegetative cover.

The satellite-based approaches for estimating ET are attracting attention in several parts of the world. Using satellite products, it is feasible to estimate the ET at scales of a few square kilometers. Hence, establishing a good ground network of observations will facilitate the calibration and validation of satellite estimates.

Figure 5.10 A flux tower showing the sensors used to measure ET. The white sensor inside the dotted circle is the gas analyzer that measures ET.

5.11 Application of Evapotranspiration in Hydrology

Evaporation is small during a runoff event because the vapor pressure gradient is low. This is because when rainfall occurs, the vapor pressure of the air is high and the RH is high. The amount of water vapor contained in the atmosphere may be high relative to the amount of moisture that it can hold at saturation at the prevailing temperature. As a result, little evaporation occurs. In general, a runoff event from a convective storm lasts only a few hours. Coupled with a low vapor-pressure gradient, such short-time events result in very small (even negligible) evaporation. Although a cyclonic storm has a longer duration, evaporation occurring over its duration is small.

Evaporation becomes important when long-term storage for irrigation or municipal and industrial water supply is the primary purpose of a study. Evaporation from storage reservoirs, particularly in arid climates, can be large and represent a direct loss for the purpose of the project. Evaporation from soil in the drainage basin is important because this antecedent moisture deficit must be satisfied before any significant runoff will occur. On the other hand, irrigation of agricultural lands is scheduled, on the basis of moisture deficit, in order to alleviate plant stress. The rate of evaporation is related to the soil type, as well as the rate of replenishment of moisture in the soil from rainfall or water delivery. For the health of soil, alkalinization, and salinization, evaporation plays a major role. Thus, ET is a primary determinant for the health of ecosystems. Knowledge of ET is important for the design of aquaculture ponds.

By suppressing evaporation, the loss of usable water can be prevented. Oil and chemicals (such as cetyl alcohol) have been used to reduce the rate of evaporation from water surfaces. Less than 10 g of this chemical can make a film covering about 4,000 m^2 of water surface. If the film can remain unbroken over the entire surface of a reservoir, as much as 70% reduction in evaporation can be effected. In water-scarce areas, chemical films can be used to suppress evaporation. In some areas (e.g., South West Africa), sand storage dams have been used to save runoff water that would evaporate otherwise.

Knowledge of the evaporative process can be used beneficially to dispose of unwanted waste and to cool water heated during industrial processes. Evaporation ponds can serve as an efficient means to dispose of waste. If the pond is shallow, the rate of evaporation is generally high. Of course, ponds must be properly lined to prevent the wastewater from percolating down to the underlying groundwater. In the San Joaquin Valley in California, evaporation ponds are commonly used to dispose of oil-field wastewater, as well as agricultural wastewater. Evaporative cooling is widely used in industry to cool water.

Knowledge of ET is needed to compute crop water requirements, in catchment modeling, reservoir management, low-flow management, and other applications.

Example 5.12

Compute ET_0 by the FAO 56 PM method by using daily meteorological data for Roorkee (latitude 29°N, elevation 270 m) for July 6: $T_{max} = 34°C$, $T_{min} = 22°C$, elevation = 250 m, $RH_{max} = 88\%$, $RH_{min} = 60\%$, number of daylight hours $n = 9$, wind speed at 2-m height = 2 m/s.

Solution

Here, the solution follows the steps given in FAO-56:

$$\text{Mean air temperature } T = (34 + 22) \times 0.5 = 28°C.$$

The atmospheric pressure, *P*, is the pressure exerted by the weight of the Earth's atmosphere. Evaporation at high altitudes is high due to low atmospheric pressure. A simplification of the ideal gas law, assuming 20°C for a standard atmosphere, can be employed to calculate *P*:

$$P = 101.3\left(\frac{293 - 0.0065\,z}{293}\right)^{5.26} \tag{5.45}$$

where P = atmospheric pressure (kPa) and z = elevation above mean sea level (m). For $z = 250$ m, $P = 100.7382$. The psychrometric constant, γ (kPa/°C) is given by

$$\gamma = \frac{c_p P}{\varepsilon\lambda} = 0.665 \times 10^{-3} \times P \tag{5.46}$$

where P = atmospheric pressure (kPa), c_p = specific heat at constant pressure, 1.013 × 10^{-3} (MJ kg^{-1}/°C), and ε = ratio molecular weight of water vapor/dry air = 0.6222.

For the example data, $\gamma = 0.665 \times 10^{-3} \times 110.7382 = 0.06699$.

Mean Saturation Vapor Pressure (e_s) As saturation vapor pressure is related to air temperature, it can be calculated from the air temperature by the following relationship:

$$e^0(T) = 0.6108 \times \exp\left[\frac{17.27\,T}{T + 237.3}\right] \tag{5.47}$$

where $e^0(T)$ = saturation vapor pressure (kPa) at the air temperature T (°C). This equation is nonlinear. Hence, the mean saturation vapor should be computed as the mean of the saturation vapor pressure at the mean daily maximum and minimum air temperatures for that period:

$$e_s = \frac{e^0(T_{max}) + e^0(T_{min})}{2} \tag{5.48}$$

For the given data,

$$e^0(T_{max}) = 5.3193, e^0(T_{min}) = 2.6439; e_s = 3.9816.$$

Slope of Saturation Vapor Pressure Curve (Δ) The slope of saturation vapor pressure curve at air temperature T, Δ (in kPa/°C), is required to calculate ET. It can be calculated by

$$\Delta = \frac{4{,}098\left[0.6108\exp\left(\frac{17.27\,T}{T + 237.3}\right)\right]}{(T + 237.3)^2} \tag{5.49}$$

For $T = 28$°C, $\Delta = 0.2201$.

The actual vapor pressure can be calculated from RH data. If RH_{max} and RH_{min} are known, then

$$e_a = \frac{e^0(T_{min})\frac{RH_{max}}{100} + e^0(T_{max})\frac{RH_{min}}{100}}{2} \tag{5.50}$$

where e_a = actual vapor pressure (kPa), $e^0(T_{min})$ = saturation vapor pressure at daily minimum temperature (kPa), $e^0(T_{max})$ = saturation vapor pressure at daily maximum temperature (kPa), RH_{max} and RH_{min} = maximum and minimum RH (%). For this example, $e_a = 2.7591$.

Radiation The radiation striking a surface perpendicular to the Sun's rays at the top of the Earth's atmosphere, called the solar constant, is about 0.082 MJ m^{-2} min. The solar radiation received at the top of the Earth's atmosphere on a horizontal surface is called the *extraterrestrial (solar) radiation*, R_a. The local intensity of radiation, however, is determined by the angle between the direction of the Sun's rays and the normal to the surface of the atmosphere. R_a changes with the season and the length of the day. Extraterrestrial radiation is a function of latitude, date, and time of day. As the radiation penetrates the atmosphere, some of it is scattered, reflected, or absorbed by the atmospheric gases, clouds, and dust. The amount of radiation reaching a horizontal plane is known as the *solar radiation*, R_s. Because the Sun emits energy by means of EM waves characterized by shortwave lengths, solar radiation is also referred to as *shortwave radiation*. For a cloudless day, R_s is about 75% of extraterrestrial radiation (Allen et al., 1998).

The relative shortwave radiation (R_s/R_{so}) is the ratio of the solar radiation (R_s) to the clear-sky solar radiation (R_{so}). R_{so} is the solar radiation that would reach the same surface during a given period under cloudless conditions. This ratio varies between about 0.33 (dense cloud cover) and 1 (clear sky).

The relative sunshine duration (n/N) is the ratio of the actual duration of sunshine (n) to the maximum possible duration of sunshine or daylight hours (N). It expresses the cloudiness of the atmosphere. In the absence of a direct measurement of R_s, n/N is often used to derive solar radiation from extraterrestrial radiation. The day length N is a function of latitude and date.

Albedo is the fraction, α, of the solar radiation reflected by the surface. It depends on the type of the surface and the angle of incidence. Albedo is very large (say, 0.95) for freshly fallen snow and very small (say, 0.05) for wet, bare soil. For the green grass reference crop, α is assumed to be 0.23. The net solar radiation, $R_{ns} = (1 - \alpha)R_s$ is the fraction of R_s that is not reflected from the surface.

The Earth emits radiation as longwave energy. Part of the radiation is absorbed by the atmosphere, and some is reflected back to the Earth. The difference between outgoing and incoming longwave radiation is the net longwave radiation, R_{nl}. As the outgoing longwave radiation is almost always more than incoming longwave radiation, R_{nl} represents the energy loss.

The net radiation R_n is the difference between incoming and outgoing radiation of both short and long wavelengths. It is the difference between the incoming net shortwave (R_{ns}) and the net outgoing longwave (R_{nl}) radiation. The total daily value for R_n is usually positive over a period of 24 h, except in extreme conditions at high latitudes. The soil heat flux, G, is the energy that is utilized in heating the soil. It is small compared to R_n and often may be ignored.

The extraterrestrial radiation, R_a, for each day of the year and for different latitudes, can be estimated from the solar constant, the solar declination, and the time of the year by

$$R_a = \frac{24(60)}{\pi} G_{SC} d_r \left[\omega_s \sin(\varphi)\sin(\delta) + \cos(\varphi)\cos(\delta)\sin(\omega_s)\right] \tag{5.51}$$

where R_a = extraterrestrial radiation (MJ m^{-2}/day), G_{sc} = solar constant (0.0820 MJ m^{-2}/min), d_r = the inverse of the relative distance between the Earth and the Sun, ω_s = sunset hour angle (rad), φ = latitude (rad), and δ = solar declination (rad). In Eq. (5.24), R_a is expressed in MJ m^{-2}/day. The corresponding equivalent evaporation in mm/day is obtained by multiplying R_a by 0.408 [Eq. (5.23)]. The latitude, φ, is commonly expressed in decimal degrees. Decimal degrees can be converted to radians as follows:

$$[\text{Radians}] = \frac{\pi}{180}[\text{decimal degrees}] \tag{5.52}$$

The inverse of the relative Earth-Sun distance, d_r, and the solar declination, δ, are given as follows:

$$d_r = 1 + 0.033\cos\left(\frac{2\pi}{365}J\right) \tag{5.53}$$

$$\delta = 0.409\sin\left(\frac{2\pi}{365}J - 1.39\right) \tag{5.54}$$

where J is the number of the day in the year between 1 (January 1) and 365 or 366 (December 31). In this example, $J = 187$. Hence, $d_r = 0.967$ and $\delta = 0.3956$.

The sunset hour angle, ω_s, is given by

$$\omega_s = \arccos[-\tan(\varphi)\tan(\delta)] \tag{5.55}$$

where $\varphi = \frac{3.1415}{180}\left(28 + \frac{0}{60}\right) = 0.48868$.

So,

$$\omega_s = \arccos[-\tan(0.48868)\tan(0.3956)] = 1.794709$$

The daylight hours (N) are computed by

$$N = \frac{24}{\pi}\omega_s = \frac{24}{3.1415}0.1794709 = 13.71 \text{ hours}$$

Hence, we compute $R_a = 40.6787$.

Solar Radiation (R_s) If the solar radiation, R_s, is not measured, it can be calculated with the Angstrom formula, which relates solar radiation to extraterrestrial radiation and relative sunshine duration:

$$R_s = \left(a_s + b_s\frac{n}{N}\right)R_a \tag{5.56}$$

where R_s = solar or shortwave radiation (MJ m^{-2}/day), n = actual duration of sunshine (h), N = maximum possible duration of sunshine or daylight hours (h), n/N = relative sunshine duration, R_a = extraterrestrial radiation (MJ m^{-2}/day), a_s = the regression constant, expressing the fraction of extraterrestrial radiation reaching the Earth on overcast days ($n = 0$), $a_s + b_s$ = fraction of extraterrestrial radiation reaching the Earth on clear days ($n = N$). Where no actual solar radiation data are available and no calibration has been carried out for improved a_s and b_s parameters, the recommended values are $a_s = 0.25$ and $b_s = 0.50$:

$$R_s = \left(0.25 + 0.5 \times \frac{9.00}{13.71}\right) = 23.52099$$

$$R_{so} = (0.75 + 2 \times 10^{-5}z)R_a \tag{5.57}$$

$$= (0.75 + 2 \times 10^{-5} \times 250)40.6787 = 30.71243.$$

The net shortwave radiation resulting from the balance between incoming and reflected solar radiation is given by

$$R_{ns} = (1 - \alpha)R_s \tag{5.58}$$

where R_{ns} = net solar or shortwave radiation (MJ m^{-2}/day), α = albedo, which is 0.23 for the hypothetical grass reference crop (dimensionless), and R_s = incoming solar radiation (MJ m^{-2}/day).

So, from Eq. (5.58)

$$R_{ns} = (1 - 0.23) \times 23.52099 = 18.11117 \text{ MJ m}^{-2}\text{/day}$$

According to the Stefan-Boltzmann law, the rate of longwave energy emission is proportional to the absolute temperature of the surface raised to the fourth power. The net energy flux leaving the Earth's surface is less, however, due to absorption and downward radiation from the sky. As humidity and cloudiness play an important role, correction is to be applied for these two factors. Thus, the net outgoing flux of longwave radiation is

$$R_{nl} = \sigma\left[\frac{T_{max,K}^{\ 4} + T_{min,K}^{\ 4}}{2}\right]\left(0.34 - 0.14\sqrt{e_a}\right)\left(1.35\frac{R_s}{R_{so}} - 0.35\right) \quad (5.59)$$

where R_{nl} = net outgoing longwave radiation (MJ m^{-2}/day), σ = Stefan-Boltzmann constant (4.903×10^{-9} MJ/K^4/m^2/day), $T_{max,K}$ and $T_{min,K}$ = maximum/minimum absolute temperature during 24-h period (K = °C+273.16), e_a = actual vapor pressure (kPa), R_s/R_{so} = relative shortwave radiation (limited to ≤ 1.0), R_s = measured or calculated [Eq. (5.56)] solar radiation (MJ m^{-2}/day), and R_{so} = calculated [Eq. (5.57)] clear-sky radiation (MJ m^{-2}/day).

$$\sigma T_{max,K}^{\ 4} = 43.6436 \text{ and } \sigma T_{min,K}^{\ 4} = 37.21276$$

Hence, $R_{nl} = 2.970891$.

The net radiation (R_n) is the difference between the incoming net shortwave radiation (R_{ns}) and the outgoing net longwave radiation (R_{nl}):

$$R_n = R_{ns} - R_{nl} \quad (5.60)$$

$$= 18.11117 - 2.970891 = 15.14027$$

We rewrite the FAO-56 PM equation as follows:

$$ET_0 = \frac{0.408\Delta(R_n - G) + \gamma\dfrac{900}{T + 273}u_2(e_s - e_a)}{\Delta + \gamma(1 + 0.34u_2)} \quad (5.61)$$

Hence, the first term of equation $\dfrac{0.408\Delta(R_n - G)}{\Delta + \gamma(1 + 0.34u_2)} = \dfrac{0.408 \times 0.22008(15.14027 - 0)}{0.22008 + 0.06699 * 1.68} =$ 4.087147.

The second term is $= \dfrac{\gamma\dfrac{900}{T + 273}u_2(e_s - e_a)}{\Delta + \gamma(1 + 0.34u_2)} = \dfrac{0.06699\dfrac{900}{28 + 273}2.0(3.9816 - 2.7591)}{0.22008 + 0.06699 * 1.68} =$ 1.472351

Thus, $ET_0 = 4.087147 + 1.472351 = 5.559$ mm/day.

5.12 Questions

5.1 Select one day from each month and note the following weather variables for any city (say, Houston, Texas): maximum temperature, minimum temperature, RH, pan evaporation, and sunshine hours.

	Temperature (°C)				
Date	Maximum	Minimum	RH	Pan Evaporation (mm)	Sunshine Hours

Note the variability and the range of the variables.

5.2 Select a place (say, Beaumont, Texas) and find its latitude and longitude. Compute the mean temperature in degrees Celsius and mean RH for June 1, 2017.

5.3 Explain the physics behind the evaporation process.

5.4 Explain how the solar radiation is measured.

5.5 Explain, with the help of a diagram, how evaporation is measured by using a pan.

5.6 On a given day, the depth of evaporation from a pan was 3.2 mm. On that basis, estimate the evaporation depth in a nearby lake. Assume any missing data/coefficient.

5.7 Using the data from Question 5.6, what volume of water will evaporate from the lake if the lake is 120 m long and 71 m wide?

5.8 For Beaumont, Texas, compute the slope of the saturation vapor pressure curve and the psychrometric constant γ.

5.9 For Beaumont, Texas, compute the shortwave radiation in mm/day on June 1, 2017.

5.10 For Beaumont, Texas, compute the net outgoing solar radiation in mm/day on June 1, 2017.

5.11 For Beaumont, Texas, compute the net radiation with an appropriate value of α on June 1, 2017.

5.12 For Beaumont, Texas, compute the vapor pressure deficit on June 1, 2017.

5.13 Briefly explain the transpiration and ET processes.

5.14 For Beaumont, Texas, compute the weighting factors in the Penman equation: $\frac{\Delta}{\Delta + \gamma}$ and $\frac{\gamma}{\Delta + \gamma}$.

5.15 Write down the PM equation and explain its terms.

5.16 Compute the amount of evaporation using the Penman method on June 1, 2017, in Beaumont, Texas.

5.17 Compute the amount of evaporation using the Taylor-Priestley method on June 1, 2017, in Beaumont, Texas.

5.18 What is the Bowen ratio? How is it used to estimate ET?

5.19 What is the Budyko curve, and what is its practical utility?

5.20 Explain any two reduced-set methods to compute ET. What are their limitations?

5.21 Compute ET_0 by the FAO 56 PM method for your city (find latitude and elevation) for June 1, 2018. If this information is not available, use the following meteorological data: $T_{max} = 30°C$, $T_{min} = 20°C$, $RH_{max} = 85\%$, $RH_{min} = 60\%$, number of daylight hours $n = 8$, wind speed at 2 m height = 2 m/s.

5.13 References

Allen, R. G. *Evaluation of Procedures for Estimating Grass Reference Evapotranspiration Using Air Temperature Data Only.* Report prepared for the Water Resources Development and Management Service, Food and Agriculture Organization (FAO), Rome, 1995.

Allen, R. G., L. S. Pereira, D. Raes, and M. Smith. *Crop Evapotranspiration: Guidelines for Computing Crop Requirements.* Irrigation and Drainage Paper No. 56. Rome: Food and Agriculture Organization (FAO), 1998.

Allen, R. G., W. O. Pruitt, J. A. Businger, L. J. Fritschen, M. E. Jensen, and F. H. Quinn. Evaporation and transpiration. In *Hydrology Handbook (Manual and Reports on Engineering Practice No. 28)*, Task Committee on Hydrology Handbook of Management Group D of ASCE, New York, pp. 125–252, 1996.

Allen, R. G., M. Smith, L. S. Pereira, and A. Perrier. An update for the calculation of reference evapotranspiration. *ICID Bulletin*, 43(2):35–92, 1994.

Bastiaanssen, W. G. M. *Remote Sensing in Water Resources Management: The State of the Art.* Colombo, Sri Lanka: International Water Management Institute, 1998.

Blaney, H. F., and W. D. Criddle. Determining water requirements in irrigated áreas from climatological and irrigation data. In *USDA Soil Conservation Service, SCS-TP-96, Hargreaves and Other Reduced-Set Methods for Calculating Evapotranspiration*, 77, 1950.

Bowen, I. S. The ratio of heat losses by conduction and by evaporation from any water surface. *Physical Review,* 27:779–787, 1926.

Burman, R. D., J. L. Wright, and M. E. Jensen. Changes in climate and estimated evaporation across a large irrigated area in Idaho. *Transactions of the ASAE*, 18(6):1089–1093, 1975.

Doorenbos, J., and W. O. Pruitt, W. O. *Crop Water Requirements.* Irrigation and Drainage, Paper 24. Rome: Food and Agricultural Organization (FAO), 1977.

Garcia, M., D. Raes, R. Allen, and C. Herbas. Dynamics of reference evapotranspiration in the Bolivian highlands (Altiplano). *Agricultural and Forest Meteorology*, 125:67–82, 2004.

Hargreaves, G. H. Moisture availability and crop production. *Transactions of the ASAE*, 18(5), 980–984, 1975.

Hargreaves G. H., and Z. A. Samani. Estimation of potential evapotranspiration. *Journal of Irrigation and Drainage Division, Proceedings of the American Society of Civil Engineers,* 108:223–230, 1982.

Hargreaves, G. H., and Z. A. Samani. Reference crop evapotranspiration from temperature. *Applied Engineering in Agriculture,* 1(2):96–99, 1985.

Henggeler J. C., Z. Samani, M. S. Flynn, and J. W. Zeitler. Evaluation of various evapotranspiration equations for Texas and New Mexico. *Proceeding of Irrigation Association International Conference,* San Antonio, TX, 1996.

Horton, R. E. Evaporation maps of the United States. *Transactions of the American Geophysical Union,* 24:750, 1943.

Irmak, S., A. Irmak, R. G. Allen, and J. W. Jones. Solar and net radiation-based equations to estimate reference evapotranspiration in humid climates. *Journal of Irrigation and Drainage Engineering,* 129(5):336–347, 2003.

Itenfisu, D., R. L. Elliott, R. G. Allen, and I. A. Walter. Comparison of reference evapotranspiration calculations as part of the ASCE standardization effort. *Journal of Irrigation and Drainage Engineering,* 129(6):440–448, 2003.

Jain, S. K., P. C. Nayak, and K. P. Sudheer. Models for estimating evapotranspiration using artificial neural networks and their physical interpretation. *Hydrologic Processes Journal,* 22(13):2225–2234, 2008.

Jensen, M. E., and H. R. Haise. Estimating evapotranspiration from solar radiation. *Journal of Irrigation and Drainage Division, Proceedings of ASCE,* 89:15–41, 1963.

Jensen, M. E., R. D. Burman, and R. G. Allen. *Evapotranspiration and Irrigation Water Requirements.* ASCE Manuals and Reports on Engineering Practice No. 70. New York: ASCE, 1990.

Makkink G. F. Testing the Penman formula by means of lysimeters. *Journal of the Institution of Water Engineers,* 11:277–228, 1957.

Masoner, Jason R., David I. Stannard, and Scott C. Christenson. Differences in evaporation between a floating pan and Class A pan on land. *Journal of the American Water Resources Association,* 44(3), 552–561, 2008.

Michael, T., and J. Huntington. Evapotranspiration and evaporative demand. In V. P. Singh (Ed.), *Handbook of Applied Hydrology,* 2nd ed. New York: McGraw-Hill, 2017.

Monteith, J. L. The state and movement of water in living organisms. *Proceedings of, Evaporation and Environment, XIXth Symposium, Society for Experimental Biology.* Swansea, U.K. and New York: Cambridge University Press, pp. 205–234, 1965.

Penman, R. L. Natural evaporation from open water, bare soil and grass. *Proceedings of the Royal Society of London,* A193:120–146, 1948.

Priestley, C. H. B., and R. J. Taylor. On the assessment of the surface heat flux and evaporation using large-scale parameters. *Monthly Weather Review,* 100: 81–92, 1972.

Samani, Z. Estimating solar radiation and evapotranspiration using minimum climatological data. *Journal of Irrigation and Drainage Engineering,* 126(4):265–267, 2000.

Saxton, K. E. and J. L. McGuinness. Evapotranspiration. In *Hydrologic Modeling of Small Watersheds,* edited by C. T. Haan. ASAE Monograph#5, American Society of Agricultural Engineers, Michigan.

Schulz, G. A., and E. T. Engman. *Remote Sensing in Hydrology and Water Resources.* Berlin: Springer-Verlag, 2000.

Shahidian, Shakib, R. Serralheiro, J. Serrano, J. Teixeira, N. Haie and F. Santos. Hargreaves and Other reduced-set methods for calculating evapotranspiration. In *Evapotranspiration–Remote Sensing and Modeling*, edited by Dr. Ayse Irmak, 2012. Published on-line at www.intechopen.com.

Shuttleworth, W. J. Evaporation. In *Handbook of Hydrology,* edited by D. R. Maidment (pp. 4.1–4.53). New York: McGraw-Hill, 1992.

Singh, V. P., and C. Y. Xu. Evaluation and generalization of 13 mass transfer equations for determining free water evaporation. *Hydrological Processes,* 11:311–323, 1997.

Thornthwaite, C. W. An approach toward a rational classification of climate. *Geographical Review,* 38:55–94, 1948.

Turc, L. Estimation of irrigation water requirements, potential evapotranspiration: A simple climatic formula evolved up to date. *Annals of Agronomy,* 12, 13–46, 1961.

Xu, C.-Y., and V. P. Singh. Dependence of evaporation on meteorological variables at different time-scales and intercomparison of estimation methods. *Hydrological Processes,* 12:429–442, 1998.

Xu, C.-Y., and V. P. Singh. Evaluation and generalization of radiation-based methods for calculating evaporation, *Hydrological Processes,* 14:339–349, 2000.

Xu, C.-Y., and V. P. Singh. Evaluation and generalization of temperature-based methods for calculating evaporation. *Hydrological Processes,* 15:305–319, 2001.

WMO. Guide to Meteorological Instruments and Methods of Observation, WMO-No. 8. World Meteorological Organization, Geneva, 2008.

CHAPTER 6

Infiltration and Soil Moisture

Infiltration is the process of the entry of water into the soil at the surface and is a vital natural process. Runoff, one of the most important components of the hydrologic cycle, is determined by subtracting from precipitation the abstractions that occur in the catchment. At short timescales, infiltration is the dominant abstraction for individual rainfall events. For longer timescales, after evaporation, it is usually the most important abstraction affecting runoff. Infiltration replenishes the soil water that is used up by plants. In the absence of infiltration, streams would cease to flow shortly after rainfall, frequent floods would inundate vast lands adjoining rivers, wells would go dry, springs would disappear, soils would erode from crop areas, crop production would fall, and droughts would be more frequent. Infiltration forms the basis to determine agricultural irrigation requirements. This chapter presents basic concepts of infiltration and soil moisture, and the methods for determining them.

6.1 Infiltration Process

Soil is the weathered and fragmented layer of material on the surface of Earth, which is formed by disintegration and decomposition of rocks. Soil is the unconsolidated granular material, including organic matter above the hard bedrock. Soil depth can range from a few centimeters to hundreds of meters.

Consider a soil profile with uniform initial water content. Let there be a shallow depth of water at the soil surface (ponding) at time $t = 0$ and water is being continuously added to maintain the depth. The rate at which water enters the soil at its surface is defined as the infiltration rate. As shown in Fig. 6.1 (ponded case), the infiltration rate f is initially high but it falls with time. After a sufficiently long time, the infiltration rate asymptotically approaches a value f_c, which is nearly equal to the saturated hydraulic conductivity of the soil. Since water is always ponded at the soil surface, the rate of infiltration is limited only by soil property. This rate is the infiltration capacity of soil, often called the potential infiltration rate, which is the maximum rate at which water can enter the soil.

Consider a different case, where water is applied to the dry soil surface (by rain or irrigation) at a constant rate R. In this case, in the beginning, the infiltration rate will be equal and limited by R. Until the time the application rate is less than the infiltration capacity, the entire amount of applied water will infiltrate the soil and $f = R$. This time is denoted by t_p in Fig. 6.1. After this time, the infiltration capacity is less than the

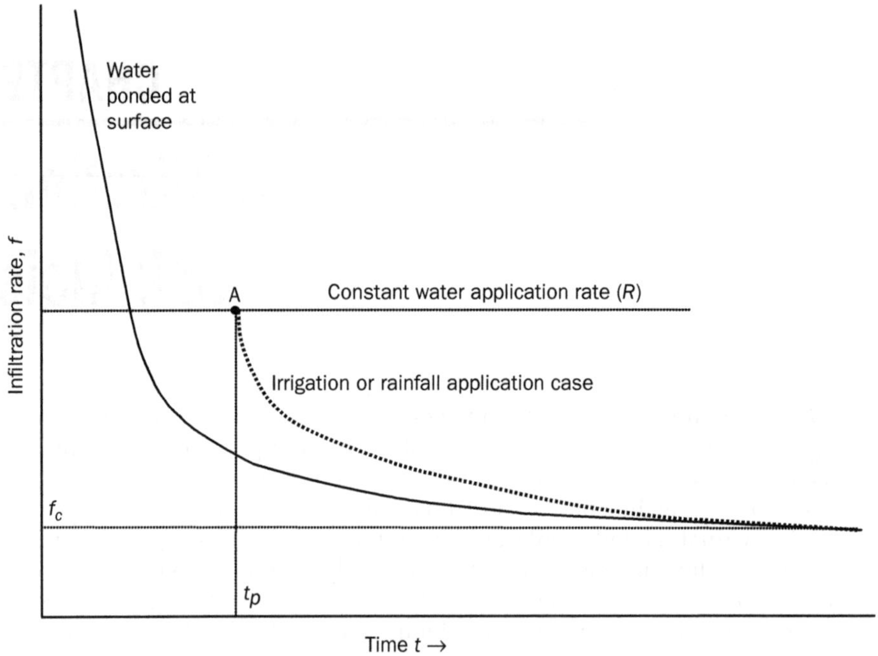

FIGURE 6.1 Variation in soil infiltration for two cases.

application rate R and water will begin to pond on the surface. After this time, the infiltration will be controlled by the soil properties. Water applied in excess of infiltration capacity will contribute to runoff.

When water from rain, snow melt, or irrigation comes in contact with the soil surface, a part of it enters the pores and moves downward. The water is pushed into the soil pores by the force of gravity and by the capillary attraction of soil pores. Cohesive forces arise due to the attraction of water molecules to each other. Adhesive forces cause water molecules to attach to the surface of soil particles and form a thin film of water around them. As more water is added to the already moist soil, water molecules are attracted to the existing molecules due to cohesive forces and the water film around the soil particles thickens. The adhesive forces attracting water and soil material act along with the capillary (surface tension) forces. Capillary action is a combination of cohesion/adhesion and surface tension forces and is demonstrated by the upward movement of water through a narrow tube against the force of gravity. Capillary action occurs when the adhesive intermolecular forces between a liquid, such as water, and the solid surface of the tube are stronger than the cohesive intermolecular forces between water molecules. Further, surface tension is defined as the property of the surface of a liquid that allows it to resist an external force, due to the cohesive nature of the water molecules. As more water enters the soil, the result of these forces is to cause the water to penetrate the openings in the soil surface and migrate downward, with numerous menisci forming at the interfaces between air and water. The pores through which water moves may range from almost microscopic size to small, tubelike pathways between soil grains. Macropores, which are large pores in the soil (usually between aggregates), permit the easy movement of water and air through soil.

As more water enters into the soil, the menisci are destroyed and the force of gravity takes over, pushing the water downward. When the supply of water to the soil (e.g., rain) stops, the water percolates (or moves down) by gravity until the menisci begin to reappear, and the forces of surface tension start dominating again, holding water in the pores. The infiltration rate depends on soil and water properties. Under most conditions, the infiltration rate is at its greatest when water first contacts the soil surface, because the capillary forces are the strongest when the soil is the driest. Hence dry soils have a high initial infiltration rate. As water penetrates the soil to some depth, the resistance to the forces acting on the water increases and the rate of infiltration falls. A fairly constant rate of infiltration is attained after a lapse of some time.

Moisture content in the soil at the beginning of a rainfall event has an important effect on the infiltration rate. Soils with high moisture content will have lower initial infiltration rates than drier soils. The amount of runoff from a given area depends on the rainfall intensity and the infiltration rate. If the infiltration rate exceeds the rainfall intensity, no runoff will occur. On the other hand, if the rainfall intensity exceeds the infiltration rate, then runoff will occur. The portion of rainfall that infiltrates the soil first fills the soil-moisture deficit; when the soil-moisture deficit is replaced, the remaining infiltrated water will continue migrating downward until it joins the groundwater table. Runoff hydrograph and infiltration characteristics of a catchment are closely related.

6.2 Soil and Its Hydrologic Properties

Soil is a heterogeneous three-phase system (Fig. 6.2). The solid phase constitutes the soil matrix, the water with dissolved matter constitutes the liquid phase, and empty space is the air phase. Figure 6.2 shows the three phases: solids, liquid, and gas, and the notations for masses and volumes of the three phases of soil.

6.2.1 Soil Profile

This term denotes the vertical cross section of the soil column. A typical soil profile consists of a number of layers or strata which depend upon the history of soil formation

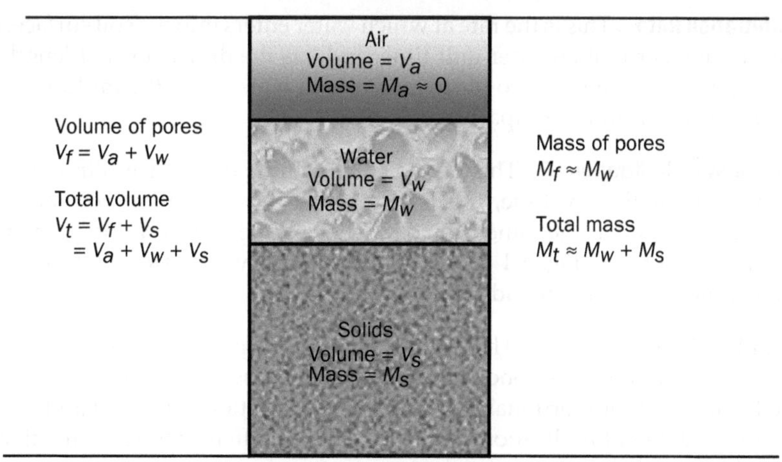

FIGURE 6.2 Schematic depiction of a soil column showing the three phases.

and deposition. In particular, the top soil layer where major biological activities take place is termed as the A horizon. This layer is rich in organic matter and is usually darker in color. Below the A horizon comes the B horizon, which is an intermediate horizon and is followed by the C horizon, which consists of the parent material of the soil. When the soil is formed by weathering, it will consist of fragmented rock matter and, in other cases, soil may be composed of material transported by alluvial or glacial means. All three profiles are not always present in soils and are not always distinct.

6.2.2 Soil Texture

Soil texture describes the range of size of soil particles. It is a qualitative way of telling whether the soil is coarse and rough or fine and smooth. Traditionally, soil is said to be sand, silt, or clay, depending on the size of particles. There is no universal definition of these terms and different organizations follow different criteria. Conventionally, the particles whose size is smaller than 2 mm are classified as soil. According to the U.S. Department of Agriculture, the term sand denotes soil particles whose diameter ranges from 2 mm to 50 μm, silt consists of particles whose diameter is between 50 and 20 μm, and clays have particles smaller than 2 μm in diameter. Clay fraction has the largest influence on the soil behavior.

Frequently used terms related to soils are defined as follows:

Infiltration The process by which water enters the soil at the soil surface. Infiltration involves change in medium, for example, from air to soil.

Percolation The process of water movement within the soil profile in the vertical direction is called percolation. It is evident that infiltration precedes percolation.

Infiltration Capacity, f_c When the availability of water is not a limiting factor, the maximum rate of infiltration at any time is called the infiltration capacity of the soil (Horton, 1933). The infiltration capacity asymptotically approaches a constant value after a certain period of time has passed, and is called the constant infiltration capacity, the steady-state infiltration capacity, the time-invariant infiltration capacity, or the minimum infiltration capacity. This is the maximum rate at which soil can absorb water through its surface.

Infiltration Rate This is the rate at which water enters into the soil surface. It is expressed as volume per unit area per unit time, and has the dimensions of length per unit time $[L/T]$. If there is no limit to the availability of water, then the infiltration rate (f) is the same as the infiltration capacity or $f = f_c$.

Cumulative Infiltration, F This denotes the volume of water infiltrated from the beginning of rainfall, t_0, to time, t. It is also called the infiltration volume or accumulated infiltration and has the dimension of L. The volume of infiltration at any time is the area under the curve in Fig. 6.1 up to that time. Clearly, the rate of infiltration depends on how much water has already infiltrated.

Relation Between f and F The infiltration rate, f, can be regarded as flux and the cumulative infiltration, F, as concentration. As F increases, f decreases. Thus f can be shown to be inversely proportional to F and an exact relation between f and F will depend on the soil. Different infiltration equations assume different types of relationship between f and F.

Density: The density of solids in a soil is defined as

$$\rho_s = \frac{M_s}{V_s} \tag{6.1}$$

For most mineral soils, the density of solids is about 2.6 g/cm³. The dry bulk density of a soil is the ratio of the mass of dry soil to its total volume.

$$\rho_b = \frac{M_s}{V_t} = \frac{M_s}{(V_s + V_a + V_w)} \tag{6.2}$$

Since ρ_b includes the volume of air and water in the denominator, $\rho_b < \rho_s$; ρ_b is high for sandy soils and small for clayey soils.

Porosity: This is the ratio of the volume of fluids to the total volume of soil.

$$f = \frac{V_f}{V_t} = \frac{V_a + V_w}{(V_s + V_a + V_w)} \tag{6.3}$$

Generally, porosity varies between 0.3 and 0.6.

Degree of saturation: This is the ratio of the volume of water present in the soil to the total volume of pores.

$$s = \frac{V_w}{V_f} = \frac{V_w}{(V_a + V_w)} \tag{6.4}$$

For completely dry soil, $s = 0$ and for completely saturated soil, $s = 1$.

Field Capacity (FC) The maximum amount of water that soil (or a field) can hold in storage is called *field capacity (FC)*. Any water at or above the FC will flow through soil under gravitational force. FC is the moisture content of a soil that has been saturated with water and allowed to drain for 24 hours. It is assumed that after 24 hours, no more moisture is moving through the soil under the force of gravity.

FC is a useful term, because it is a reference for comparing moisture-holding capacities of soils. At FC, a medium-textured soil will hold more than 5 cm of water per 30 cm of soil, a clayey soil holds about 3.2 cm of water per 30 cm, and sand will hold about 4.5 cm per 30 cm. These values represent the potential soil-moisture replacement that must be supplied to bring moisture-depleted soil up to FC. At FC, the capillary potential is approximately equal to a pressure of 0.36 atmospheres. FC plays an important role in agricultural irrigation scheduling.

Permanent Wilting Point (PWP) This is the lower limit of water available in the soil for the use by plant roots. The soil water in storage is also depleted by plant roots. The depletion occurs due to osmotic pressure differences between water in the plant roots and in soil storage. *Permanent wilting point (PWP)* is the term applied to the soil moisture content at which plants cannot extract water and permanently wilt; it is approximately equivalent to 0.15 atmosphere pressure.

Water flows from soil to plant roots as long as the suction force in the roots exceeds the capillary force in the soil pores. When the capillary force exceeds the roots' suction force, the plants are no longer able to extract soil moisture, and the residual amount of soil water is called the PWP. Thus the difference between FC and PWPs represents the soil moisture availability in the root zone that a plant can use.

Capillary Potential This is the hydraulic head caused by capillary forces and has the dimension of L. It is also known as capillary pressure, pressure head, pore pressure, moisture tension, moisture suction, or negative pressure. Capillary potential is always less than the atmospheric pressure.

Capillary Suction, *S* This is the capillary potential with an opposite sign. A positive suction will represent a negative hydraulic head. It has the dimension of L.

Hydraulic Conductivity, *K* This is defined as the volume rate of flow of water through a unit area of soil under a unit gradient. This term is dependent on the available moisture and has the dimensions of $[L/T]$.

Saturated Conductivity, K_s This represents the soil conductivity when the soil is saturated and has the dimensions of $[L/T]$.

6.2.3 Soil Water Content

The amount of water present in the soil in terms of mass or volume denotes soil wetness. Soil wetness can be defined in terms of mass or volume as

Mass wetness
$$w = \frac{M_w}{M_s} \tag{6.5}$$

Volume wetness
$$\theta = \frac{V_w}{V_t} = \frac{V_w}{(V_s + V_a + V_w)} \tag{6.6}$$

Relation between mass and volume wetness
$$\theta = w\frac{\rho_b}{\rho_w} \tag{6.7}$$

Methods for determining soil wetness can be categorized as direct and indirect methods. In the conventional direct method, a soil sample is collected from the field and is weighed. Next, the sample is dried in an oven at about 105°C for 24 hours. The difference in weight gives the amount of water which can be expressed as a fraction of the dry weight of the sample. The electrical resistance method uses a pair of electrodes embedded in gypsum blocks, which are placed in soil to equilibrate with its suction. This indirect method requires calibration of soil wetness (or suction) with electrical resistance. Neutron moisture probe and gamma-ray absorption are other indirect methods to determine soil wetness.

Water in soil is subject to a number of forces due to which its potential is different than that of the free water. For determination of the state and movement of water in soil, its potential energy (the energy due to its position or condition) is of prime importance. If we know the potential energy of water in the soil and in the plant at a given location, we can estimate the quantum of work that the plant must do to extract one unit of water from the soil. Soil-water potential represents the potential energy of soil water relative to water in a standard reference state.

The total potential of soil water is composed of three terms, these are

$$\varphi_t = \varphi_g + \varphi_o + \varphi_p \tag{6.8}$$

where φ_t is the total potential, φ_g is the gravitational potential, φ_o is osmotic potential, and φ_p is the pressure potential. The gravitational potential of soil water at any location depends upon the elevation of the location with respect to some datum. If the soil water is at a hydrostatic pressure greater than atmospheric, its pressure potential is positive and if it is at a pressure below atmospheric, the pressure potential is negative and soil is considered to have tension or suction. Negative pressure potential is also termed as matric potential. In fact, matric potential denotes the total effect resulting from the affinity of the water to the whole soil matrix, including its pore and particle surfaces (Hillel, 1971). Soil scientists use the term pF, which is defined as the logarithm of negative pressure (tension or suction) head in a centimeter of water. Accordingly, a pF of one denotes tension head of 10 cm of water and pF-2 is tension head of 100 cm of water. The forces of soil water retention are known as matric forces because they result from the soil matrix. The matric suction is a function of soil water content.

6.3 Soil Water

The water present in the soil is called subsurface water and is of two kinds: soil water and groundwater, and their zones are known, respectively, as unsaturated zone and saturated zone. These two zones behave in fundamentally different ways. In the groundwater zone, all the voids are completely filled with water and the top of this zone is termed as the water table. Figure 6.3 pictorially shows the soil water balance.

The soil water zone occurs from the water table up to the soil surface. Soil is capable of not only storing water, but also transmitting it. Some soil pores are completely filled with water, some are partially filled, and some are empty. Infiltration supplies water

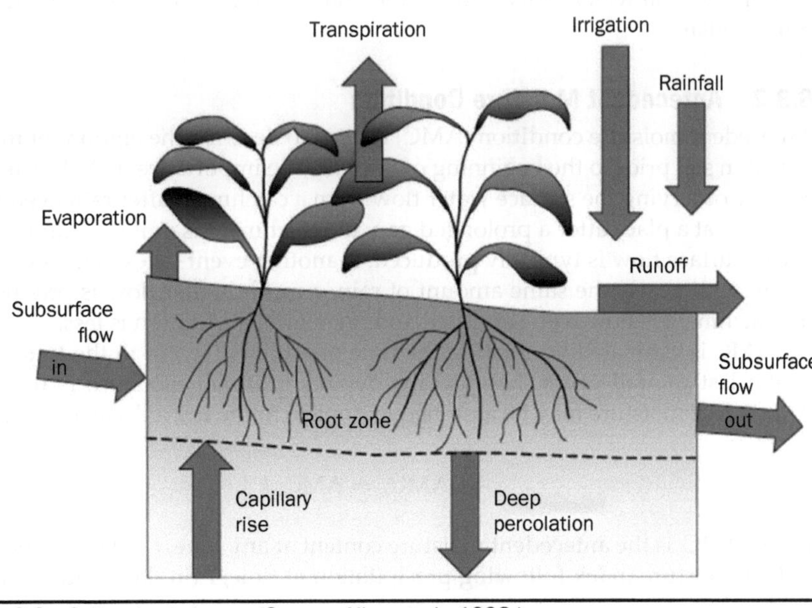

FIGURE 6.3 Soil water balance. (Source: Allen et al., 1998.)

to this zone. The void space that is not occupied by water is filled by air. The unsaturated zone acts as a conduit to move water between the surface and groundwater zone or works as a storage reservoir whose water is either used by vegetation or by evapotranspiration.

6.3.1 Movement of Soil Water

During its movement in the soil, the water flows from regions of higher potential (energy) to regions of lower potential. The potential that controls the flow of soil water is principally made up of two components: capillary potential and gravitational potential. The gravitational potential energy of an object depends upon its weight and its position on the Earth or its distance from the Earth's center. Capillary forces result from cohesive and adhesive molecular forces acting within the body of a liquid and between water and soil particles. Cohesive forces cause surface tension, which acts at the interface between two liquids or between a liquid surface and a gas such as air.

All liquids exhibit surface tension, but water manifests greater surface tension. Adhesive forces result from the attraction between water molecules and solid surface. An example is adsorbed water, which is a thin film on a solid surface that is held strongly against gravity. The cause of these cohesive and adhesive forces is hydrogen bonding. The capillarity depends upon molecular forces of adhesion with a solid surface and surface-tension forces within the liquid. It raises water in a tube against the action of gravity. In soil with interconnected granular pores or tubes, capillarity may move water many centimeters above a water level such as a water table. Since capillarity operates in the Earth's gravitational field, the capillary rise of water in a soil pore will continue until the gravitational force balances it. Thus the capillary potential energy depends on molecular forces within a liquid and on pore diameter. The capillary potential gradient will usually run from larger pores to smaller ones. The direction of the gradient can be up, down, or horizontal, with water moving from a wetter soil to a drier one. Soil water moves due to the total potential energy gradient (gravitational plus capillary).

6.3.2 Antecedent Moisture Condition

Antecedent moisture condition (AMC) is used to describe the amount of moisture contained in soil prior to the beginning of rainfall. The importance of AMC can be appreciated by observing the surface water flow from a catchment after rain events. When the rain falls at a place after a prolonged gap, the catchment is quite dry and a small quantity of surface flow is typically produced. If another event takes place shortly after this event, with nearly the same amount of rain, a much higher flow is produced because the catchment is now well-saturated and, therefore, infiltration is less.

AMC is estimated by relating the time of a rainfall event to the time and amount of the last rainfall event, because the depletion of soil moisture is related to time. Antecedent moisture may be assumed to deplete following a decay of the form

$$\mathrm{AMC}_t = \mathrm{AMC}_{t0} k^{-t} \tag{6.9}$$

where AMC_t is the antecedent moisture content at any time, t, AMC_{t0} is the initial antecedent moisture index following precipitation at time, t (in days), and k is a recession constant (a measure for soil drying). The values of k range between 0.85 and 0.95; an average value of 0.90 is often used. This states that at the end of one day, AMC1 will be 90% of the AMC of the previous day. Likewise, at the end of the second day, AMC2 will be 90% of AMC1 or 81% of AMC0. In hydrologic applications, t is taken as no more than 5 days. Equation (6.9) plays an important role in rainfall-runoff modeling for ungagged watersheds as well as in determining irrigation scheduling.

6.4 Measurement of Soil Water Properties

Here, we cover two properties of soils that are important in the field of hydrology: particle size distribution and soil moisture characteristic curve.

6.4.1 Particle Size Distribution

The distribution of the size of particles present in a soil sample can be summarized by the particle size distribution curve. The Y axis indicates the percentage of soil particles with a diameter smaller than the value shown on the X axis (typically on log scale). This graph gives a cumulative presentation of the size of soil particles. A soil sample containing particles of a few distinct sizes is considered poorly graded. A soil with a smooth distribution containing particles of different sizes is considered well graded. Usually, sieve analysis is carried out to determine the particle size distribution. A series of sieves are used with sieve opening gradually decreasing. Soil sample is placed on the topmost sieve (having the largest holes) and the whole assembly is mechanically shaken. Soil particles retained on a different sieve are those whose size is bigger than the size of the holes in that sieve. These are weighed and accumulated to determine the fraction of soil particles whose size is bigger than the sieve size. To classify particles of diameter smaller than 0.05 mm, the method of sedimentation is followed, wherein settling velocities of particles in an aqueous suspension are determined.

6.4.2 Soil Moisture Characteristic Curve

When a soil sample is saturated, the pressure is the same as atmospheric and the suction or tension is 0. As suction is applied to the soil, some water is drained out from the larger pores. With increase in suction, progressively smaller pores also yield water. Thus the amount of water retained in the soil at a given suction depends on the size and volumes of pores. If the suction (expressed in centimeter of water column) is plotted on a logarithmic scale against the water content, the resulting curve is called a soil moisture characteristic curve or moisture retention curve or h-θ curve, where h refers to the suction head and θ is the volumetric soil moisture content. The soil water retention property signifies the water storing capacity of soils. Whether water transmission actually takes place through soil pores depends on the hydraulic conductivity or permeability of the soil.

The relationship between h and θ is a soil property called the soil water characteristic. However, h depends on θ and also on the drying or wetting history of the soil. That is, the soil water characteristic exhibits hysteresis, shown in Fig. 6.4. For a given pressure head value, the volumetric moisture content in the soil is smaller if soil is wetting and more if it is drying. The shape of the curve depends, among other things, on soil texture. The greater the clay content in the soil, the more water will be retained at a given suction and the slope of the curve will be smaller.

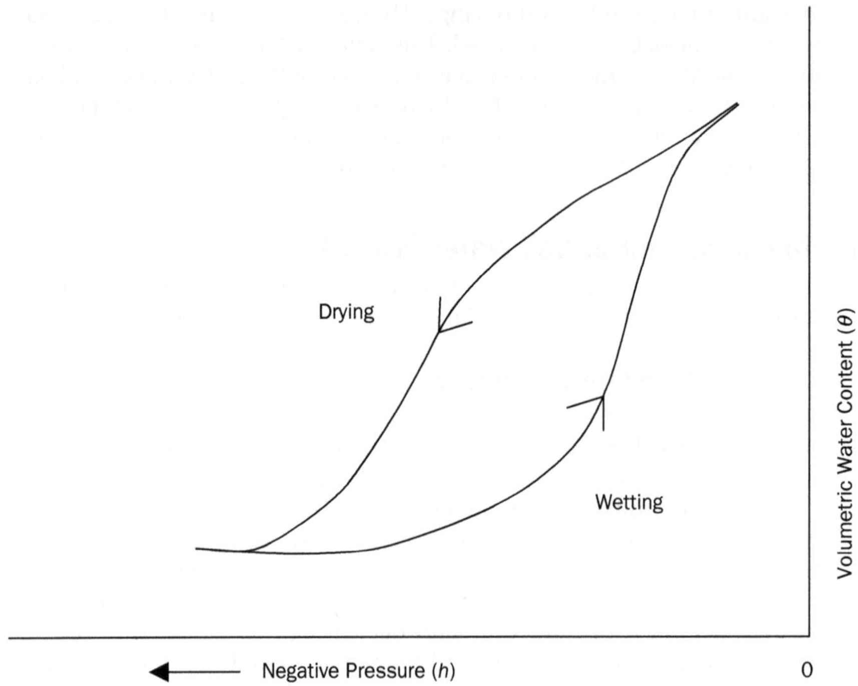

Figure 6.4 Curves of pressure versus water content, showing hysteresis.

6.5 Estimation of Soil Water Properties

A number of techniques have been developed for the estimation of soil moisture retention curve and hydraulic conductivity as a function of soil moisture by using the soil properties. Most popular among these are the Brooks and Corey (1964), Campbell (1974), and van Genuchten (1980) methods and the equations developed are shown in Table 6.1.

A two-parameter equation for a water retention curve was proposed by Assouline et al. (1998) as

$$\theta = \theta_L + (\theta_s - \theta_L)\{1 - \exp[-\xi(h^{-1} - h_L^{-1})^{\eta}]\} \quad 0 \le h \le h_L \tag{6.10}$$

where h_L is a suction head that corresponds to a very low water content, θ_L, where the hydraulic conductivity becomes negligible, and ξ and η are empirical shape parameters. Artificial neural networks (ANNs) have also been applied to determine characteristic curves.

6.6 Governing Equations of Soil Water Movement

As noted earlier, infiltration is governed by factors that control movement of water in the soil. The fundamental equation for describing movement of water in soil was

Hydraulic Soil Characteristic	Parameters	Parameters Estimation
	Brooks and Corey (1964)	
Soil water retention $\frac{\theta-\theta_r}{\varphi-\theta_r}=\left(\frac{\psi_b}{\psi}\right)^{\lambda}$ Hydraulic conductivity $\frac{K(\theta)}{K_S}=\left(\frac{\theta-\theta_r}{\varphi-\theta_r}\right)^{3-2/\lambda}$	λ = pore size index ψ_b = bubbling pressure$ θ_r = residual water φ = total porosity K_S = saturated conductivity	$\lambda=\lambda$ $\psi_b=\psi_b$ $\theta_r=\theta_r$ $\varphi=\varphi$ $K_S=K_S$
	Campbell (1974)	
Soil water retention $\frac{\theta}{\varphi}=\left(\frac{H_b}{\psi}\right)^{1/b}$ Hydraulic conductivity $\frac{K(\theta)}{K_S}=\left(\frac{\theta}{\varphi}\right)^{2b+3}$	φ = porosity H_b = pressure b = constant	$\varphi=\varphi$ $H_b=\psi_b$ $b=1/\lambda$
	van Genuchten (1980)	
Soil water retention $\frac{\theta-\theta_r}{\theta_s-\theta_r}=\left(\frac{1}{1+(\alpha h)^n}\right)^m$ Hydraulic conductivity $\frac{K(\theta)}{K_S}=\left(\frac{\theta-\theta_r}{\varphi-\theta_r}\right)^{1/2}\left[1-\left(1-\left(\frac{\theta-\theta_r}{\varphi-\theta_r}\right)^{1/m}\right)^m\right]^2$	φ = porosity θ_r = residual water α = constant n = constant m = constant	$\varphi=\varphi$ $\theta_r=\theta_r$ $n=\lambda+1$ $m=1-\frac{1}{\lambda+1}$

$Bubbling pressure is the matric suction from which air starts to penetrate into the soil. It is also called air entry value (AEV).

Table 6.1 Summary of Models to Estimate Soil Hydraulic Characteristic by Using Soil Properties (Source: Rawls and Brakensiek, 1982; van Genuchten, 1980; and others.)

derived by Darcy (1856), who found this rate to be directly proportional to the hydraulic gradient. According to Darcy's law

$$q_s = -K\frac{\partial H}{\partial s} \tag{6.11}$$

where q_s is the rate of water movement through the soil in the s-direction per unit area per unit time ($m^3/m^{-2}/t$); $\partial H/\partial s$ is the hydraulic gradient in the s-direction; K is the hydraulic conductivity (L/T); H is the total potential head (L). Here, K depends on both the properties of fluid and porous medium. Head H may be considered equal to the hydraulic head which is the sum of the pressure head (h) and the elevation head (z). If the datum lies at the soil surface then

$$H = h - z \tag{6.12}$$

where z is the distance from the surface, measured positively downward. When pressure head is positive, the soil is assumed saturated and water content is constant. For unsaturated soils, the pressure head is negative and is nonlinearly related to the volumetric water content.

For saturated soils, the hydraulic conductivity does not vary with respect to h. In heterogeneous soils, hydraulic conductivity at a given water content varies with location. In the homogeneous soils, hydraulic conductivity does not vary with location. Further, if the hydraulic conductivity does not depend on the flow direction, the soil is said to be isotropic; if the hydraulic conductivity depends on the flow direction, the soil is said to be anisotropic.

For regions of the soil mass that is only partially saturated and in which the water content varies with both time and position, the equation for flux of water, q_s, is

$$q_s = -K(\theta)\frac{\partial H}{\partial s} \tag{6.13}$$

Note that the hydraulic conductivity (K) is a function of water content (θ). Since moisture content (θ) at the given head is a property of soil (it also depends on the wetting and drying history due to hysteresis), the hydraulic conductivity is also a function of head and we may write Eq. (6.13) as

$$q_s = -K(h)\frac{\partial H}{\partial s} \tag{6.14}$$

When soils are unsaturated, water moves through space in small pores and through thin films created around and between soil particles. As the water content falls, the cross-sectional area of the films also falls and the flow paths reduce. Due to this, the hydraulic conductivity reduces rapidly with water content.

For the soil water system, the differential form of the equation of conservation of mass can be written as

$$\frac{\partial \theta}{\partial t} = -\nabla \overline{q} \tag{6.15}$$

where θ is the volumetric soil water content, $\overline{q}$ is the flux vector, and ∇ is the differential operator. If we consider flow in the z-direction only, Eq. (6.15) may become

$$\frac{\partial \theta}{\partial t} = -\frac{\partial q_z}{\partial z} \tag{6.16}$$

Combining Eqs. (6.16) and (6.14) and taking the datum at the surface, or $H = h - z$, yields the equation for the flow in the vertical direction as

$$C(h)\frac{\partial h}{\partial t} = \frac{\partial}{\partial z}\left[K(h)\frac{\partial h}{\partial z}\right] - \frac{\partial K}{\partial z} \tag{6.17}$$

where $C(h)$ is the soil water capacity, defined as

$$C(h) = \frac{\partial \theta}{\partial h} \tag{6.18}$$

Equation (6.17) was derived by Richards (1931) and it is referred to as the Richards equation.

6.7 Solving the Richards Equation

The Richards equation contains $C(h)$ and $K(h)$ terms. These parameters vary with water content or pressure head and display pronounced nonlinearity. Due to high nonlinearity of these parameters, exact analytical solutions of the Richards equation are not available except for a few limited cases. However, many numerical techniques have been developed to solve the Richards equation subject to various boundary conditions. Finite difference methods are commonly used to solve the Richards equations for two- and three-dimensional cases. Frequently used finite difference methods are the explicit schemes, implicit schemes, and the Crank-Nicolson approximation. While solving the Richards equation numerically, finer discretization in time and distance space is necessary. Otherwise, the solution may become unstable (there are numerical oscillations).

6.8 Generalized Infiltration Model

The infiltration process can be represented by a systems approach. Singh and Yu (1990) employed the systems approach to derive a generalized infiltration model and show that the various infiltration equations are special cases of the generalized model.

Consider a column of soil with unit area of vertical infiltration (Fig. 6.5). The initial storage space available in the soil column is denoted as S_o. If the soil is initially dry, then S_0 equals effective porosity multiplied by the total volume of the column. At time $t = 0$, a thin layer of water starts to cover the upper surface of the column. At any time, t, let the infiltration rate be denoted as $f(t)$, seepage rate as $f_s(t)$, and the potential water storage space as $S(t)$.

The spatially lumped continuity equation for the soil column can be written in the integral form as

$$S(t) = S_o + \int_0^t f(t)dt - \int_0^t f_s(t)dt \tag{6.19}$$

where S is the storage or empty space at time t. The differential form of continuity equation can be obtained by differentiating Eq. (6.19) with respect to t as

$$\frac{dS(t)}{dt} = f_s(t)dt - f(t)dt \tag{6.20}$$

The amount of water stored in the soil column from $t = 0$ to any time t is

$$W(t) = S_o - S(t) = \int_0^t [f(t) - f_s(t)]dt \tag{6.21}$$

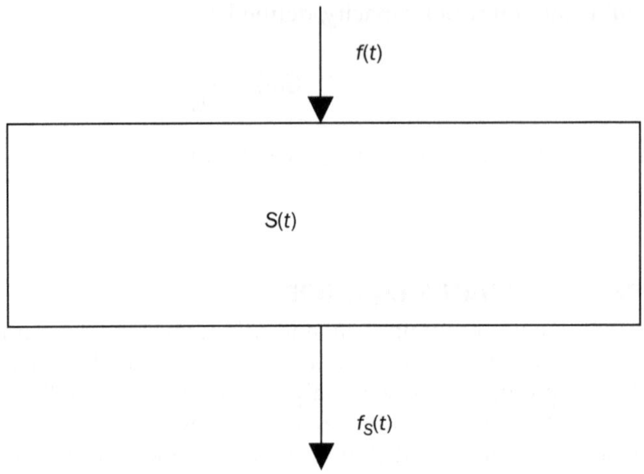

FIGURE 6.5 Systems representation of a soil column having some infiltration, seepage, and defined water storage space.

If $f_s(t) = 0$ in Eq. (6.20) (Dooge, 1973), then

$$\frac{dW}{dt} = f(t)dt \tag{6.22}$$

Equation (6.20) contains three unknowns: S, f, and f_s. To solve this equation, two relationships among the variables, $S(t)$, $f(t)$, and $f_s(t)$, are required. Analogous to development of a generalized rainfall-runoff model by Kulandaiswamy (1964), a possible relationship among these variables may be expressed as

$$S(t) = \sum_{n=1}^{m} a_n(f, f_s, t)\frac{d^n f}{dt^n} + b_n(f, f_s, t)\frac{d^n f_s}{dt^n} \quad M \geq 0 \tag{6.23}$$

where a_n and b_n are the coefficients, which may be functions of time t, and M is an integer. The third relationship could be linear or nonlinear to represent the relation among the three variables. However, the resulting model [Eqs. (6.20) and (6.23) and a third relationship] may be too general and complicated to be of any practical value (Singh and Xu 1990).

Overton (1964) has shown that several infiltration models are based on a relationship between infiltration rate (or excess infiltration rate) and the volume of either actual or potential infiltration (or excess infiltration). For example, a form of the Green and Ampt model (1911) can be derived by assuming that

$$\frac{dF(t)}{dt} = f(t) = \frac{a}{F(t)} + f_c \tag{6.24}$$

where F = volume of infiltration = $\int_0^t f(t)dt$; f_c = the ultimate infiltration rate; and a = constant of proportionality. Integration of Eq. (6.24) yields a Green and Ampt type model as

$$t = \frac{1}{f_c}\left[F - \frac{a}{f_c}\ln\left(1 + \frac{f_c F}{a}\right)\right] \tag{6.25}$$

Similarly, the Kostiakov (1932), Philip two-term (1957), Horton (1938), and Overton (1964) models can be derived. Table 6.2 summarizes these models and the postulates on which they are based. In the approach followed by Overton, no consideration is given to the continuity Eqs. (6.19) and (6.20).

According to the systems approach, the potential storage $S(t)$ and the seepage rate $f_s(t)$ are postulated for each model. The relationships expressing these postulates are then substituted into Eq. (6.20) to obtain the corresponding infiltration model.

Following Singh and Xu (1990), a summary of eight popular infiltration models is given in Table 6.2 in two ways: (1) the usually reported form in literature (column 6) and (2) the form derived by systems approach (column 5). The potential storage function $S(t)$ and seepage rate function $f_s(t)$ for each model are also given in Table 6.2: a, b, A, S_p, W_0, and n are constants; f_c and K denote steady-state infiltration rate; h is the depth of ponded water on the soil surface (a constant); θ_t is volumetric water content in transmission zone; and S_c is an effective matrix suction head at the wetting front.

From column 2 of Table 6.2, it is seen that $f_s(t)$ is either equal to zero (the Kostiakov model) or equal to the steady-state infiltration rate. Comparing these models with a typical infiltration curve (Fig. 6.1), $s(f)$ actually represents the steady-state flow rate, which is either equal to a constant or zero. As seen from Fig. 6.5, the quantity $[f(t) - f_s(t)]$ actually represents the nonsteady infiltration rate, and this function, according to column 3 of Table 6.2, is either directly proportional to the potential storage $S(t)$ or inversely proportional to the amount of water stored in the soil column, $W(t) = [S_o - S(t)]$.

Singh and Yu (1990) proposed a generalized infiltration model of the following form

$$f(t) - f_s(t) = \frac{a[S(t)]^m}{[S_o - S(t)]^n} \tag{6.26}$$

where a, m, and n = positive real constants. According to Eq. (6.26), $f(t) \to \infty$ at $t = 0$, and $f(t) \to f_s(t)$ at $t = \infty$. The general form of $f_s(t)$ may be proposed as

$$f_s(t) = f_c \tag{6.27}$$

where f_c = a constant. Eqs. (6.20), (6.26), and (6.27) constitute the generalized infiltration model. Seven popular models [Green-Ampt (1911), Kostiakov (1932), Kostiakov (modified), Horton (1938), Philip (1957), Holtan (1961), Overton (1964)] are special cases of this generalized model.

Substituting Eq. (6.26) into Eq. (6.20) yields

$$\frac{dS(t)}{dt} = \frac{-a[S(t)]^m}{[S_o - S(t)]^n} \tag{6.28}$$

Eq. (6.26) can be expressed with the use of Eq. (6.27) as

$$f(t) = f_c + \frac{a[S(t)]^m}{[S_o - S(t)]^n} \tag{6.29}$$

Integrating Eq. (6.28)

$$\int_{S_o}^{S(t)} \frac{[S_o - S(t)]^n}{[S(t)]^m} dS(t) = -at \tag{6.30}$$

Model (1)	$f_s(t) =$ assumed (2)	$[f(t) - f_s(t)] =$ assumed (3)	$s(t) =$ derived from (4)	$f(t) =$ derived from (5)	$f(t) =$ reported from (6)
Horton (1938)	f_c	$[s(t)]/a$	$S_0 e^{-t/a}$ $a > 0$	$f_c + S_{0/ae^{-t/a}}$	$f_c + (f_o - f_c)e^{-t/a}$
Overton (1964)	f_c	$aS(t)^2$ $a > 0$	$1/[1/(s_0) + at]$	$f_c + [a/(1/S_0 + at)^2]$	$f_c \sec^2\left[arstan(bs_0)/\left(\sqrt{f_c/a}\right) - \sqrt{af_c}\,t\right]$
Holton (1961)	f_c	$aS(t)^m$ $a > 0, m > 0$	$\left[s_0^{1/m-} - a(1-m)t\right)^{1/(1-m)}$, $m > 0$	$f_c + a\left[S_0^{1-m} - a(1-m)t\right)^{m/(1-m)}$	No general solution
Philip (1957)	f_c	$a/[s_0 - s(t)]$	$S_0 - \sqrt{2at}$ $a > 0$	$f_c + (a/2t)^{1/2}$	$A + (1/2)s_p t^{-1/2}$
Kostiakov (1932)	o	$a/[s_0 - s(t)]^n$ $n > 0$	$S_0 - [(n+1)at]^{1/(n+1)}$ $a > 0$	$a[(n+1)at]^{-n/(n+1)}$	KAt^{A-1}
Kostiakov (modified)	f_c	$a/[s_0 - s(t)]^n$ $n > 0$	$S_0 - [(n+1)at]^{1/(n+1)}$ $a > 0$	$f_c + a[(n+1)at]^{-n/(n+1)}$	$f_c + KAt^{A-1}$
Green-Ampt (1911)	κ	$[K\theta_t(h + s_c)]/[S_0 - S(t)]$	$S_0 - \sqrt{2K\theta_t(h + s_c)t}$	$K + \{[K\theta_t(h + s_c)]/(2t)\}^{1/2}$	$t^a = (\theta_t/K)\{L - (h + s_c)in$ $(h + s_c + L)/(h + s_c)\}$
Zhao (1981)	$[s_0 - s(t)]/a$	$b/[s_0 - s(t)] - [s_0 - s(t)]/a$	$S_0 - \left[ba\left(1 - e^{-(2/a)t}\right)\right]^{0.5}$ $a > 0,\ b > 0$	$(b/a)^{1/2}\left[1 - e^{-(2/a)t}\right]^{-1/2}$	$(b/a)^{1/2}\{1 - [1 - (w_0^2/ab)] - e^{-(2/a)t}]^{-1/2}$

No explicit form obtained.

Table 6.2 Summary of Some Infiltration Models

If n is an integer, then Eq. (6.30) becomes

$$\int_{S_0}^{S(t)} \sum_{j=0}^{n} (-1)^j \binom{n}{j} S_0^{n-j} [S(t)]^{j-m} \, dS(t) = -at \tag{6.31}$$

or

$$\sum_{j=0}^{n} (-1)^j \binom{n}{j} S_0^{n-j} \int_{S_0}^{S(t)} S^{j-m} dS = -at \tag{6.32}$$

For $m \neq 0$, $n \neq 0$, and $n \neq$ m (n and m are integers), Eqs. (6.31) and (6.32) can be integrated as

$$\sum_{\substack{j=1 \\ j \neq m-1}}^{n} (-1)^j \binom{n}{j} S_0^{n-j} \frac{\left\{[S(t)]^{j-m+1} - S_0^{j-m+1}\right\}}{j-m+1} + (-1)^{m-1} \binom{n}{m-1} S_0^{n-m+1} ln\left[\frac{S(t)}{S_0}\right] = -at \tag{6.33}$$

If m is not an integer, the solution of Eqs. (6.31) and (6.32) is

$$\sum_{j=1}^{n} (-1)^j \binom{n}{j} S_0^{n-j} \frac{\left\{[S(t)]^{j-m+1} - S_0^{j-m+1}\right\}}{j-m+1} = -at \tag{6.34}$$

The generalized infiltration model is given in parametric form by Eqs. (6.33) and (6.29) or by Eqs. (6.34) and (6.29) for n being an integer. For the cases $0 \leq n \leq 2$ and $0 \leq m \leq 2$, solutions of Eqs. (6.29) and (6.30) leading to some special cases of the generalized model are shown in Table 6.2. If $m = 0$ and $n > 0$, then the generalized infiltration model reduces to the Kostiakov model; and when $n = 0$ and $m > 0$, the generalized model becomes the Holtan model. If n is not an integer and $m \neq 0$, then Eq. (6.30) may not have a general analytical solution. If the five parameters, a, m, n, S_0, and f_c are known, then the infiltration rate can be calculated numerically by using Eqs. (6.29) and (6.30) and S as a parameter ($0 \leq S \leq S_0$). Singh (2010a, b) derived these infiltration equations using the entropy theory and showed that parameters of these equations can be determined from three measurable quantities: initial infiltration, steady infiltration, and soil moisture retention capacity. The second and third quantities have been tabulated for most soils. The first quantity is known for dry soils but can be estimated with the use of API.

6.9 Empirical Infiltration Models

A multitude of infiltration models are used in applied hydrology and soil science. Some of these models are theoretically based (Green and Ampt, 1911; Philip, 1957, 1969; Dooge, 1973), while others are empirical or semiempirical (Kostiakov, 1932; Horton, 1938; Holtan, 1961; Overton, 1964). Some of the empirical models are quite popular and frequently used in various applications because they are simple and yield satisfactory results in many cases. Since the empirical models are more or less based on experimental observations, they represent the overall infiltration process. These models compute infiltration capacity rate by using soil parameters and time as inputs. Some popular models are described here.

6.9.1 Kostiakov Model

Kostiakov (1932) proposed a simple infiltration equation

$$F = at^b \tag{6.35}$$

and

$$f = (ab)t^{b-1} \tag{6.36}$$

where parameters $a > 0$ and $0 < b < 1$. These model parameters depend on the soil properties and initial conditions. Initially, f is very large and it tends to zero as time progresses; as $t \to 0, f \to \infty$ and as $t \to \infty, f \to 0$; f is the infiltration capacity, t is the time after infiltration starts. The parameters of Kostiakov equation have no physical meaning and are estimated from the data. Singh (2010a) derived an explicit expression for estimating parameter a in terms of f_s and S.

Example 6.1

Mein and Larson conducted infiltration experiment on a soil and the measurements are given in Table 6.3, which correspond to rainfall rate of 5.56×10^{-3} and initial moisture content of 0.2. The porosity of the soil is 0.518 and saturated hydraulic conductivity is 1.39×10^{-3}. Compute the parameters of the Kostiakov model.

Solution

Taking logarithm of the Kostiakov's equation (6.35)

$$\log F = \log a + b \log t$$

$$F_L = A + B \times t_L$$

Time t (sec)	F (cm)	f (cm/h)	$x = t - t_0$ $= t - 470$	Ln (F)	Ln (t)
472.2	2.62	20.02	2.2	0.9632	0.7885
489.7	2.72	19.54	19.7	1.0006	2.9806
507.5	2.81	19.21	37.5	1.0332	3.6243
531.5	2.94	18.75	61.5	1.0784	4.1190
559.8	3.08	18.32	89.8	1.1249	4.4976
582.3	3.19	17.60	112.3	1.1600	4.7212
609.4	3.32	17.00	139.4	1.2000	4.9373
644.1	4.48	16.70	174.1	1.2470	5.1596
682.9	3.66	16.42	212.9	1.2975	5.3608
714.7	3.80	15.74	244.7	1.3350	5.5000
751.1	3.95	15.43	281.1	1.3737	5.6387
783.9	4.09	14.93	313.9	1.4085	5.7491

TABLE 6.3 Data of Soil Infiltration Experiment

which is a linear equation in log domain with two parameters, A and B, whose value can be determined by the method of least squares. Before fitting the equation, we compute time from a datum, say $t_0 = 470$ sec. For the values given in Eq. (6.35),

$$A = 0.77, \quad B = 0.092$$

To find value of F according to the developed relation for $x = 2.2$ (the first value)

$$F_{LC} = 0.77 + 0.092 \times 2.2 = 0.846.$$

This value is in log domain. The value in original domain

$$F_{LC} = \exp(0.846) = 2.33 \text{ cm}.$$

6.9.2 Horton Equation

Horton (1939) presented the following infiltration equation

$$f_p = f_c + (f_o - f_c)e^{-\beta t} \tag{6.37}$$

where f_o is the maximum infiltration rate at the beginning of the storm at $t = 0$ which gradually reduces to a nearly constant rate f_c as infiltration continues, and β is a soil parameter which controls the reduction of infiltration rate. Parameters f_o and β depend on the initial water content as well as the soil properties. The Horton equation is applicable only when the effective rainfall intensity is greater than f_c. Here also, the model parameters are estimated from measured infiltration data. Singh (2010a) showed that

$$\beta = \frac{(f_o - f_s)}{S} \tag{6.38}$$

6.9.3 Philip Two-Term Method

Philip (1957) derived a model based on the Taylor power series solution. He provided a relationship having two terms that gave the model its name. The relationship given by Philip is

$$i(t) = St^{-0.5}/2 + A \tag{6.39}$$

where i is the infiltration rate (cm/h), t is time for infiltration (hours), S is the sorptivity (the capacity of the soil for capillary uptake and release of water) of soil (cm/h$^{0.5}$), and A is a constant which depends on soil properties and initial and saturated water contents (cm/h). The Philip equation states that at early stage of infiltration when time duration (t) is very small, the equation is dominated by the first term. At this phase, the vertical infiltration progresses almost at the same rate of horizontal infiltration due to the gravity component which is denoted by the second term of equation being negligible. As infiltration progresses, the second term becomes more important until it leads the infiltration process. Philip recommended the use of this two-term model in practical hydrology.

The Philip's model assumes that soil conditions and properties are homogeneous; and the distribution of water content is uniform as well as constant. These assumptions may not be valid in some situations.

6.9.4 Green-Ampt Model

Green and Ampt (1911) proposed an approximate model which was derived using Darcy's law. Their model considered infiltration from a ponded surface into a deep soil which is homogeneous and has uniform initial water content. It was assumed that water enters the soil as piston flow, which results in a sharp wetting front. This front separates wetted soil from dry soil (Fig. 6.6). The Green-Ampt model is

$$f_p = K_s \times \frac{(H_o + S_f + L_F)}{L_F} \tag{6.40}$$

where K_s is the hydraulic conductivity of the transmission zone, H_o is the depth of water that is ponded on the surface, S_f is the effective suction at the wetting front, and L_F is the distance from the surface to the wetting front (see Fig. 6.6). If M is the difference between initial and final volumetric water contents, then $M = \theta_s - \theta_i$. We assume that the cumulative infiltration $F = (\theta_s - \theta_i)L_F = ML_F$. If the depth of ponding at the surface is small, $H \approx 0$, and Eq. (6.40) can be written as

$$f_p = K_s + \frac{K_s M S_f}{F} \tag{6.41}$$

Green and Ampt originally assumed total saturation behind the wetting front (Fig. 6.6). Philip (1954) assumed that the water content θ_s was constant but it may not be equal to the total porosity. Similarly, K_s can be less than the saturated hydraulic conductivity. Substituting $f_p = dF/dt$ in Eq. (6.41) and integrating, with initial condition at $t = 0, F = 0$, yields

$$K_s t = F - S_f M \ln\left(1 + \frac{F}{MS_f}\right) \tag{6.42}$$

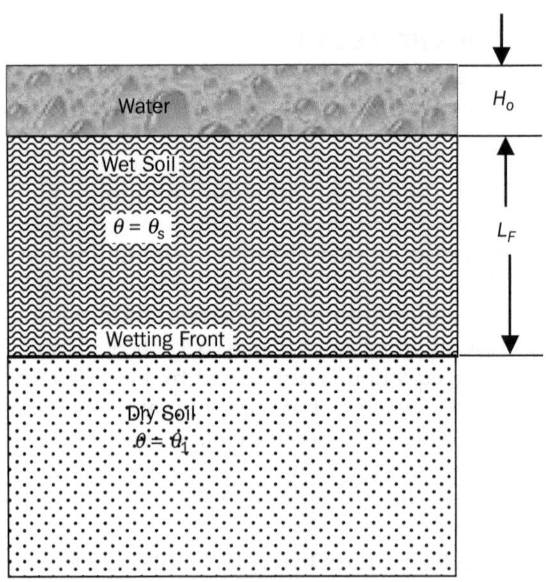

FIGURE 6.6 The Green-Ampt model assumes slug flow with a sharp wetting front between the dry and wet soil.

This equation relates infiltration volume to time from the start of infiltration. For some applications, this form is more convenient. In the derivation of Eq. (6.42), it is assumed that water is ponded so that infiltration rate is equal to soil infiltration capacity at all times. This is not the case for rainfall infiltration and Eq. (6.42) is not applicable for such cases. Singh (2010a) showed that $M = Sf_c$.

6.9.5 Green-Ampt Model for Rainfall Infiltration Case

Although the infiltration capacity-time relationship depends on the application rate, the relationship between infiltration capacity and cumulative infiltration can be assumed to be approximately independent of rainfall intensity (Skaggs and Khaleel, 1982). Mein and Larson (1973) applied the Green-Ampt model for rainfall conditions. They determined cumulative infiltration ($F = F_p$) at the time of surface ponding ($t = t_p$) from Eq. (6.41)

$$f = f_p = R = K_s + K_s MS_{av}/F_p \tag{6.43}$$

where S_{av} is the average suction at the wetting front. Substituting S_{av} for S_f and solving for F_p, Eq. (6.43) may be written as

$$F_p = \frac{S_{av}M}{R/K_s - 1} \tag{6.44}$$

Since $f = R$ prior to surface ponding, $F_p = R\,t_p$, where t_p is the time of surface ponding. Then for steady rainfall the infiltration rate may be expressed as

$$f = R \quad \text{for} \quad t < t_p \tag{6.45}$$

$$f = f_p = K_s + K_s MS_{av}/F_p \quad \text{for } t > t_p \tag{6.46}$$

where $t_p = F_p/R$ and F_p is defined by Eq. (6.44). If $R < K_s$, surface ponding will not occur if the soil profile is deep and homogeneous and $f = R$.

An equation for rainfall infiltration analogous to Eq. (6.42) was obtained by Mein and Larson (1973) expressed as

$$K_s(t - t_p + t_p') = F - MS_{av}\ln\left(1 + \frac{F}{MS_{av}}\right) \tag{6.47}$$

where t_p' is the equivalent time to infiltrate volume F_p under initially ponded surface conditions and may be calculated directly from Eq. (6.42). The Green-Ampt model can be applied by using either Eq. (6.44) or (6.45). Because Eq. (6.46) is implicit in F, it may be desirable to increment F and solve directly for time, t, and then for f from Eq. (6.44).

Example 6.2

This example is for a steady rain case and follows Skaggs and Khaleel (1982). Rainfall at a constant intensity of 0.5 cm/h occurs on a clayey soil which has an initial uniform water content of $\theta_i = 0.20$. Assume no surface depression storage or runoff begins when ponding is initiated. Use the following soil property data: $\theta_s = 0.50$, $K_s = 0.05$ cm/h, and $S_{av} = 21$ cm. Predict time of surface ponding, t_p.

Solution

The soil water deficit $M = \theta_s - \theta_i = 0.50 - 0.20 = 0.30$. Volume of infiltration at the time of surface ponding, F_p, is obtained from Eq. (6.44) as

$$F_p = \frac{S_{av}M}{R/K_s - 1} = \frac{21 \times 0.30}{\frac{0.5}{0.05} - 1} = 0.7 \text{ cm}$$

At $t = t_p$, $f = R$ and $F = Rt_p$. So, $t_p = F_p/R = 0.7/0.5 = 1.4$ h.

To determine the $F(t)$ relationship after surface ponding, Eq. (6.47) can be used. If the surface had been ponded at $t = 0$, the time required to infiltrate $F_p = 0.70$ cm is computed from Eq. (6.42)

$$t'_p = \frac{0.70}{0.05} - \frac{0.30 \times 21.0}{0.05}\ln\left(1 + \frac{0.7}{0.30 \times 21}\right) = 14.0 - 13.275 = 0.725 \text{ h}$$

Now to create table of F and f vs. t, we apply Eq. (6.47)

$$t = t_p - t'_p + F/K_s - \frac{MS_{av}}{K_s}\ln\left(1 + \frac{F}{MS_{av}}\right)$$

So

$$t = 1.4 - 0.725 + F/0.05 - \frac{0.30 \times 21}{0.05}\ln\left(1 + \frac{F}{0.3 \times 21}\right)$$

Hence when $F = 0.70$ cm

$$t = 1.4 - 0.725 + \frac{0.7}{0.05} - 126 \times \ln\left(1 + \frac{0.70}{6.3}\right) = 1.4 \text{ h}$$

From Eq. (6.44), the infiltration rate at this time can be calculated as

$$F = 0.05 + 0.05 \times 0.3 \times 21 / 0.7 = 0.5 \text{ cm/h}$$

By repeating this procedure for various values of F, the infiltration volume and rate can be obtained at different times.

6.10 Applications of Infiltration Theory

Determination of infiltration is needed in many engineering tasks, such as design of highways, design of landfills, irrigation planning, design of airport runways, and hydrological modeling.

6.11 Questions

Data on infiltration in field soils have been published in a report by the Agriculture Research Service of the U.S. Department of Agriculture. One dataset on infiltration in Robertsdale loamy sand in the Georgia Coastal Plain is given in Table 6.4.

Time from Start of Water Application (min)	Infiltration Rate (cm/h)	Accumulated Infiltration (cm)
4	12.21	0.81
5	8.55	0.96
10	5.81	1.60
15	4.89	2.03
20	4.78	2.44
25	4.27	2.82
30	3.86	3.16
35	3.51	3.48
40	3.21	3.75
45	3.23	4.03
50	3.10	4.28
55	3.18	4.56
60	3.21	4.82
65	3.36	5.10
70	3.18	5.34
75	3.31	5.62
80	3.16	5.88
85	3.11	6.14
90	2.89	6.38
95	2.51	6.58
100	2.60	6.82
105	2.68	7.04
110	2.31	7.21
115	2.44	7.43
120	2.42	7.61

Based on Rawls et al. (1976).

TABLE 6.4 Observations on Infiltration in the Robertsdale Loamy Sand (ID=09091D)

6.1 Plot the data on infiltration capacity rate and cumulative infiltration on regular graph paper and plot curves through the data.

6.2 Plot the data on infiltration capacity rate and cumulative infiltration on log-log graph paper and plot the best fit straight line through the data. Express the best-fit straight line equation and show the parameter values. The bet-fit straight line equations represent the Kostiakov equation. Determine the square of the coefficient of correlation and comment on the goodness of fit.

6.3 Determine the time from the graph when the infiltration capacity rate becomes 2.5 cm/h, 2.0 cm/h, 1.0 cm/h, and 0.5 cm/h.

6.4 Plot the infiltration rate data versus the cumulative infiltration and see if there is a relation that can be expressed mathematically.

6.5 Fit the Horton equation to the data in Table 6.3 and show its parameters.

6.6 Fit the Philip-two term equation to the data in Table 6.3 and show its parameters.

6.7 Fit the Green-Ampt equation to the data in Table 6.3 and show its parameters.

6.8 Kostiakov, Horton, Philip, and Green-Ampt equations have been fitted in Questions 6.2 and 6.5 to 6.7. Determine which equation best fits the data and under what criterion?

6.9 Differentiate between soil infiltration capacity and rate.

6.10 Write a short note on soil moisture characteristic curve. What is hysteresis and why it is observed in a soil moisture characteristic curve?

6.11 Write the Richards equation and explain the various terms. Explain how this equation is solved and the difficulties encountered in solving it?

6.12 Derive the generalized form of Singh-Yu infiltration model.

6.12 References

Allen, R. G., L. S. Pereira, D. Raes, and M. Smith. *Crop Evapotranspiration: Guidelines for Computing Crop Requirements.* Irrigation and Drainage Paper No. 56. Rome: Food and Agriculture Organization (FAO), 1998.

Assouline, S., D. Tessier, and A. Bruand. A conceptual model of the soil water retention curve. *Water Resources Research,* 34(2):223–231, 1998.

Brooks, R. H., and A. T. Corey. *Hydraulic Properties of Porous Media.* Hydrology Papers 3. Fort Collins, CO: Colorado State University, 1964.

Campbell, G. S. *A simple method for determining unsaturated conductivity from moisture retention data. Soil Science,* 117(6):311–314, 1974.

Darcy, H. *Les Fontaines Publiques de la Ville de Dijon.* Paris: Dalmont, 1856.

Dooge, J. C. I. *Linear Theory of Hydrologic Systems.* Technical Bulletin No. 1468. Washington, DC: U.S. Department of Agriculture, Agricultural Research Service, 1973.

Green, W. H. and C. A. Ampt. Studies on soil physics, I. Flow of air and water through soils. *Journal of Agricultural Sciences,* 4:1–24, 1911.

Hillel, D. *Soil and Water: Physical Principles and Processes.* New York: Academic Press, 1971.

Holtan, H. N. *ARS 41-51: A Concept of Infiltration Estimates in Watershed Engineering.* Washington, DC: U.S. Department of Agriculture, Agricultural Research Service, 1961.

Horton, R. E. The role of infiltration in the hydrologic cycle. *Transactions of the American Geophysical Union,* 14:446–460, 1933.

Horton, R. E. (1939) Analysis of runoff-plot experiments with varying infiltration capacity. Trans. Amer. Geophys. Union 20, Part IV, 693–711.

Horton, R. I. The interpretation and application of runoff plot experiments with reference to soil erosion problems. *Soil Science Society of America Proceedings,* 3:340–349, 1938.

Kostiakov, A. N. On the dynamics of the coefficient of water percolations in soils. *Sixth Commission, International Society of Soil Science* Part A, 15–21, 1932.

Kulandaiswamy, V. C. A basic study of rainfall excess surface runoff relationship in a basin system. PhD. thesis. Urbana, IL: University of Illinois, 1964.

Mein, R. G., and C. L. Larson. Modeling in filtration during a steady rain. *Water Resources Research,* 9(2):384–304, 1973.

Overton, D. E. *ARS 41-99: Mathematical Refinement of an Infiltration Equation for Watershed Engineering.* Washington, DC: U.S. Department of Agriculture, Agricultural Research Service, 1964.

Philip, J. R. Theory of infiltration, Parts 1 and 4. *Soil Science,* 85(5):345–357, 1957.

Rawls, W., P. Yates, and L. Asmussen. *ARS-S-113: Calibration of Selected Infiltration Equations for the Georgia Plain.* New Orleans, LA: U.S. Department of Agriculture, Agricultural Research Service, 1976.

Rawls, W. J. and D. L. Brakensiek. Estimating soil water retention from soil properties. *ASCE Journal of Irrigation and Drainage,* 108(IR2):166–171, 1982.

Richards, L. A. Capillary conduction of liquids through porous mediums. *Physics,* 1(5):318–333, 1931.

Singh, V. P. Entropy theory for derivation of infiltration equations. *Water Resources Research,* 46: WO3527, doi:10.1029/2009WR008193, 2010a.

Singh, V. P. Tsallis entropy theory for derivation of infiltration equations. *Transactions of the ASABE,* 53(2):447–463, 2010b.

Singh, V. P. and F. X. Yu. Derivation of infiltration equations using systems approach. *Journal of Irrigation and Drainage Engineering,* 116(6):837–858, 1990.

Skaggs, R. V. and R. Khaleel. Infiltration. In *Hydrologic Modeling of Small Watersheds,* edited by C. T. Haan, H. P Johnson, and D. L. Brakensiek. St. Joseph, MI: ASAE Monograph Number 5, American Society of Agricultural Engineers, 1982.

van Genuchten, M. Th. A closed-form equation for predicting the hydraulic conductivity of unsaturated soils. *Soil Science Society of America Journal,* 44:892–898, 1980.

CHAPTER 7

Surface Water

The terms "watershed" or "catchment" represent an area falling between ridges that separate water flowing to different river basins. Ridges of a catchment divide and direct water to a particular river or water body enclosed by them. In this chapter, we discuss the basic concepts of surface water response of a catchment.

7.1 Catchments

A watershed or a catchment is the basic geographical unit within which the hydrological processes, particularly those related with surface water, take place and are studied by hydrologists. Groundwater may flow across the boundaries of a catchment. In other words, the boundaries of surface water and groundwater basins may not always coincide. The two terms, "watershed" and "catchment," convey the same meaning; the use of the term "watershed" is more common in North America; in the United Kingdom and the Commonwealth countries, the term "catchment" is more common. Another term that is used synonymously is "drainage basin," but the term "basin" is commonly used for a large catchment.

All the water that is carried out of a catchment by rivers is known as runoff. This water moves through many different routes on the surface of a catchment (overland flow), or below the surface (interflow or groundwater flow), or the river channel. Streamflow is the discharge or flow rate of water along a river, commonly expressed in m^3/s (cumec) or liter per second. Runoff may be expressed in volume units (m^3) or as depth (mm, cm, or m); these units are easily interchangeable. Streamflow is composed of overland flow, baseflow, and interflow. Some rivers are perennial, meaning that they have flow throughout the year, while some are ephemeral, which means that they remain dry during certain periods.

A basin can be exoreic which means that its water outflows to an ocean. Examples include the Nile River basin and the Amazon River basin. There are many rivers in India that are exoreic. We also have endorheic basins whose water does not flow to an ocean. These are found mostly in arid areas and the main river may end up in a desert or marsh. Examples of such basins are the Luni River basin in India and the Okavango River basin in Africa.

7.1.1 Delineation of Catchments

In earlier times, and still in many cases, catchment areas were delineated manually by the use of topographic maps showing contours. Beginning from the specified outlet of the catchment, one looks around and marks the high hills which can be identified by small close contours. Then, a curve is drawn by moving the pen in the upstream

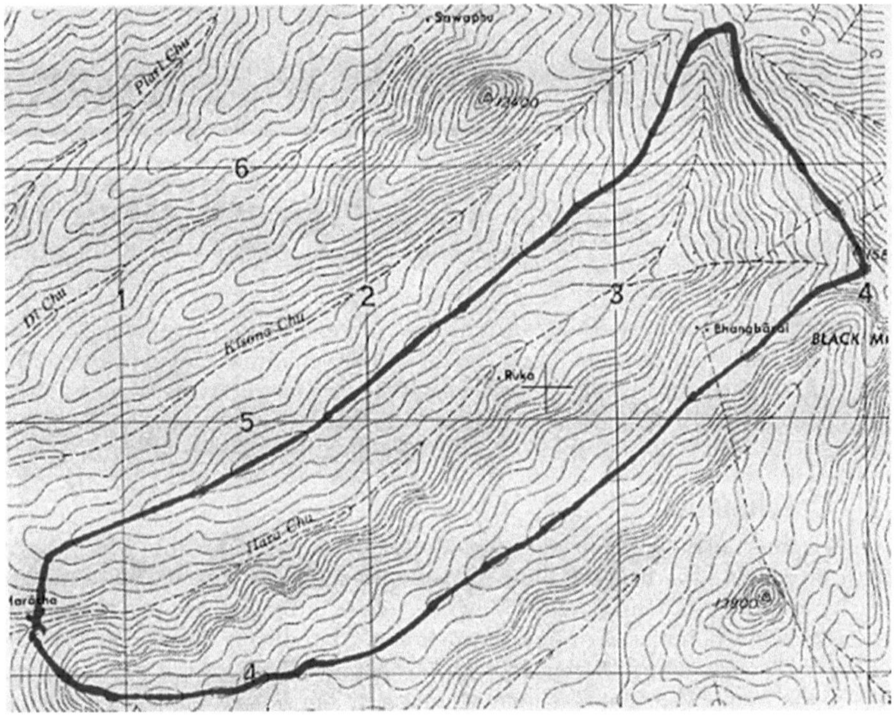

FIGURE 7.1 Contour map of a small area and the catchment delineated. Outlet is at X in the lower left corner.

direction such that the curve cuts the contour lines at right angles. Figure 7.1 shows a contour map and the catchment boundaries drawn on it. Point X in the lower left corner is the catchment outlet.

A digital elevation model (DEM) is a three-dimensional (3D) representation of the Earth's surface. Data required for preparing a DEM is commonly acquired by remote sensing. These days, the common sources of DEM data are Shuttle Radar Topography Mission (SRTM), Advanced Spaceborne Thermal Emission and Reflection Radiometer (ASTER), and Cartosat satellites from India. DEM data from SRTM, ASTER, and many other sources are freely available but they have coarse resolutions. Higher-resolution DEM data can be purchased commercially. For most hydrological applications, public domain data suffice.

Commonly used geographical information system (GIS) software packages, such as ArcGIS (developed by Environmental Systems Research Institute or ESRI) and the Geographical Resources Analysis Support System (GRASS), have tools to automatically demarcate the catchment area. To do this, depressions are to be removed from the DEM. A depression is formed when a group of raster cells in a DEM is completely surrounded by cells of higher elevations. The catchment delineation algorithm is trapped in depressions, since it cannot identify a flow direction out of the depression. Hence it is necessary to remove the depressions before using the DEM for catchment delineation. There are many ways for doing this, for example, by increasing the elevation of cells forming a depression such that there is a flow direction for water to move out. GIS software, such as ArcGIS, have built-in tools (Spatial Analysis Toolbox→Hydrology toolset→Watershed) to remove depressions. In addition to catchment delineation, GIS software can also delineate the river channel network from the DEM data products.

7.1.2 Channel Networks

Geomorphology is the study of Earth's features or landscapes. Morphology of a watershed keeps on evolving due to natural and human-induced causes, albeit over very long time horizons. Water is an important natural agent behind the evolution.

The network of river channels in a catchment is formed by geological and geomorphic processes. There is a dynamic equilibrium between the forces of moving water that try to form a catchment surface and channel network and those resisting the erosion. The shape and size of a river cross section influences the flow of water inside it, which, in turn, influences the cross section. The shape and dimensions of a cross section of a river keep changing with distance, topography, and geology. The properties of cross sections are used in hydrologic modeling and prediction; ecologists use the sections for the evaluation of habitat conditions for aquatic flora and fauna.

Water, sediment, and pollutants enter river channels from nearby areas and move downstream in these channels. As the speed of movement of water and sediment in the channel network is faster than of overland flow, a denser stream network would result in a quicker movement of water, sediment, and pollutants to the catchment outlet. Important characteristics of channel reaches are gradient (slope) and plan geometry. The gradient of a river may be steep (e.g., 1:100), mild (e.g., 1:1000), or flat (e.g., 1:10,000). The categories of planforms include meanders (tortuous, serpentine, confined, and shifting), islands, and braided.

Ordering of river channel networks is important in studies dealing with their evolution and watershed response. One of the earliest schemes for the ordering of river channel networks was developed by Horton in 1945, which was later modified by Strahler (1957). According to the Strahler system, the smallest headwater streams are called first-order streams (Fig. 7.2). The first-order streams have no tributaries. When two streams of the same order join, a new stream of one order higher is created, but no higher order stream is created when a stream is joined by a lower order stream. Hence when two first-order streams join, a second-order stream is created; when two second-order streams join, a third-order stream is created, and so on. But when a second-order stream joins a third-order stream, the resultant stream is also of third-order. The highest-order river in the world is the Amazon, which has a 12-order segment at the mouth. Figure 7.2 shows a stream network of third order and numbering as per the Horton-Strahler scheme.

Horton's law of stream numbers states that there is a geometric relationship between the number of streams of a given order, N_i, and the corresponding order, i. The parameter of this geometric relationship is the bifurcation ratio (Rb), which is the ratio of the number of stream segments of a given order N_n to the number of segments of the next higher-order, N_{n+1}, expressed as

$$Rb = \frac{N_n}{N_{n+1}} \tag{7.1}$$

A higher value of bifurcation ratio indicates more streams, implying a faster response of the catchment. Typically, Rb varies between 2 and 5.

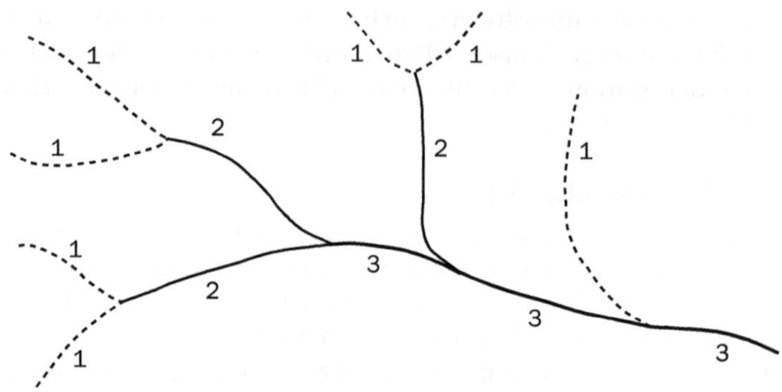

FIGURE 7.2 A stream network of third order and numbering as per the Horton-Strahler scheme. Flow direction is from left to right.

Horton's law of stream lengths states that there is a geometric relation between the average length of streams of a given order ($n + 1$) and of the order n. The stream length ratio Rl is computed as

$$Rl = \frac{L_{n+1}}{L_n} \tag{7.2}$$

where L_n and L_{n+1} are the average stream lengths of streams of order n and $n + 1$, respectively.

Schumm (1956) proposed the law of drainage areas and defined the drainage area ratio Ra as

$$Ra = \frac{A_{n+1}}{A_n} \tag{7.3}$$

where A_n and A_{n+1} are the average drainage areas of streams of order n and ($n + 1$), respectively.

To compute the three ratios, the values of N_i, L_i, and A_i are plotted against stream order on a semi-log graph and the slopes of these lines are Rb, Rl, and Ra, respectively.

Another useful parameter is the drainage density, which is the ratio of the total length of all streams (of all orders) in a catchment to its area, expressed as

$$D = \frac{\sum_{i=1}^{n} L_i}{A} \tag{7.4}$$

Sometimes, difficulties are faced in the determination of a drainage network. In many instances, catchments contain artificial channels, such as stormwater channels, waterways, and drains along highways. These channels are often difficult to locate on maps and are difficult to include in the stream order hierarchy in the natural drainage networks. Maps with scales of 1:50,000 to 1:25,000 are typically used in this analysis.

Boyd (1978) studied drainage basins in Australia and found that the basin lag time followed a law which is similar to the law of basin areas. It can be expressed as

$$L_w = L_1 R_l^{w-1} \tag{7.5}$$

where L_w is the lag time of the basin of order w, L_1 is the lag-time of the first-order basin, and R_l is the basin lag ratio. For the data used by Boyd (1978), R_l was 1.737. Boyd's equation can be helpful in hydrologic design for ungauged basins.

Example 7.1

Use the drainage network laws to determine the bifurcation ratio, stream length ratio, and drainage area ratio for a 730-ha fourth-order watershed. The drainage network characteristics are summarized in Table 7.1.

Solution

The values of *Rb*, *Rl*, and *Ra* for each set of successive stream orders can be obtained using the information in Table 7.1 and Eqs. (7.1), (7.2), and (7.3), respectively. For example, *Rb* for the ratio of stream orders 1 and 2 would simply be 80 divided by 14 or 5.7, respectively. However, what we need to obtain are values of *Rb*, *Rl*, and *Ra* that are representative of the whole network. First, we obtain natural logarithms of the N_n, L_n, and A_n values in Table 7.1.

We now plot the set of N_n, L_n, and A_n values against stream order on a semi-log graph and fit a straight line to each set of values. The slope of the line drawn through the set of N_n value will be b_1, and the slopes of lines drawn through the sets of L_n, and A_n values will be b_2 and b_3, respectively. As an example, the plot for stream lengths is shown in Fig. 7.3 and the value of b_2 is 1.01.

Stream Order, n	Number of Streams, N_n	Average Stream Length, L_n (m)	Average Drainage Area, A_n (ha)	$\text{Log}_e\,(N_n)$	$\text{Log}_e\,(L_n)$ (m)	$\text{Log}_e\,(A_n)$(ha)
1	85	400	5	4.44	5.99	1.61
2	16	1200	29	2.77	7.09	3.37
3	5	3300	150	1.61	8.10	5.01
4	1	8400	815	0.00	9.04	6.70

TABLE 7.1 Drainage Network Characteristics for Example 7.1 and Natural Logarithms of Selected Drainage Network Characteristics

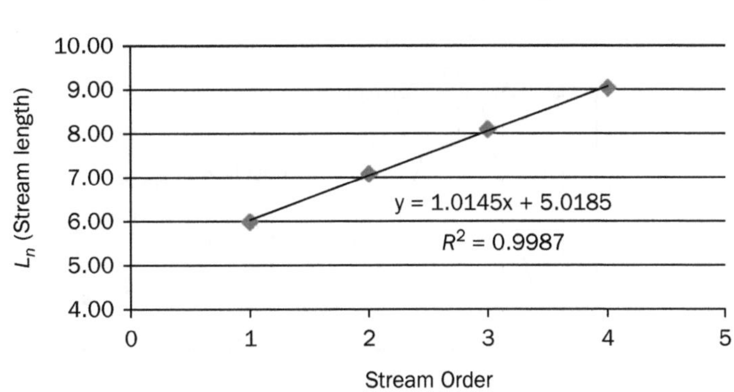

FIGURE 7.3 Plot of stream order versus length.

Thus from the plots, b_1, b_2, and b_3 turn out to be 1.44, 1.01, and 1.69, respectively. Then, by obtaining the antilogarithm (e^{b1}, e^{b2}, and e^{b3}, respectively) of each of these values, we find that *Rb*, *Rl*, and *Ra* are 4.22, 2.75, and 5.42, respectively.

Hence the bifurcation ratio *Rb* is 4.22, the stream length ratio *Rl* is 2.75, and the drainage area ratio *Ra* is 5.42.

7.2 Hydrographs

A streamflow hydrograph is a graph of the time distribution of water discharge at a location. The graph is plotted with discharge on the ordinate (Y-axis) and time on the abscissa (X-axis). A hydrograph for a given storm reflects the influence of all the physical characteristics of the drainage basin and also the characteristics of the storm. A hydrograph can be considered a thumbprint of the drainage basin. The actual shape of a hydrograph is determined by the rate at which water is transmitted from the various parts of the drainage basin to the outlet. Depending upon the topography and geology, most of this water may be carried by the channels, but some water flows overland directly to the outlet. Because of differences in factors governing runoff, such as topography, soil, and land use/cover, two drainage basins will not produce identical hydrographs for the same storm. Similarly, no two storms will produce identical hydrographs from the same basin.

A number of conceptual models are available to describe runoff generation in different catchments or at different locations in the same catchment but at different times. Based on the path taken by water, streamflow may be divided into surface flow, interflow, and baseflow. Figure 7.4 shows the components of a typical hydrograph.

7.2.1 Elements of Hydrographs

The streamflow hydrograph rises from the beginning of the wet season and with increasing catchment area at the gauging site for two reasons: (a) as we move downstream in a basin, the area contributing to flow increases and so does the amount of river flow and (b) the cumulative rainfall increases as the wet season (or storm) progresses and so does the river flow. When the wet or rain season is over, we come across the falling limb of the hydrograph and flow begins to gradually recede. The hydrograph in Fig. 7.4

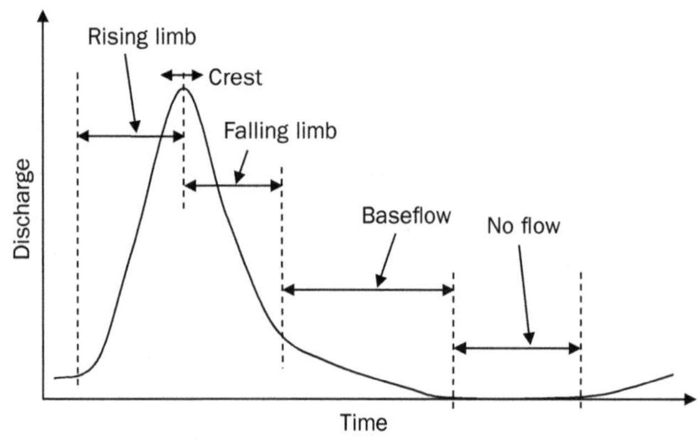

Figure 7.4 Components of a streamflow hydrograph.

corresponds to uniform rainfall and a stationary storm. If rainfall is not uniform, the hydrograph will have kinks. If the storm is moving, the hydrograph shape might be different, depending upon the direction of storm movement.

Rising Limb

As surface runoff reaches the river, the water level in it begins to rise. As rain continues, more and more surface runoff reaches the river from larger areas, and flow and water level continue to rise. After a certain amount of time, when the rain stops or its intensity or areal coverage decreases, a lesser amount of water reaches the channel and the discharge and water level begin to fall. The rising portion of the hydrograph is called the "rising limb."

Crest

The time interval bracketing the highest discharge near the peak of the hydrograph is called the crest. In the case of a sharp peak, the crest will be of short duration, while for a flat peak, the crest segment covers a fairly long time interval. Peaks produced by small catchments are sharp compared to those from large catchments, all other things remaining the same. The crest represents a subjective zone of nearly equal highest discharges. The greatest discharge within the crest is the peak discharge, which is of primary interest in hydrologic design.

Recession Limb

The recession limb is the portion of the hydrograph after the crest segment. It is also known as the falling limb or the recession curve. The recession limb represents decreasing discharge as water drains out from the catchment storage after rainfall stops. The slope of the recession limb indicates the rate at which water is drained from the basin. The lower part of the recession after the inflexion point, which has a much lower slope, is believed to represent groundwater contribution, because water is withdrawn more slowly here.

7.2.2 Hydrograph Time Characteristics

Some critical characteristics and the shape of a hydrograph can be expressed in terms of a few time parameters.

Time to Peak The time to peak of a hydrograph is the time elapsed from the beginning of the effective rainfall to the peak of direct runoff. This time depends on the drainage-basin characteristics, such as shape (elongated, circular, any other), distance from the most upstream point to the outlet, drainage density, channel slope and roughness, and soil characteristics. The time to peak somewhat depends upon the distribution of rainfall over the basin. For a given amount of runoff, hydrograph with a longer time to peak has a lower peak discharge than a hydrograph with shorter time to peak.

Lag Time It can be defined as the time interval between the centroid of effective rainfall and the centroid of the direct runoff hydrograph. Note that this term has many different definitions.

Time of Concentration (TC) It is the time required for a drop of water which falls on the most remote part of the catchment to reach the outlet. It is defined as the time interval between the end of effective rainfall and the point of contraflexture on the recession limb of the direct runoff hydrograph. If a rainfall of sufficient intensity continues for the time of concentration, then the entire drainage basin would be contributing to the hydrograph at this time and the discharge will be the maximum that can occur from a given storm intensity over the catchment. Assuming that rainfall is uniform over the entire catchment, the discharge increases as water from progressively farther distances arrives at the outlet. If the rain continues at the same intensity, at and after the time of concentration, the discharge becomes constant, since the entire basin is contributing to flow.

In a generalized manner, TC can be expressed as a function of a length measure (L_p), a slope measure (S_p), and a coefficient (C_p) reflecting the characteristics of watershed surface

$$T_c = C_p L_p^a S_p^b \tag{7.6}$$

where a and b are exponents that depend on watershed properties (Singh, 1988).

One of the commonly used formulas to compute the time of concentration was developed by Kirpich (1940) for small watersheds in Pennsylvania and Tennessee, expressed as

$$t_c = 0.06628 L^{0.77} / S^{0.385} \tag{7.7}$$

where t_c is the time of concentration in hours, L is the length of the principal watercourse from the basin outlet to the divide in kilometers, and S is the slope between maximum and minimum elevations in meters per meter (Ponce, 1989). The equation assumes uniform rainfall over the catchment. The U.S. Soil Conservation Services developed an equation for TC, expressed as

$$T_c = 27.78 \times 10^{-5} \sum_i \left((LK/S^{-0.5})_i \right) \tag{7.8}$$

where L is the length of overland flow in feet, K is a coefficient indicating land cover, and S is the slope in percent.

7.3 Components of Streamflow

Figure 7.5 shows a streamflow hydrograph and its components. The two main components of runoff are (a) direct runoff and (b) baseflow. Direct runoff can be further divided into surface runoff and quick interflow, whereas baseflow can be divided into delayed interflow and groundwater runoff. The division into quick and delayed interflows is somewhat arbitrary. The volume of total runoff corresponding to a given storm event is determined by including in the streamflow hydrograph all runoff between the baseflow discharge occurring prior to the storm up to the same baseflow discharge after the storm. Pathways of water movement can be verified by isotopic techniques or by employing a process-based hydrologic model.

7.3.1 Surface Runoff

Surface runoff or overland flow is that water which travels over the ground surface to a drainage channel. Most surface runoff flows to first-order channels which

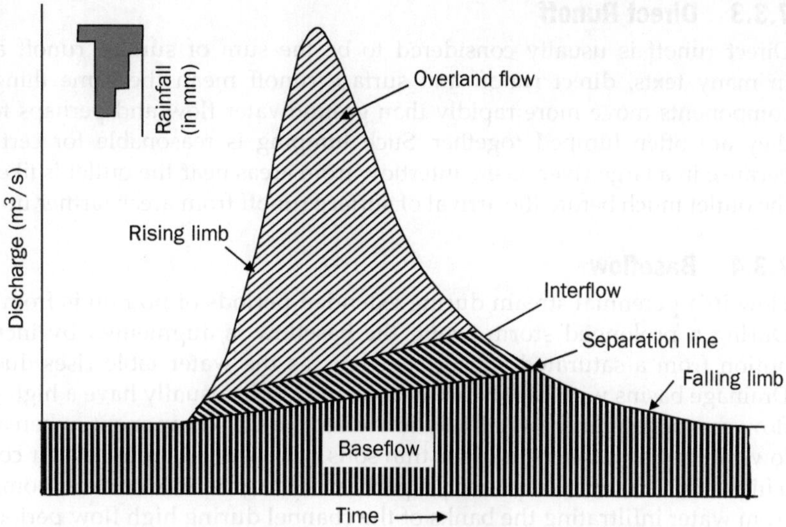

Figure 7.5 A typical streamflow hydrograph.

collectively drain the greatest area of the drainage basin. Surface runoff also includes the precipitation that falls directly on water flowing in the channel. Sheet flow usually occurs from an impervious surface, such as a paved parking lot. In a natural watershed, it can only occur when rainfall intensity uniformly exceeds the infiltration capacity. This is rarely encountered in a natural watershed. Variations in the distribution of soil type and of rainfall, uneven catchment surface, and the presence of vegetation usually result in limited sheet flow. Surface runoff is believed to be the principal contributor to the peak discharge from a storm event. Because this water rapidly runs off over the surface, it is the first to reach the channel and hence forms the rising limb and dominantly contributes to form peak of the hydrograph.

As the catchment area at a gauging site increases, the slope of the rising limb of the surface runoff hydrograph becomes flatter; discharge is typically a power function of area. According to a rule of thumb, in a typical hydrograph, one-third of the volume lies under the rising limb and two-thirds under the falling limb.

7.3.2 Interflow

Interflow, also called "subsurface storm flow," is that water which infiltrates the soil layer and moves laterally beneath the surface to a channel. Interflow can occur on forest floors, where the leaves, twigs, and other debris cover the ground. Interflow might also occur in shallow soils filled and loosened by tree roots, rock debris covering the ground surface, or surface soils loosened by any cause. During the movement as interflow, water is subject to much greater resistance than surface runoff. As a result, interflow does not move as rapidly as surface runoff and hence does not significantly add to the peak discharge; peak interflow reaches the outlet after the peak surface water discharge has passed.

7.3.3 Direct Runoff

Direct runoff is usually considered to be the sum of surface runoff and interflow. In many texts, direct runoff and surface runoff mean the same thing. These flow components move more rapidly than groundwater flow and perhaps for this reason they are often lumped together. Such lumping is reasonable for certain purposes, because in a large river basin, interflow from areas near the outlet is likely to arrive at the outlet much before the arrival of surface runoff from areas farther up in the basin.

7.3.4 Baseflow

Flow in a perennial stream during extended periods of no rain is from the baseflow. During a prolonged storm event, the baseflow is augmented by increased contribution from a saturated subsurface zone, as the water table rises due to recharge. Drainage basins with highly permeable, thick soils usually have a high groundwater-flow component and a relatively low direct-runoff component, whereas basins with low-infiltration capacity and/or thin soils have a small groundwater component and a high direct-runoff component. A portion of the groundwater flow component comes from water infiltrating the banks of the channel during high flow periods.

ASCE (1996) defined "baseflow" (also known as "groundwater flow" or "fair-weather flow") as the runoff that has reached the river by passing first through the underlying aquifer, rather than by flowing directly on the ground surface. Hence baseflow is that part of river flow that is gradually entering the river from groundwater storage. The presence of baseflow throughout the year indicates a humid climate and a shallow water table that is hydraulically connected with the stream. Baseflow is absent in (semi)arid climates and areas of deep groundwater. Contribution of baseflow to streamflow depends upon the hydrogeology of the aquifers. Sources, such as lakes, marshes, snow, and riverbanks, may also supply water for baseflow. At a given time, more than one source may be supporting baseflow. Note that the movement of groundwater in the lateral direction is slower than the vertical direction, since the hydraulic gradient for lateral movement is smaller. The catchment characteristics that affect baseflow are geology, water table position, and soil thickness and properties.

The supply of water from the aquifer to the river continues, as long as there is adequate gradient. As water moves to the stream from higher elevations, the hydraulic gradient decreases and lesser water will travel to the stream with time, unless there is additional infiltration to the aquifer. The gradual reduction of baseflow is called baseflow recession. Many different forms of baseflow recession curves have been proposed. Among these, the most common is the exponential recession, expressed as

$$Q_t = Q_0 k_r^{\,t} \tag{7.9}$$

where Q_0 is the flow at time $t = 0$; Q_t is the flow at time t; and parameter k_r is known as the recession constant. It is a dimensionless quantity which depends upon the unit of time selected. It lies in the interval [0,1] and is normally greater than 0.7. Horton (1933) suggested a nonlinear recession equation, known as Horton's double exponential, expressed as

$$Q_t = Q_0 \exp(-a_2 t^m) \tag{7.10}$$

where a_2 and m are constants.

To determine the recession constant k_r, one can plot Q_{t-1} versus Q_t on a simple graph. This plot will result in a straight line passing through the origin and the slope will be k_r.

The master baseflow recession curve is a composite recession curve, which represents the mean recession behavior. This curve can be constructed by observing discharge at a number of time intervals encompassing the entire dry-weather period. These values are plotted against time on a semi-log paper, and a best-fit straight line is drawn through the plotted points. The resulting curve is the master baseflow recession curve. Of course, the information about the recession variability is lost in the process.

7.4 Catchment Response Mechanisms

To understand how the runoff hydrograph develops in a catchment, we shall first consider a simple case involving a small elemental area.

7.4.1 Hydrograph Development: Small Elemental Area

Consider rain that is falling at a uniform rate on a catchment which is small, impervious, initially dry, and has a uniform mild slope and uniform surface roughness. Due to these properties of the surface, the abstractions from the rainfall due to interception, evaporation, and infiltration will be negligible. As rainfall begins, a thin layer of water (surface detention storage) is formed on the catchment surface. Initially, the outflow from this catchment is zero, but it begins to rise as soon as rain starts and reaches an equilibrium rate at time t_e, which is a function of rainfall rate and catchment area. Curve ① in Fig. 7.6 represents this part of hydrograph development. As the rain continues at a uniform rate, outflow also continues at a constant rate, represented by ②. When the rain stops after the time t_r, the discharge begins to fall rapidly and stops after sometime; the detention storage is rapidly emptied after the cessation of rain. Curve ③ in Fig. 7.6 shows this part of hydrograph. If the rain stops before the discharge attains a plateau, the discharge will begin to fall from that point as shown by curve ④. In Fig. 7.6, the volumes V_1 and V_2 will be equal (Huggins and Burney, 1982) since V_1 is the volume that fills up the detention and catchment storage and V_2 is the water when the same storage is emptied.

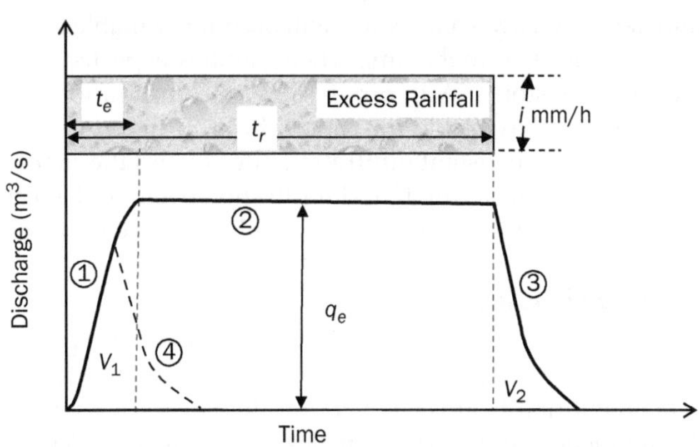

FIGURE 7.6 Hydrologic response of a small, impervious elemental area.

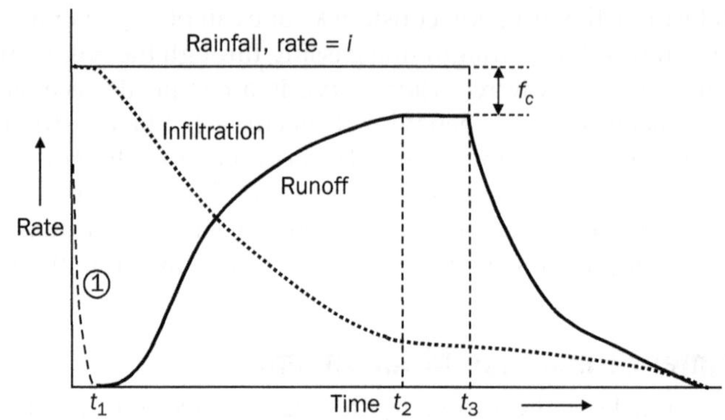

FIGURE 7.7 Hydrologic response of a small natural catchment.

7.4.2 Hydrograph Development: Small Natural Catchment

We shall now look at a small natural catchment with an uneven surface, soil of certain thickness, and vegetation. As a rainstorm begins at time $t = 0$, the interception by vegetation also begins. Since the interception storage is generally very small compared to the usual rainfall amount, this process lasts for a short time. In a vegetated catchment, part of the rainfall drips through the plant canopies (and is known as "throughfall") and some flows along branches/stem (and is known as "stemflow"). Note that evaporation has been ignored in this discussion.

Figure 7.7 shows the change in the magnitude of different components with time and the variation in interception is shown by the dotted curve ① near time zero. The rain that remains after interception loss hits the catchment surface. Some part of it infiltrates into the soil and some rain fills small depressions on the catchment surface. Depending upon the properties of the catchment surface and its saturation level, infiltration demands may be quite high and hence there may not be any runoff, particularly for the storms that fall in the beginning of the wet season. At time t_1, demands for infiltration and surface retention are equal to rainfall rate and thereafter, if the rain continues at sufficient intensity, some water will become available for runoff generation. Note that if the rain stops before the time when runoff is expected to begin, no runoff will be generated from the storm. With progress of time, infiltration rate continues to fall and more water becomes available for runoff. Beyond time t_2, infiltration and runoff rate become constant. The constant infiltration rate is f_c and the runoff rate is $i - f_c$. Rain stops at some time t_3 and hence runoff and infiltration begin to decrease, since progressively less water at the catchment surface is available for these processes.

7.4.3 Overland Flow Generation

As the term suggests, overland flow is the flow that takes place on the catchment surface in the form of thin sheets. This flow is not confined to channels. Surface runoff from a catchment first begins as overland flow and after some distance, it joins small channels. In practice, due to irregularities on catchment surface and vegetation, flow is rarely in the form of thin sheets. We describe here the important processes of overland flow generation in natural catchments.

Soils begin to saturate first at those places where the antecedent soil moisture deficit is least, since the smallest amount of rainfall is needed for saturation at these places (Beven, 2002). Overland flow is generated at all those places in the catchment where the rate of rainfall input at the catchment surface exceeds the infiltration capacity of the soil; this process is called "infiltration excess" or "Hortonian flow." This mechanism is the major reason behind overland flow in catchments in arid or semiarid regions.

On a hillslope, soils near the river channels or valley bottoms will have higher moisture content compared to the soils in the upslope areas as water converges here. As soil infiltration capacities decrease with rising soil moisture, and since water tends to accumulate in the valley bottoms near river channels, these are the areas which first contribute to overland flow. As the rainstorm continues, areas of saturation expand with time, commonly from the valley bottom toward the top of the hillslope. When the rain stops or intensity becomes small, the areas of saturation begin to contract in reverse fashion, starting from the top. Since the soil properties, rainfall (particularly in a convective storm), and snowmelt commonly vary widely in a catchment, the infiltration excess overland flow generation may have a large variation in space. Thus in a natural catchment, all the areas may not contribute to overland flow at a given time and this model of catchment response is known as the *partial area model*. Betson (1964) suggested this concept, which is a generalization of the Horton concept. Since the areas where overland flow is generated in a catchment expand and contract with the rainstorm, this concept is termed as a "dynamic contributing area concept." Areas that receive more rain are likely to become saturated first, all other things remaining the same.

Figure 7.8(a) depicts the situation where rain is falling with intensity, P, which is more than the infiltration rate, f, at some places. Overland flow velocity, soil characteristics, and infiltration rate are important in controlling the partial area response. Consequently, a layer of surface soil becomes saturated and overland flow is generated from such areas; at the given time, only a part of a catchment area may be contributing to overland flow. Further, if runoff generated on one part of a slope flows onto an area of higher infiltration capacity downslope, some of the flow will infiltrate. Beven (2002) noted that if the high- intensity rainstorm that produces overland flow is of short duration, it is likely that the flow generated at a place far from the stream channel will infiltrate before it reaches the nearest channel. Note that the soil profile at a place may also be saturated due to flow coming from the upstream area (on the surface or under the surface) in addition to rainfall.

During a rainstorm, a part of the catchment area may become saturated at the surface. As depicted in Fig. 7.8(b), due to geological set up and soil properties, soil profile in some cases becomes saturated up to the catchment surface. Irrespective of the intensity of rainfall, overland flow will be generated from these saturated areas and this mechanism is known as "saturation excess flow." Saturation rapidly occurs in those areas where soils are thin or have low permeability. In addition to the contribution to runoff from rainfall, surface runoff from such a saturated area may also be due to the return flow of subsurface water. In a given catchment, these different processes may be operating at the same time at different locations.

Surface flow response in some catchments, particularly where soils are deep and have high infiltration capacities, is controlled/dominated by subsurface flows. If subsurface areas have joints and fractures, these can also provide ways for water to flow and get stored. Forested catchments often have deep soils with high infiltration capacities and their response during storms may be governed by subsurface processes.

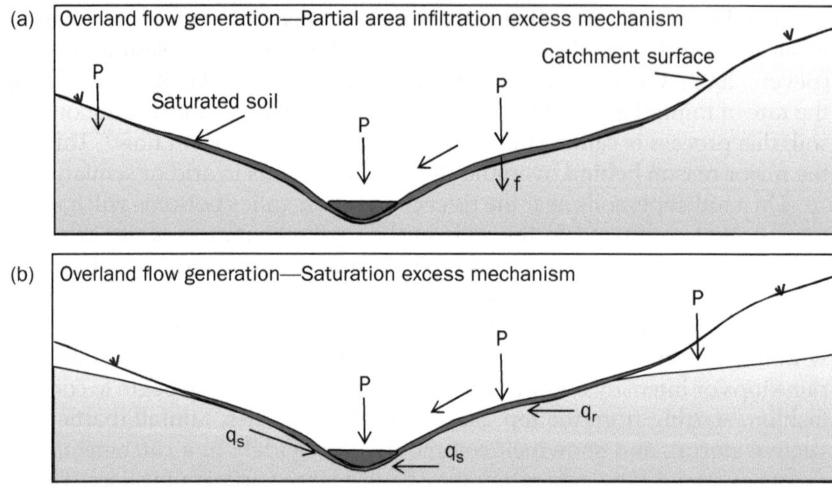

FIGURE 7.8 Two major overland flow generation mechanisms.

Sometimes, flow pathways are found below the surface due to rock fractures, macropores, and so on, and these allow water to rapidly move through the subsurface. When contribution to the river flow through subsurface pathways in a rainstorm is substantial, this is known as subsurface storm flow.

7.5 Factors Affecting Runoff

Runoff from a catchment depends upon a number of factors. Some factors affect spatial variability, while others affect temporal variability. Major factors are discussed next.

7.5.1 Properties of a Rainstorm

A rainstorm may seldom fall uniformly over the entire catchment and the storm may not be stationary. In a typical situation, the storm may be stationary, moving from upstream to downstream, moving from downstream to upstream, or moving laterally. In terms of area coverage of the catchment, the storm may be covering only a small upstream portion of the watershed, or only a small downstream portion of the watershed near the outlet, or only some intermittent portions of the watershed. During a rainfall event, the duration of the storm is also important—its duration in different parts of the catchment may be different. At a given time, rain may be falling on some parts and not on others and so on. The velocity of the storm may be the same as that of flow and faster or slower than the flow velocity. If all these factors are combined, we get a large number of combinations, which determine the hydrological response of a watershed. In a lumped model, all the combinations are merged, since the spatial variability is ignored. However, these can be explicitly considered in a distributed watershed model. Irrespective of the way these factors are considered, it allows us to take a view of the limitations of a particular modeling application.

7.5.2 Topography and Orientation

Topography of a catchment that has evolved over millions of years due to geomorphological processes plays an important role in runoff generation. In conjunction with other factors, topography also determines flow pathways, storages, and types and density of vegetation. These ultimately influence catchment water and sediment yield. In a catchment with steep slopes, water stays on the surface for a shorter time, the opportunity for infiltration is less and runoff is more, if all other things are the same. Steeply sloping catchments also have a shorter time of concentration. The average slope of a river is an important parameter in some empirical methods to determine the time of concentration. If the catchment hillslopes are steep, water swiftly moves over the land, whereas in relatively flat catchments, more water infiltrates and runoff moves slowly. The same behavior is expected for the river network.

The aspect or orientation of a catchment is important particularly in the case of snow- and glacier-fed streams. North-facing hillslopes receive less sunshine in the Northern Hemisphere, remain cooler than the south-facing slopes, and this results in less melt water to the streams from north-facing slopes. Furthermore, the orientation of the valley with respect to wind and storm direction also influences hydrologic processes in a catchment.

Some models explicitly account for the topography of a catchment through topography-based indices. For example, the TOPMODEL (Beven and Kirkby, 1979) uses a topographical index as a measure of flow accumulation at a given point on the catchment surface. It was first defined as a wetness index, expressed as

$$\lambda = \ln \frac{\alpha}{\tan \beta} \tag{7.11}$$

where λ is the topographical wetness index, α is the local upslope area draining through a certain point per unit contour length, and $\tan \beta$ is the local slope in radians. The topographical index controls the flow accumulation and soil moisture.

7.5.3 Size and Shape

The size of a catchment naturally determines the magnitude of runoff generated due to several reasons. The total amount of water received by precipitation usually increases with the size of catchment. The rational formula relates peak discharge from a catchment directly with area. However, in a large catchment, rainfall may have large variations and if the average rainfall decreases as we move downstream, the runoff per unit of catchment area may actually decrease. For example, in Krishna River basin in India, the headwaters region receives very high rainfall which rapidly decreases in the downstream direction. As a result, the runoff generated per unit area decreases with the increase in the catchment area. The time base of the hydrograph typically increases with catchment size.

The shape of the watershed controls the synchronization of flow from various parts to the outlet. Catchments with a circular or fan shape produce runoff at higher rates than those which are elongated, because runoff from different locations will reach the outlet at nearly the same time. Hydrographs from circular catchments will have sharper peaks and a shorter time base. Elongated catchments produce hydrographs with smaller peaks and a longer time base. The shape of a river valley also affects the flow velocity and hence the time to peak.

7.5.4 Soils and Geology

Soil properties chiefly influence runoff in two ways. Infiltration of water depends upon soil properties and thickness of soil profile. The soils which are highly permeable will permit higher infiltration and less runoff. The depth of soil profile determines the soil moisture storage capacity. Deeper soils will absorb more water before attaining saturation than will shallow soils. The properties of soils in a catchment also determine the type of vegetation and crops grown. Land use and vegetation impact catchment response. More vegetation will produce more resistance to overland flow and more infiltration (partly due to plant roots).

Whereas the geographical features, such as mountains and continents, evolve over long time horizons, small-scale features, such as gullies and drainage channels, form over shorter timescales. The main processes that control short-term changes are weathering, erosion, transport, and deposition of sediments. Fluvial geomorphology is the study of the formation of rivers and how they respond to anthropogenic and climate-induced changes.

Geological properties, such as types of rocks, fractures, and faults, also affect the occurrence and movement of subsurface water. Some rock types permit easy storage and movement of water and such basins are rich in groundwater occurrence. The chemical composition of rocks determines the quality of groundwater, as many minerals may get dissolved in water when water passes through these rocks.

7.5.5 Land Use and Land Cover

Land use refers to the purpose the land is serving, for example, agriculture, built-up area, or wildlife habitat. Land cover refers to the type of cover on the ground, for example, vegetation, water, bare soil, or road. The catchment properties measured by remote sensing relate to land cover. From these, land use can be determined by using additional ancillary data.

In terms of hydrologic response, infiltration is small, surface runoff is high, and the hydrograph has a sharp peak. Agricultural areas typically have small slopes and bunds, so water moves slowly. In forested areas, infiltration is high. In forested catchments, the hydrograph shows delayed peaks and baseflow lasts longer.

7.6 Baseflow Separation

Baseflow separation is the process of separating surface runoff from baseflow. Even though such separation is somewhat arbitrary and subjective, it is useful in many analyses. Three ways of separation are discussed below.

7.6.1 Hydrograph Separation

Several techniques have been developed to perform baseflow separation. The area method of baseflow separation is based upon a nonlinear relation between time and area, expressed as

$$N = b\,A^{0.2} \tag{7.12}$$

where A is the drainage-basin area in km²; b is a coefficient, equal to 0.8; and N is the time in days from the hydrograph peak. As shown in Fig. 7.9, a point N distance away (on the x axis) from the peak is marked on the recession limb of the hydrograph. A line

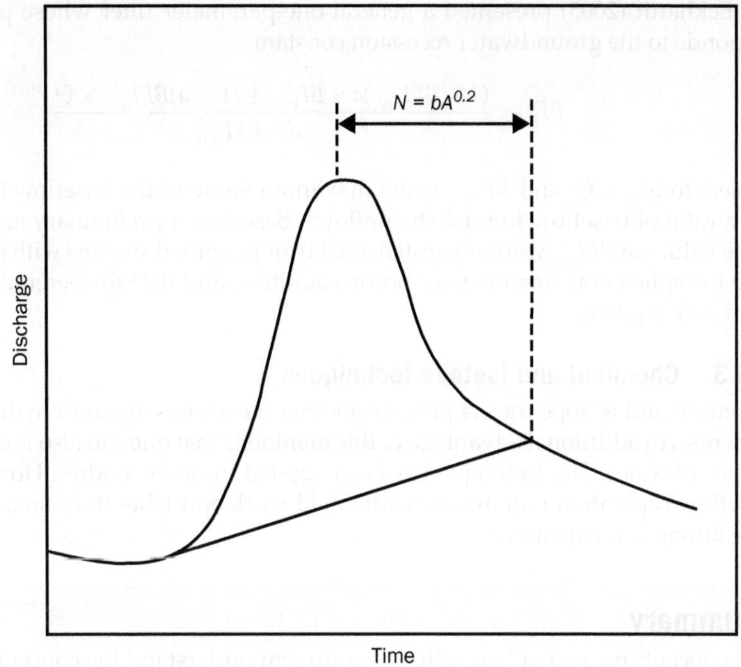

FIGURE 7.9 Baseflow separation by area method.

joining this point with the point where the hydrograph just begins to rise is drawn. All the flow below the line is baseflow. This equation is not suitable for smaller watersheds. It generally gives a longer time base. For example, if $A = 1{,}000$ km^2, then $N = 3.18$ days, that is, if rainfall occurs for 6 hours, its effect will be felt for more than 3 days.

It is convenient to draw a separation line directly from the chosen groundwater discharge on the receding limb to the point under the hydrograph peak. Although this linear separation does not represent the true boundary between direct runoff and groundwater runoff, the error may be acceptable in most cases.

7.6.2 Separation Using Digital Filters

Graphical methods of baseflow separation are inconvenient when processing a long streamflow record, however, digital filters are helpful in such situations. Basically, streamflow is partitioned into two components, expressed as

$$Q(i) = Q_1(i) + Q_h(i) \tag{7.13}$$

The first can be interpreted to be baseflow and the second to be overland and quick interflow.

A filter employing a two-parameter algorithm due to Boughton (1993) assumes that baseflow at the current time is the weighted average of direct runoff and baseflow at the previous time interval. Let $Q_b(i)$ and $Q(i)$ be the baseflow and total streamflow for the time interval i.

Eckhardt (2005) presented a general one-parameter filter whose parameter a corresponds to the groundwater recession constant

$$BF_k = \frac{(1 - BFI_{max})a \times BF_{k-1} + (1 - a)BFI_{max} \times Q_k}{1 - a \times BFI_{max}} \tag{7.14}$$

subject to $BF_k \leq Q_k$ and BFI_{max} is the maximum value of the baseflow index (the long-term ratio of baseflow to total streamflow). Based on a preliminary analysis, approximate values of BFI_{max} were suggested as 0.80 for perennial streams with porous aquifers, 0.50 for ephemeral streams with porous aquifers, and 0.25 for perennial streams with hard rock aquifers.

7.6.3 Chemical and Isotope Techniques

Chemical and isotope tracers provide another means to separate a hydrograph in components. An additional advantage of this method is that one can also identify the spatial source of water. The technique has been applied in many studies. However, its use in baseflow separation requires elaborate field work and laboratory analysis that is time consuming and expensive.

7.7 Summary

The concepts presented here will help a student understand the concept of a catchment and the processes that are responsible for the development of a river flow hydrograph.

7.8 Questions

7.1 What do you understand by the term "catchment" or "watershed"? Which is the biggest watershed in your country and how much is its area?

7.2 Obtain a topographic map of a mountain area and delineate boundaries of a catchment whose area is about 10 km^2.

7.3 Order the catchment in Question 7.2 using the Horton-Strahler ordering scheme.

7.4 The drainage network characteristics for a 1,570-ha fourth-order watershed are summarized in Table 7.2.

Stream Order, n	Number of Streams, N_n	Average Stream Length, L_n (m)	Average Drainage Area, A_n (ha)
1	96	300	4
2	29	1,000	25
3	10	2,500	195
4	1	6,900	909

Table 7.2 Drainage Network Characteristics for Question 7.4

Use the drainage network laws to determine the bifurcation ratio, stream length ratio, and drainage area ratio.

7.5 Obtain a streamflow data from a gauging station on a perennial stream during rainy season. Plot streamflow discharge against time, which will give a streamflow hydrograph. Compute the total volume of flow represented by the hydrograph.

7.6 For the hydrograph in Question 7.5, determine the time to peak, crest period, and time of recession.

7.7 Separate the baseflow for the hydrograph in Question 7.5. Compute the volume of baseflow and of surface flow. Then, compute the percentage of baseflow and surface flow.

7.8 What do you understand by the term "Hortonian flow"?

7.9 Explain the dynamic contributing area concept in the rainfall runoff process.

7.10 Compute the volume of total runoff, volume of direct runoff, and volume of baseflow if the following discharge values are given:

Time (h)	Discharge (m^3/s)
0	15
1	10
2	30
3	50
4	40
5	20
6	15

Use a straight line hydrograph separation method.

7.11 Compute the direct runoff peak discharge, time to peak, and the duration of direct runoff hydrograph for the hydrograph in Question 7.10.

7.9 References

ASCE. *Handbook of Hydrology*. New York: American Society of Civil Engineers, 1996.

Betson, R. P. What is watershed runoff? *Journal of Geophysical Research,* 69(8):1541–1552, 1964.

Beven, K. J. and M. J. Kirkby. A physically based variable contributing area model of catchment hydrology. *Hydrological Science Bulletin,* 24:43–69, 1979.

Beven, K. J. *Rainfall-Runoff Modeling—The Primer*. Chichester: John Wiley and Sons, Ltd., 2002.

Boughton, W. C. *A Hydrograph-Based Model for Estimating the Water Yield of Ungauged Catchments.* Paper presented at *Hydrology and Water Resources Symposium,* Institute of Engineering, Australia, Newcastle, N.S.W., 1993.

Boyd, M. J. A storage-routing model relating drainage basin hydrology and geomorphology. *Water Resources Research,* 14(2):921–928, 1978.

Eckhardt, K. How to construct recursive digital filters for baseflow separation. *Hydrology Process,* 19: 507–515, doi:10.1002/hyp.5675, 2005.

Horton, R. E. The role of infiltration in the hydrologic cycle. *Transactions, American Geophysical Union,* 14:446–460, 1933.

Horton, R. E. Erosional development of streams and their drainage basins: a hydrophysical approach to quantitative morphology. *Geological Society of America Bulletin,* 56:275–370, 1945.

Huggins, L. F. and J. R. Burney. Surface runoff, storage and routing. In *Hydrologic Modeling of Small Watersheds,* edited by C. T. Haan, H. P. Johnson, and D. L. Brakensiek. American Society of Agricultural Engineers, 1982.

Kirpich, Z. P. Time of concentration of small agricultural watersheds. *Civil Engineering,* 10(6):362, 1940.

Ponce, V. M. *Engineering Hydrology: Principles and Practices*. Englewood Cliffs, NJ: Prentice Hall, 1989.

Schumm, S. A. Evolution of drainage systems and slopes in Badlands at Perth Amboy, New Jersey. *Geological Society of America Bulletin,* 67:597–646, 1956.

Strahler, A. N. Quantitative analysis of watershed geomorphology. *Transactions of the American Geophysical Union,* 38:913–920, 1957.

CHAPTER 8

Groundwater

Groundwater is a vital source of freshwater. In many areas, groundwater is the only freshwater available. Excluding glaciers and icecaps, groundwater represents about 95% of the freshwater available in the world. Thus the importance of groundwater can hardly be overemphasized. This chapter discusses rudimentary aspects of groundwater, covering different types of rock formations where groundwater is found, equations governing groundwater movement, well hydraulics, stream-aquifer interaction, and recharge.

8.1 Definitions and Notation

8.1.1 Groundwater

Groundwater is the water that is stored and transported in geologic formations that are unconsolidated porous rocks made up of particles of varying sizes. These formations are located at depth below the land surface and many of these formations comprise alluvial and colluvial deposits.

8.1.2 Aquifers

An aquifer is a geologic formation that is capable of yielding significant quantities of water (Hantush, 1964). In groundwater hydrology, geologic formations, depending on their ability to transmit or yield water, are recognized to be of different types.

A geologic formation may contain sufficient quantities of water, but because of low hydraulic conductivity it may not be suitable for the development of wells or springs. Such a formation is semipermeable and is called an aquitard. It significantly influences the hydraulics of the aquifer below and it yields water slowly in comparison with the adjoining aquifer.

A geologic formation that contains water but is not capable of yielding significant quantities of water under ordinary hydraulic gradients is called an aquiclude. The formation is impermeable. An example of such a formation is clay or shale.

The terms aquifer, aquitard, and aquiclude are often used in a relative sense. For example, a sequence of gravel, sandy clay, and clay can be regarded, respectively, as an aquifer, aquitard, and aquiclude. On the other hand, a sequence of sandy clay and clay at another place may be regarded as an aquifer and aquitard. Similarly, consider two interlayered sequences: silt and clay and sand and silt. In the first sequence, silt is regarded as an aquifer and clay as an aquitard, whereas in the second sequence, sand is an aquifer but silt is an aquiclude. Most geologic formations are classified into aquifers and aquitards.

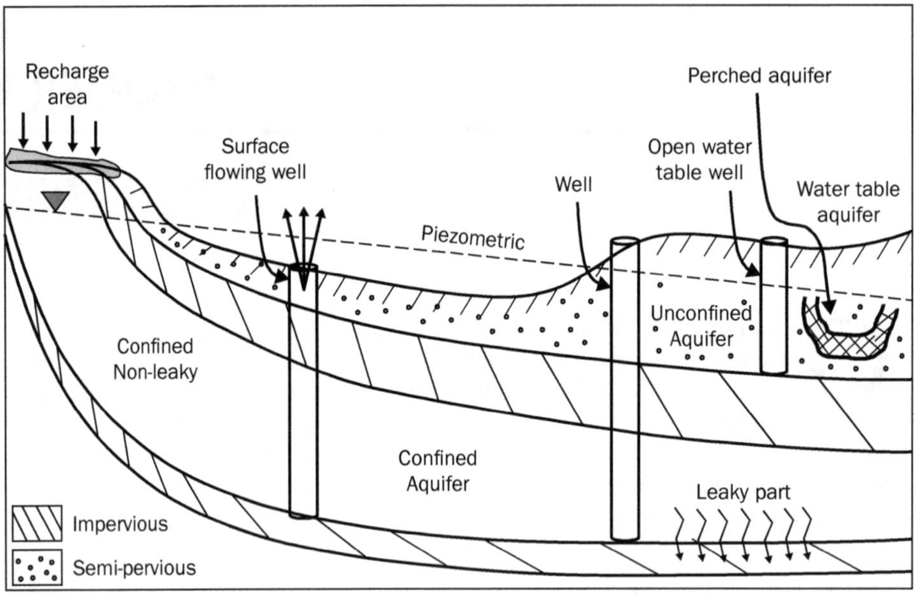

FIGURE 8.1 Confined and unconfined aquifers.

A geologic formation that can neither absorb nor transmit water is called an aquifuge. An example of such formation is solid granite or dense limestone without joints.

8.1.3 Types of Aquifers

Generally there are four types of aquifers: unconfined, confined, perched, and leaky, as shown in Fig. 8.1. An unconfined aquifer is a saturated geologic formation where the upper surface is the water table and is under atmospheric pressure and the bottom surface is a confining layer. An unconfined aquifer is also called phreatic, water table, or free aquifer. Recharge can occur over the entire spatial extent of the aquifer and would increase the volume of water storage in the aquifer. The water table can rise or fall in response to the change in the water storage caused by recharge or pumping.

8.1.4 Confined Aquifer

A confined aquifer is a saturated geologic formation that is confined at the top as well as the bottom by impervious or semipervious strata. The pressure in the formation is greater than the atmospheric pressure and hence the aquifer is also called a pressure aquifer. When a well penetrates such a formation, the piezometric head (or water table) will rise above the bottom of the upper confining layer but may not reach the land surface. The aquifer is called artesian when the piezometric head (water table) rises to the land surface and the well becomes a free flowing well. A confined aquifer has a specific and limited recharge zone. If the piezometric head falls below the upper confining layer, then the confined aquifer becomes an unconfined aquifer. However, its recharge characteristics remain unchanged.

8.1.5 Leaky Aquifer

A leaky aquifer is a geologic formation that loses or gains water through adjacent layers. It can be a confined aquifer or an unconfined one. If a confined aquifer is a leaky one, then it must have at least one semipervious stratum. Likewise, if an unconfined aquifer is a leaky one then its bottom stratum is semi-impervious. A leaky aquifer is also called a semiconfined aquifer. If the permeability of the confining layer is high, the aquifer is called semi-unconfined aquifer.

8.1.6 Perched Aquifer

A perched aquifer is an unconfined aquifer in which the water table rests over a relatively small impervious or semipervious stratum and this water table is above the main water table. If the impervious stratum is leaky, then the aquifer is a semiperched aquifer. The water table of a perched aquifer is termed a perched water table. Such water table occurs over clay lenses in sedimentary deposits. Perched aquifers yield small quantities of water and are often exploited for domestic water supply in rural areas in developing countries.

8.1.7 Water Table

A water table, also called a "phreatic surface," is defined by the surface of a zone of saturation where the pressure is atmospheric.

8.1.8 Phreatic Surface

In a confined aquifer, the piezometric or potentiometric surface coincides with the hydrostatic pressure of water in the aquifer, which is the same as the water level in a well penetrating the aquifer. The piezometric surface and the water table are the same if the potentiometric surface falls below the upper confining stratum.

8.1.9 Groundwater Basin

A watershed or drainage or river basin is a surficial area that drains into a point (often called outlet) on a stream and as this point is moved up or down, the watershed area decreases or increases accordingly. The water moves under gravity on the surface of the watershed. In a similar spirit, a groundwater basin is defined as a physiographic unit comprising one aquifer or more aquifers which may be connected or interrelated. In a physiographic region, the drainage basin and the groundwater basin may or may not coincide. In a mountain valley, they may coincide, but in limestone or sand hill areas, they will not.

8.1.10 Recharge Area

An area that supplies water to a groundwater basin is designated as a recharge area. The recharge area is limited to a specific zone for a groundwater basin containing one or more confined aquifers, whereas for an unconfined aquifer, the entire upper surface may act as a recharge zone.

8.1.11 Porosity

A porous rock or soil is made up of sediment particles or grains and spaces between the particles are called pores, voids, or interstices. Thus the ratio of the volume of interstices

(V_i) to the volume of rock or bulk volume (V_b) is defined as porosity (n). In a saturated rock formation, the pore spaces are all occupied by water, that is, the volume of interstices is the same as the volume of water (V_w). If the volume of solid particles is V_s then $V = V_i + V_s$. Thus, one can express porosity as

$$n = \frac{V_i}{V_b} = \frac{V_w}{V_b} = \frac{V_b - V_s}{V_b} = 1 - \frac{V_s}{V_b} \tag{8.1}$$

The porosity can also be expressed in terms of mass density. Let ρ_b be the mass density of a dry sample (bulk density) and ρ_s be the density of sediment or mineral particles (grain density). Let M_s define the mass of solids and M_a the mass of air. Then, the term V_b can be expressed as M_s/ρ_b, and term V_s as M_s/ρ_s. Equation (8.1) can now be expressed as

$$n = 1 - \frac{M_s/\rho_s}{M_s/\rho_b} = 1 - \frac{\rho_b}{\rho_s} \tag{8.2}$$

Clearly, the volume of pore spaces depends on the size and shape of sediment particles, the degree of compaction, and cementation, arrangement of particles, and the degree of assortment. Typical values of porosity range as 25%–40% for sand and gravel, 35%–50% for silt, and 40%–70% for clay. In rocks, the values can range from 0% to 50%. For example, porosity can range from 5%–50% for fractured basalt and karst limestone, 5% to 30% for sandstone, 0% to 20% for limestone and dolomite, 1% to 10% for shale, 0% to 10% for fractured crystalline rock, and 0% to 5% for dense crystalline rock. Based on porosity, it is hard to distinguish different types of rock formations.

Porosity is of two types: primary and secondary. Primary porosity refers to the original pore or void space when the rock or soil was formed. It is greater in a well-sorted sediment than in a poorly sorted one. Fine grained materials, such as clay, tend to be better sorted than coarse-sorted materials, such as gravel, and hence have higher porosity. Clay has a porosity of 50%, silt 40%, sand 30%, and gravel 25%. These values suggest that as the grain size decreases, porosity increases, but that is not always the case. For example, for a given volume, baseballs or marbles have the same porosity as uniformly packed bowling balls. For perfect spheres, porosity is independent of the size of the spheres. In general, it increases with sphericity and roundness. The highest primary porosity is found in unconsolidated, subangular, well-sorted sediment. Primary porosity is negligible in igneous and sedimentary rocks.

Secondary porosity developed subsequent to the formation of material. Fractures, faults, folds, cavities, and joints occur after formation and they increase secondary porosity. Effective porosity represents that void space where voids are interconnected, for the interconnection is essential for groundwater flow.

Example 8.1

Consider 2 m^3 of sand with a porosity of 0.3 and 3 cubic meters of clay having a porosity of 0.5. Sand and clay are now mixed. Determine the resulting porosity.

Solution

For clay, the volume of pore space is 3 m^3 $\times$ 0.5 = 1.5 m^3 and the volume of particles: 1.5 m^3.

For sand, the volume of pore space is 2 m^3 $\times$ 0.3 = 0.6 m^3 and the volume of solids is 1.4 m^3.

Since sand has larger pore spaces, the volume of clay that will fill the void space within sand is 0.6 m^3.

The volume of clay left over is 3 m^3 – 0.6 m^3 = 2.4 m^3

The total volume of solids is: 1.4 m^3 of sand + 1.5 m^3 of clay = 2.9 m^3

Total volume = 2 m^3 + 2.4 m^3 = 4.4 m^3

$$\text{Therefore } n = \frac{(4.4 - 2.9)\,\text{m}^3}{4.4\,\text{m}^3} = 0.341$$

8.1.12 Void Ratio

The ratio of the volume of voids or pores to the volume of solids is defined as void ratio (r_v). This is related to porosity as

$$r_v = \frac{V_i}{V_s} = \frac{V_w}{V_s} = \frac{V_b - V_s}{V_b} = 1 - \frac{V_s}{V_b} = 1 - \frac{M_s/\rho_s}{M_s/\rho_b} = \frac{n}{1-n} \tag{8.3a}$$

The void ratio is related to the strength of the porous medium. A decrease in the void ratio corresponds to an increase in the consolidation or compaction of the material. This is the reason that it is used frequently in geotechnical engineering. Consider a saturated porous medium of thickness b and initial void ratio r_{vi}. This medium gets consolidated or subsides so its void ratio reduces to r_{vf}. Because of consolidation, the thickness reduces by an amount of Δb. This amount of compaction can then be determined as (Green, 1962)

$$\Delta b = \frac{r_{vi} - r_{vf}}{1 + r_{vi}} b \tag{8.3b}$$

Example 8.2

Compute the void ratio of the sand-clay mixture in Example 8.1.

Solution

The porosity of the mixture computed in Example 8.1 is 0.341. Therefore using Eq. (8.3a), the void ratio is

$$r_v = \frac{0.341}{1 - 0.341} = 0.517$$

Example 8.3

Consider a confined aquifer of thickness 30 m and porosity 35%. Due to pumping an excessive amount of water, the aquifer porosity decreases to 25% and so does its thickness. Determine the amount of reduction in thickness or subsidence.

Solution

First, the initial void ratio and the final void ratio are to be computed, expressed as

$$r_{vi} = \frac{n}{1-n} = \frac{0.35}{1-0.35} = 0.538$$

$$r_{vf} = \frac{n}{1-n} = \frac{0.25}{1-0.25} = 0.333$$

The amount of subsidence is

$$\Delta b = \frac{r_{vi} - r_{vf}}{1 + r_{vi}} b = \frac{0.35 - 0.25}{1 + 0.35} \times 30 = 2.222 \text{ m}$$

The aquifer thickness now reduces to 27.778 m.

8.1.13 Permeability

Permeability is a characteristic of the porous medium, indicating its ability to transmit fluid under energy gradient. It does not depend on the fluid properties and is often expressed as a function of mean grain diameter as

$$k = Cd^2 \tag{8.4}$$

where k (L^2) is the intrinsic or specific permeability, d is the mean grain diameter, and C is a dimensionless constant that depends on the porosity range.

8.1.14 Hydraulic Conductivity

Hydraulic conductivity (K) is one of the most important parameters in groundwater hydrology and is fundamental to determining the movement of fluid in porous media. It is a function of both porous medium properties expressed through permeability and fluid properties such as weight density ($\gamma = \rho g$) and dynamic viscosity (μ) and can be expressed as

$$K = \frac{k\rho g}{\mu} \tag{8.5}$$

It has dimensions of length per unit time (L/T). Since hydraulic conductivity is a proportionality factor in Darcy's law, which will be discussed later, it can be stated that a porous medium has a hydraulic conductivity K of unit value if it transmits in unit time a unit volume through a unit cross-sectional area normal to the direction of flow under a unit hydraulic gradient.

The values of K have a very wide range spanning over 13 orders of magnitude from a very low value of 10^{-13} m/s for metamorphic and igneous rocks to 1 m/s for gravel (Freeze and Cherry, 1979). Also, K may vary from one location to another and from one direction to another in a geologic formation. This variability is defined by anisotropy and heterogeneity.

8.1.15 Anisotropy and Heterogeneity

A geologic formation is homogeneous and isotropic if its hydraulic conductivity is constant throughout the formation, that is, K does not vary with location and with direction. It is heterogeneous and anisotropic if K varies from one point to another in any direction, that is, it varies with direction as well as with location. It is homogeneous but anisotropic if K is constant from location to location but varies from one direction to another at the same location. It is heterogeneous but isotropic if K varies from location to location but is constant from one direction to another at the same location. Real formations are usually heterogeneous and anisotropic. For simplicity they are assumed homogeneous and isotropic.

Transmissivity	(m²/day) Class	Meaning	Supply
≥1500	I	Exceptional	A national resource
≥1000	II	Very high	Of great regional importance
≥100	III	High	Of regional importance
≥10	IV	Intermediate	Of local importance
≥1	V	Low	Of local value
≥0.1	VI	Very low	Limited private supplies
<0.1	VII	Imperceptible	Probably not an aquifer

TABLE 8.1 Krasny's Classification of Transmissivity

8.1.16 Transmissivity

Transmissivity (T) of an aquifer is also called transmissibility or coefficient of transmissivity and is a characteristic of the ability of an aquifer to transmit water. For a confined aquifer of approximately uniform thickness b, transmissivity is defined as the product of hydraulic conductivity K and aquifer thickness b

$$T = bK \tag{8.6}$$

Here, K is the mean hydraulic conductivity value in the vertical section of the aquifer. Equation (8.6) states that transmissivity $T(L^2T^{-1})$ of an aquifer is defined by the rate of transmission of water through its unit width under a unit hydraulic gradient.

Kasenov (1997) presented a useful modification of Krasny's classification of transmissivity by magnitude (Krasny, 1993), shown in Table 8.1.

These values illustrate the significance of T in groundwater development. From pumping test data, the aquifer parameters are calculated and then using the values in Table 8.1, one can get an idea about the potential value of the aquifer.

8.1.17 Specific Yield

The specific yield (S_y) of an aquifer is the ratio of the volume of water that the aquifer will yield (V_y) under gravity to its own volume (V) (Meinzer, 1923), expressed as

$$S_y = \frac{V_y}{V} \tag{8.7}$$

Sometimes, specific yield is also referred to as "effective porosity" and is a dimensionless quantity.

When water is added to an aquifer (e.g., through recharge) or extracted (e.g., through pumping), there will be a change in the volume of the aquifer below the water table. Then, apparent specific yield (S_{ya}) is defined as the ratio of the volume of water added to or extracted from the aquifer to the change in the aquifer volume. Usually, $S_{ya} \leq S_y$. It may be noted that in the field, reasonable results can be obtained by using S_y as 0.2 for gravel and coarse-grained sand and S_y as 0.1 for medium- to fine-grained sand.

8.1.18 Specific Retention

The specific retention (S_r) of an aquifer is the ratio of the volume of water retained (V_r) against the force of gravity to the aquifer volume V, expressed as

$$S_r = \frac{V_r}{V} = n - S_y \tag{8.8}$$

The water will be retained mostly by cohesion and adhesion and tension. For a soil, S_r is also known as field capacity or water holding capacity.

Example 8.4

Consider an unconfined aquifer which is being pumped and the pumping causes a decline of 5 m in the water table. The aquifer has an area of about 1 km². The porosity of the aquifer is 0.45. It is assumed that the specific retention of the aquifer is 0.15. What is the specific yield of this aquifer? What is the change in storage in the aquifer?

Solution

From Eq. (8.8),

$$S_y = 0.45 - 0.15 = 0.30$$

The change in storage is the product of area multiplied by the specific yield and drop in the water table:

Change in storage = 1000 m² × 0.30 × 5 = 1500 m³

8.1.19 Compressibility and Effective Stress

It may be informative to recall the modulus of elasticity, which is defined as the ratio of change in stress to the change in strain: $(d\sigma/d\varepsilon)$, where σ is stress and ε is strain. The inverse of the modulus of elasticity defines compressibility, that is, the ratio of the change in strain to the change in stress.

For the compressibility of water, β, the stress is through pressure (p) and strain is the change in the volume of water. Hence, it can be defined as

$$\beta = -\frac{dV_w/V_w}{dp} \tag{8.9}$$

The value of β is approximately constant around 4.4×10^{-10} m²/N (or Pa⁻¹). For a given mass of water, Eq. (8.9) can also be expressed in terms of fluid density ρ as

$$\beta = -\frac{d\rho/\rho}{dp} \tag{8.10}$$

If $\rho = \rho_0$ at $p = p_0$ as an initial or reference condition, then the solution of Eq. (8.10) becomes

$$\rho = \rho_0 \exp[\beta(p - p_0)] \tag{8.11}$$

If p_0 is atmospheric pressure then Eq. (8.11) can be written as

$$\rho = \rho_0 \exp[\beta p] \tag{8.12}$$

which is referred as the equation of state. Equation (8.12) shows that for an incompressible fluid, $\beta = 0$ and hence, $\rho = \rho_0 =$ constant.

For the compressibility of the porous medium or aquifer, the reduction in volume due to stress is caused by the compression of sediment particles, rearrangement of particles in a more packed configuration, and compression of water in the voids represented through β. The portion of the stress borne by the sediment particles is called effective stress (σ_e). The compressibility of a porous medium (α) can be expressed as

$$\alpha = -\frac{dV/V}{d\sigma_e} \tag{8.13}$$

For a confined aquifer, dV/V can also be expressed as db/b. It can be shown (Freeze and Cherry, 1979) that in the case of aquifers, the total stress remains the same and hence

$$d\sigma_e = -dp = -\rho g dh \tag{8.14}$$

Likewise, $dV = dV_w$. The value of σ can range from 10^{-9} to 10^{-11} for solid rock to 10^{-6} to 10^{-8} for clay (Freeze and Cherry, 1979).

8.1.20 Specific Storage

The specific storage S_s of an aquifer is the volume of water that is released, under a unit decline in the average hydraulic gradient, by a unit volume of an aquifer due to the expansion of water and compression of an aquifer and can be expressed (Hantush, 1964) as

$$S_s = \rho g n \beta \left(1 + \frac{\alpha}{n\beta}\right) \tag{8.15}$$

Equation (8.15) contains two parts. Part $\rho g n \beta$ corresponds to the fraction of storage derived from the expansion of water and part $\rho g \sigma$ corresponds to the fraction of storage derived from the compression of an aquifer.

8.1.21 Storativity

Storativity (S_c), also called "storage coefficient," of an aquifer is the volume of water that a vertical column of a unit cross-sectional area releases or takes in storage under a unit change in the average hydraulic head. For unconfined aquifers, $S_c \approx S_y$ because most of the water is released under gravity and only a small portion is attributed to the expansion of water and compression of the aquifer. The value of S_c ranges from 0.1 to 0.3 for most unconfined aquifers.

For confined aquifers, the release of water is entirely due to the compression of an aquifer and expansion of water. For a confined aquifer of thickness b, we can write

$$S_c = bS_s \tag{8.16a}$$

The value of S_c ranges from 0.005 to 0.00005. A value of 0.0001 or 0.0005 is reasonable for confined aquifers.

For unconsolidated and loosely consolidated geologic formations, the storage coefficient can be expressed as

$$S_c = S_y + bS_s \tag{8.16b}$$

Noting the definition of compressibility of the porous material, from Eqs. (8.13) to (8.16a), the amount of land subsidence (*db*) can be determined as (Lohman, 1972)

$$db = \Delta p\left[\frac{S}{\rho g} - nb\beta\right] \tag{8.16c}$$

Example 8.5

Consider a confined aquifer with an area of 1 km^2 and a storage coefficient of 0.003. What will be the change in storage of the aquifer if the drop in the hydraulic head is 1 m?

Solution

The change in storage is given as

$$\text{Change in storage} = 0.003 \times 1\ \text{m} \times 1000\ \text{m}^2 = 3\ \text{m}^3$$

Example 8.6

Compute the storage coefficient of an aquifer that has a porosity of 0.25, thickness of 100 m, and compressibility of 10^{-8} m^2/N. The value of β can be taken as 4.4×10^{-10} m^2/N.

Solution

From Eqs. (8.15) and (8.16)

$$\begin{aligned} S_c &= b\rho g(\alpha + n\beta) = 100\ \text{m} \times 1000(\text{kg/m}^3) \\ &\quad \times 9.81(\text{m/s}^2)(10^{-8}\ \text{m}^2/\text{N} + 0.25 \times 4.4 \times 10^{-10}) \\ &= 0.00981 \end{aligned}$$

Example 8.7

Consider a confined aquifer with porosity of 0.25, storativity of 0.0025, and thickness of 50 m. Because of pumping, the pressure drops by 5000 N/m^2. Compute the amount of subsidence that the aquifer will experience. Take the value of β as 4.4×10^{-10} m^2/N.

Solution

Using Eq. (8.16c), the amount of subsidence is

$$\begin{aligned} db &= 5000(\text{N/m}^2)\left[\frac{0.0025}{1000(\text{kg/m}^3) \times 9.81(\text{m/s}^2)} - 0.25 \times 50\ \text{m} \times 4.4 \times 10^{-10}(\text{m}^2/\text{N})\right] \\ &= 0.00127\ \text{m} = 0.127\ \text{cm} \end{aligned}$$

Example 8.8

Compute the compressibility of the aquifer having a porosity of 0.45, thickness of 50 m, and storativity of 0.0005. The value of β can be taken as 4.4×10^{-10} m^2/N.

Solution

From Eq. (8.15)

$$\alpha = \frac{S_c}{\rho g} - n\beta = \frac{0.0005}{1000(\text{kg/m}^3) \times 9.81(\text{m/s}^2)} - 0.45 \times 4.4 \times 10^{-10}\ \text{m}^2/\text{N}$$
$$= 5.02 \times 10^{-8}\ \text{m}^2/\text{N}$$

8.1.22 Hydraulic Diffusivity

The hydraulic diffusivity (D) of an aquifer combines transmission as well as storage properties and is defined as

$$D = \frac{T}{S_c} = \frac{K}{S_s} \tag{8.17}$$

8.2 Geologic Formations

Groundwater occurs in geologic formations or rocks. The nature, distribution, and structure of these formations govern the occurrence, movement, quality, volume, and availability of groundwater. From the perspective of groundwater storage, movement, and yield, rocks can be broadly distinguished into three classes: unconsolidated, semiconsolidated, and consolidated.

8.2.1 Unconsolidated Rocks

Unconsolidated rocks include nonindurated sediments of fluvial, aeolian, lacustrine, glacial, and marine origin that comprise boulders, gravel, sand, silt, and clay. These sediments are not bound or hardened by mineral cement. Examples of unconsolidated rocks include alluvial fans and piedmont deposits, valley fills, flood plain deposits, coastal deposits, glacial deposits, and aeolian sediments. Fluvial deposits, also known as alluvial deposits, are laid down by river flow processes. Streams originating from mountainous terrains transport materials varying from boulders to clay, which are deposited in the form of cones or fans in the foothills region. That is the reason for their designation as alluvial fans. The coarsest sediments are usually deposited near the apex of the cone and the finer material near the periphery. During floods, streams overflow their banks and form new channels emanating from the apex. Sediments get stratified away from the apex. The individual strata have different inclinations controlled by stream profiles. Near the apex, individual aquifers merge to form common recharge areas. Most alluvial deposits are coarse grained, having high permeability and specific yield, and the textural composition of these rocks varies widely, resulting in a wide range of porosities, permeabilities, and specific yields. The porosity may vary from 20% to 40%, except for fine-grained sediment, and hydraulic conductivity from 1 m/day to 2000 m/day.

8.2.2 Semiconsolidated Rocks

The difference between semiconsolidated, unconsolidated rocks, and consolidated rocks is in the degree of compaction, cementation, and recrystallization. Sedimentary rocks range from clatic sedimentary rocks, that are composed of silicate, carbonate, or clay minerals, to chemically precipitated sedimentary rocks, that are primarily limestone, dolomite, salt, or gypsum, to coal and lignite, whose original sediment is organic matter. In any stage of consolidation, sedimentary rocks can occur from loose granular materials like silt and sand to firmly held granular material like siltstone and sandstone. They lose their primary porosity but develop secondary porosity due to weathering and fracturing. The thickness of sedimentary rocks can vary from a few meters to thousands of meters extending over thousands of square kilometers.

Among semiconsolidated rocks, sandstone is the most productive rock. Its porosity can be as high as 40%, hydraulic conductivity from less than 1 m/day to 100 m/day, and transmissivity 5000 m^2/day. Shale is formed from the induration and consolidation of clayey sediment. During its evolution, it develops fractures, layers, bedding planes, and weathered zones in which water is found. The shale aquifers are among the least productive aquifers. The hydraulic conductivity is less than 1 m/day.

Limestone ranges from extremely fine grained, dense, massive, and fragmented to fragmental, fossiliferous, or granular types with high porosity, and constitutes about 5%–10% of sedimentary rocks. Limestone comprises predominantly calcium carbonate and some magnesium. It is interesting to note that with increasing dominance of clay, limestone grades into shale, and with the dominance of sand it turns into sandstone. Similarly, with increasing proportion of magnesium it grades into dolomite limestone and then into dolomite. Extensive limestone deposits are marine in origin and the sequence of sediment deposition in the ocean bed follows as: sand to sandstone, clay to shale, and calcareous mud to limestone. Limestone rocks develop caverns and karst formation, depending on the solution activity. They may represent about 5%–10% of sedimentary rocks.

8.2.3 Consolidated Rocks

Consolidated rocks are also called hard rocks and consist of igneous and metamorphic rocks, such as granite, basalt, schist, and indurated sedimentary rocks, including sandstone, shale, and limestone whose grains are held tightly by cementation, compaction, and recrystallization. The process of induration is incomplete in the case of semiconsolidated rocks.

Intrusive igneous rocks and metamorphosed rocks hardly have any primary porosity, if any, but openings develop through fracturing, faulting, or weathering. Fractures develop during cooling of these rocks and these can be nearly vertical or highly angular but can occasionally be horizontal. Anisotropic permeability is one of the characteristics of fractured rocks. Igneous and metamorphic rocks form as a result of cooling and solidification of molten material originating from the Earth's interior. Volcanic rocks include basalt, rhyolite, and tuff. The porosity of consolidated rocks can vary widely from almost a negligible value (e.g., basalt) to over 60% (e.g., pumice).

8.2.4 Geologic Controls

The nature, distribution, and lateral and vertical extent of water-bearing formations are controlled by the lithology, stratigraphy, and structure of the geologic deposits and formations. Lithology is defined by the physical makeup, such as mineral composition, grain size, and grain packing of sediments. Thus the storage and yield properties of the geologic formation are reflected by its lithological characteristics. Stratigraphy describes the chronological order in which rock formations occurred on the Earth's surface, which helps predict the presence or absence of an aquifer below a rock formation. Structural controls, such as cleavages, fractures, faults, folds, and unconformities, are the geometrical properties which are produced by deformation after deposition or crystallization. These are manifested in the displacement, termination, or recurrence of water-bearing formations.

8.2.5 Formations as Aquifers

Groundwater is found in both consolidated and unconsolidated geologic formations but the largest percentage of aquifers in the world is found in sedimentary materials that include coarse unconsolidated materials and hard sedimentary rocks, such as limestone and dolomite. Different types of rock formations have different types of openings, so the groundwater potential varies with the rock type. In general, these openings are (1) openings between individual particles which makeup the primary porosity; (2) crevices, joints, and fractures in hard rocks that makeup secondary porosity and result from breaking; and (3) solution channels and caverns in limestone and openings that result from shrinkage and gas bubbles in lava. Sandstone formations form the best aquifers.

8.3 Groundwater Flow Equations

In groundwater flow, there are two fundamental equations: Darcy's law (flux equation) and law of conservation of mass or continuity equation. These equations are described below.

8.3.1 Darcy's Law

Consider a porous bed of length Δx (or L) and cross-sectional area A, as shown in Fig. 8.2. Darcy's law states that the specific discharge or flux v through the bed is directly proportional to the difference in the piezometric head ($-\Delta h$) and inversely proportional to Δx, where h is the piezometric head. The negative sign with the head difference is because the head is declining along x in the downstream direction. Then

$$v = -K\frac{dh}{dx} \tag{8.18a}$$

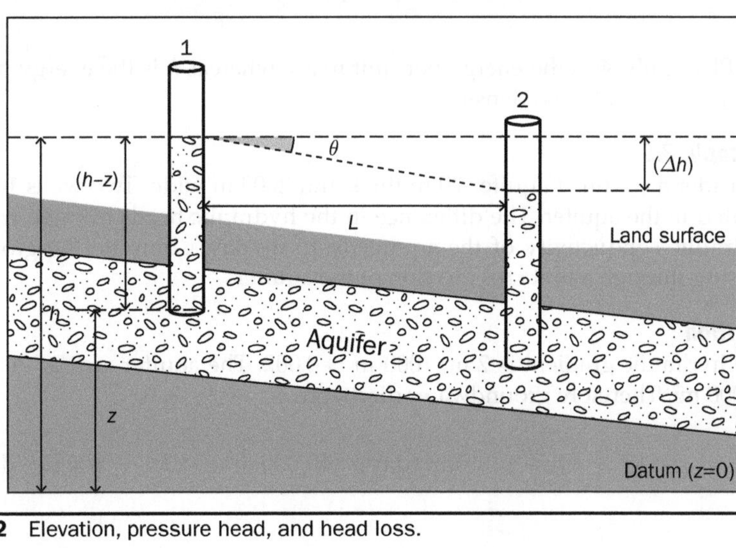

FIGURE 8.2 Elevation, pressure head, and head loss.

where K is the coefficient of proportionality or hydraulic conductivity, and dh/dx is the hydraulic gradient. The specific discharge is also referred to as Darcian velocity or bulk velocity. It may be noted that specific discharge equals the porosity multiplied by the pore velocity (also called seepage velocity), v_p, that is,

$$v_p = \frac{v}{n} = -\frac{K}{n}\frac{dh}{dx} \tag{8.18b}$$

The discharge from the bed can be written as

$$Q = -KA\frac{dh}{dx} \tag{8.19}$$

The piezometric or hydraulic head h can be expressed as

$$h = z + \frac{p}{\gamma} + C \tag{8.20}$$

where z is the elevation above a datum, p is the hydrostatic pressure (piezometric head), and C is an arbitrary constant that can be taken as zero. The hydraulic head is the sum of the elevation head and the pressure head. In theory, the hydraulic head also includes the kinetic energy head, which is given by the square of velocity divided by twice the acceleration due to gravity. Because groundwater velocity is quite low, the kinetic energy head is comparatively negligible. With the substitution of Eq. (8.20), Eq. (8.19) can be expressed as

$$Q = -KA\frac{d}{dx}\left(z + \frac{p}{\gamma}\right) \tag{8.21}$$

Darcy's law is valid in all directions.

The hydraulic head at any point in the porous medium is related to the fluid potential Φ, which can be expressed as the product of the hydraulic head at that point and the acceleration due to gravity, written as

$$\Phi = gh = gz + \frac{p}{\rho} \tag{8.22}$$

Physically, Φ is the energy per unit mass, whereas h is the energy per unit weight, and ρ is the fluid mass density.

Example 8.9

Consider a confined aquifer 20 m thick and 2000 m wide. Two wells 1000 m apart are located in the aquifer. The difference in the hydraulic heads of these wells is 2 m. The hydraulic conductivity of the aquifer is 10 m/day. Compute the groundwater flow passing through a unit cross section of the aquifer.

Solution

The hydraulic gradient is 2 m/1000 m = 0.002. The aquifer area is: 20 m × 2000 m = 40,000 m^2. Therefore groundwater discharge is

$$Q = KA\frac{dh}{dx} = 10(\text{m/day}) \times 40{,}000\ \text{m}^2 \times 0.002 = 800\ \text{m}^3/\text{day}$$

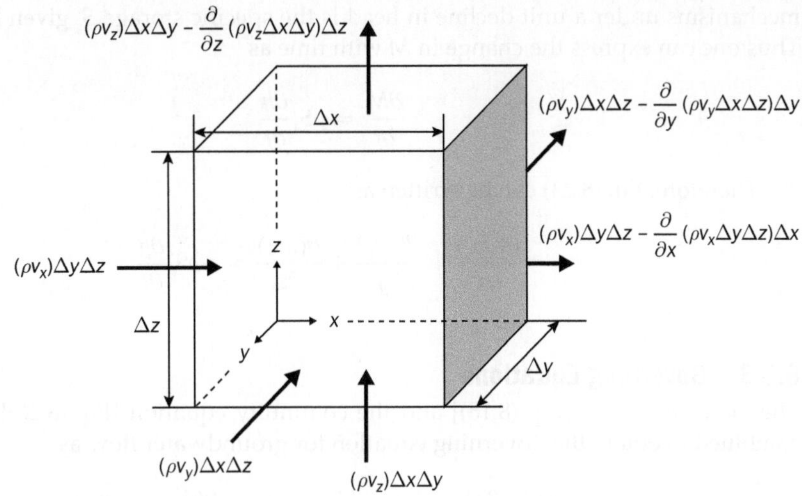

FIGURE 8.3 Volume element.

8.3.2 Continuity Equation

Consider a control volume $\Delta x\ \Delta y\ \Delta z$ as shown in Fig. 8.3. The conservation of mass states that the total mass flux into the control volume minus the total mass flux out of the control volume equals the rate of change of mass in the control volume. The volumetric flow in any direction is the product of the Darcy flux and the area normal to that direction. The net mass flux in the x-direction can be expressed as

$$\text{mass flux out-mass flux in} = \frac{\partial}{\partial x}(\rho Q_x)\Delta x$$

Similarly, the net mass flux can be expressed for the y- and z-directions, as shown in Fig. 8.3. Considering all directions, the continuity equation can be expressed as

$$\frac{\partial(\rho v_x)\Delta y\Delta z\Delta x}{\partial x} + \frac{\partial(\rho v_y)\Delta x\Delta z\Delta y}{\partial y} + \frac{\partial(\rho v_z)\Delta x\Delta y\Delta z}{\partial z} = -\frac{\partial M\Delta x\Delta y\Delta z}{\partial t} \tag{8.23}$$

where M is the fluid mass per unit volume in the control volume. The negative sign on the right side is needed to make the net outflow positive.

The term on the right side needs explanation. Here, $M = \rho n$. Therefore,

$$\frac{\partial M}{\partial t} = n\frac{\partial \rho}{\partial t} + \rho\frac{\partial n}{\partial t} \tag{8.24}$$

The first term on the right side indicates the change in mass due to the change in the density for water, which is controlled by β and the second term arises due to the compaction of the porous medium which is controlled by α. Both these terms result from the change in the hydraulic head h and the volume of water produced due to these two mechanisms under a unit decline in head is the specific storage S_s given by Eq. (8.15). Thus one can express the change in M with time as

$$\frac{\partial M}{\partial t} = \rho S_s \frac{\partial h}{\partial t} \tag{8.25}$$

Therefore, Eq. (8.23) can be written as

$$\frac{\partial(\rho v_x)}{\partial x} + \frac{\partial(\rho v_y)}{\partial y} + \frac{\partial(\rho v_z)}{\partial z} = -\rho S_s \frac{\partial h}{\partial t} \tag{8.26}$$

8.3.3 Governing Equations

The Darcy equation [Eq. (8.18)] and the continuity equation [Eq. (8.26)] can now be combined to obtain the governing equation for groundwater flow as

$$\frac{\partial}{\partial x}\left(K_x \frac{\partial h}{\partial x}\right) + \frac{\partial}{\partial y}\left(K_y \frac{\partial h}{\partial y}\right) + \frac{\partial}{\partial z}\left(K_z \frac{\partial h}{\partial z}\right) = S_s \frac{\partial h}{\partial t} \tag{8.27}$$

Equation (8.27) describes the space-time distribution of piezometric head in an anisotropic heterogeneous porous formation or aquifer.

If the porous medium is isotropic but heterogeneous, Eq. (8.27) becomes

$$\frac{\partial}{\partial x}\left(K \frac{\partial h}{\partial x}\right) + \frac{\partial}{\partial y}\left(K \frac{\partial h}{\partial y}\right) + \frac{\partial}{\partial z}\left(K \frac{\partial h}{\partial z}\right) = S_s \frac{\partial h}{\partial t} \tag{8.28}$$

If the porous medium is homogeneous but anisotropic, then Eq. (8.27) becomes

$$K_x \frac{\partial^2 h}{\partial x^2} + K_y \frac{\partial^2 h}{\partial y^2} + K_z \frac{\partial^2 h}{\partial z^2} = S_s \frac{\partial h}{\partial t} \tag{8.29}$$

If the porous medium is homogeneous and isotropic, then Eq. (8.27) becomes

$$\frac{\partial^2 h}{\partial x^2} + \frac{\partial^2 h}{\partial y^2} + \frac{\partial^2 h}{\partial z^2} = \frac{S_s}{K}\frac{\partial h}{\partial t} = \frac{\rho g(\alpha + n\beta)}{K}\frac{\partial h}{\partial t} \tag{8.30}$$

Equation (8.30) is the same as the diffusion equation. In order to compute the hydraulic head at any point in the flow field at any time, the values of hydrogeological parameters α, n, and K, and fluid parameters β and ρ are needed.

8.3.4 Governing Equations for Confined Aquifers

If a confined aquifer is of constant thickness b and flow is assumed two-dimensional (2D) (x, y), noting that $S_c = bS_s$, then Eq. (8.30) reduces to

$$\frac{\partial^2 h}{\partial x^2} + \frac{\partial^2 h}{\partial y^2} = \frac{bS_c}{bK}\frac{\partial h}{\partial t} = \frac{S_c}{T}\frac{\partial h}{\partial t} \tag{8.31}$$

Equation (8.31) also can be expressed in terms of drawdown $s = h_0 - h$, where h_0 is a reference piezometric head, usually an initial value, as

$$\frac{\partial^2 s}{\partial x^2} + \frac{\partial^2 s}{\partial y^2} = \frac{S_c}{T}\frac{\partial s}{\partial t} \tag{8.32}$$

If the flow field is steady, then Eq. (8.30) becomes

$$\frac{\partial^2 h}{\partial x^2} + \frac{\partial^2 h}{\partial y^2} + \frac{\partial^2 h}{\partial z^2} = 0 \tag{8.33a}$$

Equation (8.33a) is sometimes written simply as

$$\nabla^2 h = 0 \tag{8.33b}$$

where ∇^2 is the Laplacian operator and Eq. (8.33) is called the "Laplace equation."

8.3.5 Governing Equations for Unconfined Aquifers

In unconfined aquifers, water is primarily derived from the drainage of pores, and water expansion and compaction of the porous medium contribute relatively little. If a pumping well draws water from an unconfined aquifer then a cone of depression will form around the well. Most of the pumped water will be due to the dewatering of the aquifer within the cone of depression and the volume of water coming from outside of the cone will be small. This suggests that the change in the aquifer storage can be determined from the change in the volume of cone of depression multiplied by apparent specific yield.

In unconfined flow, Dupuit's theory is often invoked for mathematical simplicity. The theory makes two basic assumptions. First, if the line of seepage or water table has only a small gradient then streamlines are approximately horizontal and equipotential lines are vertical. Second, the slope of the line of seepage will equal the hydraulic gradient.

Consider a 2D (x, y) steady-state flow in a homogeneous isotropic formation. Note that flow will be equal to Darcian velocity times area. The velocity will be given by the Darcy equation and the flow cross-sectional area will be the hydraulic head times the length dimension. For example, consider a one-dimensional (1D) (x-direction) flow. The discharge will be

$$Q_x = -K\frac{\partial h}{\partial x}(hdy) = -\frac{K}{2}\frac{\partial h^2}{\partial x} \tag{8.34}$$

where y is the transverse direction. Following this line, the continuity equation combined with the Darcy equation can now be expressed as

$$K\frac{dx\,dy}{2}\frac{\partial}{\partial x}\left(\frac{\partial h^2}{\partial x}\right) + K\frac{dx\,dy}{2}\frac{\partial}{\partial y}\left(\frac{\partial h^2}{\partial y}\right) = 0 \tag{8.35}$$

Equation (8.35) reduces to

$$\frac{\partial^2 h^2}{\partial x^2} + \frac{\partial^2 h}{\partial y^2} = 0 \tag{8.36}$$

Equation (8.36) is the Laplace equation in h^2.

Now consider the case when the unconfined aquifer is subject to recharge rate R (LT^{-1}). Then, the amount of recharge on the element $dx\,dy$ will be $R\,dx\,dy$. Equation (8.35) can then be modified as

$$K\frac{dx\,dy}{2}\frac{\partial}{\partial x}\left(\frac{\partial h^2}{\partial x}\right) + K\frac{dx\,dy}{2}\frac{\partial}{\partial y}\left(\frac{\partial h^2}{\partial y}\right) + R\,dx\,dy = 0 \tag{8.37}$$

Equation (8.37) reduces to

$$\frac{\partial}{\partial x}\left(\frac{\partial h^2}{\partial x}\right) + \frac{\partial}{\partial y}\left(\frac{\partial h^2}{\partial y}\right) + \frac{2}{K}R = 0 \tag{8.38}$$

Equation (8.38) is also written as

$$\nabla^2 h^2 + \frac{2}{K}R = 0 \tag{8.39}$$

It may be interesting to consider just a 1D case. Equation (8.36) becomes

$$\frac{d^2h^2}{dx^2} = 0 \tag{8.40}$$

Integration of Eq. (8.40) yields

$$\frac{dh^2}{dx} = 2h\frac{dh}{dx} = b \tag{8.41}$$

Integration of Eq. (8.41) yields

$$h^2 = a + bx \tag{8.42}$$

in which a and b are constants which can be obtained from the boundary conditions. Let $h = h_0$ at $x = 0$. Then

$$a = h_0^2 \tag{8.43}$$

In order to determine b, Eq. (8.41) is utilized wherefrom

$$h\frac{dh}{dx} = -\frac{Q}{K} \tag{8.44}$$

Therefore b is given as

$$b = -\frac{2Q}{K} \tag{8.45}$$

Inserting Eqs. (8.43) and (8.45) in Eq. (8.42), the result is

$$h^2 = h_0^2 - 2\frac{Q}{K}x \tag{8.46}$$

Equation (8.46) describes the free surface and is often referred to as "Dupuit's parabola."

Now consider the length of the aquifer as L. At $x = 0$, $h = h_0$. Let there be no seepage surface at $x = L$. Then, Eq. (8.46) gives

$$h_L^2 = h_0^2 - 2\frac{Q}{K}L \tag{8.47}$$

Equation (8.47) yields

$$Q = \frac{K}{2L}(h_0^2 - h_L^2) \tag{8.48a}$$

Equation (8.48) is referred to as the "Dupuit equation."

If a constant recharge rate is considered over the entire length L, then Eq. (8.48a) will become

$$Q = \frac{K}{2L}(h_0^2 - h_L^2) - \frac{RL}{2} \tag{8.48b}$$

Example 8.10

Consider an unconfined aquifer that has a hydraulic conductivity of 10 m/day. The hydraulic head at a point upstream is 20 m and that downstream is 18 m and the distance between the two points is 500 m. Compute the discharge at the downstream point.

Solution

The Dupuit equation can be used here. Using Eq. (8.48b)

$$Q = \frac{K}{2L}(h_0^2 - h_L^2) - \frac{RL}{2} = \frac{10\ \text{m/day}}{2 \times 1000\ \text{m}}(20^2\ \text{m}^2 - 18^2\ \text{m}^2) = 0.38\ \text{m}^3/\text{day/m}$$

Using the Dupuit theory, Eq. (8.28) can be extended for unsteady flow in a 2D (x, y) heterogeneous isotropic formation as

$$\frac{\partial}{\partial x}\left(Kh\frac{\partial h}{\partial x}\right) + \frac{\partial}{\partial y}\left(Kh\frac{\partial h}{\partial y}\right) = S_{ya}\frac{\partial h}{dt} \tag{8.49}$$

where S_{ya} is the apparent specific yield. If the aquifer is homogeneous, then Eq. (8.49) reduces to

$$\frac{\partial}{\partial x}\left(h\frac{\partial h}{\partial x}\right) + \frac{\partial}{\partial y}\left(h\frac{\partial h}{\partial y}\right) = \frac{S_{ya}}{K}\frac{\partial h}{dt} \tag{8.50}$$

Equation (8.50) is known as the "Boussinesq equation."

Example 8.11

An unconfined aquifer 1000 m wide has a thickness of 40 m measured up to the water table. The difference between the water table and the ground surface is 5 m. Its hydraulic conductivity is 10 m/day. Two wells, 200 m apart, are pumping water. The distance from the surface to the water level in well one is 7 m and that in well two is 8 m. Determine the groundwater discharge passing through the section between the wells. Note the section is 1 km wide.

Solution

Here, $h_0 = 35 - 2 = 33$ m and $h_L = 35 - 3 = 32$ m. Using Eq. (8.28), the discharge per unit width is

$$Q = \frac{K}{2L}(h_0^2 - h_L^2) = \frac{10\,(\text{m/day})}{2 \times 200\ \text{m}}(33^2\ \text{m}^2 - 32^2\ \text{m}^2)$$

$$= 0.025/\text{day} \times (1089 - 1024)\text{m}^2 = 1.625\ \text{m}^2/\text{day}$$

$$\text{Discharge } Q = 1.625\ \text{m}^2/\text{day} \times 1000\ \text{m} = 1625\ \text{m}^3/\text{day}$$

Example 8.12

Consider the aquifer in Example 8.11. If the porosity of the aquifer is 0.25, then what will be the average linear groundwater velocity?

Solution

Recalling Eq. (8.18b), the velocity is

$$v = \frac{K}{n}\frac{dh}{dx} = \frac{10\,(\text{m/day})}{0.25}\frac{1\,\text{m}}{200\ \text{m}} = 2\,\text{m/day}$$

8.3.6 Well Hydraulics

Water wells are a common way of extracting groundwater. Wells are holes or shafts, normally vertical, excavated in the Earth in order to extract groundwater by pumping to the surface. A well system is comprised of three components: well structure, pump, and discharge piping. The well comprises an open section and a casing. The open section is a perforated casing or slotted metal screen allowing water to enter and at the same time prevent collapse of the hole. Sometimes, gravel is placed at the bottom of the well casing around the screen. During pumping, water is removed from the aquifer and the water level first drops around the well below its original level. This drop implies the loss of piezometric head due to the resistance offered by the soil to flow. The drop is the highest at the well and the lowest far away. The difference between the original water level and lowered water level is called drawdown and at any time it is different at different locations away from the well and defines the drawdown curve. This curve seems to have a cone-like appearance in three dimensions and hence called the cone of depression. The spatial extent of this depression defines the area of influence of pumping from the well. The cone of depression spreads until equilibrium is reached. One of the key questions is to determine the drawdown curve under various conditions of pumping and recharge. Simple methods are discussed here for determining the drawdown curve.

Steady Unconfined Radial Flow Toward a Well

Consider a well in an unconfined aquifer where the water table is horizontal, the flow is radial, the well fully penetrates the aquifer, and a steady-state condition exists. At any distance r away from the well, the flow toward the well is equal to the flow velocity times the cylindrical area at the section ($2\pi rh$). Thus, the Darcy equation can be expressed as

$$Q = 2\pi rhK\frac{dh}{dr} \tag{8.51}$$

Integration of Eq. (8.51) with the boundary conditions: $h = h_0$ at $r = r_0$, and $h = h_w$ at $r = r_w$, where subscripts w and 0 denote, respectively, the well and far away distance where the piezometric head is the same as the original head, yields

$$Q = \pi K\frac{{h_0}^2 - {h_w}^2}{\ln(r_0/r_w)} \tag{8.52}$$

Equation (8.52) can be expressed for estimating K as

$$K = \frac{Q\ln(r_0/r_w)}{\pi({h_0}^2 - {h_w}^2)} \tag{8.53}$$

Equation (8.52) is satisfactory if the drawdown in the well does not go beyond one half the original aquifer thickness.

Example 8.13

Consider an unconfined aquifer of 30 m thickness and a hydraulic conductivity of 10 m/day. There are two observation wells 30 m and 70 m away from the pumping well. These observation wells have drawdowns, respectively, of 5 m and 4.5 m. Compute the discharge from the pumping well, assuming steady-state flow.

Solution

The hydraulic heads in the two wells are: (30 − 5) = 25 m and (30 − 4.5) = 25.5 m. Equation (8.52) can be used to compute the discharge, written as

$$Q = \pi K\frac{{h_0}^2 - {h_w}^2}{\ln(r_0/r_w)} = 3.14 \times 10\,(\text{m/day})\frac{(25.5^2 - 25^2)\text{m}^2}{\ln(70\,\text{m}/30\,\text{m})} = 935.74\ \text{m}^3/\text{day}$$

Steady Confined Flow Toward a Well

Now consider a well fully penetrating a homogeneous isotropic confined aquifer of thickness b, as shown in Fig. 8.4. The flow is considered radial and it is assumed that the Dupuit assumption holds. The flow to the well can be expressed as

$$Q = Av = (2\pi rb)K\frac{dh}{dr} \tag{8.54}$$

Let the boundary condition be defined as: $h = h_w$ at $r = r_w$ and $h = h_0$ at $r = r_0$, where subscripts w and 0, respectively, denote the well and far away distance. Integration of Eq. (8.54) leads to

$$Q = 2\pi Kb\frac{h_0 - h_w}{\ln(r_0/r_w)} \tag{8.55}$$

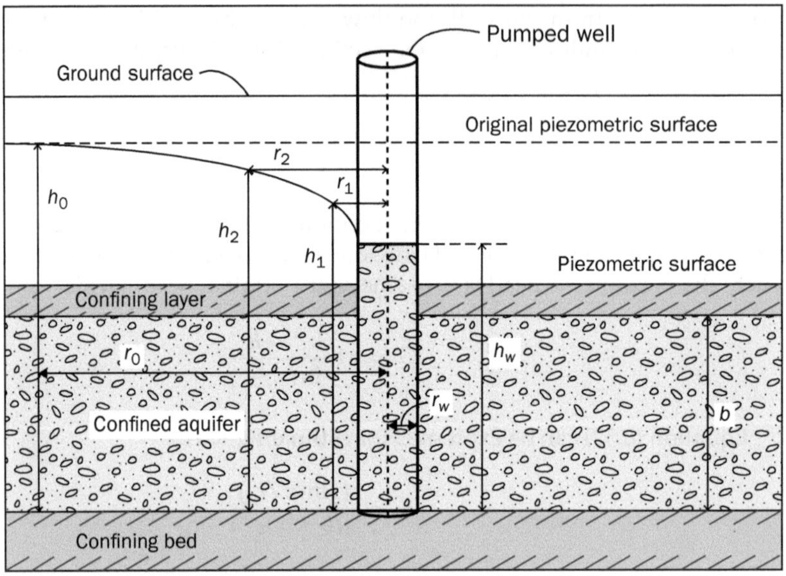

FIGURE 8.4 Radial flow to a well in a confined aquifer.

Knowing the head at any two points (1, 2) at any radial distances, Eq. (8.55) can be used to estimate K as

$$K = \frac{Q}{2\pi b(h_2 - h_1)} \ln\left(\frac{r_2}{r_1}\right) \tag{8.56}$$

If r_0 tends to infinity (assuming the aquifer to be extensive), then integration of Eq. (8.54) would yield

$$Q = 2\pi Kb \frac{h_0 - h_w}{\ln(r/r_w)} \tag{8.57}$$

Equation (8.55) is known as equilibrium or Thiem equation. One can also express the variation of head in terms of radial distance by equating Eqs. (8.55) and (8.57) as

$$h - h_0 = (h_0 - h_w) \frac{\ln(r/r_w)}{\ln(r_0/r_w)} \tag{8.58}$$

Example 8.14

Consider a confined aquifer 30 m thick is having a steady-state pumping rate of 100 m^3/day. The drawdown in one well 15 m away from the pumping well is 2 m and that in the second well 200 m away is 0.5 m. Compute the hydraulic conductivity of the aquifer.

Solution

Equation (8.56) can be used, written as

$$K = \frac{Q}{2\pi b(h_2 - h_1)}\ln\left(\frac{r_2}{r_1}\right) = \frac{100(\mathrm{m}^3/\mathrm{day})}{2\times 3.14\times 30\ \mathrm{m}\times(29.5-28)\ \mathrm{m}}\ln\left(\frac{200\ \mathrm{m}}{15\ \mathrm{m}}\right) = 0.92\ \mathrm{m/day}$$

Well Fields

Consider an isotropic homogeneous unconfined aquifer which has a number of pumping wells. It is assumed that the flow is steady. The wells may be pumped at the same rate or different rates and wells may be located at equal distances or unequal distances. When these wells are being pumped, the cones of depression will overlap, resulting in a composite effect on the water table. To determine the composite effect, let N be the number of wells and the total drawdown at any point be s_T, which will be the sum of drawdowns as

$$s_T = \sum_{i=1}^{N} s_i \tag{8.59}$$

Now consider Eq. (8.52) from which the head at any radial distance from a single well can be written as

$$h_0^2 - h^2 = \frac{Q\ln(r_0/r)}{\pi K} \tag{8.60}$$

Equation (8.60) shows that the drawdown at any radial distance r can be computed for constant Q if h_0, r_0, and K are known. Likewise, Eq. (8.59) shows that the total drawdown will depend on the arrangement of wells and pumping rates. For N pumping wells, let r_i be the radial distance from well i to the point at which the drawdown is to be determined, r_{0i} be the radial distance of the ith well to a location where the water table is almost horizontal or drawdown is negligible, h_0 be the original height of water table, h be the height due to the combined effect of N pumping wells, and Q_i be the flow rate from the ith well. Keeping Eq. (8.59) in mind, one can write

$$h_0^2 - h^2 = \sum_{i=1}^{N}\frac{Q_i}{\pi K}\ln\left(\frac{r_{0i}}{r_i}\right) \tag{8.61}$$

Equation (8.61) should be used when drawdowns are relatively small. The value of r_0 can, in practice, be taken as 500 to 1000 ft (Todd, 1980).

If the aquifer is confined, then for N pumping wells, the drawdown at any point can be computed in a similar manner:

$$h_0 - h = \sum_{i=1}^{N}\frac{Q_i}{2\pi Kb}\ln\left(\frac{r_{0i}}{r_i}\right) \tag{8.62}$$

Unsteady Radial Flow

For unsteady groundwater flow, the piezometric head changes with time, which is often the case in practice. Let the aquifer be of infinite areal extent. The well being pumped fully penetrates the aquifer. Mathematically, the treatment of unsteady flow for confined aquifers is similar to that for unconfined aquifers. Therefore, separate treatments will not be given. The hydraulic head at any radial distance from the pumping well in a confined aquifer can be expressed in polar coordinates as

$$\frac{\partial^2 h}{\partial r^2} + \frac{1}{r}\frac{\partial h}{\partial r} = \frac{S_c}{T}\frac{\partial h}{\partial t} \tag{8.63}$$

Equation (8.63) can be expressed in terms of drawdown ($s = h - h_0$) as

$$\frac{\partial^2 s}{\partial r^2} + \frac{1}{r}\frac{\partial s}{\partial r} = \frac{S_c}{T}\frac{\partial s}{\partial t} \tag{8.64}$$

The initial and boundary conditions for solving Eq. (8.64) can be assumed as

$$s(r,0) = 0;\; s(\infty,t) = 0,\; \lim_{r\to 0} r\frac{\partial s}{\partial r} = -\frac{Q}{2\pi T} \tag{8.65a, b, c}$$

Equation (8.64) can be solved using the Boltzmann variable, written as

$$u = \frac{r^2 S_c}{4Tt} = \frac{r^2}{4Dt} \tag{8.66}$$

Equations (8.64) and (8.65) can be written with the use of Eq. (8.66) as

$$\frac{d^2 s}{du^2} + \left(\frac{1}{u} + 1\right)\frac{ds}{du} = 0 \tag{8.67}$$

$$\lim_{r\to 0} u\frac{ds}{du} = -\frac{Q}{4\pi T} \tag{8.68}$$

Solution of Eq. (8.67), subject to Eq. (8.68), follows

$$s = \frac{Q}{4\pi T}\int_u^{\infty} \exp(-x)dx \tag{8.69}$$

The exponential integral in Eq. (8.69) is known as the Theis function or well function (Theis, 1935) and is often symbolically written as $W(u)$. Then, Eq. (8.69) can be written as

$$s = \frac{Q}{4\pi T}W(u) \text{ or } T = \frac{Q}{4\pi s}W(u) \tag{8.70a}$$

It may also be noted that if u and $W(u)$ are known, then T and S can be determined as

$$T = \frac{Q}{4\pi s}W(u);\; S = \frac{4Ttu}{r^2} \tag{8.70b}$$

Equation (8.70b) can be employed to the limit of the cone of depression, denoted as R, at which point the drawdown is theoretically zero but for practical purposes it can be taken as a small value of about 2.5–3 mm.

The values of well function for practical ranges of u have been tabulated and are available in textbooks. The well function can be computed from its expansion as

$$W(u) = -0.577216 - \ln u + u - \frac{u^2}{2\times 2!} + \frac{u^3}{3\times 3!} - \frac{u^4}{4\times 4!} + \ldots \quad (8.71)$$

Looking at Eq. (8.70), a log-log plot of u (on abscissa) versus $W(u)$ (on ordinate) can be constructed and this plot is known as a "type curve." Likewise, looking at Eq. (8.66), a log-log plot of r^2/t (abscissa) versus s (on ordinate) can be constructed. Keeping the axes parallel, the two curves are then superimposed and moved until segments match. Then, a coincident point is selected on the matched curves where the type curve provides the values of u and $W(u)$ for the desired point. The corresponding values of s and r^2/t are obtained from the plot of observed data. Then, the values of T and S_c are obtained from Eqs. (8.66) and (8.70).

If u is small, $u \le 0.01$ (for a large time at a given distance), then the well function $W(u)$ can be approximated by the first two terms (Cooper and Jacob, 1946; Jacob, 1949) in Eq. (8.71). Then, the drawdown can be determined as

$$s = \frac{Q}{4\pi T}(-0.577216 - \ln u) = \frac{Q}{4\pi T}\ln\frac{2.25Dt}{r^2} \quad (8.72a)$$

This is often referred to as the Jacob method. Equation (8.72a) can also be expressed as

$$s = \frac{Q}{2\pi T}\ln\frac{1.5(Dt)^{0.5}}{r} \quad (8.72b)$$

If s tends to 0 at $r = R$, then from Eq. (8.72b)

$$R = 1.5(Dt)^{0.5} = \left(\frac{2.25Tt}{S}\right)^{0.5} \quad (8.73)$$

For very large times, the difference in drawdowns at two different points (1, 2) can be computed from Eq. (8.73) as

$$s_2 - s_1 = \frac{Q}{2\pi T}\ln\frac{r_1}{r_2} \quad (8.74)$$

Equation (8.75) is for a steady-state condition.

The Cooper-Jacob simplification yields for time drawdown, expressed as

$$T = \frac{2.3Q}{4\pi\Delta s}; \; S = \frac{2.25Tt_0}{r^2} \quad (8.75a)$$

where t_0 is the time corresponding to zero drawdown on the semilog plot of drawdown versus time, and Δs is the difference in drawdown. Likewise, for distance drawdown

$$T = \frac{2.3Q}{4\pi\Delta s}; \; S = \frac{2.25Tt}{r_0^2} \quad (8.75b)$$

where r_0 is the distance when drawdown becomes zero on the semilog plot of drawdown versus distance.

Example 8.15

The following pump test data is available from McWhorter and Sunada (1977) for a confined aquifer near Fort Collins, Colorado (Table 8.2). Compute the transmissivity and storativity of the aquifer using the Theis curve method. The discharge was 1.872 m^3/min and time was 502 minutes.

Solution

The solution is obtained using the following steps.

(1) The drawdown curve [(r^2/t) versus s] is plotted on the logarithmic paper; (2) the Theis curve is then superimposed on the drawdown curve; (3) the match point is located and its coordinates are determined, as shown in Fig. 8.5; and

(4) from the match point, the coordinates are

$$W(u) = 1.0; u = 0.01$$

$$s = 0.19 \text{ m}; r^2/t = 0.46 \text{ m}^2/\text{min}$$

Distance (m)	Drawdown (m)	r^2/t (m^2/min)
4.5	1.35	0.042
10.4	0.94	0.215
20.0	0.65	0.797
34.5	0.41	2.371
65.8	0.20	8.625
91.8	0.14	16.787

TABLE 8.2 Pumping Test Data

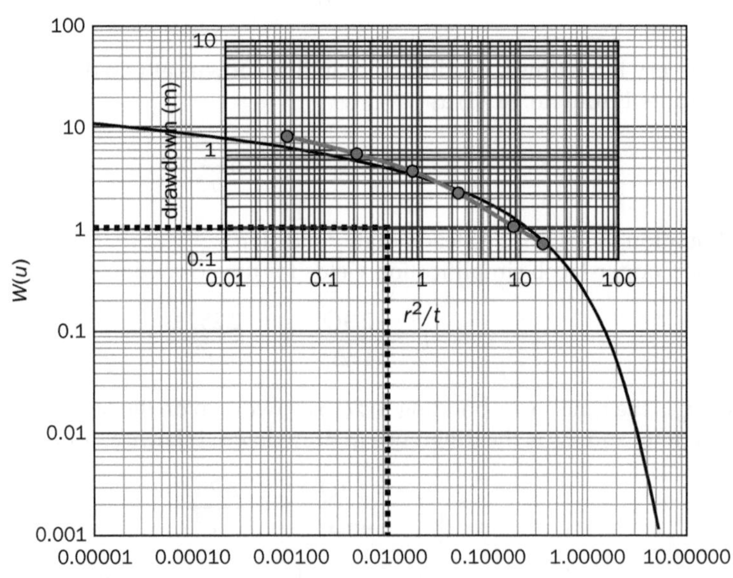

FIGURE 8.5 Match point for determining storativity and transmissivity.

(5) Transmissivity T is computed as

$$T = \frac{Q}{4\pi s}W(u) = \frac{1.872\,\text{m}^3/\text{min}}{4 \times 3.14 \times 0.19\,\text{m}} = 0.784\,\text{m}^2/\text{min}$$

(6) Storativity is computed as

$$u = \frac{r^2 S_c}{4Tt} = \frac{r^2}{4Dt}S_c = \frac{4Tu}{r^2/t} = \frac{4 \times 0.784(\text{m}^2/\text{min}) \times 0.01}{0.46\,\text{m}^2/\text{min}} = 0.068$$

Example 8.16

For a given set of drawdown data in an observation well, it is found that for the match point, the value of $u = 2.8$, $W(u) = 0.02$, $T = 140$ m²/day, $S = 0.00025$, and $t = 0.17$ day. The discharge is 135 m³/h. Assuming a drawdown of 0.25 cm at the limit of the cone of depression, compute the limit of the cone of depression.

Solution

From Eq. (8.70b), the limit of cone of depression can be computed as

$$R = \left[\frac{4Ttu}{S}\right]^{0.5} = \left[\frac{4 \times 140(\text{m}^2/\text{day}) \times 0.17(\text{day}) \times 2.8}{0.00025}\right]^{0.5} = 1032.59\ \text{m}$$

Example 8.17

Determine the drawdown at a distance of 1000 m away from the pumping well. The aquifer parameters are known as: $Q = 135$ m³/h, $T = 140$ m²/day, $S = 0.00025$, and $t = 0.17$ day.

Solution

First the value of u is computed as

$$u = \frac{r^2 S}{4Tt} = \frac{(1{,}000)^2\,\text{m}^2 \times 0.00025}{4 \times 140(\text{m}^2/\text{day}) \times 0.17\ \text{day}} = 2.63$$

Second, for this value of u, the well function $W(u)$ is obtained from the graph (Fig. 8.6), $W(u) = 0.018$.

Third, the drawdown at $r = 1000$ m is computed as

$$s = \frac{QW(u)}{4\pi T} = \frac{135(\text{m}^3/\text{h}) \times (0.018)}{4 \times 3.14 \times 140(\text{m}^2/\text{day}) \times (1/24)} = 0.033\ \text{m}$$

Example 8.18

An aquifer, having $T = 140$ m²/day and $S = 0.00025$, is being pumped at a constant rate. A drawdown of 1 m is observed at a distance of 100 m from the pumped well after pumping for 5 hours. Compute the rate of pumping.

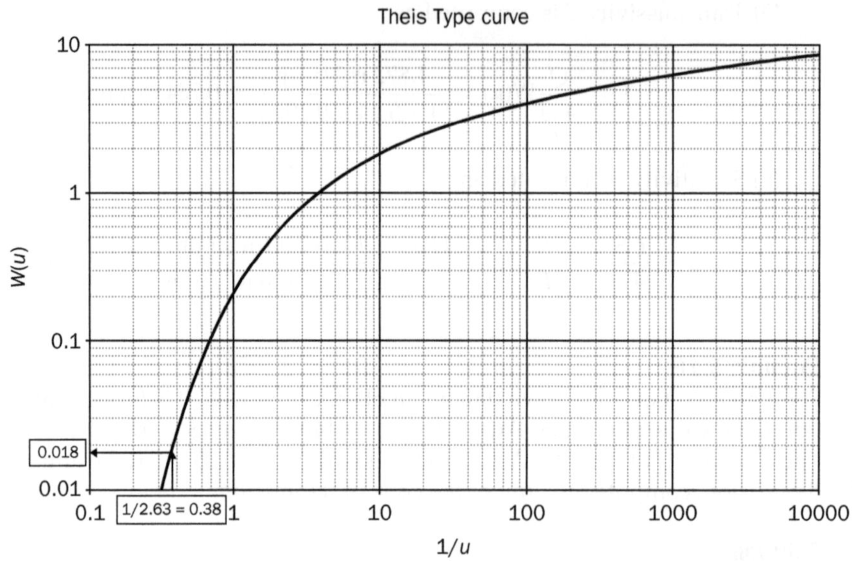

FIGURE 8.6 Theis curve for Example 8.17.

Solution

First, the value of u is computed as

$$u = \frac{r^2 S}{4Tt} = \frac{(100)^2\ \text{m}^2 \times 0.00025}{4 \times 140(\text{m}^2/\text{day}) \times 5\ \text{hours}/24\ \text{hours}/\text{day}} = 0.02$$

Second, the value of well function is obtained from the well function graph (Fig. 8.7), $W(u) = 3.1$

Third, rate of pumping is computed as

$$Q = \frac{4\pi Ts}{W(u)} = \frac{4 \times 3.14 \times 140\ \text{m}^2/\text{day} \times 1\ \text{m}}{(3.1)} = 567.23\ \text{m}^3/\text{day}$$

Example 8.19

An aquifer, having $T = 140\ \text{m}^2/\text{day}$ and $S = 0.00025$, is being pumped at a constant rate of $Q = 135\ \text{m}^3/\text{h}$. A drawdown of 1 m is observed at a distance of 100 m from the pumping well after pumping. Compute the time of pumping.

Solution

First, the well function is computed as

$$W(u) = \frac{4\pi Ts}{Q} = \frac{4 \times 3.14 \times 140(\text{m}^2/\text{day}) \times 1\ \text{m}}{135(\text{m}^3/\text{h}) \times (24\ \text{h}/1\ \text{day})} = 0.54$$

From the well function graph (Fig. 8.8), the value of $u = 0.5$.

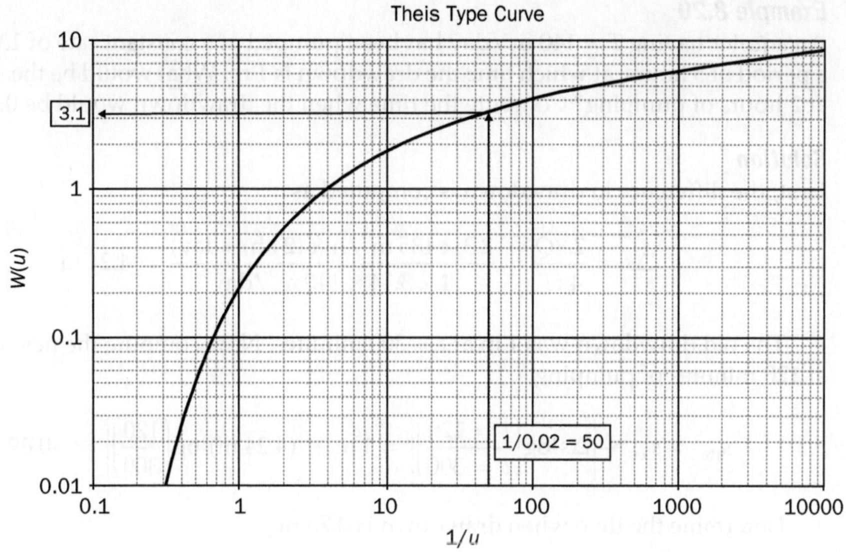

FIGURE 8.7 Theis curve for Example 8.18.

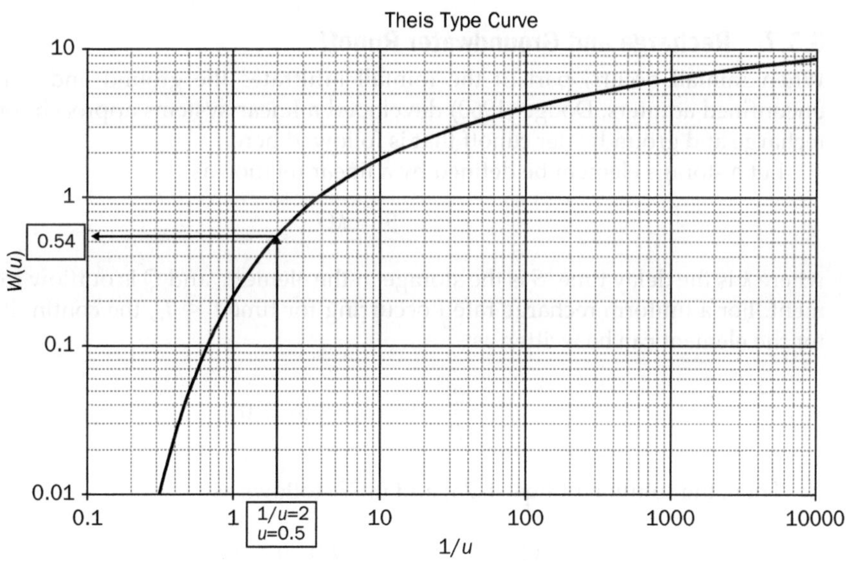

FIGURE 8.8 Theis curve for Example 8.19.

The time of pumping is now computed as

$$t = \frac{r^2 S}{4Tu} = \frac{100 \times 100 \text{ m}^2 \times 0.00025}{4 \times 140(\text{m}^2/\text{day}) \times (0.5)} = 0.01 \text{ day}$$

Example 8.20

An aquifer having $T = 140\ m^2/day$ has been pumped at a constant rate of $135\ m^3/h$ for a period of 5 hours, at which time the drawdown is 1 m. What would be the drawdown at 2 hours of pumping? Compute the time when the drawdown would be 0.75 m.

Solution

First, the difference in drawdown is computed as

$$\Delta s = \frac{2.3Q}{4\pi T} = \frac{2.3 \times 135\ m^3/h \times (24\ h/1\ day)}{4 \times 3.14 \times 140\ m^2/day} = 4.24\ m$$

The old drawdown was 1 m at $t = 300$ minutes. Now compute the new drawdown at 120 minutes of pumping

$$s_{120} = s_{300} + \left[\Delta s \log\left(\frac{t = 120}{t = 300}\right)\right] = 1\ m + \left[(4.24\ m)\log\left(\frac{120}{300}\right)\right] = -0.69\ m$$

Determine the time when drawdown is 0.75 m

$$\log t = \frac{s - s_{300}}{\Delta s} + \log(t = 300\ min) = \frac{0.75\ m - 1\ m}{(4.24\ m)} + \log(300\ min) = 2.42\ min$$

8.3.7 Recharge and Groundwater Runoff

When rainfall occurs, part of the rainfall infiltrates the ground and may recharge unconfined aquifers. Dooge (1960) developed a linear systems approach for modeling recharge and groundwater runoff that is followed here.

Let a storage element be defined by a linear relation as

$$S = kQ \tag{8.76}$$

where k is the delay time, S is the storage in the element, and Q is outflow from the element. For a uniform recharge rate I occurring for time $t \leq T_0$, the continuity equation for the element can be written as

$$\frac{dS}{dt} = I - Q, S(0) = 0 \tag{8.77}$$

Then, the solution of Eqs. (8.76) and (8.77) follows as

$$Q = I[1 - \exp(-t/k)],\ 0 < t \leq T_0 \tag{8.78}$$

$$Q = I[\exp(-T_0 k) - 1]\exp(-t/k),\ T_0 \leq t < \infty \tag{8.79}$$

The volume of outflow during T_0 can be computed as

$$V_0 = \int_0^{T_0} Q dt = \int_0^{T_0} I[1 - \exp(-t/k)]dt = I[T_0 + k\exp(-t/k) - k] \tag{8.80}$$

Let $R_0 = IT_0$. Then, Eq. (8.80) can be written as

$$\frac{V_0}{R_0} = 1 - \frac{k}{T_0}\left[1 - \exp\left(-\frac{T_0}{k}\right)\right] \tag{8.81}$$

In a similar manner, the volume of outflow during any time interval mT_0 and $(m + 1)T_0$, $m = 1, 2, \ldots$, can be computed using Eq. (8.79) as

$$V_m = \int_{mT_0}^{(m+1)T_0} I[\exp(-T_0/k) - 1]\exp(-t/k)dt$$

$$= Ik[\exp(T_0/k) - 1]\{\exp(-mT_0/k) - \exp(-(m + 1)T_0/k]\} \tag{8.82}$$

Then

$$\frac{V_m}{R_0} = \frac{k}{T_0}[\exp(T_0/k) - 1]^2 \exp[-(m + 1)(T_0/k)] \tag{8.83}$$

Our interest is more in computing the outflow during a given period due to recharge in that period and the preceding periods. This can be obtained by computing the contributions of past recharges, R_n, R_{n-1}, …, to a given volume V_n. Then, we can write

$$\begin{aligned} V_n &= R_n\left\{1 - \frac{k}{T_0}\left[1 - \exp\left(-\frac{T_0}{k}\right)\right]\right\} + R_{n-1}\frac{k}{T_0}\left[\exp\left(\frac{T_0}{k}\right) - 1\right]^2 \exp\left(-\frac{2T_0}{k}\right) \\ &\quad + R_{n-2}\frac{k}{T_0}\left[\exp\left(\frac{T_0}{k}\right) - 1\right]^2 \exp\left(-\frac{3T_0}{k}\right) + \ldots \\ &= \sum_{i=0}^{\infty} a_i R_{n-i} \end{aligned} \tag{8.84}$$

where

$$a_0 = 1 - \frac{k}{T_0}\left[1 - \exp\left(-\frac{T_0}{k}\right)\right] \tag{8.85}$$

$$a_i = \frac{k}{T_0}\left[\exp\left(\frac{T_0}{k}\right) - 1\right]\exp\left[-\frac{(i + 1)T_0}{k}\right],\ i \geq 1 \tag{8.86}$$

Equation (8.84) can be expressed in the coefficient routing form in terms of recharge in the present period and recharge and outflow in the immediately preceding period as

$$V_n = C_0R_n + C_1R_{n-1} + C_2V_{n-1} \tag{8.87}$$

where

$$C_0 = 1 - \frac{k}{T_0}\left[1 - \exp\left(-\frac{T_0}{k}\right)\right] \tag{8.88}$$

$$C_1 = \frac{k}{T_0}\left\{\left[1 - \exp\left(-\frac{T_0}{k}\right)\right] - \exp\left(-\frac{T_0}{k}\right)\right\} \tag{8.89}$$

$$C_2 = \exp\left(-\frac{T_0}{k}\right) \tag{8.90}$$

Coefficients C_0, C_1, and C_2 depend only on T_0 and k.

Example 8.21

An unconfined aquifer is being recharged monthly as follows: The recharge is 5 cm for December, 7 cm for January, 4 cm for February, 5 cm for March, 3 cm for April, and 2 cm for May. Take the value of k as 1 month. Compute the groundwater runoff.

Solution

$T_0 = 6, k = 1$

$R_1 = 5, R_2 = 7, R_3 = 4, R_4 = 5, R_5 = 3, R_6 = 2$

$$C_0 = 1 - \frac{k}{T_0}\left[1 - \exp\left(-\frac{T_0}{k}\right)\right] = 1 - \frac{1}{6}\left[1 - \exp\left(-\frac{6}{1}\right)\right] = 0.8337$$

$$C_1 = \frac{k}{T_0}\left\{\left[1 - \exp\left(-\frac{T_0}{k}\right)\right] - \exp\left(-\frac{T_0}{k}\right)\right\} = \frac{1}{6}\left\{\left[1 - \exp\left(-\frac{6}{1}\right)\right] - \exp\left(-\frac{6}{1}\right)\right\} = 0.1658$$

$$C_2 = \exp\left(-\frac{T_0}{k}\right) = \exp\left(-\frac{6}{1}\right) = 0.0025$$

$$V_1 = C_0R_1 + C_1R_0 + C_2v_0 = (0.8337 \times 5) + (0.1658 \times 0) + (0.0025 \times 0) = 4.17 \text{ cm}$$

$$V_2 = C_0R_2 + C_1R_1 + C_2v_1 = (0.8337 \times 7) + (0.1658 \times 5) + (0.0025 \times 4.17) = 6.68 \text{ cm}$$

$$V_3 = C_0R_3 + C_1R_2 + C_2v_2 = (0.8337 \times 4) + (0.1658 \times 7) + (0.0025 \times 6.68) = 4.51 \text{ cm}$$

$$V_4 = C_0R_4 + C_1R_3 + C_2v_3 = (0.8337 \times 5) + (0.1658 \times 4) + (0.0025 \times 4.51) = 4.84 \text{ cm}$$

$$V_5 = C_0R_5 + C_1R_4 + C_2v_4 = (0.8337 \times 3) + (0.1658 \times 5) + (0.0025 \times 4.84) = 3.34 \text{ cm}$$

$$V_6 = C_0R_6 + C_1R_5 + C_2v_5 = (0.8337 \times 2) + (0.1658 \times 3) + (0.0025 \times 3.34) = 2.17 \text{ cm}$$

8.3.8 Stream-Aquifer Interaction

Unconfined aquifers and streams may interact in three ways: (1) the aquifer supplies water as baseflow to the stream; (2) the stream recharges the aquifer; and (3) the aquifer and stream alternate in supplying water to each other. In the first way, the stream surface is below the water table and the stream is usually perennial because of sustained baseflow. In the second way, the stream surface is higher than the water table and stream provides recharge to the aquifer whenever it flows. Ephemeral streams are examples of such streams. In the third way, stream surface and water table in the aquifer alternately become high or low. For describing stream-aquifer interaction, simple models are presented here for both steady and unsteady cases.

Storage Models

Consider an aquifer-stream system. Inflow to the aquifer may entail natural recharge, artificial recharge, percolation, streamflow, drainage, and irrigation return flow. Outflow from the aquifer may entail pumping, subsurface flow, artificial drainage, phreatic transpiration, and spring. Let A be the aquifer area, h the piezometric head, Q_i inflow to the aquifer, and Q_0 the outflow from the aquifer. The volume balance for the aquifer can be written as

$$S_c \frac{dV}{dt} = Q_i - Q_0 \tag{8.91}$$

where V is the storage in the aquifer and S_c is the coefficient of storage. The storage V can be expressed in terms of hydraulic head as

$$V = Ah \tag{8.92}$$

Therefore, Eq. (8.91) can be written as

$$S_c \frac{dh}{dt} = \frac{1}{A}(Q_i - Q_0) \tag{8.93}$$

Equation (8.93) can be solved by hypothesizing relations between flow Q and V or h. In this equation, one of Q_i and Q_0 is known and the other is unknown to be determined.

Now consider the case where the aquifer-stream interact, as shown in Fig. 8.6. Then, one can write the net inflow to the stream per unit area, q, as

$$q = a(H - h) \tag{8.94}$$

in which a is the subsurface outflow constant, H is the mean water level of the stream, and h is the piezometric head in the aquifer. Parameter a denotes the stream-aquifer flow per unit aquifer area due to a unit difference of mean head between aquifer and stream, which may be either way. Using Eq. (8.94), the volume balance Eq. (8.93) becomes

$$S_c \frac{dh}{dt} + ah = aH = I \tag{8.95}$$

Using the initial condition $h(0) = h_0$, the solution of Eq. (8.95) follows

$$h(t) = h_0 \exp(-at/S_c) + \frac{1}{S_c}\int_0^t I(\tau)\exp\left[-\frac{a(t-\tau)}{S_c}\right]d\tau \tag{8.96}$$

Steady-State Flow

Consider an unconfined aquifer, subjected to constant recharge I, that is contributing flow to the stream and flow is steady. Then, under the Dupuit approximation,

$$\frac{d}{dx}\left(Kh\frac{dh}{dx}\right) + I = 0 \tag{8.97}$$

For solving Eq. (8.97), the boundary conditions can be specified as

$$h = H, x = 0; \quad \frac{dh}{dx} = 0, \quad x = L \tag{8.98}$$

where H is the head in the stream. Integration of Eq. (8.97) yields

$$h = H\left[1 + \frac{I(2L - x)}{KH}\right]^{0.5} \tag{8.99}$$

Unsteady-State Flow

Using the Dupuit approximation, the 1D governing equation for aquifer-stream interaction in linearized form can be written as

$$\frac{T}{S_c}\frac{\partial^2 h}{\partial x^2} + \frac{I}{S_c} = \frac{\partial h}{\partial t} \tag{8.100}$$

where recharge I occurs at a constant rate before the start of the interaction. For solving Eq. (8.100), the following initial and boundary conditions can be employed

$$h - H = Ix\frac{2L - x}{2T}, \; t < 0 \tag{8.101}$$

$$h - H = 0, \quad x = 0; \quad \frac{\partial(h - H)}{\partial x} = 0, \quad x = L \tag{8.102}$$

The solution of Eq. (8.100), subject to Eqs. (8.101) and (8.102), follows as

$$h - H = \sum_{i=1}^{\infty} C_i \exp\left[-\left(\frac{i\pi}{2L}\right)^2\left(\frac{T}{S_c}\right)t\right]\sin\left(\frac{i\pi x}{2L}\right) \tag{8.103}$$

where

$$C_i = 1 - \cos\left(i\pi\frac{8IL}{Ti^3\pi^3}\right) \tag{8.104}$$

The average head in the aquifer can be computed as

$$\bar{h} = \frac{1}{2L}\int_0^{2L} h dx \tag{8.105}$$

Using Eq. (8.103) in Eq. (8.105), the average head can be expressed as

$$\bar{h} - H = \sum_{i=1}^{\infty} C_i \exp\left[-\left(\frac{i\pi}{2L}\right)^2\left(\frac{T}{S_c}\right)t\right]\frac{1 - \cos(i\pi x)}{i} \tag{8.106}$$

Noting that

$$q = \frac{T}{L}\frac{d\bar{h}}{dx}\Big|_{x=0} \tag{8.107}$$

Therefore

$$q = \sum_{i=1}^{\infty} C_i \exp\left[-\left(\frac{i\pi}{2L}\right)^2\left(\frac{T}{S_c}\right)t\right]\left(\frac{i\pi T}{2L^2}\right) \quad (8.108)$$

If only the first term of the series is retained then

$$\bar{h} - H = C_1\frac{2}{\pi}\exp\left(-\frac{T\pi^2}{4S_cL^2}t\right) \quad (8.109)$$

and

$$q = C_1\exp\left(-\frac{\pi^2 Tt}{4S_cL^2}\right)\frac{T\pi}{2L^2} \quad (8.110)$$

8.4 Summary

Study of groundwater hydrology is extremely important since these days the use of groundwater in providing drinking water and food security has surpassed that of surface water. This chapter covers fundamentals of groundwater hydrology as well as interaction between surface and subsurface water. Practical application of the basic laws discussed here will be clearer in some subsequent chapters.

8.5 Questions

8.1 Consider 2 m³ of sand having a porosity of 0.3, 2 m³ of silt with a porosity of 0.4, and 2 m³ of clay with a porosity of 0.5. If the sand, silt, and clay are mixed, what will be the resulting porosity?

8.2 Compute the void ratio of the mixture in Question 8.1.

8.3 A confined aquifer is of thickness 40 m and has a porosity of 0.4. Due to excessive pumping, the aquifer thickness reduces (or subsides) and so does its porosity to 0.3. Compute the reduction in the aquifer thickness.

8.4 Compute the specific yield of the unconfined aquifer with an area of 2 km² and porosity of 0.4 whose water table declines by 4 m due to excessive pumping. Assume the specific retention of the aquifer as 0.1. Also, compute the change in storage in the aquifer.

8.5 A confined aquifer has an area of 2 km² and a storage coefficient of 0.0025. Compute the change in storage of the aquifer if the hydraulic head drops by 2 m.

8.6 An aquifer has a porosity of 0.35, thickness of 200 m, and compressibility of 10^{-8} m²/N. The value of β can be assumed as 4.5×10^{-10} m²/N. Compute the storage coefficient of the aquifer.

8.7 A confined aquifer has a porosity of 0.35, storativity of 0.003, and thickness of 100 m. Because of pumping, the pressure drops by 4000 N/m^2. What will be the subsidence of the aquifer? Assume the value of β as $4.5 \times 10^{-10}\ m^2/N$.

8.8 What is the compressibility of an aquifer that has a porosity of 0.40, thickness of 100 m, and storativity of 0.004? Assume the value of β as $4.5 \times 10^{-10}\ m^2/N$.

8.9 Two wells that are 500 m apart are located in a confined aquifer 50 m thick and 1000 m wide. The difference in the hydraulic heads of these wells is 3 m and the hydraulic conductivity of the aquifer is 5 m/day. What is the groundwater discharge passing through a unit cross section of the aquifer?

8.10 An unconfined aquifer has a hydraulic conductivity of 5 m/day. The hydraulic head at a point upstream is 15 m and that downstream is 14 m and the distance between the two points is 400 m. What is the groundwater discharge at the downstream point?

8.11 Consider the aquifer in Question 8.10. If the porosity of the aquifer is 0.3, then what will be the average linear groundwater velocity?

8.12 Consider an unconfined aquifer (2000 m wide and 40 m thick) up to the water table and having a hydraulic conductivity of 5 m/day. The aquifer has two pumping wells 300 m apart. The distance from the surface to the water level in well one is 4 m and that in well two is 5 m. Consider the difference between the water table and the ground surface as 3 m. What is the groundwater discharge passing through the 2-km-wide section between the wells?

8.13 An unconfined aquifer (50 m thick), with a hydraulic conductivity of 5 m/day, has two observation wells 50 m and 200 m away from the pumping well. The drawdown observed is 4 m in the nearer well and 3 m in the farther well. If the steady-state flow is assumed, then what will be the discharge?

8.14 A confined aquifer (50 m thick) is having a steady-state pumping rate of 250 m^3/day. The drawdown in one well 20 m from the pumping well is 1 m, and in the second well 300 m away, it is 0.5 m. What is the hydraulic conductivity of the aquifer?

8.15 Lohman (1972) reported values of drawdown observed in a well 200 ft (r) away from the pumping well in a confined aquifer, which was being pumped at a steady rate of 96,000 ft^3/day as given in Table 8.3. The values of r^2/t are also computed as given in the table. Compute the transmissivity and storativity (coefficient of storage) of the aquifer.

8.16 Compute the transmissivity and coefficient of storage (or storativity) of the aquifer in Question 8.15 using the Cooper-Jacob method.

8.17 An aquifer that is 30 m thick has a hydraulic conductivity of 10 m/day and storativity of 0.005. The aquifer is being pumped at a rate of 3000 m^3/day. What will be the drawdown after a week of pumping 10 m away from the pumping well?

8.18 Consider a confined aquifer with transmissivity of 200 m^2/day and storativity of 0.005. It has a fully penetrating well that is being pumped at a rate of 2000 m^3/day. Determine the drawdown 5 m away from the pumping well after 5 hours and at a distance of 200 m after 1 day of pumping.

8.19 Compute the approximate values of radius of influence after 5, 10, 24, 36, and 48 hours of pumping.

Time from Start of Pumping, *t* (min)	Observed Drawdown, s (ft)	r^2/t (ft²/day)
1.0	0.66	5.76×10^7
1.5	0.87	3.84×10^7
2.0	0.99	2.88×10^7
2.5	1.11	2.30×10^7
3	1.21	1.92×10^7
4	1.36	1.44×10^7
5	1.49	1.15×10^7
6	1.59	9.06×10^6
8	1.75	7.20×10^6
10	1.86	5.76×10^6
12	1.97	4.80×10^6
14	2.08	4.10×10^6
18	2.20	3.20×10^6
24	2.36	2.40×10^6
30	2.49	1.92×10^6
40	2.65	1.44×10^6
50	2.78	1.15×10^6
60	2.88	9.60×10^5
80	3.04	7.20×10^5
100	3.16	5.76×10^5
120	3.28	4.80×10^5
150	3.42	3.84×10^5
180	3.51	3.20×10^5
210	3.61	2.74×10^5
240	3.67	2.50×10^7

TABLE 8.3 Drawdown of Water Table in an Observation Well ($r = 200$ ft) (Source: Lohmann, 1972)

8.20 Two wells located 500 m apart fully penetrate a 30-m-thick aquifer having a hydraulic conductivity of 10 m/day and storage coefficient of 0.20. A well is being pumped at a rate of 10,000 m³/day. What will be the decline in the other well after 5 days of pumping?

8.21 An unconfined aquifer is being recharged monthly as follows: The recharge is 3 cm for July, 6 cm for August, 4 cm for September, 3 cm for October, 2 cm for November, and 1 cm for December. Assuming the value of *k* as 1 month, compute the groundwater runoff.

8.6 References

Cooper, H. H., and C. E. Jacob. A generalized graphical method for evaluating formation constants and summarizing well-filed history. *Transactions of the American Geophysical Union*, 27(4):526–534, 1946.

Dooge, J. C. I. The routing of groundwater recharge through typical elements of linear storage. *International Association of Scientific Hydrology (IASH)*, 52:286–300, 1960.

Green, J. H. *Compaction of the Aquifer System and Land Subsidence in the Santa Clara Valley, California*. U.S. Geological Survey Professional Paper 450-D, Governing Printing Office, Washington, DC, 1962.

Freeze, R. A., and J. A. Cherry. *Groundwater*. Englewood Cliffs, NJ: Prentice-Hall, 1979.

Hantush, M. S. Hydraulics of wells. In *Advances in Hydroscience, Vol. 1*, edited by V. T. Chow, 281–432. New York: Academic Press, 1964.

Jacob, C. E. Flow of groundwater. In *Engineering Hydraulics*, edited by H. Rouse, 321–386. New York: Wiley, 1949.

Kasenov, M. *Applied Ground-Water Hydrology and Well Hydraulics*. Highlands Ranch, CO: Water Resources Publications, 1997.

Krasny, J. Classification of transmissivity magnitude and variation. *Groundwater*, 31(2): 230–236, 1993.

Lohman, S. W. *Groundwater Hydraulics*. Geological Survey Professional Paper 708, U.S. Department of the Interior, Washington, DC, 1972.

McWhorter, D. B., and D. K. Sunada. *Groundwater Hydrology and Hydraulics*. Highlands Ranch, CO: Water Resources Publications, 1977.

Meinzer, O. E. *The Occurrence of Groundwater in the United States, with a Discussion of Principles*. U.S. Geological Survey Water Supply Paper 489, Government Printing Office, Washington, DC, 1923.

Theis, C. V. 1935. The relation between the lowering of the piezometric surface and the rate and duration of discharge of a well using groundwater discharge. *Transactions, American Geophysical Union*, 16:519–524.

Todd, D. K. 1980. *Groundwater Hydrology*. 2nd ed. New York: John Wiley & Sons.

CHAPTER 9

Water Quality

Water quality plays a fundamental role in the biosphere in general and water resources development and management in particular. Natural waters have a variety of contaminants that define water quality characteristics. The sources of these contaminants vary, depending on where, when, and who the water comes in contact with as well as atmospheric pollution resulting in acid precipitation. Water quality characteristics can be grouped into physical, chemical, and biological characteristics. These characteristics do not occur in isolation and many of them occur simultaneously and are interactive. Furthermore, these characteristics are dynamic variables and vary with space and timescales. They also interact with the components of the hydrologic cycle and are strongly influenced by human activities. The discussion in the preceding chapters has been on water quantity and related aspects. This chapter discusses water quality and its characteristics.

9.1 Introduction

Water quality usually implies the materials contained in the water, but may also include heat or nuclear radiation. Water quality can be defined in terms of tens of parameters. The quality of water is affected by pollution, which may be point or nonpoint source. Pollution implies that water has been degraded by human activities, although water quality degradation can be caused by natural events, such as heavy rainstorms, floods, droughts, fires, and volcanic eruptions. Some of the contaminants dissolve in water and some may be carried by water in suspension. Some of the contaminants are more hazardous than others. The significance of a contaminant depends on the use of water. As an example, salinity is important for water supply but not for recreation. The focus in this chapter is on nonpoint source pollution.

Since water is an effective solvent, it picks up constituents when in contact with rocks, soil, and organic matter. There are tens of contaminants and some of them are more harmful than others. The organic matter in natural water is usually low and is derived from the decay of plants, weeds, and leaves. When the organic matter is present in water, it may cause color formation, taste and odor problems, oxygen depletion, formation of halogenated compounds when chlorine is added, and interference with treatment facilities. These contaminants define water quality characteristics, may be different, and may be interactive. The quality of water in natural streams, that are an integral part of the watershed and its ecosystem, is influenced by the interaction with the atmosphere, soil, and biota. Standards have been established to define the limits of these characteristics for different uses. This chapter provides a qualitative discussion of the water quality characteristics.

9.2 Sources of Contaminants

The sources of generation of contaminants can be classified as (1) domestic, (2) industrial, (3) agricultural, (4) energy-related, and (5) environmental. Domestic contaminants occur due to the leakage of sewer and septic tanks, pets, garbage, and waste disposal. Industrial contaminants are generated by waste disposal in streams, lakes, or seas. Depending on the type of industry, contaminants may be heavy metals, nondeteriorating compounds, radioactive materials, highly toxic materials, and so on. Agricultural contaminants are generated from the dissolution of fertilizers, herbicides, weedicides, pesticides, animal waste, chemicals in irrigation water, and others. Some or part of these contaminants are washed off in surface water and pollutant receiving streams. Other or part of the contaminants infiltrate into the soil, travel through the unsaturated zone, and eventually pollute groundwater. Contaminants are generated by the energy generation industry. For example, if energy is generated by nuclear power then radioactive waste is generated. Likewise, if coal is used for the generation of electricity then coal residue is a waste material and this material may contain a lot of chemicals. Some contaminants are produced by nature. For example, during geological evolution, soils may have excess salt or excess alkalinity. There are also areas where groundwater is contaminated by arsenic.

9.3 Quality of Water

Pure water is found only in the vapor state. Impurities start to accumulate when condensation occurs as water droplets that form clouds, because the atmosphere contains gases that dissolve in the droplets. Prior to the industrial revolution, oxygen (O_2) and carbon dioxide (CO_2) were the two most important dissolved gases. In the subsequent decades, there has been an increasing use of fossil fuel and internal combustion engines that have led to the production of sulfur dioxide (SO_2) and oxides of nitrogen (NO_2), which pollute the atmosphere. On dissolution of these gases into water droplets, the result is the generation of mineral acids and the increase of the hydrogen ion content, which may be one to three orders of magnitude. Therefore when precipitation occurs, it is acid rain. Observations in Canada, Europe, and the northeastern United States have shown the rainwater pH value of as low as 4. Another important point to note is that gases can travel long distances prior to precipitation, meaning the cause of acid rain is in one country but another country may receive the actual acid rain. During the Iraq War, the effect of the burning of oil wells was observed in the darkening of snow and glaciers in the far away Himalayas in India and Nepal.

Water is often characterized as freshwater, brackish water, and saline water, depending upon the concentration of dissolved constituents. The concentration is usually prescribed in terms of total dissolved solids (TDS). Freshwater is usually the water that has the concentration of TDS below 1500 g/m^3. Brackish water is the water that has the TDS concentration of up to 5000 g/m^3. The water that contains the TDS concentration exceeding 5000 g/m^3 is characterized as saline water. Sea water has a TDS range of 30,000 to 34,000 g/m^3.

Another term that is commonly used in day-to-day language to characterize water is hardness. This property will be discussed later on in the chapter, but for practical purposes it is represented by the sum of calcium and magnesium concentrations in water. When one cleans hands with soap and water that contains these concentrations in significant quantities, then these ions are easily precipitated and react with the soap and form a scum that becomes difficult to remove under normal temperature. Such a water is termed as hard, otherwise the water is soft. Based on the concentration of calcium carbonate ($CaCO_3$), water can be regarded as soft if $CaCO_3$ is less than 50 mg/L, moderately hard if the concentration is between 50 and 150 mg/L, hard if the concentration is between 150 and 300 mg/L, and very hard if the concentration is greater than 300 mg/L. Hard water hampers washing because soap does not easily lather in hard water. Also, it leaves scaly mineral deposits in pipes, conduits, tanks, and boilers. It may not necessarily be harmful to human health and may, in fact, be even better than soft water (Tchobanoglous and Schroeder, 1985).

In general, wherever there is a body of water, such as a lake, reservoir, lagoon, pond, tank, or stream, its habitat is supported and certain microorganisms are maintained. The diversity of the species of organisms provides an indicator of water quality and aquatic health. Pollution reduces the variety of organisms, because certain organisms do not survive in a polluted environment. Thus biological indicators provide a better portrayal of the overall quality of water than do individual chemical characteristics, such as dissolved oxygen, temperature, and so on.

9.4 Chemical Contaminants

In general, chemical contaminants found in water in significant amounts are calcium, magnesium, sodium, potassium, bicarbonate, chloride, sulfate, and nitrate. Also present in the water are trace amounts of other ions, such as lead, copper, arsenic, manganese, and a wide range of organic compounds ranging from humic materials to synthetic organics used as detergents, pesticides, herbicides, and solvents. There are four main sources that lead to the generation of these compounds and these are agricultural runoff, wastewater, decaying vegetation and animal matter, and inadequate hazardous waste discharge.

9.5 Surface Water Quality

Surface water, vadose zone water, and groundwater usually have different water quality characteristics, partly because these waters have different environments and are differently influenced by anthropogenic impacts. Surface waters are derived from precipitation and are stored in lakes, reservoirs, ponds, lagoons, rivers, streams, and bayous, and as snow, ice, and glaciers. Although abundantly available, surface water stored in seas, oceans, and bays is saline and cannot be used without treatment. Otherwise, surface waters are usually fresh but tend to exhibit turbidity, which is a property caused by the presence of clays and other light scattering colloidal particles. Except for agricultural irrigation, it is essential to treat surface water for the removal of turbidity before use. Surface water gets more contaminated as it picks up minerals from surface runoff. Besides silt and debris being transported by surface water, plants and algae may also grow if the water is moving slowly. However, the greatest contamination of surface waters is caused by the disposal of liquid waste and this is happening all over the world, especially in developing countries, sometimes without any discrimination. The contaminants become concentrated upon evaporation of water and the residues get deposited in sinks. Of course, oceans and seas are the major sinks, but there are also significant inland sinks. Examples of noteworthy inland sinks are the Caspian

Sea on the Iranian–Russian border; the Dead Sea in Palestine; the Great Salt Lake in Utah; the Salton Sea, formed by a major break in the Alamo Canal, in the Imperial Valley in Southern California; and the Dull Lake in Kashmir and the Bhopal Lake in India, among others. Surface waters, if available, are preferred for use, because of low treatment requirements. Glacial melt and snow melt waters are excellent for drinking. However, their dependability is becoming questionable in many areas, because of the uncertainty owing to climate change.

9.6 Subsurface Water Quality

Vadose zone water, also called soil water, is in the pedosphere and groundwater is found in the lithosphere. The vadose zone water is temporarily stored in the soil for consumption by plants and is impacted by what happens on the land surface. Pollutants applied on the surface, such as fertilizers or waste disposal, are transported by water infiltrating the soil. The water also picks up minerals from soil and has higher concentration of TDS than does the surface water. As such, this water is in the transient zone and extracted for other uses. Percolation from this zone adds contaminants to groundwater. Groundwater also picks up minerals from soil and rocks and may usually have higher TDS concentration. In many cases, groundwater is found to have high concentrations of calcium, magnesium, boron, and fluoride. Leaching of salts from soil, application of fertilizers, and evapotranspiration are contributing to the increase in the concentration of contaminants in groundwater. Increasing nitrate concentrations in groundwater in the western United States are the result of leaching of fertilizers applied to crops. Owing to the low cost of treatment and dependability, groundwater is, in general, the preferred source for domestic consumption and is being increasingly utilized these days for agricultural irrigation. Water from igneous and metamorphic rocks has excellent quality for drinking, except in arid areas where recharge water may have a high salt concentration due to high evaporation.

In general, groundwater is of better quality than surface water and has greater dissolved mineral salts, such as sodium, calcium, magnesium, and potassium cations as well as chloride, sulfate, and bicarbonate anions. However, if the concentration of these salts exceeds 1000 parts per million (*PPM*), the water becomes saline and has limited use for domestic, municipal, and industrial uses. The use of groundwater may be limited by the presence of iron. When the iron concentration exceeds 0.3 mg/L, the taste and odor of water are affected and its use for drinking, cooking, and washing becomes limited.

9.7 Characteristics of Water Quality

Water quality characteristics are classified in two main ways (Tchobanoglous and Schroeder, 1985, p. 765). The first way includes grouping characteristics into three classes: physical, chemical, and biological. The other way includes two classes: gross and specific. The gross class does not distinguish individual physical, chemical, and biological characteristics. The gross characteristics are commonly used as measures of water quality, partly because they are easily measured and interpreted. Suspended solids, color, odor, hardness, alkalinity, and biochemical oxygen demand are examples of gross measures. On the other hand, specific measures, as the name suggests, consider these characteristics individually. These characteristics are employed to describe water quality for specific applications. For example, a particular use may need specification of a toxic organic compound or heavy metal ion.

Solutes are dissolved substances, including artificial tracers, natural constituents, and others, and exist in liquid and/or gaseous phases, and can be reactive or nonreactive, and hazardous or innocuous. Because of increasing dependence on chemicals, pollution of surface and subsurface waters is growing, especially in developing countries.

Water quality characteristics are closely connected to the components of the hydrologic cycle that is discussed in Chap. 1. Since these components are dynamic and have different timescales, the water quality characteristics at any point in the hydrologic cycle are also dynamic. Before discussing these characteristics, it will be useful to note units used for their expression.

9.7.1 Units of Measurement

It is important to define the units of measurement that are employed for doing physical and chemical water quality analyses. Several units are used for expressing concentration but only the most commonly used units are defined here. First, symbols commonly used are g = gram, mg = milligram, PPM = parts per million, L (or l) = liter, mol = mole, m = molality, M = molarity, eq = equivalent, meq = milliequivalent, and N = normality.

1. Parts per million (PPM): PPM is defined as

$$\text{PPM} = \frac{\text{mass of solute (g)}}{10^6\,\text{g of solution}} \quad \text{or} \quad \frac{\text{concentration}\,(\text{g/m}^3)}{\text{specific gravity of liquid}} \tag{9.1}$$

   Sometimes, parts per billion (PPB) is also used. In that case, 10^6 in the denominator of Eq. (9.1) is replaced by 10^9.

2. Mass concentration (conc): Noting that 1 gram per cubic meter (g/m^3) is equal to 1 milligram per liter (mg/L), conc is defined as

$$\text{conc}(\text{g/m}^3) = \frac{\text{mass of solute (g)}}{1\,\text{m}^3\ \text{of solute}} \tag{9.2}$$

3. Percent by mass: This is expressed as

$$\%\ \text{by mass} = \frac{\text{mass of solute} \times 100}{\text{mass of solute} + \text{mass of solvent}} \tag{9.3}$$

4. Density (ρ): This is written as

$$\rho(\text{kg/m}^3) = \frac{\text{mass of solution (kg)}}{1.0\,\text{m}^3\ \text{of volume}} \tag{9.4}$$

5. Molality (m): Noting that one mole of a compound is equal to the molecular mass in grams, molality (m) can be expressed as

$$m(\text{mol/kg}) = \frac{\text{mole of solute}}{1\,\text{kg of solvent}} \tag{9.5}$$

6. Molarity (M): This defined as

$$M(\text{mol/L}) = \frac{\text{mass of solute (g)}}{1.0\ \text{l of solution}} \tag{9.6}$$

7. Normality (N): Noting that 1.0 eq/m³ = 1.0 mg/L, N is defined as

$$N(\text{eq/L})(\text{meq/L}) = \frac{\text{equivalent of solute (eq)(meq)}}{1.0\ \text{l of solution}} \tag{9.7}$$

where

$$\text{equivalent mass (g/eq)} = \frac{\text{atomic mass (g)}}{z,\ \text{eq}} \tag{9.8}$$

or

$$\text{equivalent mass (mg/meq)} = \frac{\text{molecular mass (mg)}}{z,\ \text{meq}} \tag{9.9}$$

in which z equals the number of replaceable hydrogen atoms or their equivalent, and equals the change in valence for the case of oxidation-reduction reactions.

9.7.2 Chemical Relationships

It is useful to now define those chemical relations that are often used in chemical analysis of water and interpret its results.

1. **Mole fraction:** It is defined as the ratio of the number of moles of a specified solute to the total number of moles of all compounds in a solution, written as

$$r_i = \frac{n_i}{\sum_{i=1}^{N} n_i} \tag{9.10}$$

where r_i is the ratio of mole fraction, n_i is the number of moles of the i-th solute, and N is the number of compounds in the solution.

2. **Electroneutrality:** The principle of electroneutrality states that in any solution the sum of positive ions or cations (i.e., positively charged species in eq/L) is equal to the sum of negative ions or anions (negatively charged species in eq/L), written as

$$\sum \text{cations} = \sum \text{anions} \tag{9.11}$$

3. **Equilibrium Constant:** Consider two reactants A and B which combine to produce two products C and D in a reversible chemical reaction. Then, one can write

$$aA + bB \Leftrightarrow cC + dD \tag{9.12}$$

where a, b, c, and d are weight constants. The chemical reaction will eventually reach a state of equilibrium. Then, following the law of mass action, the ratio of products to reactants is called the equilibrium constant (K) expressed as

$$\frac{[C]^c[D]^d}{[A]^a[B]^b} = K \qquad (9.13)$$

The use of brackets in Eq. (9.13) indicates molar concentration. The value of K will vary with temperature and ionic strength of the solution.

4. **Ionic Strength:** In a given solution, the behavior of ions depends on the ionic concentration. In a dilute solution, ions act independent of each other, but with the increase of concentration, ions start to interact with each other. The ionic strength is a measure of this interaction and can be defined as

$$\mu = \frac{1}{2}\sum_{i=1}^{N}(C_i \times z_i^2) \qquad (9.14)$$

where μ is the ionic strength, C_i is the concentration of ion species i (M), and z_i is the charge of ion species i.

5. **Solubility Product:** The equilibrium constant for reaction that entails precipitate and its constituent ions is termed as solubility product. As an example, the reaction for calcium carbonate ($CaCO_3$) can be expressed as

$$CaCO_3 \Leftrightarrow Ca^{+2} + CO_3^{-2} \qquad (9.15)$$

The solubility product can be expressed, considering the activity of solid phase as 1, as

$$[Ca^{+2}][CO_3^{-2}] = K_{sp} \qquad (9.16)$$

where K_{sp} denotes the solubility product.

Example 9.1

Consider a 1.0 L solution that contains 30 g of calcium carbonate ($CaCO_3$) whose density can be taken as 1.12 g/mL. Determine the molarity and mole fraction of the solution.

Solution

First, we need to determine the moles of $CaCO_3$ as

$$\text{mol}\,(CaCO_3) = \frac{30\text{ g}}{100\text{ g/mol}} = 0.30\text{ mol}$$

Now the molarity of the $CaCO_3$ solution can be written as

$$M\,(CaCO_3) = \frac{30\text{ g}}{1.0\text{ L}} = 0.30\text{ M}$$

For mole fraction of $CaCO_3$, from Eq. (21.10), we obtain

$$r(CaCO_3) = \frac{n(CaCO_3)}{n(H_2O) + n(CaCO_3)}$$

Now we determine the volume (V) of $CaCO_3$ and that of water. The volume of $CaCO_3$ in the 1.0 L solution is

$$V(CaCO_3) = \frac{30\ \text{g}}{1.12\ \text{g/mL}} = 26.79\ \text{mL}$$

Now the volume of water is

$$V(\text{water}) = 1000\ \text{mL} - 26.79\ \text{mL} = 973.21\ \text{mL}$$

The moles of H_2O can be written as $n(H_2O) = \dfrac{973.21}{18\ \text{g/mol}} = 54.07\ \text{M}$

Now, n $(CaCO_3) = 0.30$ M and n $(H_2O) = 54.07$ M
The mole fraction can be expressed as

$$r(CaCO_3) = \frac{0.30\ \text{M}}{54.07\ \text{M} + 0.30\ \text{M}} = 5.52 \times 10^{-3}$$

9.7.3 Physical Characteristics

In real life, it is the physical characteristics, such as color, clarity, and odor, not chemical or biological, that are used to base our impressions of water quality. Natural waters are seldom colorless and clear because of the impurities due to human influences and interactions with their natural environments. Additional important physical characteristics include suspended solids, temperature, and turbidity. Each of these characteristics is now briefly discussed.

Color The color of water that we observe is not the true color but it results from colloidal suspension. True colors are a product of dissolved material, often organic, produced by decaying plants. Industries produce waste that is often colored and colors the receiving stream water.

Odor Odors in water are caused by a variety of factors, such as decaying of organic matter, transformation of sulfates by bacteria into H_2S gas, ammonia, chlorine, and so on. A number of different smells are encountered as a result and these odors often occur together. It is not uncommon to find plant debris, dead animals, and microorganisms in the water. These and wastewater discharge are the sources of organic matter. Upon decay, these organic materials create an appropriate environment for anaerobic bacteria to produce noxious gases that smell bad.

Turbidity Clarity of water is associated with the ability of light to penetrate into a body of water. When water is clear, light penetrates without getting scattered. When it has colloidal matter, light gets scattered and water is regarded as turbid. Thus turbidity causes scattering or absorption of light and is hence an indicator of the property of water. Lower turbidity allows light to penetrate deep, thus providing greater opportunity for photosynthesis and higher oxygen levels. Suspended clays, silts, organic matter, plankton, and other inorganic and organic particles in the water are the cause of turbidity.

Solids Solids encompass all contaminants, except dissolved gases. Of all the solids in hydrology, sediment constitutes the largest portion of the solid load (Brooks et al., 2003, p. 575). The quality of surface water is principally determined by the amount of sediment it carries. From a water quality perspective, suspended sediment is more important, because it affects turbidity and aquatic ecosystems. Furthermore, sediment is also the carrier of many nutrients and heavy metals that in turn affect water quality. Land use changes in drainage areas, urbanization, road construction, agricultural cultivation, forestation, overgrazing by livestock, logging operations, fires, floods, and landslides cause erosion, which in turn leads to suspended sediment.

Sediment transport is an important source of nutrients and heavy metal losses from upland watersheds. These contaminants strongly impact water quality. For example, sediment transport of phosphorous reduces chemical quality of surface water, which in turn impacts aquatic ecosystems. Eutrophication, leading to algae growth, accelerates when phosphorous increases in water bodies. Some land use practices increase sediment production and in turn lead to an increase in nutrient and heavy metal loss. In some rivers, variations in heavy metals, such as zinc (Zn), iron (Fe), copper (Cu), manganese (Mn), lead (Pb), and cadmium (Cd) levels are correlated with variations in sediment concentration.

Solids can be classified in different ways, depending on (1) size, (2) chemical characteristics, and (3) size distribution. Based on the size and state, solid and semisolid materials can be distinguished as settleable, suspended, colloidal, and dissolved. Often, solids are classified as dissolved and suspended. Any particle passing a membrane filter having a pore size of about 1.2 μm is regarded as dissolved. Dissolved solids include colloidal and small suspended particles, whereas suspended solids include settleable. An example of a typical colloidal suspended solid is clay. The total solid load is the sum of dissolved and suspended solids.

Solids are grouped as volatile and nonvolatile (fixed). Volatile solids are typically organic and volatilize at a temperature of 550°C. By doing sieve analysis, the distribution of sediment sizes can be computed. In rivers and streams, the total mass and the distribution of solids vary, depending on the time of the day.

Temperature Temperature plays a fundamental role in determining the water chemistry as well as the health of aqua ecosystems. In deep lakes, the temperature varies vertically from the bottom to the surface and hence fish populations vary. Likewise, water releases from a lake impact downstream populations. With the increase in temperature, gas solubility decreases, mineral solubility increases, and chemical and biochemical reaction rates increase. For example, the solubility of oxygen in water is related inversely to temperature. Organisms have specific temperature ranges that determine their ability to reproduce and compete. Changes in temperature can lead to changes in the aquatic diversity.

Dissolved Oxygen Dissolved oxygen (DO) is one of the most important water quality characteristics in the aquatic environment. It is often used as a measure of the health of a water body. Depending on the pollutant load, each water body has a certain oxygen requirement in order for the aquatic ecosystem to remain viable. The DO concentration decreases with decreasing pressure and increasing salt content. The solubility of oxygen, which is inversely proportional to water temperature, pressure, and biological activity, determines the DO concentration. Each water body in the context of its pollutant load requires a certain level of oxygen. If a water body is highly polluted, the DO concentration may become very low, even approaching zero; in that case, no living species can survive and the water body may be declared dead. For fish to survive, the DO concentration must be at least 4–6 mg/L. In pristine waters, which are hard to come by these days, the DO concentration would be at the saturation level, which would be about 14.6 mg/L at 0°C, 7.56 mg/L at 30°C, and 6.41 mg/L at 40°C.

DO is a transient property and fluctuates in time, and depends on temperature, elevation, and salt content. The fluctuation of DO in a water body stems from two processes: oxygenation and deoxygenation. Oxygenation is a process of adding oxygen to the water body, which occurs through photosynthesis, from tributaries having greater DO concentration, and transfer of oxygen from atmosphere through air-water interface called reaeration. On the other hand, deoxygenation is the process of oxygen depletion, which occurs through the oxidation of carbonaceous biochemical oxygen demand (CBOD), nitrogenous biochemical oxygen demand (NBOD), sediment oxygen demand (SOD), and algal respiration. CBOD represents the reduction of organic carbon to CO_2 when microorganisms, such as bacteria, are present; NBOD represents the biological oxidation of ammonia (NH_3) to nitrates (NO_3^-); and SOD represents the aerobic decay of organic benthic material, which is insignificant in streamflow.

The solubility of oxygen in water can be determined as (Churchill et al., 1962)

$$O_s = 14.652 - 0.41022T + 0.007991T^2 - 0.000077774T^3 \tag{9.17}$$

where O_s is the solubility of oxygen (mL/L), and T is temperature (°C).

Biochemical Oxygen Demand and Chemical Oxygen Demand The biochemical oxygen demand (BOD) is an indication of the oxygen-demanding property of biodegradable material in water. During degradation, oxygen is demanded for oxidization that occurs in two stages. In the first stage, oxygen is required to oxidize carbon compounds and this stage is related to carbonaceous demand. The second stage, referred to as nitrification, requires oxygen to oxidize nitrogen compounds. BOD consists of three components: CBOD, NBOD, and SOD.

Chemical oxygen demand (COD) is an indication of the oxygen demand for the complete oxidation of the pollutant load in water. Unlike BOD, COD does not depend on bacteriological activity and is not regarded as a good measure of oxygen-demanding properties of materials in water. Hence BOD is preferred to COD.

Bacteria and other organisms break down organic material that is in water, such as decaying plants, dead animals, human sewage, livestock waste, logging debris, to more stable compounds. When there is sufficient dissolved oxygen in water and the organic loading is not excessive, oxidation can occur without causing any harmful reduction in DO. If oxygen is insufficient or pollutant loading is too large, then anaerobic processes occur that impact the aquatic ecosystem and lead to an inefficient oxidation process with aerobic and anaerobic end products.

pH (Hydrogen Ion Concentration) The pH of water can be defined as the negative logarithm of positive hydrogen ion. The pH at any time reflects the balance of chemical equilibria in water and the availability of certain nutrients for uptake by plants. It influences and is influenced by chemical reaction in aquatic systems. A pH of >7 is an indicative of alkaline water which happens when carbonate and bicarbonate ions are present, a value of <7 is an indicative of acidic water, and values of <4.5 and >9.2 are indicative of toxic limits. Algae cannot survive when the pH values are >8.5 and freshwater fish can tolerate values <4.5. Carbon dioxide reactions are the most important in determining the pH level in water.

Acidity This is linked to pH, and is caused by free hydrogen ion from carbonic, organic, sulfuric, nitric, and phosphoric acids. Acidity influences chemical and biological reactions. Its importance stems from its ability to contribute to the corrosiveness of water. Acid precipitation in the northeastern United States was a major environmental concern in the 1980s.

Alkalinity Alkalinity is opposite of acidity, that is, it neutralizes acidity and is linked to pH. A high value of alkalinity is related to a high value of pH and excessive dissolved solids. When carbon dioxide is dissolved, carbonate, bicarbonate, and hydroxide are formed, which then cause alkalinity. A well-diversified aquatic life is supported if the alkalinity range is 100–120 mg/L. If water has high alkalinity then large amounts of acid will be needed to change the pH value.

Specific Conductance This specifies the ability of water to conduct electrical current. Since specific conductance increases with dissolved solids, it can be used to estimate total dissolved solids (TDS) and this indicates its significance. Directly, it is not related to water quality.

9.7.4 Chemical Characteristics

All natural waters have dissolved ionic constituents and the principal constituents present in most waters include calcium (Ca^{+2}), magnesium (Mg^{+2}), sodium (Na^{+}), and potassium (K^{+}) representing cationic species; and bicarbonate (HCO_3^{-}), sulfate (SO_4^{-2}), chloride (Cl^{-}), and nitrate (NO_3^{-}) representing anionic species. When water comes in contact with various mineral deposits, it extracts the ionic species. The bicarbonates, sulfates, and chlorides of calcium, magnesium, and sodium are the most abundant species. On the other hand, the principal nonionic mineral observed in all natural waters is silica (SiO_2). Silica causes severe scaling problems in boilers and heat exchangers. There are also many minor ionic species that are found in the water. These include under cations aluminum (Al^{+3}); ammonium (NH_4^{+}); copper (Cu^{+2}); iron, ferrous (Fe^{+2}); iron, ferric (Fe^{+3}); manganese (Mn^{+2}); barium (Ba^{+2}); boron, as borate (BO_4^{-3}); and arsenic (As^{+}); and under anions sulfide (S^{-2}); sulfite (SO_3^{-2}); bisulfate (HSO_4^{-}); bisulfite (HSO_3^{-}); carbonate (CO_3^{-2}); phosphate, mono ($H_2PO_4^{-}$); phosphate, di (HPO_4^{-2}); phosphate, tri (PO_4^{-3}); hydroxide (OH^{-}); and fluoride (F^{-}). In addition, a variety of inorganic constituents are generated by anthropogenic activities and find their way into water, including under cations arsenic (As^{+3}), barium (Ba^{+2}), cadmium (Cd^{+3}), chromium (Cr^{+3}), lead (Pb^{+2}), mercury (Hg^{+2}), selenium (Se), silver (Ag^{+2}), and zinc (Zn^{+2}); and under anion cyanide (CN^{-}). These constituents are toxic to plants, animals, and microorganisms. The anthropogenic activities include inadequately treated industrial waste and wastewater sludge. Nitrogen and phosphorous are applied as nutrients to crops and they also are found in water. Some of the important chemical characteristics will now be discussed.

Nitrogen Nitrogen (N) is of main interest in watershed management. Its sources are organic matter dumped in water bodies, fixation of nitrogen gas by bacteria and plants, and weathering of rocks. It occurs in seven states of oxidation and the nitrogen-containing compounds of interest in water quality are organic nitrogen, ammonia (NH_3), gaseous N (N_2), nitrate-N (NO_3^-), urea [$CO(NH2)_2$], and dinitrogen. Nitrogen is recycled between its organic and inorganic forms. Bacteria and plants produce proteins (organic compounds containing nitrogen) from inorganic forms of nitrogen. Animals, humans included, cannot use nitrogen from the atmosphere or from inorganic compounds to produce protein and hence must obtain nitrogen in organic form. Proteins are broken down into urea and ammonia by bacterial activity. Urea is then converted enzymatically to ammonia. Ammonia from these two sources is then oxidized bacterially, first to nitrite and then to nitrate. Nitrite and nitrate are also converted to nitrogen gas by bacteria. This sequence of conversions constitutes what is called the nitrogen cycle. Agricultural activities and urban development increase nitrate-N and total N concentration in streams. Higher concentrations of nitrate-N in water promote the growth of algae and aquatic plants. In the presence of phosphorous, even 0.30 mg/L of nitrate-N leads to algal bloom. For fish life, the nitrate-N concentration should not exceed 4.5 mg/L and for drinking water it is restricted to 10 mg/L by the U.S. Environmental Protection Agency (EPA).

Phosphorous Phosphorous is very important in water supply systems and aquatic ecosystems. It is essential for the growth of algae and other aquatic organisms. Natural waters are normally deficient in phosphorous but agricultural activities and urbanization greatly contribute to phosphorous loading in water bodies. The sources of phosphorous are weathered igneous rocks, soil leaching, and organic matter. Phosphorous compounds are employed for the production of synthetic detergents and control of corrosion in water supply systems. The phosphorous concentration in treated municipal wastewater is about 10–20 mg/L.

Calcium Calcium is a major constituent of many rocks and is abundantly available in most waters, except for acid peat and swamp waters. It is a major ion that contributes to the hardness of water, TDS, and specific conductance. It dissolves when CO_2 is present and pH is $< 7–8$. High concentrations of calcium are normally not harmful to fish and aquatic life.

Magnesium It exists abundantly in igneous and carbonate rocks, such as limestone and dolomite. It can be toxic to fish if its concentration is >100–400 mg/L and supports fish fauna if the concentration is <14 mg/L.

Sodium It is abundant in igneous and sedimentary rocks and leaches easily in surface water and groundwater and remains in solution. It is not usually harmful to fish fauna, except when both sodium concentration and potassium concentration exceed 85 mg/L. Some levels of sodium reduce toxicity due to aluminum and potassium salts and hence can be beneficial to fish.

Potassium Potassium is fundamental to plant growth and is recycled by aquatic vegetation. Igneous rocks, clays, and glacial material are sources of potassium. In clear waters, the concentration of potassium is found to be less than 1.5 mg/L, but in nutrient rich and eutrophic waters it can be more than 5 mg/L. Concentrations of potassium greater than 400 mg/L can kill fish and greater than 700 mg/L can kill invertebrates.

Manganese It is essential to plant metabolism, and is found in igneous rocks. It leaches from the soil, is circulated organically, and becomes soluble upon decay of organic material. Its concentration for drinking water is 0.05 mg/L and seldom exceeds 1 mg/L in natural waters not subject to human interference. When pH is less than 7, the most dominant form of manganese is Mn^{++}.

Sulfur The sources of sulfur are gypsum and igneous rocks, and sedimentary rocks. Air pollution results in greater concentrations of atmospheric sulfate leading to acid precipitation. Rainfall may have a sulfate concentration of 1–10 mg/L. Sulfate concentrations of less than 90 mg/L are favorable for fish fauna, drinking water has 250 mg/L, and algal growth is not supported when the concentration drops below 0.5 mg/L. Organic sulfur can transform to organic sulfide. The hydrogen sulfide H_2S gas smells awful.

Pesticides and Fertilizers Today's agriculture heavily depends on pesticides and fertilizers which find their way into water systems and impact their water quality. These can be toxic to nontarget organisms when they come in contact. The degree of hazard depends on the amount of exposure and the toxicity of chemicals to the organism.

9.7.5 Biological Characteristics

Biological characteristics of water are related to the aquatic populations of microorganisms. The principal microorganisms found in water include animals, plants, bacteria, algae, and fungi. The important impacts on water quality are transmission of disease by pathogenic organisms, development of tastes and odors, and corrosion and biofouling in cooling and water supply systems.

Waterborne pathogenic bacteria cause gastrointestinal illnesses in humans, livestock, and wildlife species. *Escherichia coli* (*E. coli*) is found in the gut of all warm-blooded animals (e.g., rodents, deer and elk, and livestock). Several strains of *E. coli* pose a health risk. *E. coli* is often used as an indicator for determining if water quality is suitable for human consumption and its number present is a measure of bacteriological water quality. Waterborne protozoa parasites also cause gastrointestinal illnesses in humans, livestock, and wildlife species. Some of these parasites have been found to cause large-scale outbreaks of gastroenteritis in humans. They are carried by a variety of warm-blooded animals.

Fungi, in conjunction with bacteria, are primarily responsible for the decomposition of carbon in the biosphere. They can grow in low-moisture areas as well as in low-pH environments. They decompose organic material in both terrestrial and aquatic environments and release carbon dioxide to the atmosphere and nitrogen to the terrestrial environment. Thus they play a fundamental role in the carbon cycle.

Algae are important for water quality and form a symbiotic relationship with bacteria in an aquatic environment. They can influence the DO balance by creating anaerobic conditions at night. They can increase pH by extracting bicarbonate from water when bacteria or other sources of CO_2 do not exist, and calcium carbonate may then be precipitated. Algae contain chlorophyll and are hence different from fungi and are associated with odor and taste problems. Various colors of algae are observed in nature.

9.8 Analytical Concepts

In water quality analysis, the mass conservation or continuity equation contains terms that describe the chemical or biological reaction. The reaction may entail production and decay of chemical or biological constituents. The mechanism of reaction depends on the order of reaction which is often described by zero order, first order, or second order. In reaction kinetics, it is important to determine the time required for the constituent concentration to become half from its initial state, often termed as half-life.

If c is concentration of a constituent in water then the zero-order reaction states that the change in c with time (negative), that is the rate of decline, is constant, expressed mathematically as

$$\frac{dc}{dt} = -k \tag{9.18}$$

where k is the rate constant having the dimension of T^{-1}, indicating the speed of reaction. Let the initial condition be defined as

$$c = c_0 \text{ at } t = 0 \tag{9.19}$$

where c_0 is the initial concentration. Integration of Eq. (9.18) with the condition given by Eq. (9.19) yields

$$c(t) = c_0 - kt \tag{9.20}$$

The rate constant k is the slope of a line and can be computed by plotting the values of c against time and fitting a straight line and determining its slope.

In a similar vein, the first-order reaction states that the change in c with time (negative), that is the rate of decline, is directly proportional to c, expressed mathematically as

$$\frac{dc}{dt} = -kc \tag{9.21}$$

Integration of Eq. (9.21) with the condition given by Eq. (9.19) yields

$$c(t) = c_0 \exp(-kt) \tag{9.22}$$

The value of k can be determined by plotting the values of c against time on a semi-log paper and fitting a straight line. The slope of this line will yield the value of k.

The second-order reaction states that the change in c with time (negative), that is the rate of decline, is directly proportional to the square of concentration, expressed mathematically as

$$\frac{dc}{dt} = -kc^2 \tag{9.23}$$

Integration of Eq. (9.23) with the condition given by Eq. (9.19) yields

$$c(t) = \frac{c_0}{1 + ktc_0} \tag{9.24}$$

The value of k can be obtained by plotting $1/c$ against time, fitting a straight line, and computing the slope of the line.

Example 9.2

Determine the half-life for zero-order reaction, of first-order reaction, and of second-order reaction. Also, determine the second half-life of the first half-life in each case.

Solution

For half-life, $c = 0.5\, c_0$. Therefore for the zero-order reaction from Eq. (9.20)

$$\frac{1}{2}c_0 = c_0 - kt_{1/2}$$

This gives the half-life as

$$t_{1/2} = \frac{c_0}{2k}$$

The second half-life is now given as

$$t_{1/4} = \frac{c_0}{4k}$$

Similarly, for the first-order reaction from Eq. (9.22)

$$\frac{1}{2}c_0 = c_0 \exp(-kt_{1/2})$$

This yields the first half-life as

$$t_{1/2} = \frac{\ln 2}{k}$$

Similarly, the second half-life is given as

$$t_{1/4} = \frac{\ln 4}{k}$$

For the second-order reaction from Eq. (9.24), the half-life can be written as

$$\frac{1}{2}c_0 = \frac{c_0}{1 + kc_0 t_{1/2}}$$

This yields the half-life as

$$t_{1/2} = \frac{1}{kc_0}$$

Likewise, the second half-life is given as

$$t_{1/4} = \frac{3}{kc_0}$$

9.8.1 Biochemical Oxygen Demand

As indicated earlier, BOD comprises three components: CBOD, NBOD, and SOD, where CBOD is further divided into particulate CBOD and dissolved CBOD. CBOD denotes the oxygen demand for biodegradable pollutants and oxidizable nonbiodegradable pollutants. The effects of both CBODs are normally lumped but, if treated, the effects can be added. The particulate CBOD becomes part of SOD. BOD represents the total amount of oxygen consumed for complete oxidation of all organic wastes. It tells how fast biological organisms consume oxygen in a water body. In a waste that entails both CBOD and NBOD, CBOD occurs first, almost immediately. As time increases, the process of oxidation slows down, decreasing BOD and the final BOD may be taken as the one corresponding to 30 days, for example. Often, BOD_5, that is oxygen consumed over a 5-day period, is taken as a reference, but Nordic countries use 7-day BOD and France uses both 5-day and 21-day BOD.

The value of BOD_5 varies from one river to another, depending on the quality of its water. Thomann and Mueller (1987, p. 644) have given values of BOD_5 for different cases. It is typically less than 1 mg/L in pristine rivers and 2–8 mg/L in moderately polluted rivers. Raw sewage may have a BOD_5 value of 600 mg/L and sewage after tertiary treatment may have a value of 20 mg/L. In the United States, the value of BOD concentration in municipal waste ranges from 100 to 450 mg/L for $CBOD_5$, 120 to 580 mg/L for CBOD (untreated), 220 mg/L for NBOD, 5 to 35 mg/L for organic nitrogen, and 10 to 60 mg/L for ammonia nitrogen. For combined sewer outflows, $CBOD_5$ ranges from 40 to 503 mg/L, CBOD (untreated) 220 mg/L, organic nitrogen 0.08 to 2.5 mg/L, and ammonia nitrogen 0 to 11.5 mg/L. In urban runoff, $CBOD_5$ ranges from 2 to 84 mg/L, organic nitrogen from 0.2 to 4.8 mg/L, and ammonia nitrogen from 0.1 to 1.9 mg/L. Thomann and Mueller (1987) have also given reaction rates for different cases. In the United States, the reaction rates approximately range from 0.3 to 0.4 for CBOD at 20°C for untreated wastes, and 0.1–0.3 for wastes after primary and secondary treatment, and the ultimate CBOD to BOD_5 ratio is 1.2 for untreated waste and 1.6 for treated waste.

The CBOD kinetics can be briefly described. Let L denote the concentration of oxidizable carbonaceous material (ML^{-3}). Assuming a first-order reaction, one can express

$$\frac{dL}{dt} = -K_1 L; \; L(t = 0) = L_0 \tag{9.25}$$

where K_1 is constant denoting the rate of oxidization of the carbonaceous material (T^{-1}), and L_0 is the initial concentration of the oxidizable carbonaceous material (MT^{-3}). The solution of Eq. (9.25) follows

$$L(t) = L_0 \exp(-K_1 t) \tag{9.26}$$

The average value of untreated municipal waste at 20°C is about 0.35/day (Thomann and Mueller, 1987). Equation (9.26) can be used to compute the CBOD on any day and hence the amount of reduction from the initial value by taking the difference between L_0 and $L(t)$ or the amount of oxygen consumed during oxidization as

$$\text{Amount of oxygen consumed} = L_0 - L(t) = L_0[1 - \exp(-K_1 t)] \tag{9.27}$$

Now we consider NBOD, which denotes the oxygen required for nitrification, which represents the biochemical oxidation of ammonia to nitrates, which occurs in two stages, written as

$$\text{Stage 1: } NH_4 + 1.5O_2 \rightarrow 2H^+ + H_2O + NO_2^-$$

$$\text{Stage 2: } NO_2^- + 0.5O_2 \rightarrow NO_3^-$$

For expressing the reaction kinetics for these stages, let L^N represent the overall concentration of oxidization nitrogenous material. Assuming a first-order reaction and initial concentration as L_0^N, one can write

$$\frac{dL^N}{dt} = -K_N L^N,\ L^N(t=0) = L_0^N \tag{9.28}$$

where K_N denotes the overall NBOD oxidation rate. The solution of Eq. (9.27) can be written as

$$L^N(t) = L_0^N \exp(-K_N t) \tag{9.29}$$

Hence the amount of oxygen consumed during nitrification can be written as

$$\text{Amount of oxygen consumed} = L_0^N - L^N(t) = L_0^N[1 - \exp(-K_1 t)] \tag{9.30}$$

The value of K_N ranges from 0.1 to 0.5 per day but it depends on the temperature and so does K_1. Both these reaction rates can be expressed as a function of temperature as

$$K_1(T) = K_{1(20)}\theta^{T-20} \tag{9.31}$$

$$K_N(T) = K_{N(20)}\theta^{T-20} \tag{9.32}$$

where K_1 and K_N are, respectively, the CBOD and NBOD reoxygenation coefficients at a temperature in centigrade, $K_{1(20)}$ and $K_{N(20)}$ are the values of these coefficients at 20°C, and θ is the temperature correction factor (1.024).

Example 9.3

Compute and plot CBOD and NBOD at a 2-day interval from the beginning ($t = 0$) to $t = 48$ days. Assume $L_0 = 250$ mg/L, $K_1 = 0.32$/day; and $L_0^N = 140$ mg/L and $K_N = 0.28$/day. Also compute the amount of oxygen consumed during oxidization and nitrification. Also plot the computed values.

Solution

The CBOD is given by Eq. (9.26) for which $L_0 = 250$ mg/L and $K_1 = 0.32$/day are given. Therefore the values of CBOD at a 2-day interval are as follows:

Time (day)	CBOD (mg/L)	Time (day)	CBOD (mg/L)	Time (day)	CBOD (mg/L)	Time (day)	CBOD (mg/L)	Time (day)	CBOD (mg/L)
0	250	10	10.19	20	0.42	30	0.017	40	0.0007
2	131.82	12	5.37	22	0.22	32	0.009	42	0.0004
4	69.51	14	2.83	24	0.12	34	0.005	44	0.0002
6	36.65	16	1.49	26	0.06	36	0.002	46	0.0001
8	19.33	18	0.79	28	0.03	38	0.001	48	5.34E-05

The amount of oxygen consumed during oxidization is given by Eq. (9.27). Therefore the values of consumed oxygen at a 2-day interval are as follows:

Time (day)	Oxy. Con. (mg/L)	Time (day)	Oxy. Con. (mg/L)	Time (day)	Oxy. Con. (mg/L)	Time (day)	Oxy. Con. (mg/L)	Time (day)	Oxy. Con. (mg/L)
0	0	10	239.81	20	249.58	30	249.98	40	250.00
2	118.18	12	244.63	22	249.78	32	249.99	42	250.00
4	180.49	14	247.17	24	249.88	34	250.00	44	250.00
6	213.35	16	248.51	26	249.94	36	250.00	46	250.00
8	230.67	18	249.21	28	249.97	38	250.00	48	250.00

The NBOD is given by Eq. (9.29) for which $L_0^N = 140$ mg/L and $K_N = 0.28$/day are given. Therefore at a 2-day interval, the values of NBOD are as follows:

Time (day)	NBOD (mg/L)	Time (day)	NBOD (mg/L)	Time (day)	NBOD (mg/L)	Time (day)	NBOD (mg/L)	Time (day)	NBOD (mg/L)
0	140	10	8.51	20	0.52	30	0.031	40	0.0019
2	79.97	12	4.86	22	0.30	32	0.018	42	0.0011
4	45.68	14	2.78	24	0.17	34	0.010	44	0.0006
6	26.09	16	1.59	26	0.10	36	0.006	46	0.0004
8	14.90	18	0.91	28	0.06	38	0.003	48	0.0002

The amount of oxygen consumed during nitrification is given by Eq. (9.30). Therefore the values of consumed oxygen at a 2-day interval are as follows:

Time (day)	Oxy. Con. (mg/L^3)	Time (day)	Oxy. Con. (mg/L^3)	Time (day)	Oxy. Con. (mg/L^3)	Time (day)	Oxy. Con. (mg/L^3)	Time (day)	Oxy. Con. (mg/L^3)
0	0	10	131.49	20	139.48	30	139.97	40	140.00
2	60.03	12	135.14	22	139.70	32	139.98	42	140.00
4	94.32	14	137.22	24	139.83	34	139.99	44	140.00
6	113.91	16	138.41	26	139.90	36	139.99	46	140.00
8	125.10	18	139.09	28	139.94	38	140.00	48	140.00

The above computed values are plotted in Fig. 9.1.

9.8.2 Nitrification

Nitrification is an important component of the nitrogen cycle and is a microbial process of conversion of ammonia nitrogen to nitrate nitrogen. In other words, nitrogen compounds, especially ammonia (NH_3) are oxidized to nitrites (NO_2) and then to nitrates (NO_3). In the oxidization process, an electron is taken from one atom of nitrogen and added to the oxygen atom. In the presence of two types of bacteria, a two-stage chemical reaction occurs which can be expressed as

$$NH_3^+ + O_2 + \text{bacteria } (\textit{Nitrosomonas}) \rightarrow NO_2^- + 3H^+ + 2e^-$$

$$NO_2^- + H_2O + \text{bacteria } (\textit{Nitrobacter}) \rightarrow NO_3^- + 2H^+ + 2e^-$$

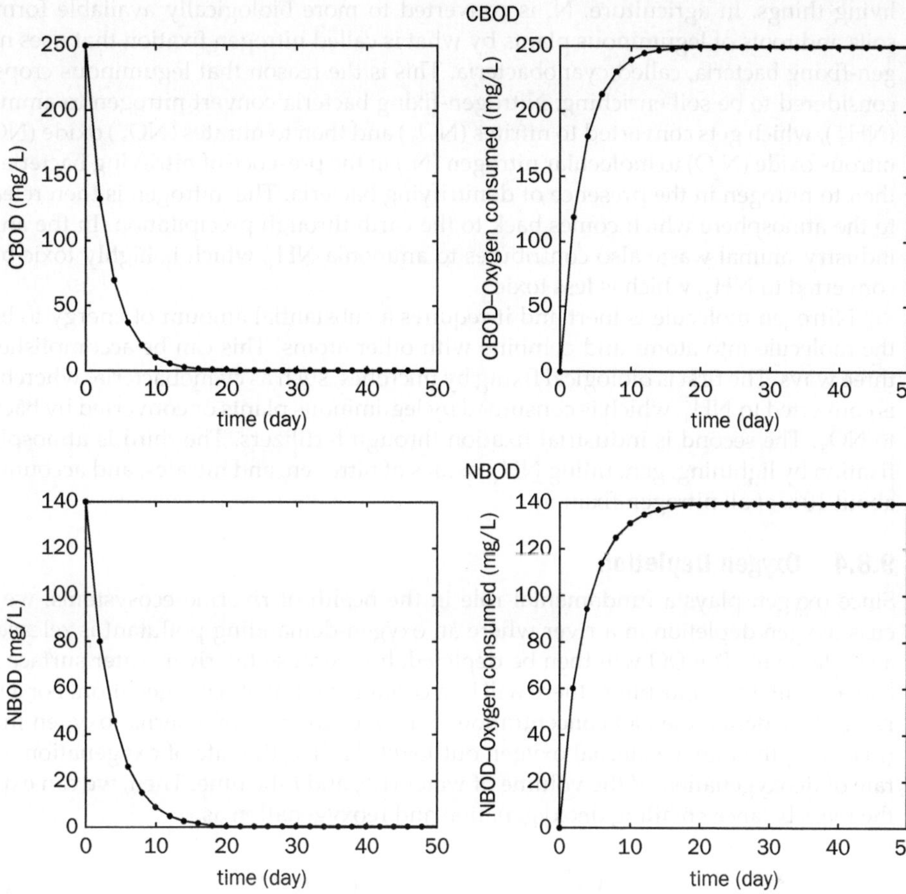

Figure 9.1 CBOD and NBOD as well as oxygen consumed during oxidization and nitrification as a function of time.

Nitrification is important in wastewater treatment and especially for fish. Toxic ammonia generated by fish excretion, food waste, algae, and other organic material is converted to less toxic nitrates and the efficiency of this conversion is vital for fish survival.

9.8.3 Denitrification

Denitrification is an anaerobic process in which nitrites are converted into molecular nitrogen (N_2) in the presence of denitrifying bacteria, called *Paracoccus denitrificans*. The stages of conversion are nitrates (NO_3) to nitrites (NO_2) to nitric oxide (NO) to nitrous oxide (N_2O) to molecular nitrogen (N_2) and can be expressed as

$$NO_3^- \rightarrow NO_2^- \rightarrow NO \rightarrow N_2O \rightarrow N_2$$

Like nitrification, denitrification is an important component of the nitrogen cycle. Molecular nitrogen is abundantly available in the atmosphere but is not accessible to living things. In agriculture, N_2 is converted to more biologically available forms in soils and roots of leguminous plants by what is called nitrogen fixation that uses nitrogen-fixing bacteria, called cyanobacteria. This is the reason that leguminous crops are considered to be soil enriching. Nitrogen-fixing bacteria convert nitrogen to ammonia (NH_4^+), which gets converted to nitrites (NO_2^-) and then to nitrates (NO_3^-) oxide (NO) to nitrous oxide (N_2O) to molecular nitrogen (N_2) in the presence of nitrifying bacteria and then to nitrogen in the presence of denitrifying bacteria. This nitrogen is then released to the atmosphere which comes back to the earth through precipitation. In the animal industry, animal waste also contributes to ammonia NH_3, which is highly toxic but is converted to NH_4^-, which is less toxic.

Nitrogen molecule is inert and it requires a substantial amount of energy to break the molecule into atoms and combine with other atoms. This can be accomplished in three ways. The first is biological fixing by microbes, such as cyanobacteria, whereby N_2 is converted to NH_4^-, which is consumed by leguminous plants or converted by bacteria to NO_3^-. The second is industrial fixation through fertilizers. The third is atmospheric fixation by lightning, generating NH_3, oxides of nitrogen, and nitrates, and accounts for about 10% of all nitrogen fixation.

9.8.4 Oxygen Depletion

Since oxygen plays a fundamental role in the health of riverine ecosystems, we discuss oxygen depletion in a river where an oxygen-demanding pollutant is released at a single point. The DO will then be depleted, but because the river water surface is in contact with the atmosphere there will be a certain amount of reoxygenation. For a river reach, let c define the DO concentration (ML^{-3}), O_i the rate of external oxygen inflow (MT^{-1}), O_o the rate of external oxygen outflow (MT^{-1}), r_o the rate of oxygenation, r_d the rate of deoxygenation, V the volume of water (L^3), and t the time. Then, we can express the mass balance entailing deoxygenation and reoxygenation as

$$V\frac{dc}{dt} = O_i - O_o + r_o \times V - r_d \times V \tag{9.33}$$

On the right side (*RS*) of Eq. (9.33), the rate of external oxygen inflow and reoxygenation can be combined into the rate of change of DO due to reaeration; likewise, the external oxygen outflow and deoxygenation can be combined into the rate of change in DO due to decay. Thus the RS can be expressed as

$$RS = \left[\frac{dc}{dt}\right]_{\text{decay}} V + \left[\frac{dc}{dt}\right]_{\text{reaeration}} V \tag{9.34}$$

Thus, Eq. (9.33) can be written as

$$V\frac{dc}{dt} = \left[\frac{dc}{dt}\right]_{\text{decay}} V + \left[\frac{dc}{dt}\right]_{\text{reaeration}} V \tag{9.35}$$

This yields

$$\frac{dc}{dt} = \left[\frac{dc}{dt}\right]_{\text{decay}} + \left[\frac{dc}{dt}\right]_{\text{reaeration}} \tag{9.36}$$

Each term on the right side of Eq. (9.36) can be expressed by invoking a first-order reaction. For the rate of reaeration or reoxygenation, let k_r be the reaeration coefficient and c_s be the saturation value of the DO concentration. Then, we can write

$$\left[\frac{dc}{dt}\right]_{\text{reaeration}} = k_r(c_s - c) \tag{9.37}$$

Likewise, for the rate of decay or deoxygenation, let k_d be the reaeration coefficient reflecting both oxidation of settled and soluble BOD, and L be the BOD remaining in the water at time t. Then, we can write

$$\left[\frac{dc}{dt}\right]_{\text{decay}} = -k_d L \tag{9.38}$$

The value of c_s (mg/L) can be estimated from its relation with temperature (°C) (Committee on Sanitary Engineering Research of the Sanitary Engineering Division, ASCE, 1960) as

$$c_s = 14.652 - 0.41022T + 0.0079910T^2 - 0.000077774T^3 \tag{9.39}$$

With the use of Eqs. (9.37) and (9.38), Eq. (9.26) can be expressed as

$$\frac{dc}{dt} = -k_d L + k_r(c_s - c) \tag{9.40}$$

Equation (9.40) describes the change of DO concentration in a water body that is subjected to BOD loading.

Example 9.4

Compute the value of c_s for temperature equal to 0, 4, 8, 12, 16, 20, 24, 28, 32, 36, 40, 44, 48, and 52°C and plot the values.

Solution

The values of c_s are given by Eq. (9.39) and for given temperature values, these are plotted in Fig. 9.2 and are given as follows:

Temp. (°C)	C_s (mg/L)	Temp (°C)	C_s (mg/L)	Temp (°C)	C_s (mg/L)
0	14.65	20	9.02	40	6.05
4	13.13	24	8.33	44	5.45
8	11.84	28	7.72	48	4.77
12	10.75	32	7.16	52	3.99
16	9.82	36	6.61		

Equation (9.40) can also be expressed in terms of oxygen deficit (D) defined as

$$D = c_s - c \tag{9.41}$$

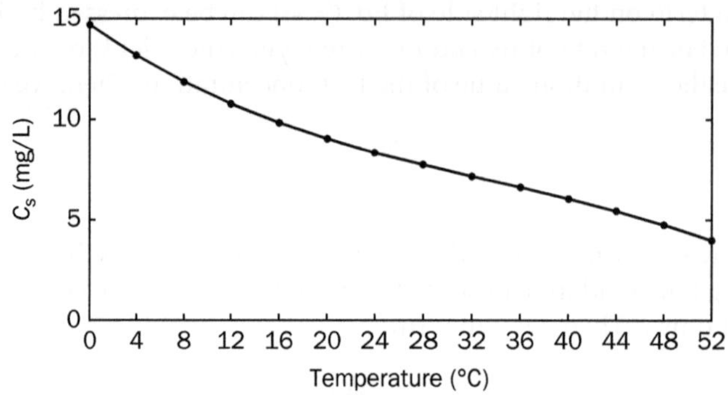

FIGURE 9.2 Plot of c_s versus temperature.

Then

$$\frac{dD}{dt} = k_d L - k_r D \tag{9.42}$$

If a first-order decay is invoked, then L can be expressed as

$$L(t) = L_0 \exp(-k_d t) \tag{9.43}$$

and Eq. (9.42) as

$$\frac{dD}{dt} = k_d L_0 \exp(-k_d t) - k_r D \tag{9.44}$$

where L_0 represents the BOD remaining in water at time t. The solution of Eq. (9.44), subject to the condition $D(t) = D_0$, follows

$$D(t) = \frac{k_d L_0}{k_r - k_d}[\exp(-k_d t) - \exp(-k_r t)] + D_0 \exp(-k_r t) \tag{9.45}$$

Differentiating Eq. (9.45) and equating to 0, we obtain the minimum or critical oxygen deficit as

$$t_c = \frac{1}{k_r - k_d} \ln\left\{\frac{k_r}{k_d}\left[1 - \frac{D_0(k_r - k_d)}{k_d L_0}\right]\right\} \tag{9.46}$$

Equation (9.45) can be written in terms of DO concentration $c(t)$ as

$$c(t) = c_s - \frac{k_d L_0}{k_r - k_d}[\exp(-k_d t) - \exp(-k_r t)] - D_0 \exp(-k_r t) \tag{9.47}$$

Equation (9.46) expresses $c(t)$ as a function of time, taking into consideration only reaeration and deoxygenation. If the flow in a river is assumed steady state with an average velocity u, then time can be converted to the space coordinate as $x = ut$, where x is the distance from the outfall. Then, Eq. (9.47) can be expressed as

$$c(t) = c_s - \frac{k_d L_0}{k_r - k_d}\left[\exp\left(-k_d \frac{x}{u}\right) - \exp\left(-k_r \frac{x}{u}\right)\right] - D_0 \exp\left(-k_r \frac{x}{u}\right) \quad (9.48)$$

Equation (9.48) is the famous Streeter-Phelps (1925) equation, also referred to as the oxygen sag curve. Clearly, this equation does not hold if $k_r = k_d$.

Example 9.5

Compute the values of DO and DO deficit at time equal to 2, 4, 6, 8, 10, 12, 14, 16, 18, and 20 days, assuming $k_r = 0.4$/day, $k_d = 0.25$, $L_0 = 25$ mg/L, $D_0 = 2.5$ mg/L, and $c_s = 12$ mg/L. Plot these values.

Solution

The DO concentration is given by Eq. (9.47) or Eq. (9.48) for which $k_r = 0.4$/day, $k_d = 0.25$, $L_0 = 25$ mg/L, $D_0 = 2.5$ mg/L, and $c_s = 12$ mg/L are given. Thus at a 2-day interval, the DO concentration is plotted in Fig. 9.3 and is given as follows:

Time (day)	DO (mg/L)	Time (day)	DO (mg/L)	Time (day)	DO (mg/L)
0	14.50	8	8.16	16	11.31
2	6.57	10	9.39	18	11.57
4	5.59	12	10.29	20	11.73
6	6.71	14	10.91		

The DO deficit is given by Eq. (9.45). Therefore the DO deficit values are plotted in Fig. 9.4 and are given as follows:

Time (day)	DO deficit (mg/L)	Time (day)	DO deficit (mg/L)	Time (day)	DO deficit (mg/L)
0	2.50	8	4.04	16	0.70
2	7.67	10	2.70	18	0.43
4	7.42	12	1.75	20	0.27
6	5.74	14	1.11		

9.8.5 Reoxygenation Coefficient

The reaeration coefficient in a river has been found to depend on a number of factors, such as temperature, flow velocity, turbulence, and flow depth and empirical relations have been developed. Some of these relations are given here. Taking $k_{r(20)}$ as the value of the coefficient at 20°C temperature and θ as the temperature correction factor (=1.024), the reaeration coefficient as a function of temperature (°C) can be expressed as

$$k_r(T) = k_{r(20)}\theta^{T-20} \quad (9.49)$$

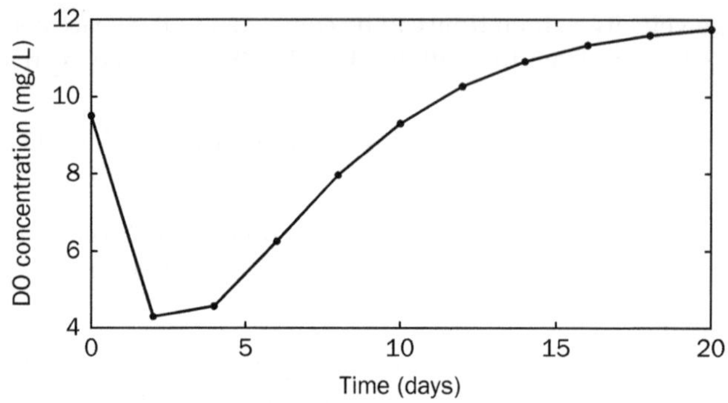

FIGURE 9.3 Plot of DO concentration as a function of time.

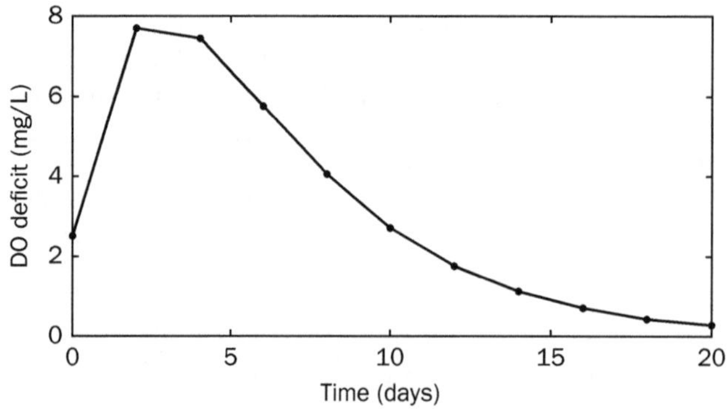

FIGURE 9.4 Plot of DO deficit as a function of time.

In terms of flow velocity and depth, two popular equations are due to O'Connor and Dobbins (1956) and Churchill et al. (1962). If velocity u is ft/s, h in ft, and k_r in 1/day, then the Churchill-Elmore-Buckingham equation can be expressed as

$$k_{r(20)} = \frac{11.6u}{h^{1.67}} \tag{9.50}$$

and if u is in m/s, and h in m, then

$$k_{r(20)} = \frac{26.7u^{0.969}}{h^{1.673}} \tag{9.51}$$

The O'Connor-Dobbins equation in FPS units can be written as

$$k_{r(20)} = \frac{(D_L u)^{0.5}}{h^{1.5}} \tag{9.52}$$

and in the SI units, it can be written as

$$k_{r(20)} = \frac{3.28u^{0.5}}{h^{1.5}} \tag{9.53}$$

Bachelor and Lazo (1999) considered the effect of bed slope and obtained values of 0.39–0.98 for mild slopes and 0.84–3.22 for medium slopes. They reported the following equations in SI units:

For mild slopes up to 0.05%,

$$k_r = \frac{10.046u^{2.696}}{h^{3.902}} \tag{9.54}$$

and for medium slopes from 0.5% to 0.5%

$$k_r = \frac{1.923u^{1.328}}{h^{2.006}} \tag{9.55}$$

9.8.6 Deoxygenation Coefficient

The value of decay coefficient is dependent on the pollutant. For example, raw sewage has a value of 0.5/day and the value decreases for treated sewage. Litton (2003) reported the value of the decay coefficient as 0.087 for BOD, 0.11 for CBOD, and 0.057 for NBOD for the San Joaquin River in California at 20°C; and as 0.094 for BOD, 0.11 for CBOD, and 0.076 for NBOD for Stockton Deep Water Ship Channel.

9.9 Nonpoint Source Pollution

Nonpoint source pollution is one of the major causes of environmental pollution and pollution occurring in agricultural, forest, rural, and urban watersheds is primarily nonpoint source.

9.9.1 Pollutants in Urban Runoff

In urban watersheds, pollutants are generated by motor vehicles, littering, animal fecal droppings, road pavement material, construction activities, garbage disposal, sewage, gas leakage, hospital waste, chemicals applied in lawns and parks, and other kinds of waste. These pollutants are carried by stormwater. Some of the contaminants are washed off in surface water and some infiltrate the pervious areas.

9.9.2 Pollutants in Agricultural Runoff

Chemicals applied in agriculture are either short-lived such as pesticides, or long-lived such as chlorinated hydrocarbons. Short-lived chemicals cause a pollution hazard in weeks upon their application. Long-lived chemicals cause pollution year round. Owing to their persistence and biological concentration, their application has been banned (Bailey et al., 1974).

Depending on crop, soil, season, and tillage practices, different types of chemicals are applied. For example, pesticides applied to agriculture may have five types of formulations, such as (1) wettable powder, (2) solution, (3) emulsion, (4) dust, and (5) granules and pellets. Pesticides are sprayed or dusted over crops, applied to the soil surface, or incorporated in the soil. The type of formulation and method of application affect soil-pesticide interaction, equilibria, and attenuation processes. Pollutants transported by agricultural runoff depends on the chemical type and formulation, soil properties, climatic conditions, watershed characteristics, and agricultural practices.

9.9.3 Pollution in Subsurface Waters

The buildup and transport of contaminants in a subsurface environment depend on the physical, chemical, and microbiological processes and their interactions. Pollutants interact with soil and may undergo chemical or microbial degradation. Chemical sorption and desorption, precipitation, and dissolution may occur, governing the mobility of compounds.

9.10 Water Quality Management

In order to determine the water quality of a water body and evaluate the impacts on water quality, water quality monitoring is of profound importance. From a water quality management perspective, it is important to identify water quality problems, abide by established water quality standards, and comply with laws and regulations governing water quality management. Water quality problems emanate from increasing discharges of untreated or inadequately treated wastewater, untreated waste, discharge of agrochemicals, discharge of hazardous or toxic or other kinds of industrial wastes, or drainage of saline water from croplands, overexploitation of groundwater and saltwater intrusion, and improper management of watersheds. Proper watershed management ensures high infiltration rates, controlled erosion, sustenance of baseflow, and controlled streamflow response to high precipitation.

An important issue in water quality management is the maintenance of water quality standards. This is particularly a problem in developing countries where water quality standards are often ignored or not attended to. Local, state, and federal regulatory agencies are responsible for ensuring that water quality standards are followed at all levels, but this is often not the case. It may partly be that such standards have been passed down from the top and the people who need to follow them were not consulted and have therefore little appreciation or they are not well informed of their significance and the consequences for them if not followed. Water quality problems arise at the local level and that is where action has to be taken.

9.11 Summary

One of the principal goals of watershed management is to secure flows of high water quality for all people in the river basin of interest. The quality of water flowing from upland areas to downstream users and how the quality is impacted by land use activities are receiving an increase in worldwide attention. This is because water quality management is fundamental to watershed management. All western countries and most developing countries have established environmental protection agencies, legislated laws for regulating water quality, and have established standards for water quality for different uses and sectors. Future water engineers will need to have a sound understanding of the principles of water quality.

9.12 Questions

9.1 Consider a 1.0 L of solution that contains 25 g of sodium carbonate (Na_2CO_3) whose density can be taken as 2.54 g/mL. Determine the molarity and mole fraction of the solution.

9.2 Determine the half-life for zero-order reaction, of first-order reaction, and of second-order reaction. Also, determine the second half-life of the first half-life in each case.

9.3 Compute and plot CBOD and NBOD at a 3-day interval from the beginning ($t = 0$) to $t = 18$ days. Assume $L_0 = 300$ mg/L, $K_1 = 0.35$/day; and $L_0^N = 145$ mg/L and $K_N = 0.27$/day. Also compute the amount of oxygen consumed during oxidization and nitrification.

9.4 Compute the value of c_s for temperature equal to 0, 3, 6, 9, 12, 15, and 18°C and plot the values.

9.5 Compute the values of DO and DO deficit at time equal to a 2-day interval from the beginning ($t = 2$) to 12 days, assuming $k_r = 0.5$/day, $k_d = 0.28$, $L_0 = 30$ mg/L, $D_0 = 3$ mg/L, and $c_s = 12$ mg/L. Plot these values.

9.6 Consider a 0.5 L of solution that contains 16 g of calcium carbonate ($CaCO_3$) whose density can be taken as 2.83 g/mL. Determine the molarity and mole fraction of the solution.

9.7 Compute the value of c_s for temperature equal to 0, 2, 4, 6, 8, 10, 12, 14, and 16°C and plot the values.

9.8 Compute the value of DO and DO deficit at a steady-state condition with uniform velocity (along the x axis) of 1, 1.25, 1.5, and 2 m/s with a distance of 86.4 km from the outfall.

9.9 What is critical oxygen deficit? Write down the equation for critical oxygen deficit. What is the Streeter-Phelps equation? Under what condition does this equation not hold?

9.10 What is reoxygenation coefficient? Find the reaeration coefficient for the variation in temperature at an interval of 2°C from $T = 25$ to 45°C with a slope of 0.03% (assume $u = 1.25$ m/s and $h = 2$ m).

9.13 References

Bachelor, H. V., and O. I. Lazo. Evaluation of water quality modeling parameters: Reaeration coefficient. *Proceedings, IAHR Congress*, Graz, Austria, 1999.

Bailey, G. W., R. R. Swank, and H. P. Nicholson. Predicting pesticide runoff from agricultural land: A conceptual model. *Journal of Environmental Quality*, 3(2):95–102, 1974.

Brooks, K. N., P. F. Ffolliott, H. M. Gregerson, and L. F. DeBano. *Hydrology and the Management of Watersheds*. Ames, IA: Iowa State University Press, 2003.

Churchill, M. A., R. A. Buckingham, and H. L. Elmore. *The Prediction of Stream Reaeration Rates*. Chattanooga, TN: Tennessee Valley Authority, 1962.

Committee on Sanitary Engineering Research of the Sanitary Engineering Division, ASCE, Solubility of atmospheric oxygen in water: Twenty-ninth Progress Report of the Committee on Sanitary Engineering Research of the Sanitary Engineering Division. *Journal of the Sanitary Engineering Division*, 86(4):41–54, 1960.

Litton, G. *Deposition Rates and Oxygen Demands in the Stockton Deep Water Ship Channel of The San Joaquin River*. Stockton, CA: San Joaquin River Dissolved Oxygen TMFL Steering Committee, 2003.

O'Connor, D. J., and W. E. Dobbins. The mechanisms of re-aeration in natural streams. *Journal of the Sanitary Engineering Division, ASCE,* 82(12):1115–1130, 1956.

Streeter. H. W., and E. B. Phelps. *A Study of the Pollution and Natural Purification of the Ohio River.* U.S. Public Health Bulletin No. 146, Government Printing Office, Washington, DC, 1925.

Tchobanoglous, G., and E. D. Schroeder. *Water Quality*. Menlo Park, CA: Addison-Wesley, 1985.

Thomann, R. V., and J. A. Mueller. *Principles of Surface Water Quality Modeling and Control.* New York: Harper & Row, 1987.

PART 3

Measurements

CHAPTER 10

Hydrologic Data Observation Networks

Hydrologic and hydrometeorologic variables vary with space and time, that is, their values are not constant in space and time. Therefore they need to be monitored and observed and this is the only way to communicate with nature. Observations or data form the basis of hydrologic analysis. These observations are made at a number of locations and measurement devices at locations of these observations form a network. The network requires a careful design, which depends on the objective of measurement or data. Also, a network should be regarded as a dynamic system, evolving with time and space. This chapter discusses fundamentals of network design under different types of environment. The World Meteorological Organization (WMO) (2008) has published *The Guide to Hydrological Practices*. It contains useful guidelines to set up networks for various types of data.

10.1 Hydrometeorological Data Networks

Assessment of freshwater resources, their magnitude, variability, dependability, and quality entail the collection and analysis of hydrologic data. Networks of hydrometeorological stations are established to capture hydrologic data. Data collection is the first step in sound and sustainable water resources management. Some typical objectives of hydrological observation networks are water resources assessment for basins, sub-basins, or administrative units; planning of water resources projects for irrigation, domestic and industrial water supply, hydroelectric power generation, environmental requirements, navigation, tourism, recreation, and so on; flood management including forecasting; and assessing impacts of climate change on water resources.

Design of a hydrological observations network is based upon two considerations: (1) the observation objectives and (2) physical characteristics of the hydrologic system to be monitored. To know the objectives, the potential data users and their data needs are to be identified. Priorities need to be set if there are more objectives than one. Objectives also determine the way data are analyzed. Physics of hydrologic processes must be known to estimate the variabilities with respect to space and time.

A good hydrometeorological network should be able to provide the desired information for different users. For example, to plan a reservoir project, planners need river flow data, rainfall in the catchment and downstream areas, and meteorological data, such as temperature, radiation, wind velocity, and so on. Good data give many long-term benefits, including sound investment decisions, in water supply, hydropower, irrigation projects, and the ability to provide timely warnings of droughts and floods. Coordination of data collection activities among associated agencies is essential to optimize the use of resources. It may also be noted that measurements are subject to random and systematic errors, leading to some uncertainty in results.

Optimal design of a monitoring network needs the following inputs: (1) sampling variables (which hydrological variables are to be measured?); (2) sampling locations (where these variables are to be measured?); (3) sampling frequencies (how frequently should observations be made?); (4) sampling accuracy (what technique/equipment/protocols should be used?); and (5) sampling duration (how long should measurements continue?).

Answers to these questions determine the costs of network establishment and operation and costs of land acquisition, construction, equipment procurement and installation, operation, maintenance, and staffing of field stations and data centers. Once the relationship between the objectives, effectiveness, and costs has been established, the optimal network configuration can be found by weighing these in cost-effectiveness analysis, as depicted in Fig. 10.1. HP (2003), WMO (2008), and Jain (2011) provide detailed material on this topic.

10.1.1 Types of Hydrometeorological Networks

Hydrometeorological networks can be classified into two types: basic network and minimum network.

Basic Network A basic or primary or principal network in a region has a low density of stations where measurements are taken over a long period of time. The basic network provides hydrological information that would have multiple and long-term uses.

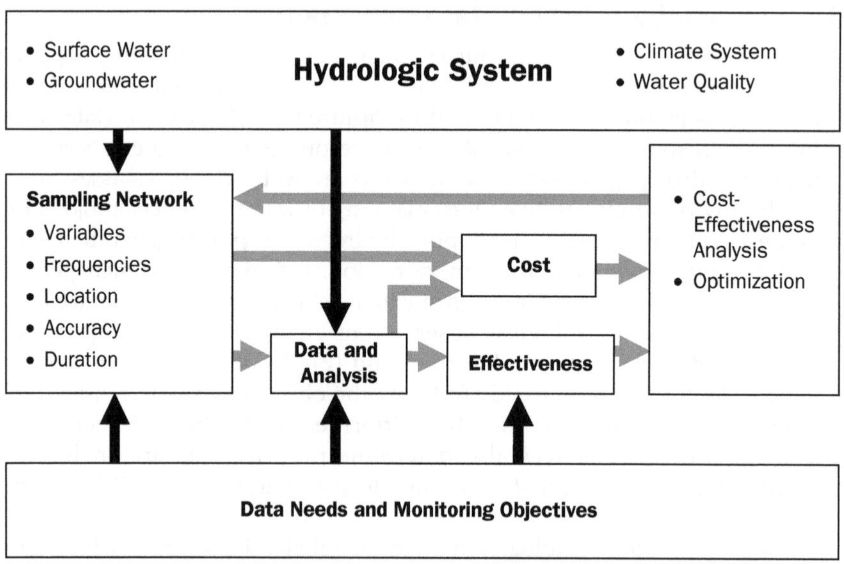

Figure 10.1 Network design process for hydrometeorological data. (black arrows represent flow of data and gray arrows represent results of analysis). (Adapted from HP, 2003.)

Minimum Network The first step in setting up a hydrological network is to establish a minimum network which contains the least number of stations necessary to initiate water resources development (WRD). The minimum network is intended as a first step to obtain the key data for WRD. The minimum network is developed rapidly by incorporating existing stations as appropriate. A minimum network will be inadequate to prepare detailed development plans.

A minimum network should include at least one primary climate and streamflow station in each climatological and physiographic area. The minimum network should also include special stations. Since a minimum network is not typically very dense, it is important that the data measured at all stations are of good quality.

Special purpose stations are required for specific purposes and are discontinued when the purpose is served. The purpose could be design and management of a project to monitor and fulfill court decisions/agreements between co-basin states or special investigations. Sometimes, the specific purpose to be served may require observations on only one variable, or be limited to one season of the year. For example, a storage gauge for measuring the total precipitation during a month or season. Networks for representative basins are established for research purposes.

Part of the hydrometeorological network should have a permanent character to ensure that some basic information is continually gathered. In addition to the basic network, stations may be established to better understand the spatial variability of a variable. Once sufficient data have been collected at the secondary stations to establish relations with the primary stations, the added value of operating the secondary station should be reexamined.

10.1.2 Steps in Network Design

The steps for network review and design include:

1. **Identify data needs:** The existing and potential future data users are approached to find out their data needs, typically by using a questionnaire.
2. **Identify the network objectives:** Based on the outcome of step 1, hydrological information needs are to be identified by listing objectives of the Hydrological Information System (HIS). The consequences of not meeting the target are to be indicated.
3. **Prioritization:** A priority ranking among the objectives is made. This helps in the allocation of resources.
4. **Determine network density:** Based on the objectives, the required network density is determined, taking in view the spatial (and temporal) correlation structure of the variable(s).
5. **Review the existing network:** The review covers the existing network density versus the required one as worked out in step 4, covering the stations in meteorological, surface water, and groundwater networks; available equipment and their adequacy for collecting the required information; and adequacy of operational procedures and possible improvements.
6. **Select the site and equipment**: If the existing network is inadequate to meet data needs, additional sites with appropriate equipment are installed.

7. **Estimate the cost:** Costs involved in developing, operating, and maintaining the existing and new sites are estimated. Then costs and effectiveness are compared. Steps 6–7 are repeated if the budget is insufficient to cover the anticipated costs and sites or equipment are revised.
8. **Implement:** Once the design is approved, the network is to be implemented. It involves executing civil works, procuring and installing equipment, and recruiting staff and training them.
9. **Review:** The network should be reviewed after about 5 years and the above procedure should be repeated as necessary.

Figure 10.1 depicts the design process for hydrometeorological networks.

10.1.3 Assessment of Data Needs

Step 1 in the network design/review is an assessment of the data requirements of the user. Needs of a user may change with time, as new problems emerge and hence an HIS would better fulfill its purpose if the needs of the user are periodically reviewed. Table 10.1 lists typical objectives of WRD and the data needed from the HIS.

A practical approach to assess data needs of the users is through small interview teams who explore the mandates and data needs of potential data users. In the teams,

Objective/Function	Data Needed from HIS
Flood—management	• Design of embankments along rivers, culverts, and bridges to pass floods requires data on temporal and spatial distribution of extreme rainfall, and high river stages and discharges. • Flood early warning systems require the data of rainfall, river flows, and meteorological variables at short intervals.
Ecological sound water management	• Environmental flow assessment requires data on the natural river stage, flow velocities, sediment movement, variation of groundwater levels, water quality, and of anthropogenic effects. • Channel cross-sections, slopes, roughness, and so on.
Design of municipal and industrial water supply systems	• Data on water quantity and quality, for example, temporal distribution of river flows and groundwater levels.
Agriculture-irrigated and rain-fed systems	• Data on water quantity and quality (surface and groundwater), climate and rainfall, and cropping data. • For operation of irrigation projects, data on water demands, rainfall, river stages, and flows (quantity and quality) are needed.
Hydropower production	• Design and operation of hydropower systems, often in combination with water supply for municipal and industrial uses, irrigation, and flood mitigating measures, require data on river flows, water quality, and sediment transport, irrigation demands, probable maximum precipitation, and so on. • For operation of the system, data on water demands, rainfall, river stages, and flows (quantity and quality) in real-time and forecasts are needed.
Recreation	• Data on water quantity and quality, flow conditions, and water levels in lakes and reservoirs and their variation.

Table 10.1 Major Objectives of WRD and Data Requirements from HIS (Adapted from HP, 2003)

both surface and groundwater organizations should be represented. To guide and streamline the discussions, a questionnaire may be prepared which is filled in during interaction with data users. It may contain the following items:

- Description of data user (name, sector, mandate, services provided, staffing, and financing).
- Water system use (present and future) with respect to quantity and quality, and responsibility.
- Data use and requirements (variables, type, location, frequency of measurement, form, accuracy, consequences if not available, summary of present status of data supply).

A list of potential users of hydrologic data, adapted from HP (2003), is given in Table 10.2.

10.2 Integration of Networks

An HIS aims to integrate the networks for measuring hydrometeorological data, streamflow data, geohydrological data, and surface and groundwater quality data. Commonly, these networks are operated by different agencies. To avoid duplication and save resources, the networks of various agencies should be integrated, technically and organizationally. Integration of networks implies that these are complimentary and there is regular exchange and reconciliation of data to produce authenticated high-quality data.

The hydrological cycle is a continuum, and its inter-connections permit the partial transfer of information obtained from one component of the cycle to another. The possibility of such transfer is a function of hydrological dependence of a component on the other. For example, if a precipitation-runoff model has been developed for a basin, it can be used to reconstruct streamflow records during the periods when observed stream-gauge data are missing but precipitation records are present.

Governmental Departments	**Nongovernmental Organizations**
– Surface Water Departments: Federal/Local	• Chambers of Commerce
– Groundwater Department: Federal/Local	• Environmental Protection Organizations
– Agricultural Department	• Farmers Associations/ Cooperative Bodies
– Environment and Forest Department	• Water Users Associations
– Fisheries Department	**Private Sector**
– Geology and Mines Department	– Contractors
– Hydro and Thermal Power Producers	– Energy Utilities
– Industries and Commerce Department	– Engineering Consultants
– Meteorological Department	– Industries: for example, Beverages, Capital Goods, Consumer Goods, Cotton Mills, Metals, Paper Mills
– Pollution Control Department	
– Public Health Department	
– Tourism Department	
– Transport/Railway Department	
– Water Supply and Sewerage Department	
– Research and Academic Institutes	

Table 10.2 List of Potential Users of Hydrological Data

10.3 Precipitation Data Networks

Precipitation data are used in a wide range of activities in WRD and management. Network designers need to know which statistical parameter(s) are to be estimated and with what accuracy. Given the variability in space and time, this determines the number of required stations in the network and the frequency of measurements.

Since the area from which a gauge catches rainfall is very small compared to the areal extent of a storm, a very large number of rain gauges would be required to get a correct estimate of storm rainfall over a catchment. However, factors, such as topography, accessibility, budget, security, and so on, limit the number of rain gauges that can be installed and operated. Hence there is a need to determine an optimum density of gauges to yield reasonably accurate information about areal rainfall.

For design of structures, data of short-duration high-intensity rainfall (e.g., hourly or daily) and their areal extent are needed. For actions, such as reservoir operation and flood forecasting, we require real-time short-duration data. Clearly, different objectives lead to different information needs and different network densities.

WMO (2008) presented a general guide for the minimum density of precipitation stations for different physiographic regions (Table 10.3) and suggested that at least 10% of the network stations should be equipped with recording rain gauges. WMO (2008) has not mentioned the criteria applied in the preparation of this table and the accuracy of results from this network.

10.3.1 Design of Optimal Rain Gauge Network

In design of an optimal rain gauge network, the aim is to determine the smallest number of rain gauge stations to get the required accuracy and to suggest suitable locations for the stations. To this end, the coefficient of variation (C_v) based technique is a popular method.

	Precipitation		
Physiographic Unit	**Non recording**	**Recording**	**Evaporation**
Coastal	900	9,000	50,000
Mountains	250	2,500	50,000
Interior plains	575	5,750	5,000
Hilly/undulating	575	5,750	50,000
Small islands	25	250	50,000
Urban areas	–	10–20	–
Polar/arid	10,000	100,000	100,000

Table 10.3 Recommended Minimum Densities of Precipitation Stations (area in km^2 per station) (Source: WMO, 2008)

10.3.2 C_v-Based Technique

For the area where some rain gauges already exist, the optimal rain gauge network can be designed by using the coefficient of spatial variation of rainfall. The optimum number "N" of rain gauges is obtained by using the relation

$$N = (C_v/p)^2 \tag{10.1}$$

where C_v = the coefficient of variation of rainfall based on existing rain gauge stations and p = the percentage error in the estimation of rainfall. Alternatively, the existing coefficient of variation of rainfall (C_{ve}) and the required coefficient of variation (C_{vr}) can also be utilized for determining the optimal network. If "n" and "N" are the existing and required number of rain gauges, then

$$N = n\frac{C_{ve}}{C_{vr}} \tag{10.2}$$

If there are m stations in the catchment, each recording values $P_1, P_2, \ldots, P_i, \ldots P_m$ in a known time, the coefficient of variation C_v is calculated as

$$C_v = \frac{100 \times \sigma}{\overline{P}} \tag{10.3}$$

where $\sigma = \sqrt{\left[\dfrac{\sum_1^m (P_i - \overline{P})^2}{m-1}\right]}$ = the standard deviation of precipitation values, P_i = the precipitation depth at the i-th station, and $\overline{P} = \dfrac{1}{m}\left(\sum_1^m P_i\right)$ = the mean precipitation.

In calculating N from Eq. (10.1), it is usual to take $p = 10\%$. It is seen that the required number of rain gauge stations increases with value of p.

Example 10.1

In the catchment (area = 2800 sq. km) shown in Fig. 10.2, six rain gauge stations have been set up. The average annual precipitation at the stations is given below:

Station	A	B	C	D	E	F
Rainfall (mm)	700	830	650	710	600	400

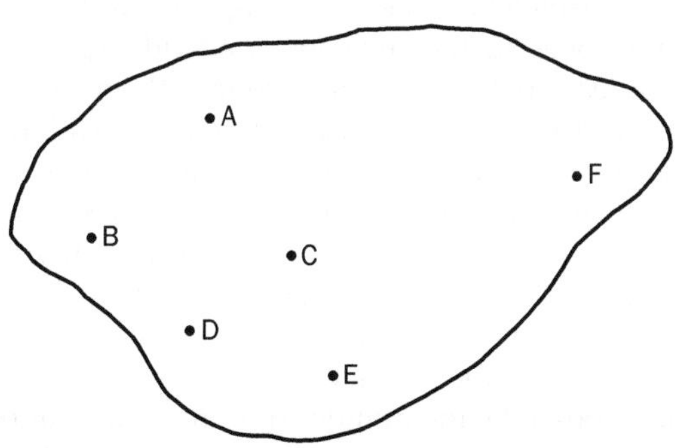

FIGURE 10.2 Map showing locations of rain gauge stations for Example 10.1.

Estimate the minimum number of rain gauge stations for the catchment. The allowable error may be taken as 5%.

Solution

$$P = \text{mean of the rainfall} = \frac{\sum P}{n}$$

$$= \frac{700 + 830 + 650 + 710 + 600 + 400}{6} = \frac{3890}{6} = 648 \text{ mm}$$

$$\text{Standard deviation } \sigma = \sqrt{\frac{\sum (P - \bar{P})^2}{n - 1}} = \sqrt{\frac{103484}{6 - 1}} = 143.86 \text{ mm}$$

$$\text{Coefficient of variation} = C_v = \frac{\sigma}{P} = \frac{143.86}{648} = 0.222$$

$$\text{Therefore } N = \left(\frac{0.222}{0.05}\right)^2 = (4.44)^2 = 19.7 \text{ say } 20.$$

Hence 20 rain gauge stations should be provided in the catchment, if accuracy up to 5% in estimation of rainfall is desired.

Example 10.2

For the problem of Example 10.1, compute the required number of rain gauges if required accuracy is 10%. Compare the results.

Solution

$$\text{If } p = 0.1, N = \left(\frac{0.222}{0.1}\right)^2 = 4.84 = 5.$$

Thus if the acceptable error is doubled, the required number of stations reduces by a quarter. Note that the WMO norm for plain areas is one gauge per 575 km^2 (Table 10.3). For this basin, the density requirement matches with the required density for monthly values if a 10% error in monthly areal rainfall estimates is acceptable.

10.3.3 Location of Rain Gauge Stations

To determine the location of stations in the catchment, the following procedure is adopted: (1) draw isohyets as shown in Fig. 10.3. In this case, the isohyets divide the basin area into 10 zones; (2) the rain gauges should be placed such that an equal number of rain gauge stations are located in each zone. If the network is to have 20 stations, each zone should have 20/10 = 2 rain gauge stations. Additional stations in a zone will be placed, depending upon the existing stations in that zone. For example, in zone (I), one rain gauge station already exists and hence one more station is to be established. In this case of zone (II), two rain gauge stations are to be established. Similarly, the requirement of rain gauges in various zones can be determined.

The exact location of stations should be decided, keeping in view the following: (1) for logistical reasons, the rain gauge stations should be accessible throughout the year; (2) as far as possible, the number of rain gauge stations in each of the subcatchment should be proportionate to the area of that subcatchment.

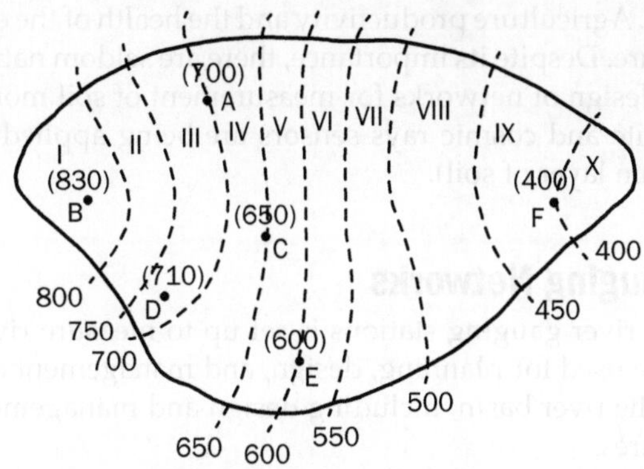

FIGURE 10.3 Isohyets for the example catchment.

Example 10.3

A catchment has six rain gauge stations. In a year, the annual rainfall recorded by each of the gauges is as follows:

Station	A	B	C	D	E	F
Rainfall (cm)	82.6	102.9	180.3	110.3	98.8	136.7

For a 10% error in the estimation of mean rainfall, calculate the optimum number of stations in the catchment.

Solution

For this data

$$m = 6, \bar{P} = 118.6, \sigma_{m-1} = 35.04, \varepsilon = 10$$

$$C_v = \frac{100 \times 35.04}{118.6} = 29.54$$

$$N = \left(\frac{29.54}{10}\right)^2 = 8.7, \text{say 9 stations}$$

The optimal number of stations for the catchment is nine. Hence three more additional stations are needed.

10.4 Networks for Other Hydrometeorological Data

Data pertaining to temperature, relative humidity, sunshine hours, radiation, wind velocity, and so on, are also frequently used in hydrologic analysis. For these variables, we do not require a dense network since their spatial variability is small.

Soil moisture is an important variable which is often termed as the "gateway to hydrology." It interfaces with land surface and subsurface processes. Soil moisture determines percolation and groundwater recharge as well as leaching of nutrients and contaminants. Agriculture productivity and the health of the ecosystem largely depends on soil moisture. Despite its importance, there are seldom national networks and guidelines for the design of networks for measurement of soil moisture. These days, remote sensing satellite and cosmic rays sensors are being applied to estimate soil moisture (for the top thin layer of soil).

10.5 River Gauging Networks

A network of river gauging stations is set up to measure river stages and discharges. These data are used for planning, design, and management of water and other natural resources of the river basin, including design and management of flood/drought protection measures.

Every major stream should be gauged near its mouth and its major tributaries should also be gauged as feasible, depending on the existing and likely development in the basin. According to WMO (2008), the first gauging station on a river is established at a location where the drainage area is about 1300 km^2. The second station is located at a point in the downstream direction where the drainage area is approximately doubled. Stations are also established at the sites where significant changes in the flow are noticed.

The distance between two consequent stations on the same river may vary from about 50 km to several hundred kilometers. Two consecutive observation sites on a large river should be established such that the difference in flow at the stations is significant. Sometimes, gauging stations are set up to answer a hydrological question, for example, to determine the flow contribution from a tributary.

A different approach is necessary for regions having small independent rivers which flow directly into the sea. In such cases, the first hydrological observation station should be established on a stream that is typical of the region and further stations are added to the network to cover the area and obtain information about the variability. Those streams whose flows are low should not be excluded from the network. If the surface water potential of the area is assessed on the basis of the data from the high flow streams only, it may lead to wrong assessment. Care is to be exercised to ensure that all important hydrological features are adequately monitored by the gauging network.

The WMO recommendations regarding the minimum density of river gauging networks in different physiographic regions are reproduced in Table 10.4. Regarding the

Physiographic Unit	Streamflow	Sediments	Water Quality
Coastal	2,750	18,300	55,000
Mountains	1,000	6,700	20,000
Interior plains	1,875	12,500	37,500
Hilly/undulating	1,875	12,500	47,500
Small islands	300	2,000	6,000
Urban areas	–	–	–
Polar/arid	20,000	200,000	200,000

Table 10.4 Recommended Minimum Densities of River Gauging Stations (area in km^2 per station) (Source: WMO, 2008)

accuracy desired in measuring river water depth and discharge, WMO recommends that the water depth measurement should have an accuracy of about 2%, velocity of flow 2%–5%, and discharge about 5%. Suspended sediment concentration should be estimated with an accuracy of 10%.

Note that the WMO recommendations are general in nature and serve only as guidelines. The density of a particular network will ultimately be based on the network objectives, the temporal and spatial variability of river flows, availability of finance and other resources, and so on.

10.5.1 Design of River Gauging Networks

Design of networks is not a one-time affair. Factors affecting river gauging networks keep on evolving with time and thus the networks also require periodic review and adjustments (HP, 2003). Design of networks to measure river stage and discharge involves the following steps:

1. Network design begins with collection of maps and background information about the area/region. Usually 1:250,000 scale (this will depend upon the size of area) topographical maps of the river basin showing basin boundaries form the base map for the network design. Additionally, the following maps should also be collected:
   i. Map showing existing precipitation and gauge-discharge gauging stations operated by various departments.
   ii. Map showing river network, canals, existing and proposed water projects and command areas of irrigation projects, infrastructure (road/rail network), and inhabited localities. Demarcate coastal, mountainous, interior plains, hilly, and plain regions.
   iii. Land use map, also showing forests, main industries, and population centers.
   iv. Communications map showing roads, rails, power transmission lines, and so on.
   v. Map showing soil classes, geological formation, and mining areas.
2. Define the objectives of the network: Who are the data users? What will be the use of data? At what times data will be collected each day?
3. Evaluate the existing network and how well it meets the objectives?
4. Review existing databases to identify gaps and ascertain variability in catchment behavior.
5. Identify gaps and over-design in the existing network; new stations may be proposed and existing stations deleted/shifted.
6. For the rivers originating within the country, identify the first site on each river where the catchment area is nearly equal to the value suggested for that physiographic unit.
7. For the rivers coming from neighboring countries, identify one site on each important river close to the international border considering accessibility/logistics and determine the catchment area up to that place.
8. Find out the next location where the catchment area has increased as per the applicable density norms. Check the location for the ease of communication/

connectivity and suitability regarding the existing/planned water resources projects. Determine the catchment area at this location.

9. Repeat the above two steps until the international border downstream or the outfall into sea.
10. A reconnaissance survey is necessary before finalizing the location of sites.
11. Identify site-specific observation methodology and instrumentation; data collection frequency for discharge, sediment, and water quality parameters.
12. Prioritize stations by following appropriate system.
13. Decide location of new sites after site surveys.
14. Estimate average capital and recurrent costs of installing and maintaining different categories of stations and overall cost of operating and maintaining the network.
15. Review revised network in relation to overall objectives and available budget; adjust as necessary.

Sometimes, special stations are required to fulfill a legal requirement, for example, to monitor releases to follow a court order.

10.5.2 Evaluation of Networks

To evaluate the networks, the existing and proposed stations should be marked on a map. Scanning the network systematically, the following questions need to be considered for each station:

1. What purpose the station will fulfill?
2. Does a better location exist nearby?
3. Have any developments (e.g., dam construction) taken place or are likely which could affect this station?
4. How close are the nearest upstream and downstream gauging stations? Stations should not be very close unless there are specific reasons.
5. Does any other organization operate a gauging station in the vicinity? If yes, could that station serve the purpose expected from this station?

Based on the answers, stations which need to be added, deleted, or relocated are identified. Before removing/shifting an existing station in the network, the main users of the data should be consulted.

Financial Aspects

Besides technical, financial considerations are also important in network evaluation and design, because large finances and manpower are required to set up and operate river gauging stations. Hence after the preliminary design of the network is completed, the cost of establishing and operating stations should be estimated. These costs should match with the budget so that the proposed network is sustainable. In case the funds are inadequate, the network should be revisited or additional budget should be arranged. Stations in the network may be prioritized to best attain the objectives, given the constraints.

10.5.3 Site Selection Surveys

Site selection surveys are carried out to select the most appropriate location for a station. These surveys can be divided into three distinct phases: (1) desk study, (2) reconnaissance surveys, and (3) topographic and other surveys.

Locations for the gauging stations are first identified on a map; large-scale topographic maps should be checked to identify possible sites. Reconnaissance surveys should be undertaken by an experienced hydrologist along with persons familiar with the area. As the hydraulic conditions of rivers vary considerably from dry to wet season, reconnaissance surveys should be undertaken in both seasons to arrive at the correct decision about the suitability of the site.

At sites of interest, ownership of the land and accessibility should be ascertained. The site should be accessible in all seasons. It is important to use updated maps, since catchments and site conditions may be subject to large changes. Recent situations can be obtained from remote sensing images. Information on the historical high flood level should also be collected during the site visit.

On the completion of reconnaissance surveys, one or more locations are shortlisted for further consideration. After this, field investigations are carried out and the river cross section of the proposed site is surveyed. If artificial controls (e.g., a weir) are planned, it will be helpful to survey the river for some distance upstream and downstream to demarcate submergence area and the impact of the structure on flows and water levels. It will also be important to understand the type of control that exists and to make sure that the planned gauging station will not be impacted by the backwater of any structure.

10.5.4 Hydraulic Engineering for Selection of Sites for River Gauging Stations

The locations of stream gauging stations are determined based on the following factors:

1. Places where major rivers cross state/national borders.
2. Places where flow has changed considerably after the last upstream station.
3. Based on locations of proposed dams/diversion/run-of-river schemes, including diversion/inflow points for (proposed) inter-basin water transfers links.
4. Locations whose data may be needed for flood forecasting.
5. Locations of interest from ecological and biodiveristy perspective or where environmental assessment is needed.
6. Areas expected to have significant land use changes, for example, deforestation or reforestation.

After the broad locations of stations have been decided, the general site selection guidelines are followed to determine exact locations. It is emphasized that an ideal site which satisfies all requirements cannot always be found in real-life cases. WMO (1980) have given guidelines in this regard. In practice, measurements are often taken in nonideal conditions.

1. There should be a straight, well-defined approach channel upstream of the measuring section to ensure parallel and nonturbulent flow. The approach channel should have a uniform cross section, free from irregularities and the flow should have a regular velocity distribution. For rivers less than 100 m

wide, a straight approach of four times the channel width should be preferred. For rivers more than 100 m wide, a straight approach channel of minimum 400 m is desirable. When an adequate length of straight channel is not present, the straight length upstream should be at least twice that downstream.

2. The site should not have a tendency to collect floating debris, which may hinder the working of instruments. Sites where high sediment deposition or scouring occurs or sites subjected to weed growth should be avoided, to the extent feasible.
3. Locations which are subject to high turbulence should be avoided.
4. Ideally, flow should be confined to a single channel. When this is not possible, each channel should be gauged separately to obtain the total flow. It also needs to be ensured that there is no parallel by-pass channel, natural or manmade, surface or subsurface, around the station.
5. The channel bed should be rigid, and free from large undulations, obstructions, and debris. Sites with highly erodible beds and banks shall be avoided. In some rivers, it is difficult to find such a site and the site with minimum bed and bank changes may be chosen.
6. Riverbanks at the site should be stable, and free from large vegetation and other obstructions.
7. The control shall be sensitive, which means that a significant change in discharge, even for low discharges, should result in a significant change in stage. Small errors in stage readings during calibration at a nonsensitive station can result in large errors in discharge values indicated by the stage-discharge relationship.
8. The station should be located where the flood plain is narrow and the out-of-bank flood flow is small. It is often not possible to locate a gauging station so that all flood flows are contained within the river channel. At many locations, there is an elevation after which the flow overtops banks.
9. River channel downstream to the site should preferably be stable. Sites, which are influenced by downstream confluences with other rivers, river control structures, dams, tidal conditions, should be avoided.
10. Factors, such as unhindered access to the site in all seasons, space to construct office, living space for the observers, electricity, and other services, should be available as per the needs.
11. Sites with a tendency for the formation of vortices, reverse flow, or dead water shall be avoided.
12. Human interference (out of curiosity or mala fide intention) with instruments is a problem in some places. If a choice is to be made between two hydraulically similar sites, the site with fewer problems due to human interference is preferred.
13. The measurement section should be clearly visible across its width and unobstructed by vegetation or other obstacles.
14. There should be a sufficient depth of flow across the whole cross section.

15. The site shall be sufficiently far away from the disturbance caused by rapids and falls.
16. If the site is upstream of the confluence of two rivers, it should be located sufficiently far upstream so that it is beyond backwater or any disturbance due to the joining of two rivers.
17. Velocities should be well in excess of the minimum required speed if a current meter is to be employed.

Example 10.4

Consider the basin shown in Fig. 10.4. Its flow in the main river and its two tributaries needs to be known. What would be the best strategy if the average flows at the three locations are estimated as: $A = 100$ m³/s, $B = 80$ m³/s, and $C = 20$ m³/s. Measurement procedures are such that error at each location is about 10%.

Solution

We may operate all the three stations, A, B, and C. Suppose we want to save money and operate only two stations, the following are the options for flow estimation:

1. $A = B + C$
2. $B = A - C$
3. $C = A - B$

Assume that the measurement errors made at each location are independent. For case 1

$$\sigma_A = \sqrt{\sigma_B^2 + \sigma_C^2 + 2\sigma_B\sigma_C}$$

where σ_A denotes the error in the estimation of A. Neglecting the product $\sigma_B\,\sigma_C$ term,

$$\sigma_A = \sqrt{\sigma_B^2 + \sigma_C^2}$$

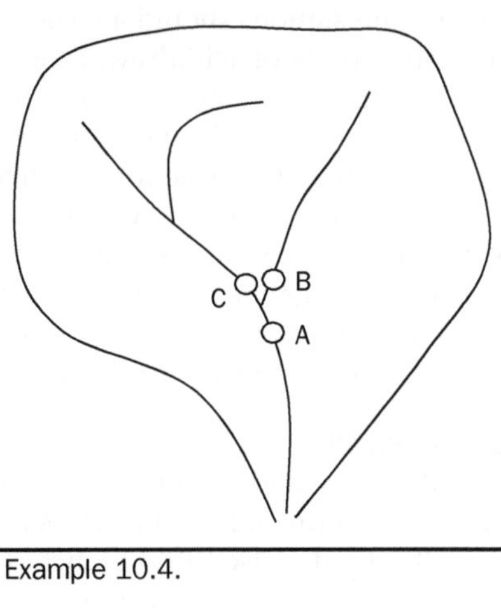

FIGURE 10.4 Catchment for Example 10.4.

Since measurement error at each place is about 10%, $\sigma_A = 10$, $\sigma_B = 8$, $\sigma_C = 2$. Hence in case of 1:

1. $\sigma_A = (8^2 + 2^2)^{1/2} = 8.25$ m^3/s; relative error in the flow at $A = 8.25/100 \approx 8\%$.
2. Case 2, $\sigma_B = (10^2 + 2^2)^{1/2} = 10.2$ m^3/s; relative error in the flow at $B = 10.2/80 \approx 13\%$.
3. Case 3, $\sigma_C = (10^2 + 8^2)^{1/2} = 12.81$ m^3/s; relative error in the flow at $C = 12.81/20 \approx 64\%$.

Clearly, if the flows at A and two tributaries (B and C) are to be determined, the best option from the point of view of accuracy and cost would be to have stations at B and C. Further, the sum of B and C will be slightly less than actual flow at A because, due to backwater effects, stations B and C cannot be located too close to the confluence. It is also observed that if flow at C is estimated as the difference of two large values ($C = A - B$), it might result in a highly inaccurate estimate.

10.5.5 General Considerations About River Gauging Networks

In general, a sufficient number of streamflow stations should be located along the main stems of large streams to permit reliable interpolation of discharge between the stations. The specific locations of these stations should be governed by topographic, hydrologic, and climatic considerations. If the difference in flow between two points on the same river is not greater than the limit of error of measurement at the station, then an additional station is unjustified (WMO, 1980). Where the tributary flow is of special interest, a station on the tributary will be required.

Wherever possible, the base stations should be located on streams with natural regimes. Where this is impractical, it may be necessary to establish additional stations near reservoirs to obtain the requisite data to reconstruct the natural flows at the base stations. Computed flows past hydroelectric plants or control dams will be useful for this purpose, but provisions will have to be made for the calibration of control structures and turbines and for the periodic checking of such calibrations.

Stations should be located on the lower reaches of major rivers of the country, immediately above their mouths (usually above tidal influence), or where the rivers cross borders. River gauging stations should also be located where rivers emerge from mountains and above the points of withdrawal for irrigation water. Stream gauging stations are placed at points where the discharge from the last gauging station has increased considerably, before the confluence points of major tributaries, at the outlets of lakes, and at locations where large structures are likely to be built.

The discharge of small rivers is strongly influenced by local factors. To ensure a correct appraisal of catchment hydrology, small streams should be adequately gauged. However, it is impracticable to gauge all small streams. In regions where small rivers are important, stations may be established even on streams having small contributing areas. WMO (2008) has discussed this topic in greater detail.

Networks for Large River Basins

A primary station should be located at a point on the main river near its mouth where the discharge attains its maximum value. This site is usually little upstream of the point where the river normally divides itself into branches before joining the sea or a lake or crosses a national border. In the case of mountainous rivers, a station should be established where water leaves the mountainous reach and enters the plain. Subsequent stations are established at sites where significant changes in the volume of flow are noticed, namely, confluence of a major tributary, near a major diversion, or at the outflow point of a lake, and so on.

The distance between two gauging stations on the same river may vary from about 25 to several hundred kilometers, depending on the size of the river. The drainage areas computed from the origin up to two consecutive gauging sites on a large river should preferably differ by more than 10% so that the difference in flow is significant. The uncertainties in discharge values, particularly for high flows in alluvial rivers, are likely to be more than ±10%. Attempt should be made to minimize these uncertainties.

Networks for Small River Basins

Some regions, for example Western Ghats in India, have many small independent rivers which flow for a short distance, say a few hundred kilometers. It is not possible to operate and maintain gauging stations on all the smaller watercourses. Therefore representative basins have to be selected and the data from those are used to develop techniques for estimating flows for similar ungauged sites. In such cases, the first hydrological observation station might be established on a stream that is typical of the region and then further stations could be added to the network to cover the entire area. Streams in a particular area having lower yields should not be ignored for inclusion in the network. Absence of a station on a low flow stream may lead to the wrong conclusions on the water potential of the area as a whole, evaluated on the basis of flow in the high flow streams. Care should be exercised in designing the network to ensure that all distinct hydrologic areas are adequately covered.

Networks for Deltas and Coastal Floodplains

Deltaic areas, where gradients are usually low and channels bifurcate, pose unique problems in stream gauging. Further, deltas are dynamic systems, that is, they are continually changing. Hence, the type of network required in deltas may differ from more conventional river basins. Due to the low gradients, it is often not possible to locate stations with stable stage-discharge relationships. In other words, variable backwater effects can occur due to tidal influences and/or changes in the aquatic vegetation growth. Stage readings should be made at all principal off-takes/bifurcations or nodes in the system. These should be supplemented by current meter gauging when required.

Sustainability of Networks

Since hydrologic analysis requires long-term data, sustainability of the network is very important. Many benefits of a gauging network will be lost if its operation is not sustained. Frequently, there are networks whose stations fail to work well due to the lack of funding support, skilled manpower, and logistic support. It is far better to operate and maintain 10 gauging stations well than to operate and maintain 20 stations poorly. Better quality data from fewer stations is preferable to poor quality of data from a greater number of stations.

Since more than one organization may be responsible for the establishment and operation of gauging stations, it is essential that the activities are coordinated so they complement each other and duplication of efforts is avoided.

Networks of river gauging stations require periodic reevaluation. The developments that take place in the basin, like construction of new WRD projects and industrialization in the area, may require addition or closure of stations. For example, river reaches are often polluted due to the discharge of effluents from industry. Therefore a need may arise to establish stations to assist with water quality monitoring and pollution assessments.

10.6 Groundwater Monitoring Networks

Water level measurements are fundamental for determining the directions of groundwater flow, areas of recharge and discharge, evaluating the effects of manmade and natural stresses on groundwater, defining the hydraulic characteristics of aquifers, and ascertaining stream-aquifer relations. Groundwater measurements depend on the complexity of hydrogeologic environment and are costly. Also, groundwater cannot be measured continuously. That may explain why groundwater networks are not as extensive as surface water networks. In many cases, especially in developing countries, groundwater data are collected for a particular site of interest. In the United States, there are federal and state agencies that are responsible for collection of data on groundwater that help efficient development, utilization, and management of groundwater resources. At each data collection site, aquifer properties, the depth of the well, the type of construction, and well elevation should be known. Water level data are fundamental and are collected from wells or piezometers. A well is essentially a large manometer tube that measures the hydrostatic pressure of water in an aquifer. The aquifer can be either confined or unconfined. Existing wells can be employed as convenient measuring sites at low cost; however, these wells must satisfy certain conditions before use, such as proper maintenance and regular use. Wells drilled especially for observation purposes are desirable.

10.7 Water Quality and Sediment Measuring Stations

The possible uses of water depend, to a large degree, on its physical, chemical, and biological quality. Sediment data cannot be continuously measured and require laboratories and trained manpower for analysis of samples. Hence the networks are not dense.

Sediment gauging stations may be designed either to measure the total sediment discharge to the ocean or to measure erosion, transport, and deposition of sediment within a country, state, basin, and so on. In designing a minimum network, emphasis should be placed on erosion, transport, and deposition of sediment within a country or river basins. All networks should contain a sediment station at the mouth of the river discharging into the sea.

Sediment transport by rivers is a major problem in arid regions, particularly in those regions underlain by friable soils and in mountainous regions where, for engineering applications, the amount of sediment loads should be known. The densities given by WMO (2008) (see Table 10.4) serve as guides in setting up a basic network.

While measuring sediment transport, emphasis should be placed on those areas where soil erosion is known to be severe. Sediment-transport data may be supplemented by surveys of sediment trapped in lakes or reservoirs.

Observations of chemical quality consist of periodic sampling and laboratory analyses. The number of sampling points and frequency of observations depend on the variability, severity of problem, and water uses. If fluctuations are more, more

ISO Number	Title
ISO 748:2007	Hydrometry—Measurement of liquid flow in open channels using current meters or floats
ISO 772:2011	Hydrometry—Vocabulary and symbols
ISO 1070:1992	Liquid flow measurement in open channels—Slope-area method
ISO 1100-2:2010	Hydrometry—Measurement of liquid flow in open channels—Part 2: Determination of the stage-discharge relationship
ISO 1438:2008	Hydrometry—Open channel flow measurement using thin-plate weirs
ISO 2425: 2010	Hydrometry—Measurement of liquid flow in open channels under tidal conditions
ISO 3846:2008	Hydrometry—Open channel flow measurement using rectangular broad crested weirs
ISO 4359:1983	Liquid flow measurement in open channels—Rectangular, trapezoidal, and U-shaped flumes
ISO 4360:2008	Hydrometry—Open channel flow measurement using triangular profile weirs
ISO 4362:1999	Hydrometric determinations—Flow measurement in open channels using structures—Trapezoidal broad-crested weirs
ISO 4373:2008	Hydrometry—Water level measuring devices
ISO 4374:1990	Liquid flow measurement in open channels—Round-nose horizontal broad-crested weirs

Table 10.5 A Partial List of ISO Standards Related to Hydrometry.

measurements would be required. As a minimum network, records of water quality should be obtained at the densities as shown in Table 10.4.

10.8 International Standards for Hydrometry

The International Organization for Standardization (ISO) has published a number of standards related to hydrometry. A partial list is given in Table 10.5. More details are given on their website (www.iso.org).

10.9 Reference Climate and Water Data Networks

Data are of primary importance in the geosciences community, which embraces the premise that geosciences are by nature induction-based, rather than deduction-based and rely, therefore, to a greater extent on historical data as the key to the future. Adequate data are crucial to understand the observed changes and to improve models. If only short hydrometric records are available, the full extent of natural variability can be understated and detection studies confounded (Kundzewicz et al., 2008).

Past data are considered to be the key to the future for at least two reasons (Koutsoyiannis et al., 2009). The first is related to the indispensable utility of data in model building. Geophysical models, including climate and hydrological models, necessarily involve conceptualizations and parameterizations of the natural processes. Hence the observed data are employed to test hypotheses and calibrate and validate mathematical models. The second utility of data is related to their crucial role in understanding past climatic and hydrological changes, which also provide an extremely useful guide in tracing possible futures. Analysis of these data provides useful clues to determine whether climate change is causing any change in the state and variation of the key processes and whether the impacts are positive, negative, or indeterminate.

One cannot rely on short-term or only recent observations to determine the magnitude of impacts or risk from climate change because there has been only small global warming so far. To detect the state of climate and changes in it, severity of changes, present and projected impacts, and so on, there is a need to continue and improve/enhance data collection efforts and expand the data observation networks. Currently, the observation networks are dense in developed countries whereas these are fairly sparse in developing countries. Since changes in the climate are observed at timescales which are much larger than those in typical hydrological studies, it is helpful to have very long measured data series. Given the need for decision-makers to have the best data possible, it is therefore critical to have an unimpeachable benchmark source of climate data with which to do this (Diamond et al., 2013). Such data can be generated by specialized network of stations. The material presented in this section has been condensed from Jain (2015).

10.9.1 Global Climate Networks

Climate observations are sourced from the numerous meteorological and related observational networks and systems throughout the world. These observations provide the basis for applications, such as weather forecasting and environmental impact assessments. However, not all climate observations are the same and some are considered of higher quality than others. According to the WMO, three factors are involved in collecting climate observations (http://www.wmo.int):

1. Climate observations need to account for the full range of elements that describe the climate system—not just those that describe the atmosphere. Extensive observations of the ocean and land-based systems are also required.
2. An observation at any point in time would not mean very much, unless it can be compared to a reference climate, that is, a reference climatological period must be selected (e.g., the current reference dates are typically from 1961 to 1990). For this reason, the observations from a station, that only exists for a short period (i.e., from days to a few years) or which relocates very frequently, will generally be of less value than those observations from a station whose records have been maintained to established standards over many years.
3. A climate observation should be associated with a set of data describing the conditions in which the climate observations were collected and that will provide users with information on how the observation should be interpreted and used.

There are over 11,000 weather stations around the world measuring land, air, and sea surface temperatures, as well as satellites, ships, and aircraft that also take measurements. These stations carefully follow specified standards and 1040 of these stations have been selected by WMO to provide high-quality climate data to quantify and detect global aspects of climate change. The more stringent requirements on observation networks and systems for monitoring climate, including the detection of climate change, has led to the development of special networks at national and global scales.

10.9.2 Country-Level Networks

Many countries and international organizations have established specialized observational networks for surface, climate, hydrometric, and water quality observational systems whose data can be used in studies related with climate change. Pilon (2000) has discussed selection criteria for such networks. Some of these networks are international in scope, such as the GSN (Peterson et al., 1997) and the WMO RCS network (WMO, 1986).

Pilon and Yue (2002) have given six criteria for selection of stations for river gauging stations. Taking a clue from these, the following criteria are suggested to select and operate stations in a reference data network:

1. The station must have at least 20 years of past data, except for data sparse areas.
2. The station must operate for the entire year.
3. The station must be located in (near) pristine conditions and those places should be preferred where no large-scale future development is anticipated. Analysis of nearby stations in pristine and nonpristine locations will provide insight in the nonstationary (Lins and Cohn, 2010) behavior of hydro-climatic data.
4. Those rivers/locations are preferable where there are minimum upstream interventions. Of course, one station each should be set up near the mouth of each major river.
5. Data for the relevant variables must be obtained by measurements which follow the national standard procedures (not allowing the use of estimated or reconstructed records).
6. Data measured at the station should be communicated to a central station in real time and should be processed and uploaded on the Internet in near real time for free download.
7. The accuracy of measurements should preferably be $\pm 5\%$.
8. For each station, the description of site conditions and records of maintenance history should be available.

Among the country level networks, those of the United States and the United Kingdom are prominent.

10.9.3 U.S. Climate Reference Network

The U.S. Climate Reference Network (USCRN) consists of 114 stations in the United States. Its purpose is to maintain a sustainable high-quality network which will detect, with high confidence, signals of climate change in the United States. The USCRN is to provide future long-term high-quality observations of surface air temperature and precipitation that can be coupled to past long-term observations for the detection and attribution of present and future climate change. The main goal of the USCRN is to answer a simple prospective question posed 50 years into the future: How has the climate of the United States changed in the last 50 years? To meet this goal, the USCRN program aims to create a set of station records that will provide a robust multidecadal climate monitoring capability (Diamond et al., 2013).

Each station is positioned at a pristine site which is expected to remain free from development over coming decades and avoid possible urban microclimate interference. These stations measure air temperature, precipitation, solar radiation, and surface wind, and are being expanded to include triplicate measurements of soil moisture and soil temperature at five depths, as well as atmospheric relative humidity.

10.9.4 Reference Climatological Stations in the United Kingdom

The purpose of the climate network in the United Kingdom is to enable monitoring and determination of the UK climate and thereby meet the requirements for international exchange, commercial applications, and research. The minimum observing requirement at a climate station is the daily measurement of maximum temperature, minimum temperature, and rainfall amount, although many stations make a wider range of observations.

The RCSs are designated to meet the United Kingdom's needs for climate monitoring. All stations have a long observational record spanning at least 30 years and their distribution aims to be representative of the major climatological areas unaffected by environmental changes.

A network of around 130 gauging stations across the United Kingdom has been designated, based on hydrometric performance, the degree of artificial influence to the flow regime, homogeneity, and length of flow series (Dixon et al., 2013).

10.10 Recent Advances in Hydrometry

Recent advances in information technologies and electronics have yielded a new generation of hydrometric instruments. An important development in streamflow measurement in recent times has been the Acoustic Doppler Current Profilers (ADCPs). ADCPs use acoustic energy to measure water velocity by measuring the shift in the frequency of the acoustic signals (using a principle of sound waves called the Doppler effect) reflected from materials suspended in, and moving with, the water. The ADCPs have several advantages over other methods. ADCP is described in greater detail in Chapter 11. CWC (2017) has described the procedure to be followed for measurement of river flow by use of ADCP.

The measuring and monitoring of surface water from space seems a realistic possibility. Space-based instruments hold promise for measuring elevations and perhaps even flow rates of rivers in different parts of the world. In the long run, space-based techniques will be highly useful for hydrologic monitoring.

Recent works have shown that weather radar can provide an estimate of rainfall rates with good spatial and temporal resolution. Rainfall estimation using satellite measurement (mainly thermal infrared images) gives satisfactory results. Satellite rainfall estimation algorithms must be calibrated and validated by using rain gauge networks.

These drones are being employed for measuring vegetative cover which then helps estimate evapotranspiration. Ground, air, and space-based microwave sensing is being employed for measuring soil moisture over large areas. As a result, the United States now has a large number of sites across the country where soil moisture is being measured in the upper soil horizons. The soil measurements of these sites are available on websites for public use.

10.11 Summary

A network is a dynamic system, not a static entity. Once it becomes operational, it needs to be evaluated periodically to see whether (revised) objectives still match with the output produced in a cost-effective manner. This implies flexibility in establishing new stations and closing down those that are no longer needed.

10.12 Questions

10.1 Indicate the main purposes of data collection.

10.2 Why are hydrometeorological data networks set up? What factors impact the design of these networks?

10.3 What is a basic hydrometeorological network? Write the main steps in hydrometeorological network design.

10.4 As per the WMO guidelines, what are the minimum densities of precipitation stations in coastal, mountains, interior plains, and small islands? What are the corresponding guidelines in your country?

10.5 Many countries define "sufficient accuracy" of measurement differently. Take five countries as examples (e.g., Australia, Canada, the United Kingdom, Japan, and the United States) and define what sufficient accuracy quantitatively means.

10.6 List which components of the hydrologic cycle are measured at what level (e.g., networks, isolated, etc.) in the countries selected in Question 10.5.

10.7 Which agencies in your country are responsible for the collection of what kinds of hydrometeorological data?

10.8 Select five countries and provide information on density of networks for measuring rainfall, discharge, and hydrometeorological variables (e.g., temperature, wind velocity, relative humidity, vapor pressure, sunshine hours, and radiation).

10.9 Different countries use different design criteria. Select any two countries and list the methodologies used in these countries for measuring precipitation, evaporation, discharge, soil moisture, and snow.

10.10 A catchment has 5 rain gauges at various locations. From the data of these, the mean annual rainfall of the catchment is computed as 1020 mm and the standard deviation as 375 mm. Estimate the minimum number of rain gauge stations needed if the allowable error is 10%. What will this number be if the standard deviation of rainfall is 523 mm?

10.11 What hydraulic and hydrologic criteria are checked to determine the exact location of a river gauging station?

10.12 Are water quantity and water quality networks coordinated? If yes, why?

10.13 Should a network be revisited for improvement from time to time? If the answer is yes, then why? If the answer is no, then why not?

10.14 Often, there are sites where no observations are available. For hydrologic design, data for such ungauged sites are obtained from nearby gauged sites. What methods can be used to obtain data at ungauged sites? Discuss the sources of error in data obtained.

10.15 What are the sources of error in measured data? What can be done to eliminate these errors?

10.16 Why are reference data networks set up and what are the key considerations in selecting stations for such networks?

10.13 References

CWC. *Handbook for Hydrometeorological Observations*. Central Water Commission, New Delhi, January, 2017.

Diamond, H. J., T. R. Karl, M. A. Palecki, C. B. Baker, J. E. Bell, R. D. Leeper, D. R. Easterling, et al. U.S. Climate reference network after one decade of operations: Status and assessment. *Bulletin of the American Meteorology Society*, 94:485–498, 2013.

Dixon, H., J. Hannaford, and M. Fry. The effective management of national hydrometric data—Experiences from the United Kingdom. *Hydrological Sciences Journal*, 58(7): 1383–1399, 2013.

HP. Field Manual—Hydrometry (SW), Part I—Network Design and Site Selection. Hydrology Project, Govt. of India, New Delhi, 2003.

Jain, S. K. Hydrological Measurements and Analysis of Data. Web-based course under National Program on Technology Enhanced Learning (NPTEL). http://nptel.ac.in/courses/105107129/, 2011.

Jain, S. K. Reference Climate and Water Data Networks for India. *J. Hydrol. Eng.*, ASCE, doi: 10.1061/(ASCE)HE.1943-5584.0001170 , 02515001, 2015.

Koutsoyiannis, D., C. Makropoulos, A. Langousis, S. Baki, A. Efstratiadis, A. Christofides, G. Karavokiros, et al. Climate, hydrology, energy, water: Recognizing uncertainty and seeking sustainability. *Hydrology and Earth System Sciences*, 13:247–257, 2009.

Kundzewicz, Z. W., L. J. Mata, N. W. Arnell, P. Döll, B. Jimenez, K. Miller, T. Oki, et al. The implications of projected climate change for freshwater resources and their management. *Hydrological Sciences Journal*, 53(1):3–10, 2008.

Lins, H., and T. Cohn. Stationarity: Wanted dead or alive. In *Workshop on Non-Stationarity, Hydrologic Frequency Analysis, and Water Management* (eds. J. Olsen, J. Rolf, J, Kiang, and R. Waskom). Colorado Water Institute Information Series No. 109. Colorado State University, 2010.

Peterson, T., H. Daan, and P. Jones. Initial selection of a GCOS surface network. *Bulletin of the American Meteorological Society*, 78(10):2145–2152, 1997.

Pilon, P. Criteria for the selection of stations in climate change detection network. In *Detecting Trends and Other Changes in Hydrological Data* (eds. Z. W. Kundzewicz and A. Robson). World Climate Programme—Water, World Climate Programme Data and Monitoring, WCDMP-45, WMO/TD No. 1013. Geneva, Switzerland: WMO, 2000.

Pilon, P. J., and S. Yue. Detecting climate-related trends in streamflow data. *Water Science and Technology (Environment Canada)*, 45(8):89–104, 2002.

World Meteorological Organization (WMO). *Manual on Stream Gauging. Volume I. Fieldwork. WMO No. 519.* Secretariat of the World Meteorological Organization, Geneva, Switzerland, 1980.

World Meteorological Organization (WMO). *Guidelines on the Selection of Reference Climatological Stations (RCSs) from the Existing Climatological Station Network. WMO/TD-No.* 130. Geneva, Switzerland: World Meteorological Organization, 1986.

World Meteorological Organization (WMO). *Guide to Hydrological Practices. Volume I: Hydrology—From Measurement to Hydrological Information. WMO No. 168.* Geneva, Switzerland: World Meteorological Organization, 2008.

CHAPTER 11

Streamflow Measurement

Rivers have served as the lifeline for mankind and will continue to do so in the future. Hence the most important hydrological data for surface water analysis pertains to streamflow. Streamflow records are records of flow depth and/or volume passing through a particular section on a stream at different times and are therefore continuous flow records. Streamflow data have multiple uses, such as planning, design, and operation of surface water projects; flood forecasting and management; flood plain delineation; water quality and environment protection; and design of bridges and culverts.

The river cross sections, where flow is measured, are known as stream gauging stations. A network of these stations is established to collect data about surface water resources of a region. "Hydrometry" is the science of water measurements. Since there is a huge network of stream gauging sites around the world, it is important to set up standards to guide how to set up a gauging site and do measurements. A technical committee (TC113) of the International Organization for Standardization (ISO) has brought out a large number of standards dealing with measurement of liquid flow in open channels. For instance, ISO 748 describes flow measurement in open channels by following the velocity-area-method. These standards have been referred while preparing material for this chapter. The World Meteorological Organization (WMO) has also brought out many publications related to streamflow measurement, including the famous *Guide to Hydrological Practices* (2008) and many countries have their own standards. The topic is covered in detail in Rantz (1982), Herschy (1978), Herschy (2008), and Boiten (2008).

11.1 River Stage and Discharge

The term "river stage" denotes the water level in a river with respect to a datum; the depth of flow is the vertical distance between the bottommost point in a cross section to the water surface (Fig. 11.1). Here, datum may be an arbitrary datum that is selected for convenience or the national reference (e.g., the mean sea level or an arbitrary level, slightly below the point of zero flow in the stream). To eliminate the possibility of negative values of the gauge height, the datum selected should be sufficiently below the elevation of zero flow. The gauge height is usually expressed in hundredths or thousandths of a meter. The

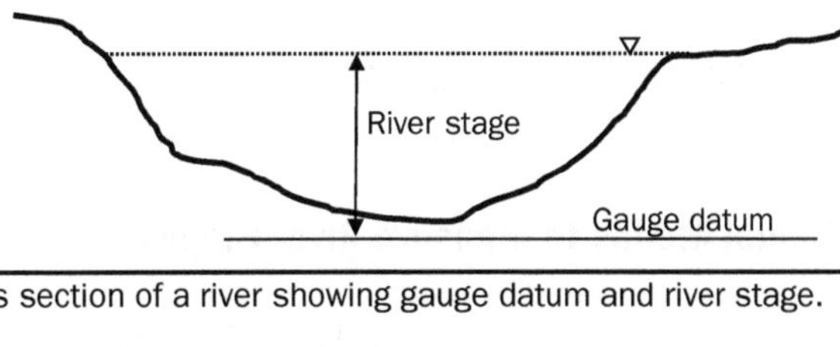

FIGURE 11.1 Cross section of a river showing gauge datum and river stage.

term discharge denotes the volume of flow passing through a cross section in the river per unit time. Its unit in the metric system is cubic meters per second (m^3/s or cumec).

Measurement of stage in a natural river is easy but frequent and direct measurement of discharge is comparatively difficult, time consuming, expensive, and requires skill. Therefore the discharge at a site is measured less often compared to stage. River stage and discharge at a cross section are related through the stage-discharge rating curve or relationship (discussed later). We will now discuss the concept of station control.

11.1.1 Station Control

For correct hydrologic measurements, a stable stage-discharge relation (or rating relation, discussed later) is desirable. To establish such a relation, the flow conditions at the gauging station must be such that for a given stage, the discharge past the station is always nearly the same. The stability, reliability, and shape of the stage-discharge relation are controlled by a section or a reach of the channel at/or downstream from the station. The combination of elements that control stage-discharge relation at a station is known as control. Station control can be of four types: (1) section and channel controls; (2) natural and artificial controls; (3) complete, compound, and partial controls; and (4) permanent and shifting controls.

In the section control, a change in the physical features of the channel downstream to the gauging station has no effect on the flow at the gauging station. Local narrowing of the flow cross section (waterfalls, rock, or gravel bar) creates a zone in which flow is accelerated and is an example of section control. A section control will always have a critical flow section some distance downstream. In a channel control, the flow at the station is controlled by the geometry and roughness of the downstream reach; the length of the downstream reach of the river that affects the flow at the station is directly proportional to discharge. This length increases with discharge.

A complete control governs the rating relation over the entire range of stages at the gauging station. In a compound control, a section control dominates lower discharges and a channel control dominates higher discharges. A weir or dam that is not submerged at high discharges is a complete control. A partial control acts along with another control to influence the rating relation (Rantz, 1982).

Natural section controls (see Fig. 11.2) can be a ledge of rock across a channel, the brink of a waterfall, or a local constriction in width. Examples of artificial controls are weirs and flumes. In a stable control, the geometry of a section and the resulting rating relationship do not change with time, but do in shifting controls. The properties of a cross section may change with time due to erosion or deposition and growth of vegetation.

River gauging efforts and cost to obtain data of good quality are much greater for shifting controls than for permanent controls. It is preferable to select a gauging site with a section or an artificial control. WMO (2008) and Braca (2008) have discussed this topic in greater detail.

11.2 Measurement of River Stages

We will now describe how the river stage is measured.

11.2.1 Manual Means to Measure River Stages

The river stage can be measured by using a variety of equipment: staff gauges, autographic water level recorders, and digital water level recorders. These can be classified in two broad

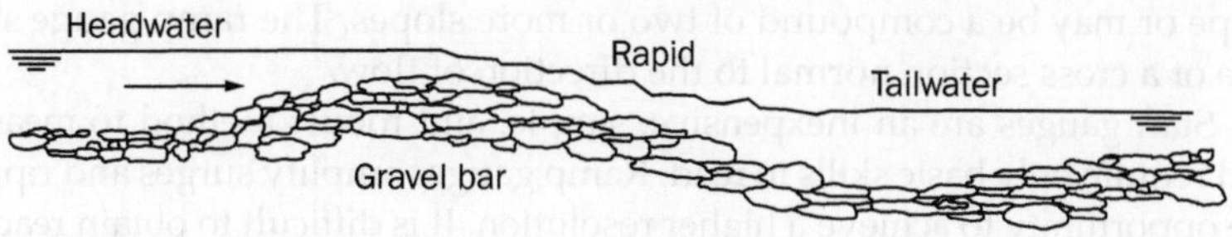

FIGURE 11.2 A section control. (Adapted from Braca, 2008.)

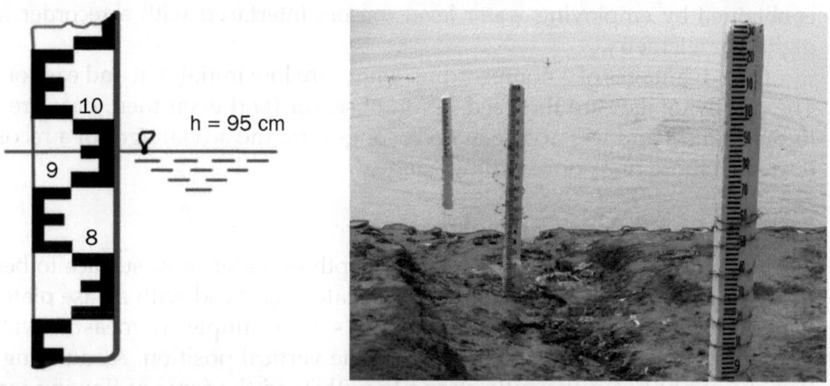

FIGURE 11.3 Staff gauge (left) and stepped staff gauges (right).

categories: (1) nonrecording manual gauges and (2) recording gauges. Manual gauges are being gradually replaced by automatic ones.

A simple way to measure river stage is by a *staff gauge*, which is an iron section (normally porcelain enameled), installed such that a part of it is always immersed in the water (Fig. 11.3). The gauge may also be a vertical scale attached (or painted) to a bridge pier, or a rock that extends into the lowest part of the stream. Manual staff gauges are simple and inexpensive but must be manually read frequently. CWC (2017) describes the steps of the procedure to manually observe river stage.

Staff gauges can be either vertical or inclined. Vertical staff gauges are normally porcelain enameled iron sections with a scale (with a graduation of 5 or 10 mm) marked on or securely attached to a suitable vertical surface graduated every 10 mm (Jain and Singh, 2003). Where the range of water level variation is large, observations using a single staff gauge are difficult. In such cases, the gauge consists of stepped sections installed at different locations in a line normal to the flow. Each of these stepped gauges should refer to the common datum and they should overlap sufficiently to ensure continuity of readings and consistency with each other.

An inclined staff gauge is usually a graduated surface anchored to a permanent foundation. Rock outcrops make a good base for inclined gauges. Inclined gauges built flush with the stream bank are less likely to be damaged by floods, floating ice, or drift than are projecting vertical staff gauges. Such gauges should be located as close as possible to the measuring section, without affecting the flow. Staff gauges are manually read generally each day in the morning in the lean season and at (multi) hourly intervals during high flows.

A ramp gauge consists of a scale marked on or securely attached to a suitable inclined surface. Throughout its length, the ramp gauge may lie on one continuous slope or may be a compound of two or more slopes. The ramp gauge should lie on the line of a cross section normal to the direction of flow.

Staff gauges are an inexpensive, simple, and robust method to measure water level and require only basic skills to read. Ramp gauges amplify surges and ripples but provide the opportunity to achieve a higher resolution. It is difficult to obtain readings in the field with an accuracy better than ±5 mm.

Manual gauges are read at fixed times, whereas a recording gauge provides a continuous record of the stage. In automatic recording stations, a continuous record of stage is obtained by employing water level sensors interfaced with a recorder (analogue or digital) or telemetry.

The advantages of a nonrecording gauge are low initial cost and ease of installation. The disadvantages are the need for an observer (and even then, data are observed at limited times) and less accuracy. Over long-term, the advantages of a recording gauge outweigh those of a nonrecording gauge.

11.2.2 Direct Depth Sounding

During field investigations, the vertical depth of water from surface to bed may have to be measured. A sounding rod is a graduated rigid rod with a base plate to measure depths up to 5 m and velocities up to 2 m/s, for example. To measure water depth by a sounding rod, the rod must be held in the vertical position. A sounding rod should be lightweight but sufficiently strong to withstand the force of flowing water without undergoing significant deflection or vibration. It should not disturb the flow of water, and should not penetrate into the channel bed.

Cable suspension equipment is generally used in deep streams, where rigid rod suspension equipment cannot be used. Depending on the depth and velocity of flow, weights are attached to the cable suspension to keep the cable as vertical as possible. Corrections are required if the cable does not remain vertical and forms large vertical angles.

11.2.3 Water Level Recorder

A water level recorder (WLR) senses and records water level. It consists of a device to keep track of time and a device to record water level. Together, these produce a time-series of observed water levels. The timer device is controlled by a clock, while the gauge height recorder is actuated by a float or a pressure actuated system. These recorders can be either analogue type or digital type.

Float-Type Water-Stage Gauges A float-type gauge consists of a float, installed in a stilling well, a graduated tape or wire, a counterweight, a pulley, and a pointer. As shown in Fig. 11.4, this device records the rise and fall of a float with river water level. The stilling well protects the float and other accessories from floating debris and dampens the waves in the river water level. Motions of the float move a pen or stylus that records the rise and fall on a chart that moves at right angles. Thus a continuous record of water levels is produced. Such gauges are more suitable for streams with narrow and deep channels, since the stilling well can be placed near the stream. But if the channel is wide and sandy, the stilling well must be installed on the bank at some distance and pipes connect the stilling well with the stream. However, silt and debris frequently tend to block these pipes, which need regular cleaning.

A float gauge provides a direct observation of river stage with good accuracy and does not require external energy. Of course, the accuracy of data largely depends on the quality of the recording mechanism. Stilling wells are easy to construct and maintain. A continuous hard copy output chart of gauge variation is produced and data is manually extracted from it.

Bubble Gauges An air reaction gauge consists of a pressure regulator, a gas flow regulating valve, and a pressure sensing system (Fig. 11.5). A small quantity of air or an inert gas flows into a pipe and then to a nozzle fixed below the water surface so that a steady stream of bubbles emerges from the nozzle. The pressure in the pipe feeding the nozzle is equal to the head of liquid above the nozzle. The tube feeding the nozzle is also connected to a pressure sensing system. This type of gauge is often referred to as a "bubbler gauge." An air reaction gauge requires a source of compressed gas, usually nitrogen or compressed air. Its advantage is that it does not need a costly stilling well like that for float-operated gauges.

The digital WLR is an instrument to record water level in rivers with time (either at preset equal intervals of time, say hourly, or when water level changes by more than a preset amount) and (optionally) displays the data. The data is stored in an electronic memory from which these can be downloaded to a computer or transmitted to a central location, if a telemetry device is attached.

Water level measuring devices are also classified in accordance with the performance classes that account for the resolution to be achieved and the limits of accuracy required over specified ranges. Typically, these can give an accuracy of 2 mm.

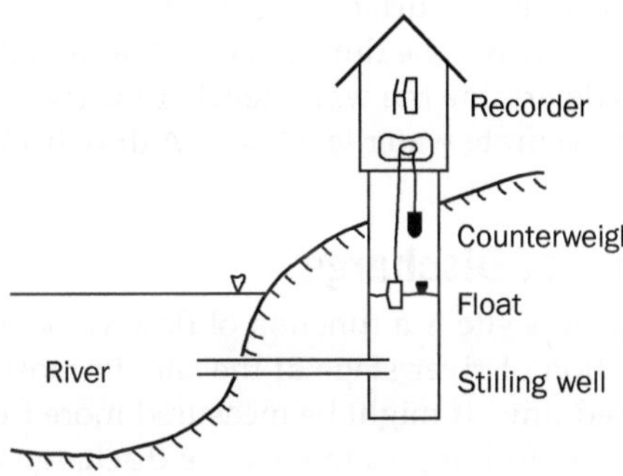

FIGURE 11.4 Float type recorder.

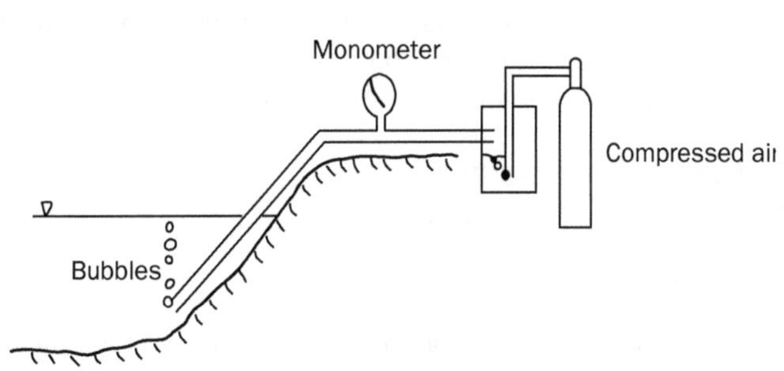

FIGURE 11.5 Bubbler gauge.

FIGURE 11.6 A radar water level recorder.

11.2.4 Radar Water Level Measurement

Radar technology is being increasingly employed for continuous noncontact measurement of water levels in rivers and reservoirs. The operation of all radar water level measurements involves sending an EM signal to the water surface by a sensor which is mounted much above the surface (Fig. 11.6). After hitting the water surface, the EM signal returns to the sensor. The time taken by the signal to return to the sensor is measured and used to determine the water level of the river. The radar measurement technique gives highly accurate water level data. A drawback is their high cost.

11.3 Measurement of Discharge

The river discharge at a site is a function of flow velocity and cross-sectional area and the latter is a function of river stage at the site. In most cases, discharge is measured once a day at a fixed time. It might be measured more frequently at important stations or during high flows. Techniques to measure discharge can be classified into two categories: (1) direct measurement and (2) indirect measurement.

11.3.1 Direct Determination of Discharge

In these methods, either discharge is measured directly or some variable on which discharge uniquely depends is measured. The commonly used methods in this category are velocity-area methods, dilution techniques, electromagnetic method, and ultrasonic method.

11.3.2 Velocity-Area Methods

Mathematically, the discharge (Q) passing through a cross section is

$$Q = A\,V \tag{11.1}$$

where A is the area of cross section of flow (m^2), and V is the mean flow velocity (m/s). In the velocity-area method, the flow area and velocity are measured and are multiplied to get discharge.

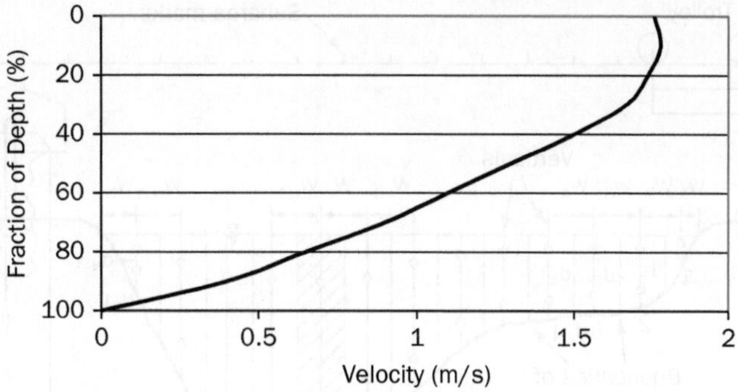

FIGURE 11.7 Typical velocity profile.

Water moving in a river cross section is subject to friction as it comes in contact with the sides and bottom of the channel. Speed of flow of water is the fastest slightly below the open water surface and center of the channel (away from the frictional influences). A typical velocity profile is shown in Fig. 11.7 where the velocity varies as a parabola from zero at the channel bottom to a maximum at (or near) the surface. Due to the variation in velocity, it is not enough to measure the velocity at a single point. Depending on the accuracy required, the cross-sectional area is divided into a number of vertical segments (Fig. 11.8). In each segment, velocity is measured at one or more points along the depth to get the average segment velocity. The area of the individual segment can be calculated by use of data of bed profile and flow depth.

Figure 11.9 shows isovels, which are the contours of equal velocity in a river cross section.

Usually, there are national guidelines on how the vertical sections should be chosen and the depth at which the velocity should be measured to determine the mean velocity in the vertical profile (see, e.g., CWC, 2017). Two methods often used to determine the mean velocity over the vertical velocity profile are discussed here.

Two-Point Measurement The variation of flow velocity with depth for most channels is such that the average of the velocities at 0.2d and 0.8d depth below the surface equals the mean velocity in the vertical (Fig. 11.10).

One-Point Measurement The velocity at 0.6d depth below the surface closely approximates the mean velocity in the vertical. The validity of this assumption for a particular stream can be tested by making a detailed vertical velocity traverse.

The number of velocity determinations is also dictated by the number of measurements that are possible in a reasonable time. If the river stage is changing rapidly, one should quickly complete the measurements before the stage changes significantly.

11.3.3 Measurement of Velocity

Flow velocity can be measured by a number of means. Some important ones are discussed here.

Floats A float is an object that does not sink and remains near or on the water surface, for example, a wooden log, or branch of a tree, or a closed bottle partly filled with

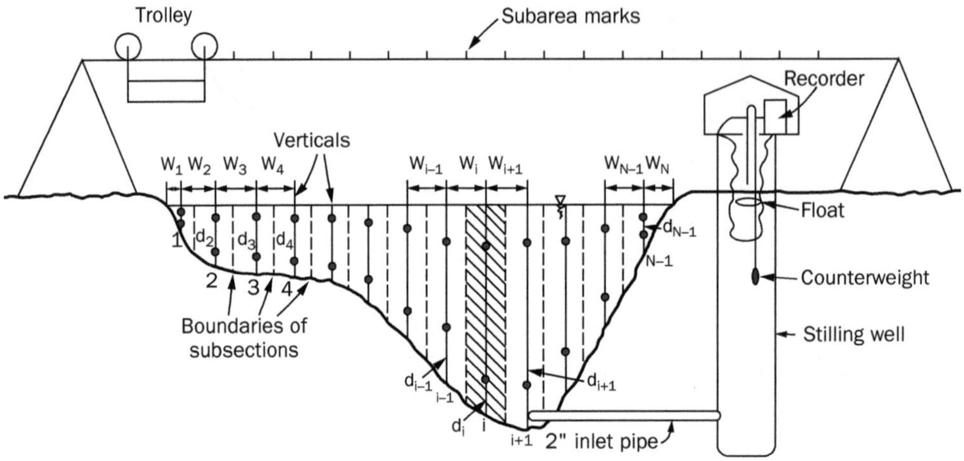

FIGURE 11.8 Schematic sketch for a velocity-area station. (Redrawn from Jain and Singh, 2003.)

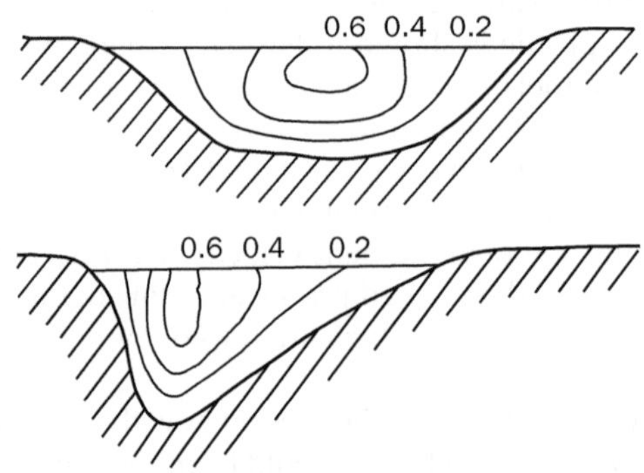

FIGURE 11.9 Velocity distribution in channel cross sections (isovels in m/s).

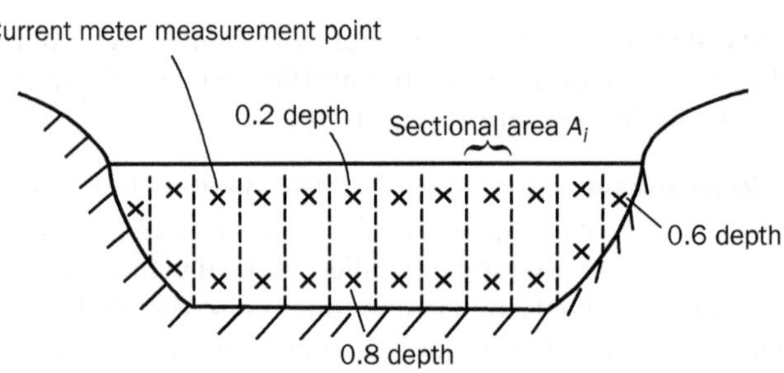

FIGURE 11.10 Scheme for a current meter measurement.

water. Floats that are used to measure streamflow velocity are wooden cylindrical rods of about 0.3 m length. They are shaped such that they remain vertical while float and sufficient length protrudes above water. It is common to paint floats in bright colors for easy identification in muddy or turbulent water. At times, a floating object, such as a branch of a tree, may serve as a float.

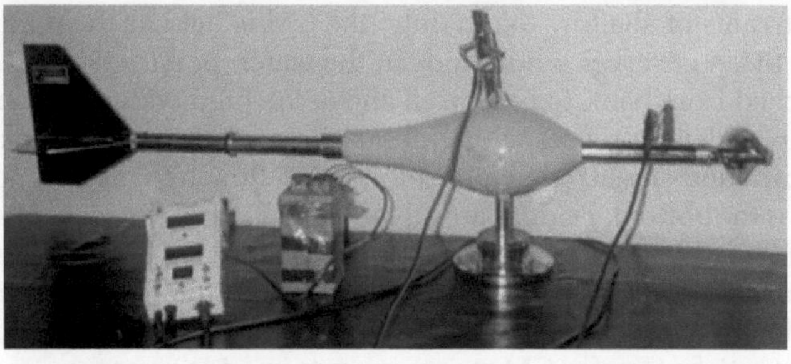

Figure 11.11 Propeller-type current meter.

To measure velocity, two cross sections sufficiently apart (for accurate assessment of float travel time) on a straight river reach are selected. Cross sections should be clearly marked so that the time when the float crosses the cross section is correctly known. The river width at the upstream cross section is divided into a number of equal segments in which floats are tossed.

Observers are placed at either end of the reach such that they can see each other. Floats are thrown uniformly (as feasible) across the cross section a short distance before the upstream section (so that they attain the velocity of water when they arrive at the upstream cross section). Floats may be tossed from the riverbank, bridge, or cableway. When the float crosses the upstream cross section, the observer placed there gives a signal to the downstream observer who acts as a timekeeper and who notes the time taken by the float to cover the reach and compute the float velocity. Next, the mean flow velocity in the vertical is the float velocity multiplied by a coefficient, which depends on the velocity profile and the depth of immersion of the float; it ranges from 0.80 to 0.85. International Standard ISO 748 may be seen for more details.

Floats can rarely be thrown at the desired locations in the river. It is difficult to throw them far from the bank manually and hence only the velocity of water near the banks can be measured. Measurements by floats are not reliable and are normally taken only when other means are not available or cannot be used.

Current Meter A current meter (CM) is commonly used to measure the velocity of flowing water. It consists of a rotating element (rotor) which moves due to the reaction of the water current and acquires an angular velocity proportional to that of water. The CM is placed at a point in the water body and the number of revolutions of the rotor during a time interval are counted to determine the velocity of water. CMs are of two types: propeller type and cup type.

A horizontal axis propeller-type CM consists of a propeller mounted at the end of a horizontal shaft (Fig. 11.11). These are available in a range of propeller diameters and pitch as per flow conditions. Propellers can be made of metal or plastic. Some CMs come with propellers of different sizes for different flow conditions. In the vertical axis CMs, a number (about 4–6) of cups are mounted on a vertical axis and rotate due to the force of moving water.

The propeller-type CM causes less disturbance to flow than vertical axis rotors. Further, due to axial symmetry with the flow direction, the rotor is less likely to be entangled by debris than vertical axis rotors and has less bearing friction.

CM measurements are usually classified based on the means used to cross the stream during observations, such as wading, cableway, bridge, or boat. Wading is feasible in small streams of shallow depth only; the CM is held at the requisite depth below the surface by an observer who stands in the water. In narrow deep channels, a cableway is stretched from bank to bank well above the high water level and a carriage moving over the cableway serves as the observation platform (Subramanya, 2013). CMs are held down and stabilized at a position in the flowing water by sounding weights connected at the bottom. Bridges are advantageous from the view of accessibility, although these are not the best locations hydraulically since the bridge piers disturb natural flow conditions. The velocity should be measured on the downstream of the bridge. If the CM is deployed upstream side of the bridge, it may drift and knock against piers. Boats are most satisfactory for CM measurements in wide rivers.

The relation between revolutions per second N of the current meter and the water velocity v is given by an equation of the following type

$$v = a + b \times N \tag{11.2}$$

where a = the starting velocity or velocity required to overcome mechanical friction, and b = a constant of proportionality. The values of a and b are determined by calibration which should be checked periodically due to wear and tear.

11.3.4 Moving Boat Method

On very wide streams and estuaries, measurement of discharge by conventional methods by CM are frequently impractical or tedious. In the moving boat method, flow data are collected by an observer who is riding a boat traversing the stream along a preselected path. During the traverse, an echo sounder records the bed profile of the river and a CM measures the combined flow and boat velocities. The observer also measures the angle between the CM, which aligns itself in a direction parallel to the movement of water and the preselected path. Data are collected at a large number of observation points in the cross section.

The velocity (v_v) measured at an observation point (Fig. 11.12) is the velocity of water past the CM resulting from both streamflow and boat movement. This velocity is the vector sum of the velocity of water (v) and the velocity of boat (v_b). The flow velocity can be obtained by measuring the angle α between the selected path of the boat and a vertical vane, which aligns itself in a direction parallel to the movement of water past it.

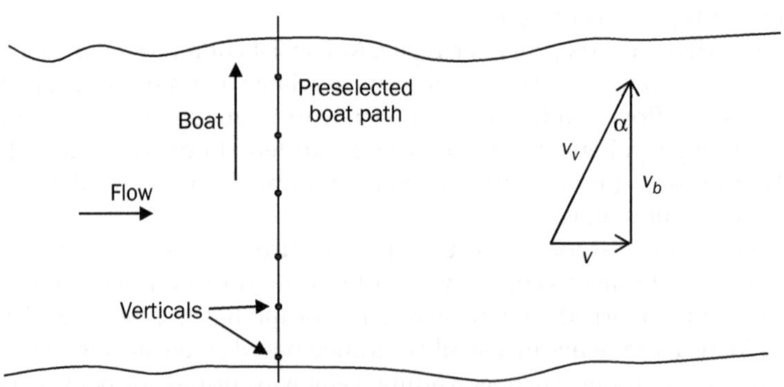

FIGURE 11.12 Moving boat method of discharge measurement.

The flow velocity v, perpendicular to the boat path (true course) at each observation point 1, 2, 3,..., can be determined from the relationship

$$v = v_v \sin \alpha \tag{11.3}$$

11.4 Computation of Discharge

Depending upon the size and variabilities, the river cross section is divided into a number of partial sections such that there is less variation between two adjacent verticals. In general, no partial section should carry more than 5%–10% of the total discharge. Let the cross section be divided into m segments. The sum of the cross-sectional area of each segment multiplied by the average velocity of each section gives the total discharge as

$$Q_{tot} = \sum_{i=1}^{m} Q_i = \sum_{i=1}^{m} A_i v_i \tag{11.4}$$

where Q_{tot} = the total discharge, m = the number of segments in the cross section, A_i = the area of segment i, v_i = the mean velocity in segment i, and Q_i = the discharge in segment i.

Example 11.1

To measure discharge of a small river, the river width was divided into seven segments of 1.5 m width each. Average depths and velocities for the segments are as given in Table 11.1. Compute the discharge in the river.

Compute the discharge in the river by using the velocity-area method.

Solution

We compute the average velocity in each segment as well as its area. These are multiplied to get the segmental discharge and these are summed to get the total discharge. Computations are shown in Table 11.2.

Hence, discharge = 0.24 + 0.77 + 2.01 + 2.67 + 1.83 + 0.80 + 0.29 = 8.61 m^3/s.

11.4.1 Indirect Methods of Discharge Determination

In these methods, discharge is not measured directly, but some easily measurable variable(s) which can be related to discharge are measured. Commonly, these methods use the relationship between discharge and flow depth at a given location. The field measurement is restricted to the measurements of depth only. Two important indirect methods are the slope-area method and the use of flow measuring structures.

11.4.2 Estimation of Discharge by the Slope-Area Method

The slope-area method is an approximate and indirect method of discharge estimation which is used when it is not possible to use more accurate methods, for example, the velocity-area method. International Standard ISO 1070:1992 describes this method. Discharge is estimated by observing the water surface slope and cross-sectional area. Although the accuracy of the slope-area method is less than that of the velocity-area methods, it is helpful when the other methods during high flow conditions are difficult to use. Such situations may arise due to non-accessibility to the site during floods, the presence of debris and floating matter in the river preventing the use of CM, and so on. In flashy streams, high floods may pass without

Segment #	Average Depth (m)	Velocity (m/s) at		
		0.2d	0.6d	0.8d
1	0.4		0.4	
2	0.7	0.76		0.70
3	1.2	1.13		1.19
4	1.4	1.25		1.25
5	1.1	1.09		1.13
6	0.8	0.69		0.65
7	0.45		0.42	

Table 11.1 Depth and Velocity Data for Example 11.1

Segment #	Average Depth (m)	Velocity (m/s) at			Average Velocity (m/s)	Area of Segment (m^2)	Discharge of Segment (m^3/s)
		0.2d	0.6d	0.8d			
1	0.4		0.40		0.40	0.60	0.24
2	0.7	0.76		0.70	0.73	1.05	0.77
3	1.2	1.13		1.19	1.16	1.80	2.01
4	1.4	1.25		1.29	1.27	2.10	2.67
5	1.1	1.09		1.13	1.11	1.65	1.83
6	0.8	0.69		0.65	0.67	1.20	0.80
7	0.45		0.42		0.42	0.68	0.29

Table 11.2 Computation of Discharge by the Velocity-Area Method for Example 11.1

being measured due to their short duration. The magnitude of such floods can be assessed shortly after their occurrence by the slope-area method. Being a less accurate method, the discharge estimated by the slope-area method should not be used to establish rating curves.

The slope-area method may provide reasonably accurate results in channels with stable bed and banks. It may also be used in alluvial channels, but there may be larger errors in results, since a correct assessment of Manning's roughness coefficient n may be difficult. However, this method is not suitable for very wide rivers, rivers with very flat slopes and high sediment concentration, and channels with significant curvature.

To use the slope-area method, a measurement reach is chosen for which the cross-sectional geometry and properties at both ends, Manning's coefficient n, and water-surface elevations at the end sections are known. The river reach should be straight (to the extent possible), should have stable bed and banks, and a uniform cross section over a length of at least five times the width of the channel. The average cross-sectional area and water slope for the reach are determined and the average velocity and discharge are computed. To the extent feasible, the reach should be long enough so that the difference between water levels at the two ends is not less than ten times the uncertainty in the difference. A contracting reach is preferred; and an expanding reach should be avoided.

The slope of water surface is computed from the gauge values at the two ends, which should be read simultaneously. If this is not possible (only one observer), the upstream gauge shall be read, then the downstream gauge, and again the upstream gauge. If the reading of the upstream gauge has changed by more than 10 mm, then the average of the two readings should be used. The data of intermediate gauge(s), if present, are used to ascertain if the slope is uniform in the reach.

Figure 11.13 shows the longitudinal section of a river between two sections, 1 and 2. The total head at a cross section is the sum of water surface elevation and the velocity head. The head loss is sum of two parts: frictional loss and energy loss due to expansion or contraction.

The water surface slope is computed by dividing the fall $(Z_1 - Z_2)$ over the reach by the distance (L) between sections 1 and 2 as

$$S = (Z_1 - Z_2)/L \tag{11.5}$$

The friction slope S_f can be computed by

$$S_f = \frac{(Z_1 - Z_2) + \left[\dfrac{V_1^2}{2g} - \dfrac{V_2^2}{2g}\right](1-k)}{L} \tag{11.6}$$

where k is the coefficient for energy loss; its value is 1 for contractions and 0.5 for expansions. According to Manning's formula, the mean velocity in reach 1 to 2 is

$$v_{1\text{-}2} = (1/n)R^{2/3}\,S^{1/2} \tag{11.7}$$

where R is the hydraulic mean depth ($= A/P$ where P is the wetted perimeter, A is the cross-sectional area), and n is Manning's roughness coefficient. Using the estimated hydraulic radius, the *conveyance* (K) of the channel at the upstream and downstream sections is computed as

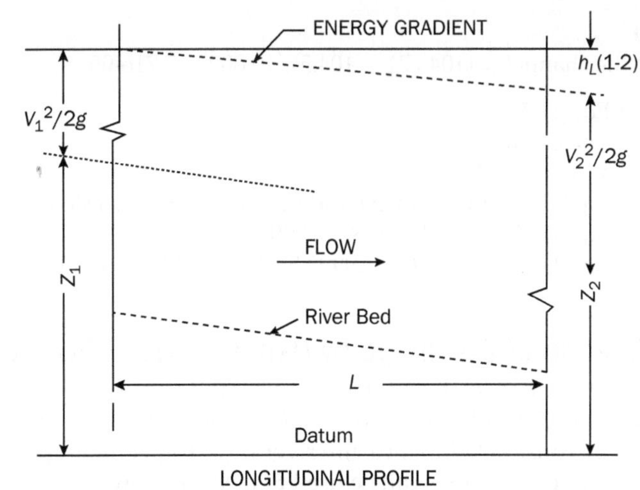

FIGURE 11.13 Channel reach for the slope-area method.

$$K = \frac{1}{n} A R^{2/3} \tag{11.8}$$

Since the flow in the reach may not be truly uniform, the average conveyance of the reach is the geometric mean of the conveyances of the two end sections (K_1 and K_2). Using the values of K_1, K_2, and the surface water slope, the following equation is applied to compute discharge

$$Q = \sqrt{\left(K_1 K_2 S_f\right)} \tag{11.9}$$

The slope-area method can be used in cases, such as alluvial channels including channels with over-bank flow or nonuniform channel cross sections. However, in such cases, the results would have large uncertainties.

Example 11.2

During a high flow, water-surface elevations of a small stream were noted at two sections A and B, 10 km apart. These elevations and other salient hydraulic properties are given below:

Section	Water-Surface elevation (m)	Area of cross section (m^2)	Hydraulic radius (m)	Remarks
A	104.771	73.293	2.733	A is upstream of B
B	104.500	93.375	3.089	$n = 0.020$

The eddy loss coefficient of 0.3 for gradual expansion and 0.1 for gradual contraction are assumed to be appropriate. Estimate the discharge in the stream.

Solution

Slope of the channel = (104.771 – 104.5)/10,000 = 2.71e-05

Using Eq. (11.7)
velocity at A, $V_A = 0.5088$
velocity at B, $V_B = 0.5521$
Using Eq. (11.6) and for an expanding section, $S_f = 2.694\text{e-}05$
Using Eq. (11.8), $K_A = 7163.5$; $K_B = 9902.5$
Using Eq. (11.9), $Q = (7163.5 \times 9902.5 \times 2.694\text{e-}05)^{0.5} = 43.71$ m^3/s.

11.5 Estimation of Discharge by Using Artificial Structures

Hydraulic structures, such as notches, weirs, and flumes, can be used to measure flow in laboratories. If the physical and hydraulic conditions in the field permit, such structures may be constructed to measure river flow. Depending upon the field conditions, different types of structures can be used to measure flow:

1. Thin-plate, sharp crested weirs consisting of a vertical metal plate: v-notch or rectangular;

2. Broad-crested weirs made up of concrete or masonry; these are typically used for big discharges;
3. Triangular or trapezoidal profile weirs; and
4. Flumes: Rectangular, trapezoidal, U-shaped, Parshall (measuring flume with a converging entrance section), or SANIIRI (measuring flume with a converging entrance section having a level floor with a vertical drop at its downstream end and perpendicular walls to join it to the downstream channel; see ISO 9826:1992 for more details). These have specified shape and dimensions and are made up of concrete, masonry, or metal sheets, depending on their use and location. Flumes depend primarily on the width constriction to produce a control section.

Weirs or flumes are frequently used gauging structures. Weirs are used for three main purposes: to partially regulate rivers, to divert river flow, and to measure discharge. They produce a critical relationship between stage and discharge by obstructing channel flow; the head of water over the weir can then be related to discharge. Since weirs have a fixed cross section (either rectangular or triangular), the computation of discharge is simple. Flow conditions are more complex in flumes; a section of critical, high-velocity flow is produced by a lateral smooth-walled constriction of the flow. In both weirs and flumes, the objective is to prevent influences of the water level downstream of the installation on flow and water depth at the point of measurement. Figure 11.14 shows a typical weir and flume.

The use of weirs and flumes to measure flow in the field is not widespread due to a number of reasons. In many rivers, the water level varies over a large range. Rivers frequently carry boulders, pebbles, sediment, and other material which tend to pile up behind these structure or choke them. Thus, it is difficult to meet the site requirements for river gauging for these structures in most cases. In addition, the cost of construction and maintenance of these structure, if built only for the purpose of flow measurement, becomes high. Due to all these reasons, the use of weirs and flumes for streamflow

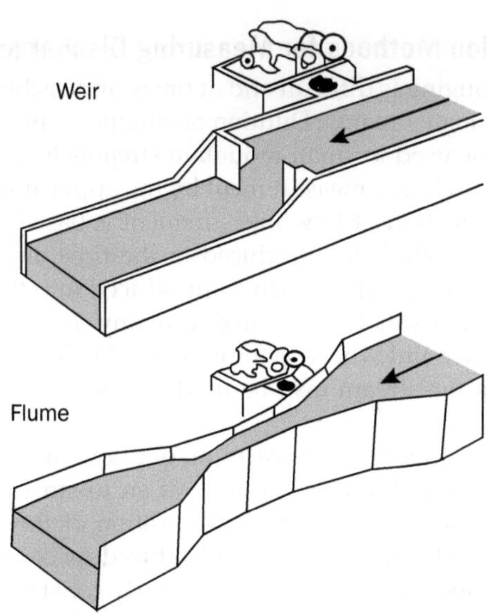

FIGURE 11.14 Typical weir and flume.

measurement is restricted to the upper and middle reaches of rivers where the size of rivers is small and site conditions are manageable to meet the requirements.

Flow measurement at structures, such as a weir or flume, is governed by the hydraulic principle; these structures produce a unique control section. Hence the discharge in the channel (Q) is a function of the head of water at the structure, written as

$$Q = f(H) \tag{11.10}$$

where Q is discharge (m^3/s), and H is the head of water (m) at the structure. For weirs, the equation becomes:

$$Q = K H_n \tag{11.11}$$

where K and n are constants. Equation (11.11) is applicable as long as the water level in the river downstream of the structure is below a certain limiting water level known as the modular limit. In the free flow condition, the flow is not affected by the water level in the river downstream of the structure. If indeed the flow is affected by the water level in the river downstream of the structure, it is called as drowned or submerged flow condition. The discharge under drowned conditions is obtained by applying a reduction factor to the free flow discharge. For a two-dimensional weir, the discharge is estimated as

$$Q = C_d \sqrt{g}\, b\, H^{1.5} \tag{11.12}$$

where C_d is the discharge coefficient, g is the acceleration due to gravity, and b is the crest width (m).

11.6 Advanced Discharge Measurement Techniques

Some advanced methods of flow measurement in open channels are finding application in the field in recent times. These are the dilution methods, the ultrasonic method, and the electromagnetic method.

11.6.1 Dilution Methods for Measuring Discharge

Current-meter gauging is difficult and at times impossible in boulder-strewn mountain rivers and very small streams. Dilution techniques which employ chemical or fluorescent tracers can be used in small mountain streams in which it is difficult to find suitable reaches for discharge measurement by the application of velocity-area methods.

The dilution method of flow measurement is based on the principle of continuity applied to a tracer which is introduced in the flow and completely mixes with it. A tracer is an ion or compound which is introduced into the river to follow the behavior of flow. Materials like salt, fluorescent dye, or any easily measurable material that is not present in the water and whose quantity is not likely to diminish by chemical reaction with materials in the stream may be used as tracers. These should not have environmental issues.

A river reach is identified for carrying out flow measurement and measured quantity of a tracer is introduced in the flow at an upstream end of a river reach. At the downstream end of the reach, the concentration of the tracer is measured at regular intervals of time. The tracer can be introduced at constant injection rate. Here, the duration of injection should be such that nearly constant concentration of the tracer is achieved for an adequate duration (about 10–15 minutes) in the downstream sampling section. Common salt (NaCl) is frequently used as a tracer and it can be detected with an error of 1% up to a concentration of 10 parts per million (Soliman, 2006). A cocktail of tracers may also be used.

Tracer of concentration C_1 is injected at a constant rate Q_t at upstream section 1. At section 2 downstream, where the tracer material is uniformly mixed, the flow is sampled continuously. The concentration of the tracer at section 2 gradually rises from a background value C_0 to a constant value C_2.

By continuity

$$Q_t C_1 = (Q + Q_t) C_2 \tag{11.13}$$

where Q is the river discharge and C_2 is the tracer concentration at section 2. If $Q >> Q_t$ and there is no tracer loss

$$Q_t C_1 = Q C_2$$

or

$$Q = Q_t C_1/C_2 \tag{11.14}$$

Dilution gauging is successful in those streams where mixing occurs readily. Complete mixing of the tracer in the flow and accurate determination of initial and final concentrations are essential. The reach selected for measurement should not have loss or gain of water and the reach length should be sufficient to achieve complete mixing.

11.6.2 Acoustic Doppler Current Profiler

The acoustic Doppler current profiler (ADCP) is an instrument that uses sound waves in the ultrasonic range and applies the Doppler principle to measure velocity profile (not velocity at a point) and compute discharge in streams. The ADCP is typically towed by a boat such that the transducers are deployed beneath the water surface and point downward (Fig. 11.15).

FIGURE 11.15 An ADCP being towed away with a boat.

The ADCP uses sound waves in the ultrasonic range to measure water velocity. The ADCP emits several pulses of sound and then measures the response. Doppler's principle relates the observed change in frequency of a sound source to the relative velocities of the source and the observer. An ADCP applies the Doppler principle by reflecting an acoustic signal off scatterers (small particles of sediment and other material present in water). The velocity measured by the Doppler principle is parallel to the direction of the transducer emitting the signal and receiving the backscattered acoustic energy (Mueller and Wagner, 2009).

Typical boat-mounted ADCPs have three or four beams pointing between 20 and 30 degrees from the vertical. Three beams are needed to obtain a three-dimensional velocity measurement. When water velocity is measured from a moving boat, it yields the velocity of the water relative to the boat. A global positioning system (GPS) keeps track of the location of the boat. The ADCP measurements are fast and accurate. In addition, the ADCP allows measurements in environments where conventional mechanical current meters are inappropriate or unreliable, such as in tidally affected flows, highly unsteady flows, and flood flows. ADCPs also provide continuous profiles of water velocity. Thus the vertical velocity distribution is no longer assumed but rather measured for all the near-bed and near surface, thereby providing more accurate measurements of streamflow. An additional advantage is that the ADCP measures the flow field in the stream channel in three dimensions, which provides a far more accurate and detailed view of velocity and flow profiles. CWC (2017) has described the procedure to measure riverflow by use of ADCP.

11.7 Streamflow Measurement Under Difficult Conditions

Extreme cold weather conditions, such as ice or frost, can hamper the operation of stage sensing or recording device and current meters. Hence, data for such periods may be missing or erroneous. Stilling well and inlet pipes should be constructed such that they continue to operate in freezing temperatures.

Many rivers carry a large amount of silt, which may clog the inlet tube and the stilling well. The area of the cross section may change due to sediment deposition/erosion, leading to shifting-control and changing stage-discharge relations. The weed growth in a channel can change the hydraulics in the measuring reach and thereby the stage-discharge relationship. In such situations, periodic weed removal is necessary.

Arid and semiarid regions typically have harsh and dusty environment and high soil erosion by strong winds. Rivers commonly carry high sediments that damage equipment and installations and cause high wear and tear of the instruments deployed for flow measurement. This leads to frequent malfunctioning of equipment and loss of data. Rivers in these areas are commonly ephemeral and flashy and channel carrying flow may shift laterally, isolating gauge and recorder from the flow. As beds are erodible, any construction in the channel is vulnerable to damage. Variations in bed are difficult to measure and lead to errors in discharge values. If measurements are manual, the likelihood of missing peak events is high (duration is typically short) unless the observer is present at the site.

Careful planning of logistics, local knowledge, and experience are often helpful to successfully tackle problems in such regions. New but robust and appropriate technology is helpful to overcome difficulties. Many equipment with minor improvements could work equally well under arid and semiarid conditions. The sensors and associated hardware currently available have been found to work well in harsh conditions with a wide range of environment. Innovative application of hardware and software to solve the problems encountered in "difficult situations" is required.

11.8 Stage-Discharge Rating Relationship

Measurement of river stage is simple and does not require very expensive instruments. However, measurement of discharge is time consuming and requires skilled staff as well expensive equipment. Hence, it is difficult to measure discharge at close time intervals, particularly in large rivers. River stage and discharge at a cross section exhibit a relation, known as rating curve or stage-discharge relation. A rating curve is developed by the use of concurrent data of stage and discharge observed over a period of time. It is important that the data covers the range of stages that are likely to occur at the gauging station. At a new site, river stage and discharge are measured and when sufficient values over the expected range of stage are available, rating relationship is established. Later, river stage only is measured frequently and the rating relationship is used to convert it to discharge values. Discharge is measured at periodic intervals only to verify the existing rating relation and update it, if necessary. As most hydrologic analyses are based on discharge data, the rating curve has important bearing on water resources planning and management.

Factors that influence the accuracy and stability of the rating curve at a gauging station can be broadly classified into two groups: natural and artificial. The natural factors include the geometry of cross section, properties of material constituting bed and banks, and channel upstream of the gauging station. The artificial factors are any structures that may regulate flow, such as a weir.

11.8.1 Simple Rating Curve

In a simple rating curve, river stage is uniquely related to discharge. When more than one curve is required depending upon the range of stage values, this is known as a compound rating curve. When variables other than stage enter the relationship, this leads to a complex rating curve.

Before establishing a rating curve, it is helpful to plot the stage and discharge data on graph paper (Fig. 11.16). Commonly, stage is plotted on the Y axis and discharge on the X axis. Plotting stage-discharge data on log-log graph paper commonly yields a straight line relationship(s). Ideally, the graph should have a sufficient number of points, well distributed over the entire stage and discharge range. In practice, however, one may get more points in lower stage ranges and less at high stages. If the points display a small scatter, a smooth curve can be drawn indicating that a simple rating curve is adequate.

A simple rating curve is commonly represented by the power equation of the form

$$Q = a\,(H - H_0)^b \tag{11.15}$$

where Q is the discharge (m^3/s), H is the river stage (m), a and b are constants to be determined, and H_0 is the stage (m) at which discharge is nil (known as the datum correction). Taking log of both sides of Eq. (11.15)

$$\text{Log}\,(Q) = \log\,(a) + b\log\,(H - H_0) \tag{11.16}$$

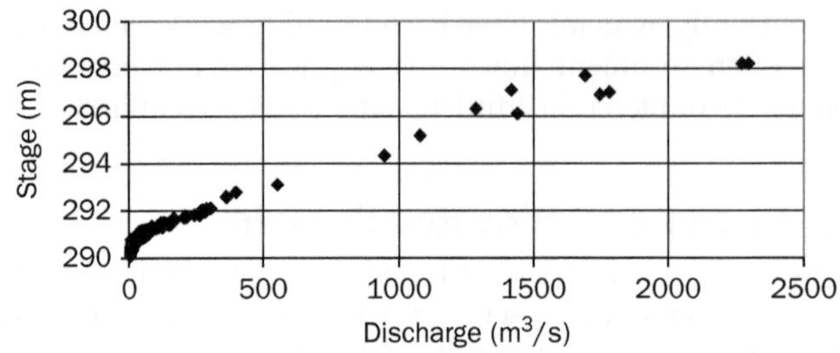

FIGURE 11.16 Plot of stage and discharge at a gauging station.

Equation (11.2) plots as a straight line in the log domain and can also be written as

$$Y = a + b\,X \tag{11.17}$$

Thus, if discharge Q and stage $(H - H_0)$ are plotted on a log-log paper, a straight line is obtained. The method of least squares is commonly used to estimate the rating curve coefficients a and b in Eq. (11.17). The datum correction H_0 is iteratively estimated; the rating relationship is established by varying H_0 and the value which yields the best relation is adopted. It should be ensured that the selected value of the datum correction is physically plausible. The slope of a rating curve plotted on a log-log paper reveals important information about the control. A slope >2.0 generally indicates a section control, while a slope <2.0 indicates that a channel control is likely (Kennedy, 1984).

Example 11.3

Forty pairs of stage-discharge data are available at a gauging site, listed in Table 11.3. Fit a linear rating curve of the type given by Eq. (11.17).

Solution

We assume $H_0 = 93.4$ m. The data are plotted in Fig. 11.17.

Computations are shown in Table 11.4 below.

As per regression rule, in Eq. (11.17)

$$b = Sxy/Sxx = (x - xm) \times (y - ym)/(x - xm)^2 = 41417.9/165.002 = 251.01$$

$$a = ym - b \times xm = 395.292 - 251.01 \times 2.571 = -250.15.$$

Here, xm denotes the mean of x, and so on.

As shown on the graph, R^2 is quite high at 0.9145.

Alternately, one can fit an equation like Eq. (11.16) by first taking log of stage and discharge values and then find the coefficients. In this, the following equation will be obtained

$$\text{Log10}\,(Q) = 1.82 + 1.62\,\text{log10}\,(H - 93.4)$$

It is helpful to fit this equation with different values of H_0 and adopt the value which gives smallest sum of squares of errors.

11.8.2 Compound Rating Curve

In many situations, the influence of roughness and boundary conditions on discharge is not be the same over the entire observed range of river stages and, therefore, a single rating curve does not properly work for the gauging site. One may, therefore, develop

SN	Q, m³/s	H, m	SN	Q, m³/s	H, m
1	79.9	94.743	21	138.6	94.65
2	34.63	94.685	22	165.2	95.485
3	87.5	94.763	23	225.12	96.067
4	64.56	94.733	24	359.62	96.285
5	272.12	94.433	25	714	96.86
6	517.5	97.65	26	742	97.72
7	980	100.84	27	610	97.105
8	416	96.798	28	279.95	95.55
9	233.02	95.47	29	203.15	95.192
10	96.06	94.865	30	190.3	95.185
11	96.06	94.864	31	170.8	95.07
12	185.05	94.978	32	136.36	94.871
13	165.45	95.08	33	159.75	94.915
14	130.25	94.723	34	170.12	94.921
15	31.75	94.61	35	148.1	94.82
16	81.98	94.537	36	632.5	96.7
17	55	94.519	37	879.91	97.21
18	101.15	94.634	38	1738	100.238
19	98.3	94.635	39	2600	103.65
20	77.92	94.56	40	1744	100.24

TABLE 11.3 Stage-Discharge Data for Example 11.3

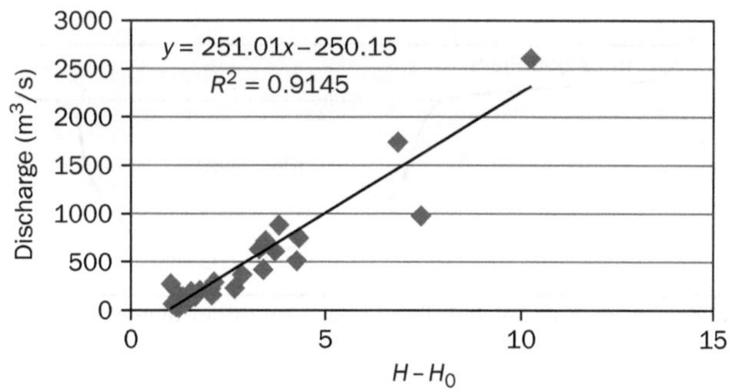

FIGURE 11.17 Plot of stage and discharge for the example problem.

up to three curves—for lower range of stages, for middle range, and for higher range. This is known as a compound rating curve. A compound rating curve is also necessary when the cross section consists of a narrow channel and flood plains (Fig. 11.18) on side(s). At low water levels, flow is confined to the main channel but the river occupies flood plains at higher levels. Since the conveyance characteristics of flood plains are quite different from those of the main channel, this causes a change in the slope in the rating curve. In such cases, two rating curves are established, the first is applicable

SN	Q, m³/s	H, m	$H - H_0$	(x – xm)^2	(y – ym)^2	(x – xm) (y – ym)	Q_c	$Q - Q_c$
1	79.9	94.743	1.343	1.509	99472.1	387.4	87.0	7.1
2	34.63	94.685	1.285	1.655	130077.1	463.9	72.4	37.8
3	87.5	94.763	1.363	1.460	94735.9	371.9	92.0	4.5
4	64.56	94.733	1.333	1.534	109383.7	409.6	84.4	19.9
5	272.12	94.433	1.033	2.367	15171.3	189.5	9.1	–263.0
6	517.5	97.650	4.250	2.818	14934.8	205.1	816.7	299.2
7	980	100.840	7.440	23.704	341883.4	2846.7	1617.4	637.4
8	416	96.798	3.398	0.683	428.8	17.1	602.8	186.8
9	233.02	95.470	2.070	0.251	26332.2	81.4	269.4	36.4
10	96.06	94.865	1.465	1.224	89539.8	331.1	117.6	21.5
— Intermediate lines deleted —								
35	148.1	94.820	1.420	1.326	61103.9	284.6	106.3	–41.8
36	632.5	96.700	3.300	0.531	56267.6	172.8	578.2	–54.3
37	879.91	97.210	3.810	1.534	234854.6	600.3	706.2	–173.7
38	1738	100.238	6.838	18.204	1802864.8	5728.9	1466.3	–271.7
39	2600	103.650	10.250	58.962	4860737.4	16929.2	2322.7	–277.3
40	1744	100.240	6.840	18.221	1819013.3	5757.2	1466.8	–277.2
	395.292	95.971	2.571	165.002	11368005.2	41417.9	15811.7	0.0

TABLE 11.4 Computations to Determine Parameters of Rating Relations

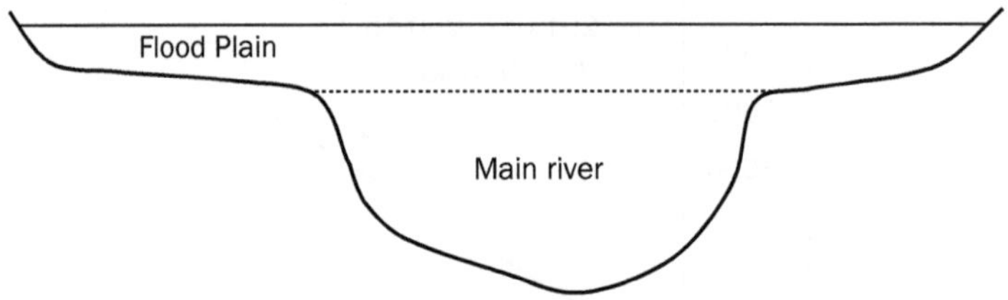

FIGURE 11.18 A river channel with flood plains on either side.

when the flow is confined to the main channel and the second for flow in the main channel plus flood plain.

Before developing a compound rating relation, the stage and discharge data are plotted on a log-log graph and examined to identify change points. The stage-discharge relation for the different segments can be developed following the procedure explained above. A compound rating curve is shown in Fig. 11.19.

The methods in which the effect of slope is accounted for in the rating relations are known as slope rating. Two commonly used methods by which the effect of slope can be taken into account are (1) constant-fall rating and (2) variable fall rating. The

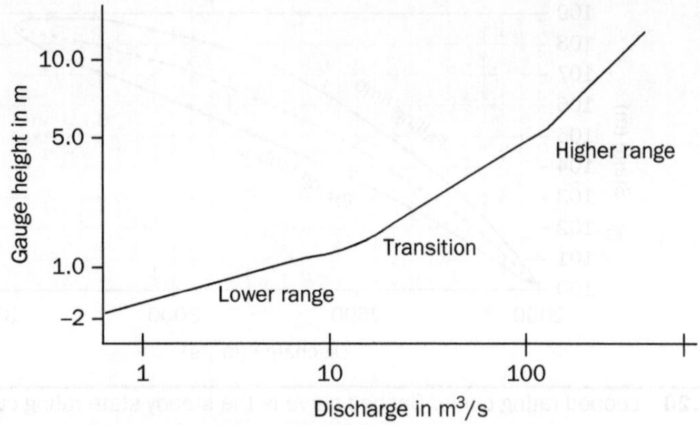

FIGURE 11.19 A compound rating curve.

first method suits uniform channels and those for which the water surface between the base and auxiliary stations does not have large curvatures. The second method is more suitable when the water surface profile has an appreciable curvature, or the reach is not uniform, or a submerged section control exists in the channel reach between two gauges, or a combination of these. Rantz (1982) has described the computational procedure.

11.8.3 Hysteresis in Rating Curves

In large rivers with flat-bed slopes and significant changes in flow rates, the impacts of unsteady flow on the rating curve may be significant. During the rising stage of a hydrograph, the velocity and discharge at a given stage are higher than the same at steady-state flow. During hydrograph recession, the situation is reverse (Jobson and Keefer, 1979). This phenomenon is known as hysteresis and it produces a loop rating curve (Fig. 11.20), that is, discharge is not a unique function of stage. This type of behavior of flow is observed because the slope of flood wave front during the rising stage is much steeper than during the steady-state condition. The situation is reverse when the discharge is reducing or river stage is falling. Such behavior of river is observed in plains where the slopes are low. Note that the shape of the loop may vary from one station to another; sometimes it can be a complicated loop.

The modeling of a looped rating curve is not possible using Eq. (11.1), which can fit an average or steady-state curve. One way in such situations is to plot the stage and discharge data on graph paper and label the points with date and time of measurement. Next, a looped rating curve is drawn.

Neglecting acceleration terms, the relation between unsteady and steady flow can be expressed by the Jones formula

$$Q_{\text{act}} = Q_{\text{steady}}\sqrt{\left(1 + \frac{1}{cS_0}\frac{dh}{dt}\right)} \tag{11.18}$$

where Q_{act} is the actual (measured) discharge in the channel, Q_{steady} is the steady-state discharge obtained from the rating curve, c is the wave celerity, S_0 is the energy slope (steady state), and dh/dt is the rate of change of stage. Equation (11.18) can also be expressed as

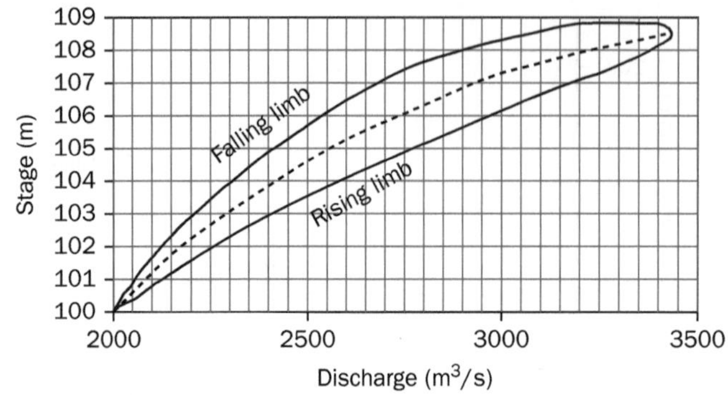

FIGURE 11.20 Looped rating curve. Dashed curve is the steady-state rating curve.

$$Q_{act} = Q_{steady}\sqrt{\left(1 + K\frac{dh}{dt}\right)} \tag{11.19}$$

where $K = 1/(S_0c)$. K may be a constant or a function of river stage. Rantz (1982) and Singh (1992) have described procedures to determine K.

11.8.4 Extrapolation of Rating Curves

Rating curves need to be extrapolated if river stage and discharge measurements are not available over the entire range of observed river stages. In simple cases, the rating curve may be smoothly extended. But this will give erroneous results if, in the extended range of stage, cross section or channel geometry changes, flow spills over flood plains, or the roughness coefficient changes significantly.

To extrapolate the rating curve in low flow ranges, it is better to plot the rating curve on a normal graph, since zero discharge cannot be plotted on a log scale. A smooth curve is drawn to join the gauge of zero flow with the lowest end of the rating curve (see Fig. 11.21).

Extrapolation of the rating curve in high flow ranges should be done with caution and should be attempted, if indirect methods of discharge estimation cannot be used. Rantz (1982) recommended the use of the conveyance-slope method to extrapolate a rating curve. The conveyance (K) can be computed by

$$K = \frac{1}{n}AR^{2/3} \tag{11.20}$$

Conveyance depends on the measurable elements of a cross section. A curve of K versus gauge height, covering the entire range, is plotted on ordinary graph paper. Now, for the measured discharges, S can be obtained from Manning's equation by using the value of K. Using these data, a curve of gauge height versus S is plotted and extrapolated to the required gauge height (see Fig. 11.22). Note that slope tends to attain a constant value at higher river stages under certain conditions. The discharge at the desired gauge can be computed by reading the values of K and S from the respective curves. Rantz (1982) recommends that the extrapolation of discharge beyond a value exceeding twice the measured discharge should be attempted only as a last resort.

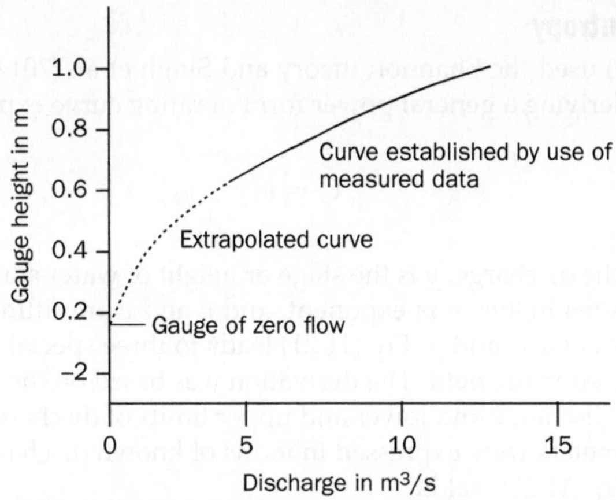

FIGURE 11.21 Extrapolation of rating curve in lower ranges.

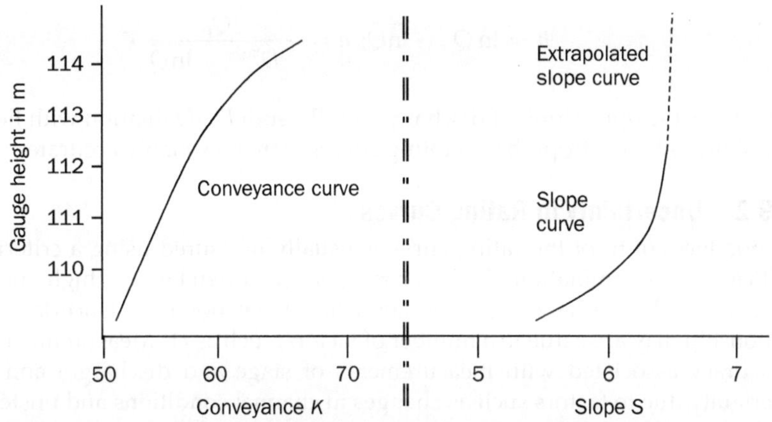

FIGURE 11.22 Conveyance slope method of extrapolation of rating curve.

11.9 New Techniques for Development of Stage-Discharge Relation

Establishing a rating curve can be viewed as a mapping problem where stage is the input variable and discharge is the output. Among the emerging data analysis techniques, the Artificial Neural Networks (ANNs) technique is a powerful procedure for nonlinear function mapping. Tawfik et al. (1997), Jain and Chalisgaonkar (2000), and Jain (2001) applied this technique to establish stage-discharge relations. Integration of ANN and fuzzy systems leads to a powerful methodology in which fuzzy systems provide a framework to represent expert knowledge, while ANNs provide learning capabilities.

11.9.1 Entropy

Singh (2010) used the Shannon theory and Singh et al. (2014) used the Tsallis entropy theory for deriving a general power form of rating curve expressed as

$$Q = a(y - y_0)^b + c \tag{11.21}$$

where Q is the discharge, y is the stage or height of water surface, y_0 is the height when discharge is negligible, b is exponent, and a and c are fitting parameters. Depending on the values of a, c, and $y_{0,}$ Eq. (11.21) leads to three special forms of rating curve that have been used in the field. The derivation was based on the knowledge of the mean of logarithmic discharge and lower and upper limits of discharge. For each form of rating curve, parameters were expressed in terms of known discharge values. For example, if $b = y_0 = 0$, Eq. (11.21) yields

$$Q = a\,y^b \tag{11.22}$$

then, Singh (2010) showed that

$$b = \ln Q_D - \overline{\ln Q};\ a = -\frac{Q_D}{D^{\ln Q_D} - \overline{\ln Q}} \tag{11.23}$$

where Q_D is the upper limit of discharge at full depth D. Evaluation with field observations showed that entropy-based rating curves were reasonably accurate.

11.9.2 Uncertainty in Rating Curves

The goodness of fit of the rating curve is usually measured using a criterion, such as coefficient of determination, R^2. This coefficient can often be very high, indicating that estimated discharges have high precision while it may not be so. Uncertainties in rating relationship may arise due to a number of factors, such as (1) Measurement uncertainty (inaccuracy associated with measurements of stage and discharge) and (2) natural uncertainty due to factors such as changes in channel conditions and unsteady flow in the rivers.

11.10 Summary

In river gauging, the choice of measuring methods and instruments depends upon the site conditions, resources (human, infrastructure, and instruments), and the desired accuracy. In emergent conditions, such as equipment breakdowns or extreme floods, photographs or floodwater marks can be used to fill data gaps (of course with much uncertainty). An important element for good measurement is motivated, well trained, and properly equipped personnel. A number of references have been provided in the chapter which may be pursued by those who want to explore this topic further. Important advances are taking place in hydrometry which are helping in collecting better data and thereby improved understanding of hydrologic processes.

11.11 Questions

11.1 What do you understand by station and section controls?

11.2 What are the various means of measuring river stages and what are their advantages and limitations?

11.3 To measure discharge of a small river, the river width was divided into 8 segments of 2 m width each. Average depths and velocities for the segments are as given in Table 11.5 below. Compute the discharge in the river.

Segment #	Average Depth (m)	Velocity (m/s) at		
		0.2d	0.6d	0.8d
1	0.35		0.4	
2	0.82	0.76		0.70
3	1.22	1.13		1.19
4	1.53	1.25		1.25
5	1.21	1.09		1.13
6	0.86	0.69		0.65
7	0.42			
8			0.42	

TABLE 11.5 Average Depths and Velocities for Segments in a Small River

Compute the discharge in the river by using the velocity-area method.

11.4 Briefly describe the slope-area method of discharge estimation and list its advantages and disadvantages.

11.5 A small stream has a trapezoidal cross section with a base width of 12 m and side slope 2 horizontal: 1 vertical in an 8 km reach. During a flood, the high water levels recorded at either ends of the reach are:

Section	Elevation of bed	Water surface elevation	Remarks
Upstream	100.20	102.70	Manning's $n = 0.030$
Downstream	98.60	101.30	

Estimate the discharge in the stream.

11.6 During a flood flow, the depth of water in a 10-m-wide rectangular channel was found to be 3.0 and 2.9 m at two sections 200 m apart. The drop in the water-surface elevation was found to be 0.12 m. Assuming Manning's coefficient to be 0.025, estimate the flood discharge through the channel.

11.7 River flow is to be measured in a hilly stream whose bed has boulders and the flow depth is usually small. Which method would you use and briefly describe it?

11.8 Thirty pairs of stage-discharge data are available at a gauging site, listed in Table 11.6 below. Fit a rating curve of the type given by Eq. (11.15).

SN	Q (m^3/s)	H (m)	SN	Q (m^3/s)	H (m)
1	77.32	80.243	16	79.4	80.037
2	32.05	80.185	17	52.42	80.019
3	84.92	80.263	18	98.57	80.134
4	61.98	80.233	19	95.72	80.135
5	269.54	79.933	20	75.34	80.06
6	514.92	83.15	21	136.02	80.15
7	977.42	86.34	22	162.62	80.985
8	413.42	82.298	23	222.54	81.567
9	230.44	80.97	24	357.04	81.785
10	93.48	80.365	25	711.42	82.36
11	93.48	80.364	26	739.42	83.22
12	182.47	80.478	27	607.42	82.605
13	162.87	80.58	28	277.37	81.05
14	127.67	80.223	29	200.57	80.692
15	29.17	80.11	30	187.72	80.685

Table 11.6 Stage-Discharge Data at a Gauging Site

11.12 References

Braca, G. *Stage-Discharge Relationships in Open Channels: Practices and Problems*. Università degli Studi di Trento, Trento, March 2008.

CWC. *Handbook for Hydrometeorological Observations*. Central Water Commission, New Delhi, January 2017.

Boiten, W. *Hydrometry*. London: CRC Press, 2008.

Herschy, R. W. (Ed.). *Hydrometry Principles and Practices*. New York: John Wiley and Sons, 1978.

Herschy, R. W. *Streamflow Measurement*. London: CRC Press, 2008.

International Organization for Standardization. *ISO 748. Liquid Flow Measurement in Open Channels—Velocity-Area-Method*. Geneva, Switzerland: International Organization for Standardization, 2007.

ISO. 1070: Hydrometry—Slope-area method. The International Organization for Standardization, Geneva, 2018.

Jobson, H. E., and T. N. Keefer. Modeling highly transient flow, mass and heat transport in the Chattahoochee river near Atlanta, Georgia. Geological Survey Professional Paper 1136, US Department of Interior, Washington DC, 1979.

Jain, S. K. Development of integrated sediment rating curves using ANNs. *Journal of Hydraulic Engineering, ASCE*, 127(1):30–37, 2001.

Jain, S. K., and V. P. Singh. *Water Resources Systems Planning and Management*. Amsterdam: Elsevier, 2003.

Jain, S. K. Hydrological Measurements and Analysis of Data. Web-based course under National Program on Technology Enhanced Learning (NPTEL). http://nptel.ac.in/courses/105107129/, 2011.

Jain, S. K., and D. Chalisgaonkar, D. Setting up stage discharge relations using ANN. *Journal of Hydrologic Engineering, ASCE*, 5(4):428–433, 2000.

Kennedy, E. J. *Techniques of Water-Resources Investigations of the United States Geological Survey: Discharge Ratings at Gaging Stations* (Book 3, Chapter A10). Washington, DC: USGS, 1984.

Mueller, D. S., and C. R. Wagner. Measuring discharge with acoustic Doppler current profilers from a moving boat. *U.S. Geological Survey Techniques and Methods* [Online] 2009, 3A–22. http://pubs.water.usgs.gov/tm3a22.

Rantz, S. E. *Measurement and Computation of Streamflow: Volume 2. Computation of Discharge*. Paper # 2175. Washington, DC: USGS, 1982.

Singh, V. P. *Elementary Hydrology*. Englewood Cliffs, NJ: Prentice-Hall Inc., 1992.

Singh, V. P. Derivation of rating curves using entropy theory. *Transactions of the ASABE*, 53(6):1811–1821, 2010.

Singh, V. P., H. Cui, and A. R. Byrd. Derivation of rating curve by the Tsallis entropy. *Journal of Hydrology*, 513:342–352, 2014.

Soliman, M. M. *Engineering Hydrology of Arid and Semi-Arid Regions*. Boca Raton: CRC Press, 2006.

Subramanya, K. *Engineering Hydrology*. New Delhi: Tata McGraw-Hill Education, 2013.

Tawfik, M., A. Ibrahim, and H. Fahmy. Hysteresis sensitive neural network for modeling rating curves. *Journal of Computing in Civil Engineering*, 11(3):206–211, 1997.

World Meteorological Organization (WMO). *Guide to Hydrological Practices: Volume I. Hydrology—From Measurement to Hydrological Information*. Report No. 168. Geneva: World Meteorological Organization, 2008.

CHAPTER 12

Remote Sensing and Geographic Information Systems

This chapter discusses remote sensing (RS) and geographic information systems (GISs), which are two tools extensively used in hydrologic analysis and modeling. The objectives of this chapter are to introduce fundamental concepts of data collection through RS and analysis, and to explain the concepts of a GIS and its analytical capabilities. We briefly describe the theoretical aspects of these tools, intermixed with their applications in hydrology.

12.1 Remote Sensing

Remote sensing (RS) is the process of acquiring information about an object without physical contact between the object and the sensor. Although it may appear to be new, this technique of acquiring information about an object is very old. Photography, which has been in use for a long time, is based on the same principles. Human beings also obtain the information by RS when they "see" an object; eyes can be considered to be sensors, mounted on the human body. Most commonly, the sensors of an RS system are mounted on a satellite, orbiting Earth and information is extracted from the electromagnetic radiation emitted or reflected by the objects. These days, the term RS mainly denotes acquisition and analysis of satellite data. Spaceborne RS provides many advantages: coverage of large areas, frequent and repetitive coverage of the area of interest, quantitative measurements of terrestrial features, no extra efforts to capture data from remote and difficult-to-access regions, and relatively low cost of data acquisition compared to field measurements. These days, drones are also being used to collect data about Earth features and they share some common characteristics with satellites.

Nuclear reactions in the Sun produce electromagnetic (EM) radiation which is transmitted through space and reaches Earth's atmosphere. The EM energy propagates in the form of waves in accordance with the wave theory. The distance from one peak of the wave to the next is the wavelength λ and the number of peaks passing through a point per unit time is known as the wave frequency, ν. The wave velocity $c\ (= \nu\lambda)$ is equal to the velocity of light ($c = 3 \times 10^8$ m/s). The spectrum of EM radiation has waves ranging from very short wavelength (10^{-7} μm) cosmic rays to very long wavelength (10^8 μm) radio waves. Figure 12.1 shows the EM spectrum. The Sun is the chief source of EM radiation on Earth.

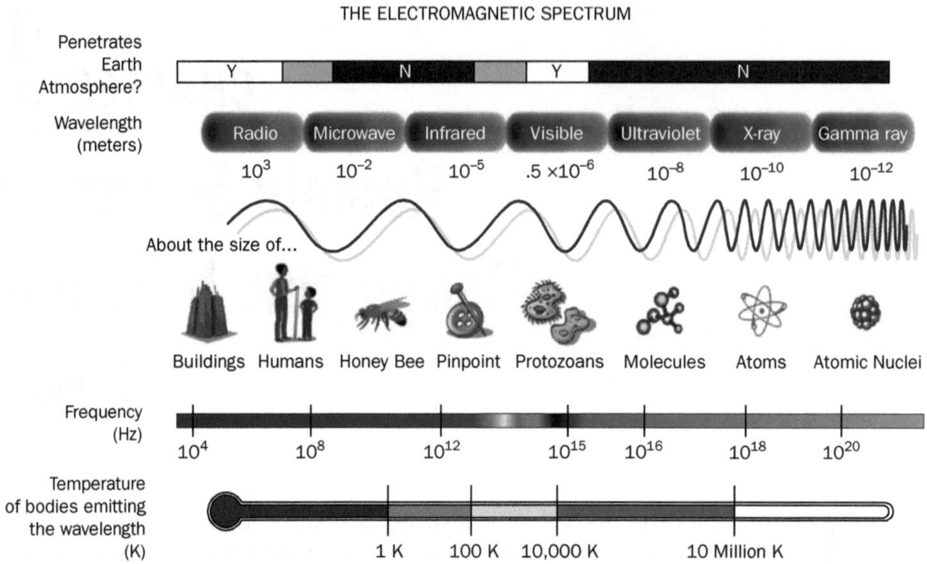

FIGURE 12.1 Electromagnetic spectrum. (Source: https://mynasadata.larc.nasa.gov/science-practices/electromagnetic-diagram.)

One RS technique is aerial photography. This is a manual method of interpreting "photographs" and basically makes use of only the visible part of the EM spectrum. As seen in Fig. 12.1, the visible part covers a very small portion of the EM spectrum. RS sensors use data from the bands of the EM spectrum beyond the visible range; these days, a very wide range of the EM spectrum is utilized.

Measurements of the EM spectrum reflected by an object reveals a wide range of information about the object: What is it? Where is it located? What are its physical properties? What spatial relationships it has with the surrounding objects, and so on? Data from one or more wavelength ranges are used to identify the object and determine its properties. From analysis of these measurements, it is possible to discriminate between different objects on the ground and determine if the object is water, dry soil, vegetation, snow, and so on. The spatial extent of each object on the ground can be mapped. Moreover, depending upon the sensor and the EM band being used, it is also possible to infer the properties and water contents of a top thin layer (say about 5 cm) of Earth's surface. Due to these properties, RS is a powerful technique to assist in exploring, mapping, and managing Earth resources.

The EM radiation emitted or reflected from an object passes through atmosphere before it is detected by the sensor. Thus the properties of atmosphere significantly determine the use of EM spectrum for RS. The most important atmospheric gases that influence the incident radiation are water vapor, oxygen (O_2), ozone (O_3), carbon dioxide (CO_2); suspended particles or aerosols also interact with radiation. The spectral characteristics of energy sources, atmospheric effects, and RS systems are shown in Fig. 12.2. We know that the human eye can detect only a small portion of the EM spectrum. This range is known as the *visible* range which lies between 0.4 and 0.7 μm. The wavelength ranges of three primary colors are blue 0.45–0.50 μm, green 0.5–0.6 μm, and red 0.6–0.7 μm. The color of an object illuminated by visible light depends upon its

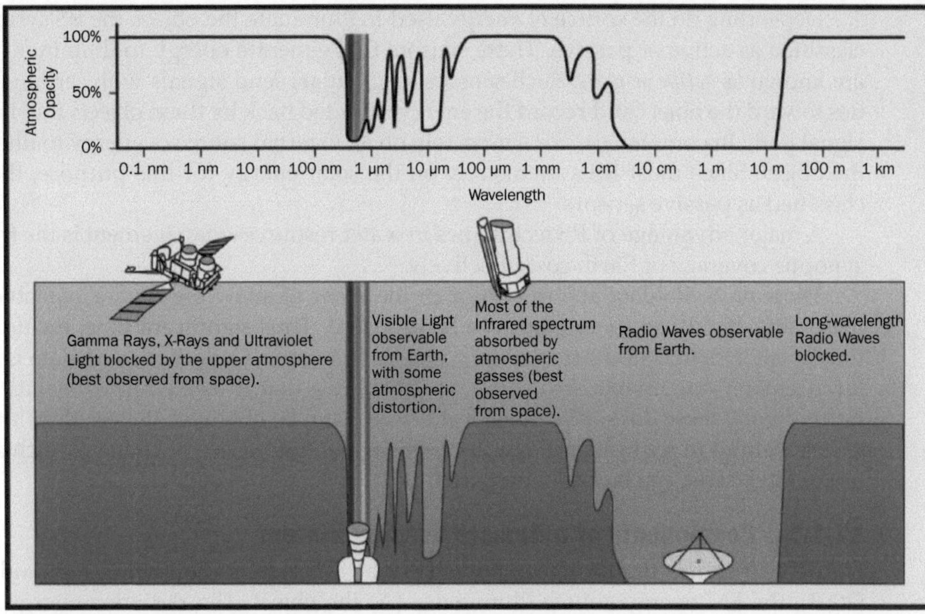

FIGURE 12.2 Atmospheric windows. (Source: https://commons.wikimedia.org/wiki/File:Atmospheric_electromagnetic_transmittance_or_opacity.jpg.)

reflectance properties. For example, chlorophyll present in healthy vegetation absorbs the blue and red portion of the EM spectrum and reflects green light. Hence, vegetation appears green.

The infrared (IR) region of the EM spectrum that covers the wavelength range from 0.7 to 100 μm is of high interest in RS. Based on radiation properties, the infrared region is divided into two categories: reflected IR and emitted or thermal IR. The reflected IR region covers wavelengths from 0.7 to 3.0 μm and is used in ways very similar to the radiation in the visible portion. The thermal IR region covering the range of 3.0–100 μm is different. In this region, energy is essentially the radiation emitted from Earth's surface in the form of heat; hence the name. Spectra from wavelengths 10^{-10}–10^{-6} are called X-rays and microwaves have a wavelength from 10 cm to about 1 mm.

Earth's atmosphere does not respond to all the wavelength ranges in the same manner. Some wavelength ranges are transmitted by the atmosphere without much loss of energy and these are known as *atmospheric windows*. The various ranges of the EM spectrum and the transmission/absorption of energy in different ranges by the atmosphere is depicted in Fig. 12.2. Naturally, the RS technique uses these wavelength ranges which are passed by the atmosphere without significant loss of energy. Useful atmospheric windows are 0.3–0.75 μm [ultraviolet (UV) and visible], 0.77–0.95 μm [near infrared (NIR)], 1.0–1.12 μm, 1.19–1.34 μm, 1.55–1.75 μm, 2.05–2.4 μm [shortwave infrared (SWIR)], 3.5–4.16 μm, 4.5–5.0 μm, 8.0–9.2 μm, 10.2–12.4 μm, 17.0–22.0 μm (middle and thermal infrared), and the microwave region. Jain and Singh (2003) noted that the design of sensors is influenced by atmospheric windows as the sensors measure energy corresponding to a particular range.

Depending on the source of energy used to illuminate the object, the RS systems are classified as active or passive. These sensors that generate energy to illuminate objects are known as *active sensors*. Such sensors (e.g., radar) send signals with certain properties toward the object and record the energy reflected back by these objects that lie in the signal path. In contrast, *passive sensors* rely on an external source of energy to illuminate the object. Since most RS systems rely on the solar energy for this purpose, these are classified as passive sensors.

A major advantage of RS techniques in water resources management is the frequent synoptic coverage of Earth cost-effectively.

These days, RS data at a resolution of the order of a few meters are available. With these data, highly accurate maps can be prepared. Thus significant time, including the time spent in field is saved by the use of the RS technique. Collection of RS data is as easy for a remote, inaccessible area as for an urban area that is located near the city center. Furthermore, these days, RS data of a study area can be obtained at close time intervals. This is helpful in hydrological studies because the time series of changes, such as land use or land cover, can be easily prepared.

12.1.1 Components of a Remote Sensing System

Figure 12.3 shows the basic components of an RS system. Following Jain and Singh (2003), the EM energy source illuminates (A) the object. The radiation energy passes through the atmosphere (B) and interacts with it. When the energy reaches the target (C) it interacts with it, depending on the properties of the target and the radiation. The radiation, after it is scattered or emitted from the target, passes through the atmosphere and travels to the sensor. The sensor (D) senses the object, records the signal, and sends it to a receiver on Earth (E), where the data are stored on a computer media. Users (F) get the data in the form of an image (digital and/or hard copy). These days, almost all images are available in digital form. At the processing centers (F), users interpret the image and produce the output usually in the form of a map (G), which is used in planning, decision-making, research, and so on.

12.1.2 Remote Sensing Sensors

The most commonly used RS sensors are discussed next.

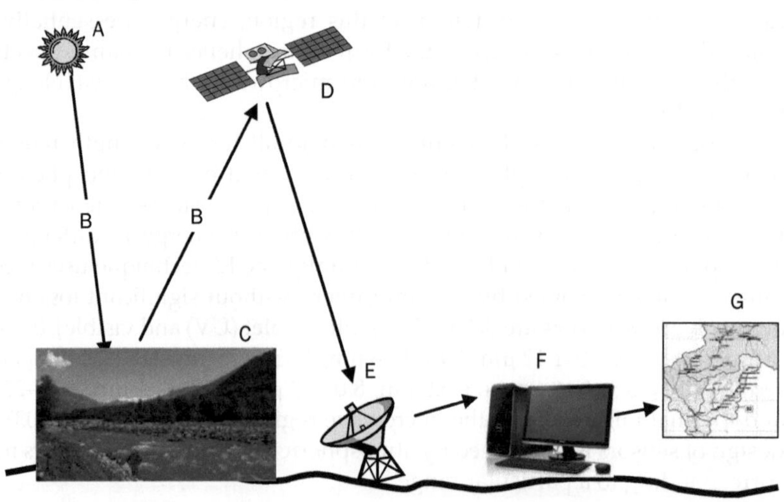

FIGURE 12.3 Basic components of a remote sensing system. (Adapted from Jain and Singh, 2003.)

Multispectral Scanners These are sensors that are sensitive to specific regions of the spectrum and measure the spectral reflectance in narrow wavelength *bands*. Commonly employed bands are Blue (0.45–0.50 μm), Green (0.50–0.60 μm), Red (0.60–0.70 μm), NIR (0.7–1.5 μm), and thermal IR range (TIR, 3–15 μm). The exact wavelength ranges of bands differ with the satellite and the sensor.

Thermal Sensors These sensors measure radiation in the thermal range of the EM spectrum (3–15 μm); TIR is useful in many hydrologic applications.

Microwave Sensors Microwave sensors measure the dielectric properties of the object. The dielectric constant is the amount of charge per unit field strength required to create an electric field in a substance. A large dielectric constant of water implies that substances whose molecules contain ionic bonds will tend to dissociate in water yielding solutions containing ions. Microwave sensors can work in cloudy weather too, since microwaves can penetrate through clouds. Hence, microwave sensors are useful in mapping areas inundated during the flood season when the sky is usually covered by clouds. Microwave sensors are also used for measurement of soil moisture.

Microwave RS can provide soil moisture information, because there is a large difference between the dielectric constants of water (~80) and soil particles (~4). Furthermore, the "dielectric constant" changes, depending on the amount of water and organic matter present in the soil. Dry soil has a low dielectric constant and hence little radar energy is reflected; saturated soil has a high dielectric constant and high radar energy is reflected.

Table 12.1 shows the applications of data of various spectral ranges in hydrologic applications.

12.1.3 Resolution of Remote Sensing Data

The RS images consist of a matrix of *pic*ture *el*ements, or *pixels,* which are the smallest units or building blocks of an image. These pixels are normally square shaped. The RS data has four types of resolutions.

Band (μm)	Applications in Hydrology and Water Resources
Blue (0.45–0.50)	Land use, soil and vegetation classification, sediment properties
Green (0.50–0.60)	Vegetation classification
Red (0.60–0.70)	Vegetation discrimination based on absorption of red chlorophyll
Panchromatic (0.50–0.75)	Land use mapping
Near-infrared (0.75–0.90)	Biomass, crop identification, soil-crop, boundaries of water bodies
Mid-infrared (1.5–1.75)	Vegetation moisture, drought, clouds- snow-ice discrimination
Shortwave-infrared (2.0–2.35)	Geology, rock formation
Thermal infrared (3–12.5)	Thermal mapping, discrimination of minerals/rocks, vegetation classification, moisture studies

Table 12.1 EM Wavelength Bands and Their Applications in Water Resources

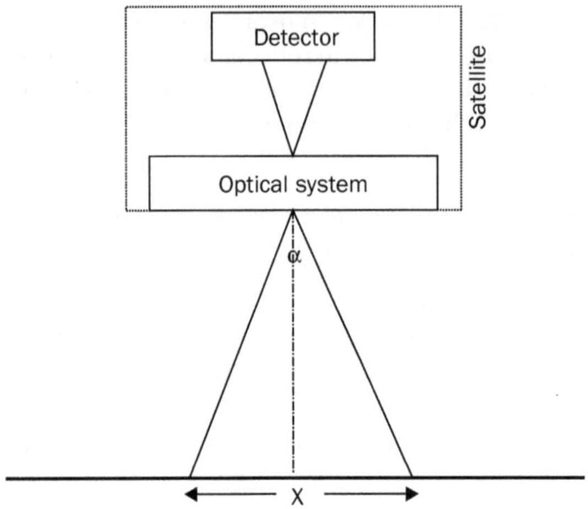

Figure 12.4 Angular IFOV and projection X on ground. X is the diameter of a circle.

Spatial Resolution This refers to the size of the smallest possible feature of an object that can be detected and distinguished by the sensor. Spatial resolution of a sensor depends primarily on its instantaneous field of view (IFOV), which is the area on the ground that is viewed by the sensor from a given altitude at a given time. As shown in Fig. 12.4, it can be either the angle α or the corresponding distance on the ground. The area viewed by the sensor is found by multiplying the IFOV by the distance of the sensor from the ground. As the altitude of a satellite may undergo small changes, IFOV is determined by considering the circular region having "high" energy. A homogeneous feature can be detected if its size is greater than or equal to the area viewed. A feature smaller than this area may remain undetectable, since the average brightness of all the features in that area will be registered. However, a smaller feature may be detectable, if its reflectance dominates.

Images where only large features can be seen are said to have a coarse (or low) resolution. If the small sized objects can be detected, the image has a fine (or high) resolution. In general, the finer the resolution is, the smaller the ground area that can be seen is.

For a map, the term scale refers to ratio of distance on a map to actual distance on the ground. If a map has a scale of 1:100,000, an object of 1 cm length on the map would actually be an object 100,000 cm or 1 km on the ground. Maps or images having small "map-to-ground ratios" are referred to as small scale and those with larger ratios are called large scale. A large-scale (e.g., 1:5000) map shows a small area but in great detail. A small-scale (e.g., 1:100,000) map shows more area but has less detail. A map depicting a large area, such as a big country would be called a small-scale map.

Spectral Resolution This is the number and width of spectral bands in the EM spectrum to which a sensor is sensitive. If the sensor is sensitive to narrow bandwidths in some regions of the EM spectrum, it will be easier to discriminate among the various objects on the ground. Therefore, the current trend is to have more number of spectral bands with narrow bandwidths. For example, the Moderate Resolution Imaging Spectroradiometer (MODIS) payload on board the Aqua satellite operates in 36 spectral bands whose wavelength ranges from 0.4 to 14.4 μm.

A panchromatic (sensitive to all visible colors) image contains data of only one band and the image is usually displayed as a grey-tone image where brightness of pixels is proportional to their digital numbers. In normal photographs, an object is shown in its true color(s) or the way a normal person will see it and the photograph is known as true color composite. An object appears green because it reflects only green band of EM spectrum. Different colors can be created by combining the three primary colors of red, green, and blue (RGB) in different proportions. In the prints of an RS image, the colors are commonly changed. The resulting product is known as *false color composite (FCC)*. There are many possible schemes to produce FCCs, depending upon which types of objects are to be highlighted. A commonly used scheme to produce FCC is R = NIR band, G = red band, and B = green band. Since vegetation is highly reflective in the NIR band, it will appear as red in the standard FCC and this helps with easy detection of vegetation in the image.

Radiometric Resolution An RS sensor measures the radiation reflected from an object. Radiometric resolution of an imaging system is the ability of the sensor to differentiate among the spectral reflectances received from various targets so that the targets can be correctly identified. This resolution is the smallest difference in reflectance that can be detected by the sensor. The strength of radiation received by the sensor is converted into a number; typically, there are 128 or 256 discrete levels.

Temporal Resolution This shows how often the satellite revisits the same area to record data. It commonly varies from a few days to weeks. Geostationary satellites have a quite high resolution in time (e.g., about 30 minutes) but have a coarse spatial resolution (e.g., of the order of several hundreds of meters) partly because they are placed in very high orbit.

The RS image of an area as received from a satellite contains the reflectance recorded at each pixel in various spectral bands which is processed to obtain useful information. A range of hardware and software systems with different features have been developed to analyze digital images or *image processing*.

We now discuss techniques to process RS images with focus on water resources applications.

12.1.4 Reflectance Characteristics of Earth Features

Reflectance is the ratio of reflected energy to total incident energy on a body. It is commonly expressed as percentage. The signal measured by an RS sensor is often called the *spectral signature* of the target, because it is unique to each object on Earth's surface, such as clear water.

Spectral reflectance of an object is the ratio of reflected radiation to incident radiation. Figure 12.5 shows spectral reflectance curves for three common objects: water, soil, and vegetation. Reflectance of water is zero when wavelength is more than 0.8 μm, whereas reflectance of soil increases with wavelength. Vegetation may reflect only about 10% in the green band of the EM spectrum but about 50% in the NIR band. Note that the reflectance characteristics of different objects may be the same under certain conditions. For example, water and wet black soil may have the same reflectance properties in a particular band. Hence, to differentiate them, it is necessary to analyze their spectral reflectance in other spectral bands in which their reflectance is different.

The mapping and identification of vegetation is important in water management. In the visible band, the green color of the leaf is due to peak reflectance at 0.54 μm. The

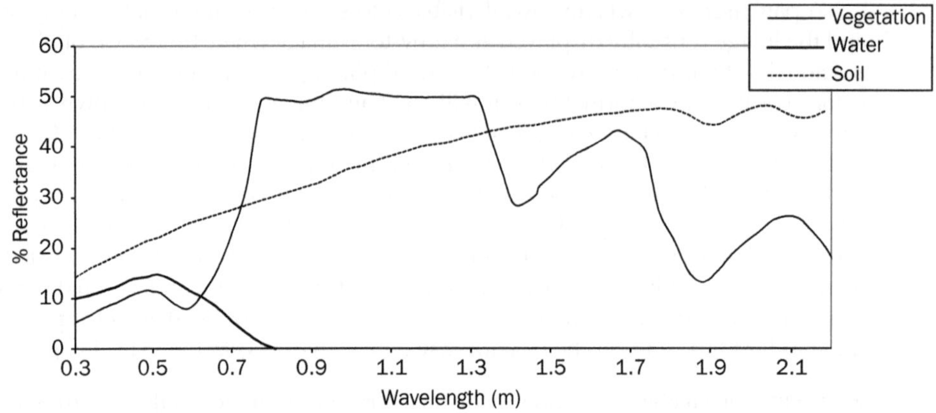

FIGURE 12.5 Typical spectral reflectance curves for water, soil, and vegetation.

presence of chlorophyll is responsible for very low reflectance (10%–30%) in the red and blue bands. Some middle infrared (MIR) bands are absorbed by water vapor.

The reflectivity rises rapidly near 0.75 μm (Fig. 12.5) and remains high in the range 0.75–1.35 μm. The peak reflectance in MIR occurs at 1.6 and 2.1 μm and measurements in 1.55–1.75 and 2.08–2.35 μm bands give useful information about moisture content of the plant canopy.

A healthy plant gives high reflectance in NIR band and low in the red band. Spectral signature of a plant also depends upon its growth state. Also, if the plant is in stress due to nutrient deficiency, insects, or a disease, the reflectance is affected and the same can be detected in the NIR band. Spectral reflectance by different vegetation species can be considerably different and this is to discriminate among the species.

The reflectance of soil depends on its texture (proportion of clay, silt, and sand), the properties of its constituents, presence of organic matter and iron oxide, moisture content, and surface roughness. In the visible bands, soils generally have higher reflectance than plants. Reflectance by plants is higher than soils in the NIR band. Moisture present in soil reduces its reflectance. Soil moisture content and texture are strongly related.

RS is also being applied for water quality mapping and monitoring although it poses difficulties because a host of factors impact the signal from the water body. Good quality water absorbs nearly all the incident energy in NIR band and the reflectance is low in the visible band. It is noted that the reflectance from a surface water body depends on floating material, suspended particles, and the depth of lake or river. Generally, the reflectance of turbid water is more than that of clear water in the visible and NIR bands; the change depends on the characteristics of the suspended material.

12.1.5 Remote Sensing Platforms

RS sensors are mounted on platforms which may be ground based; or fixed on aircrafts, satellites, and drones. To capture the data of a large area, it is necessary to take the platform high and this makes the satellites a good choice. All parts of Earth's surface can be seen from satellites. A geostationary satellite moves synchronously with Earth and is always positioned above the same location (equator) on Earth. A sensor mounted on a geostationary satellite can monitor the weather of a place. Sensors on such satellites can cover large areas but have poor resolution.

The polar orbiting satellites view a given area again after a certain fixed time and cover the entire surface of Earth. They fly at about 833 km above Earth's surface. While low orbit allows much more detailed images, the satellite cannot see the whole of Earth's surface at any one time. These RS satellites can measure a wide range of atmospheric variables, such as temperature, precipitation, and concentrations of gases. Sensors aboard some satellites can be steered sideways, which allow the cameras to also view and map off-nadir areas. McCabe et al. (2017) have given a good overview of space-based earth-observation systems.

12.2 Remote Sensing Data Analysis

Remote sensing imageries can be analyzed using either visual or digital processing techniques. These are discussed next.

Visual Image Processing In visual image processing, the analyst systematically inspects an image and other supporting material, such as a map of the area, to demarcate features of interest.

The analyst interprets an image by using their experience and knowledge about the area and its features. The key characteristics of image that are used in the visual interpretation are texture, tone/color, shape, size, pattern, and association.

Visual interpretation of images has some limitations. An imagery contains a large amount of data and it is difficult for human beings to process it. Computers can do this job efficiently. Humans can differentiate about a dozen of shades of gray and much useful data remains unused. Further, results from different analysts may not always be the same. Due to these reasons, visual interpretation methods are not popular these days.

Digital Image Processing (DIP) A digital image consists of a number of pixels which are typically of square shape. Associated with each pixel is a *digital number* (or DN) which is the integrated radiance response (in a given band) of a small area on the ground whose size depends on the resolution of the sensor on the satellite. Naturally, as the pixel size becomes smaller, the greater details of the object are mapped.

A row of pixels represents a scan line that is recorded as the sensor records the reflected energy by different pixels. Each scan line is composed of n pixels in the direction of the x axis. An image consists of a number (m) of geographically ordered scan lines that are placed adjacent to each other in the direction of the y axis. Thus a digital image is a two-dimensional (2D) array of radiance values or the DN of that pixel. Typically, the DN values range from 0 to 255.

Digital image processing (DIP) techniques manipulate one or more digital images by using specialized software. The image is first pre-processed. Next, mathematical operations are performed on image data or DN values to improve image qualities, maximize clarity, sharpness, details of the objects, and to enhance/map the features of the object that are of interest. In DIP, the results of computation are typically stored as a new image.

12.2.1 Steps in Digital Image Processing

The steps involved in DIP are described next.

Preprocessing of the Remotely Sensed Images Raw RS data received from the sensors on-board the satellite contains noise and deficiencies. Data are preprocessed to prepare the image for analysis by a range of operations: radiometric correction, geometric correction, and atmospheric correction. The image is processed to remove unwanted elements, such as atmospheric interference and noise.

Geometric Correction and Registration Remotely sensed images frequently need to be integrated with the topographic maps of the area. For this purpose, it is necessary to correct them geometrically so that they have the same scale and projection as the maps. This is achieved by *geometric correction*. When many images of the same area are to be used, it is necessary that all the images follow the same coordinate system and projections. In *registration*, the coordinate system of one of the images (called master image) is fitted with other images of the same area. The relation between the image and the coordinate system of the maps are determined by assigning map coordinates to a number of points that are identifiable and can be located accurately on the map and the image. These points are called ground control points (GCP) or tie points and are used to determine transformation coefficients. At least, 5 to 6 GCPs are identified.

In image enhancement, an image is digitally manipulated so as to optimize its appearance on the visual display. This is carried out by use of tools such as different kinds of filters, image editors, and tools that change the properties of an entire image or part(s) of it. Basic image enhancement tools may just change the brightness or contrast of the image or manipulate the grayscale of the image. Some filters may change the color of the image or add visual effects. Sophisticated image enhancement tools can edit/modify only certain parts of an image and not the complete image.

Mathematical operations, such as addition, subtraction, multiplication, and division, are also carried out under DIP. These operations are performed on two or more coregistered images of the same area. These operations are performed on data from individual bands, multiple bands, or single bands in images from different dates. More complicated operations are performed to estimate values, such as sea-surface temperature from multispectral thermal infrared data. Subtraction operation is sometimes carried out on images of the same area acquired at different dates for change detection. Band ratioing or division of images is an arithmetic operation which is applied to images in studies from different disciplines such as hydrological, ecological, and agricultural.

Image Classification The intent of the image classification process is to categorize all pixels in a digital image into broader groups of land cover classes or *themes*. Image classification is the most important part of digital image analysis. One may have a colorful image showing various terrain features using a range of colors but it is not of much use unless we know what each color means.

Image classification is the task in which information classes from a multiband raster image are extracted. Here, the spectral information represented by the DN in one or more spectral bands is used and the classification algorithm attempts to categorize each pixel in one of the themes. Typical themes or classes encountered in water resources studies are water, forest, crop, built-up area, etc. The resulting classified image will comprise of pixels, each of which will belong to a particular theme. This is nothing but the thematic "map" of a part of the study area.

Common image classification procedures can be grouped into two broad subdivisions: (1) unsupervised classification and (2) supervised classification. Both types of classifications make use of two primary methods: training and classifying.

Unsupervised Classification In unsupervised classification, a large number of unknown pixels are examined and are clustered into a number of classes based on natural groupings present in the image values. Unsupervised classification does not require analyst-specified training data or prior knowledge of the area. The basic premise is that the DN values of a particular cover type or feature class (e.g., forest) would lie close together, whereas those in different classes would be well separated. Pixels are grouped based on their DNs and the groups are called "clusters." The user specifies the bands whose data are to be used and how many clusters are to be generated. Using this data, the image classification algorithm generates the requisite number of clusters. A number of image clustering algorithms, such as the K-means method, are available. After the clusters have been generated, the user identifies each cluster with the corresponding classes.

Unsupervised classification is useful for the cases in which the user does not have knowledge of the area. It can also be used as a screening tool to analyze the scene prior to a supervised classification. These days, efficient clustering procedures have been developed that can do fairly accurate classifications. With suitable ground truth collection, this technique can rapidly produce thematic maps.

Supervised Classification This classification involves supervision by the user who specifies DN ranges in various bands of object types present in an imagery to the classification algorithm. Supervised classification requires some knowledge about the area, for example, type and location of crops. Ground truth (field data) or information from maps can be used to create training sets.

In the training stage, the user identifies representative training sites within an image, and develops a numerical description of the spectral attributes of each land cover type, for example, wheat field, water body, forest, and so on. The accuracy with which the training sites are identified determines the success of classification. It is helpful if the user is familiar with the area whose RS data are being classified. It is best to collect the ground truth data in the training areas at the time of satellite over pass.

The training areas in an image are normally demarcated by drawing polygons for each cover type. The coordinates of these polygons are used for extracting the digital numbers from the pixels in these polygons in the image file. A training class contains the sum of pixels or polygons (clusters of pixels).

The image processing software is used to develop a statistical characterization of the reflectance for each information class by *signature analysis* (Maini and Agrawal, 2014). This may involve developing a characterization simply by using mean or a range of reflectances on each band, or by a complex analysis of mean, variance, and covariance over all bands. After completing statistical characterization for each information class, each pixel is examined to find out to which of the signatures the pixel has the highest resemblance. The pixel is assigned to the class having the highest resemblance. Image pixels are categorized into the designated classes by discriminant analysis, which is commonly carried out by pattern recognition algorithms. Three popularly used algorithm are: minimum distance to means, maximum likelihood, and parallelepiped. All of these algorithms use spectral values to establish the relationship between individual pixels and the training class (or training site).

Minimum Distance Classification The minimum distance to the mean classification is based on the distance of the pixel DN from the mean of the training class. It is determined by calculating the distance of a pixel in multispectral feature space from the average spectral value of each category. A pixel is assigned to the training class for which it has a minimum distance.

Maximum Likelihood Classification Maximum likelihood classification, as the name suggests, assigns each pixel to that category to which it has the highest likelihood (probability) of being a member. It is assumed here that the training data follow the normal distribution. The algorithm establishes probability density curves using the variance and covariance for spectral response data for each category (Lillesand et al., 2008). The algorithm then calculates the probability of an unassigned pixel belonging to each class and is assigned to that class for which it has the highest probability.

Parallelepiped Classification This classifier uses the class limits to determine if a given pixel falls within the class or not. The class limits specify the range of values (in terms of standard deviations) on each side of a parallelepiped (a solid body whose each face is a parallelogram) surrounding the mean. If the pixel falls inside a particular parallelepiped, it is assigned to that class. If the pixel falls in more than one parallelepiped, it is assigned to a class arbitrarily. However, If the pixel does not fall inside any parallelepiped, it may remain unassigned. Special decision rules are used to minimize misclassification.

12.2.2 Spectral Indices

Spectral indices are very useful in image classification. A few important indices are described below.

Normalized Difference Vegetation Index (NDVI) This is an indicator of the chlorophyll content of the vegetation

$$\text{NDVI} = \frac{(\text{NIR} - \text{RED})}{(\text{NIR} + \text{RED})} \tag{12.1}$$

where NIR and RED are the reflectances in the NIR and RED bands. Chlorophyll largely absorbs radiation in the red band and hence reflectance (DN) is low. But, in the NIR band, reflectance (DN) is high. NDVI varies from -1 to $+1$; a high positive value indicates healthy vegetation and a low value indicates water stressed or deceased vegetation. NDVI is used to identify vegetation, disease afflicted plants, crop yield, and so on.

Normalized Difference Wetness Index (NDWI) This index was developed to identify irrigated crops with a view that it is better than NDVI for this purpose. NDWI is defined as

$$\text{NDWI} = \frac{(\text{NIR} - \text{SWIR})}{(\text{NIR} + \text{SWIR})} \tag{12.2}$$

NDWI is a ratio image of difference between reflectance in NIR and SWIR bands, normalized to fall between -1 and 1. Nearly all NIR radiation is absorbed by water. In visible bands, the transmittance is high and absorption is less. This results in a gradual reduction in the reflectance of water with the increase in the wavelength from blue to NIR bands. The NDWI value is high for water compared to other objects and thus water can be identified by using this index. NDWI is sensitive to changes in the liquid water content of vegetation canopies (Gao, 1996). NDWI is less sensitive to atmospheric effects than NDVI, NDWI does not remove completely the background soil reflectance effects.

Soil Adjusted Vegetation Index (SAVI) SAVI was developed to minimize soil influences on canopy spectra by incorporating a soil adjustment factor L in the NDVI equation. Huete (1988) proposed the following equation to account for first-order soil background variations and compute SAVI

$$\text{SAVI} = \frac{(\text{NIR} - \text{RED})}{(\text{NIR} + \text{RED} + L)}(1 + L) \tag{12.3}$$

where L is a soil adjustment factor which varies with vegetation density. Huete (1988) used a constant $L = 0.5$ since this reduced soil noise considerably throughout a wide range of vegetation amounts. Many attempts have been made to relate SAVI with features of crops.

Surface Energy Balance Algorithm for Land (SEBAL) The SEBAL developed by Bastiaanssen et al. (1998) uses the surface energy balance to estimate evapotranspiration. SEBAL carries out energy balance by using satellite data. Land surface characteristics, such as surface albedo, leaf area index, vegetation index, and surface temperature, are derived from satellite imagery. SEBAL also requires meteorological data, such as wind speed, humidity, solar radiation, and air temperature.

Density Slicing The histogram of an image is divided into a number of intervals or slices. All the pixels that fall in a given interval are likely to belong to a particular category. Density slicing has been used in many tasks, for example, to identify the water spread area of a lake.

Soft computing techniques, such as fuzzy classification or ANN, have also been used in image classification. Texts, such as Lillesand et al. (2008), contain detailed discussions on DIP.

12.2.3 Change Detection

Change detection involves analysis of a series of images of different dates typically to identify changes in ground features. Commonly the analyst wants to know the short-term (e.g., flood inundation or snow cover) or long-term (e.g., deforestation or expansion of urban area) changes. It is better to use the data of the same sensor, the same band, and the same resolution. All the images should be accurately registered so that overlays, that help visualize the change, can be prepared.

Changes can also be detected by subtracting the DN values of one image from the other. Ratio of two DN values of two images may also be used to identify changes. If no major changes has taken place in an area the ratio will be close to unity; at places of change, the ratio will be some higher or lower number. It may be necessary to fix a threshold above which the change is considered significant. Any change below the threshold is considered insignificant or due to error in analysis.

12.3 Selected Space Programs

Some satellite systems whose data are most commonly used include the Landsat [including the Thematic Mapper (TM)] of the National Aeronautics and Space Administration

(NASA), the French SPOT (Systeme Probatoire d'Observation de la Terre) satellites, the Indian Remote Sensing (IRS) satellites, the ERS and JERS satellites, and GOES, GMS, and so on (Jain and Singh, 2003). The data from Landsat program are available since 1972, the data from SPOT program since 1986, and from the IRS satellites since 1988. The spectral bands whose data are most common in hydrological applications are visible (VIS) bands, IR, NIR, thermal infrared, and passive and active microwave (radar). In addition to satellites, ground-based RS data collection systems are also important in water-related studies. For instance, ground-based weather radar is found to be very useful for areal estimation of precipitation. Drones are also being employed to capture data, for example, land use, for hydrological modeling and water management.

The Landsat program has provided probably the most extensively utilized RS images. Two popular sensors on-board the Landsat satellites are the Multispectral Scanning System (MSS) and the TM. The website of Landsat (NASA) can be visited to obtain more details about this program. It is available at http://landsat.gsfc.nasa.gov/.

The National Oceanic and Atmospheric Administration (NOAA) operates a system of satellites for meteorological purposes. The French SPOT satellite system commenced operation in 1986. Details of this system and data are available at https://spot.cnes.fr/en/SPOT/index.htm.

Under the Indian Remote Sensing Satellite (IRS) program, the Indian Space Research Organization (ISRO) has launched a series of land observation satellites. The operational first-generation RS satellites IRS-1A and IRS-1B were launched in 1988 and 1991, respectively. The second generation, operational, multisensor satellites IRS-1C and IRS-1D were launched in 1995 and 1997. Recent launches include CARTOSAT for cartographic applications, Resourcesat series for mapping of resources, SARAL for satellite altimetry and so on. Indian Space Research Organization (ISRO) has a very ambitious program to launch satellites in near future which will provide high resolution data, say at a topographic resolution of 0.25 m and satellite-based altimetry data. A new concept of sensor web aims to make available, at one place, data from a multitude of RS products and ground-based sensors. Further details of satellite data products are available at http://www.nrsc.gov.in.

The Gravity Recovery and Climate Experiment GRACE was a joint mission of NASA and the German Aerospace Center (www.nasa.gov\mission_pages\Grace). Its twin satellites took detailed measurements of Earth's gravity field anomalies. Grace satellites were launched in 2002 and the science mission ended in October 2017. By measuring gravity anomalies, GRACE determined how mass is distributed on Earth and how it varies with time. Data collected by GRACE satellites had numerous path-breaking applications. This helped understand the thinning of polar ice sheets and changes in water stored in aquifers. A follow-up GRACE mission is planned.

12.4 Geographic Information Systems

Maps are routinely used in all stages of water management—from data measurement to operation of the projects. Water professionals use many types of maps: topography, land use/cover, soils, irrigated areas, forest, and so on. Geographic information systems (GISs) are software systems to input, store, retrieve, analyze, and produce hard/soft display of spatial data. Conventionally, paper maps are used to represent spatial data; GISs work with digital maps. Figure 12.6 pictorially shows the GIS concept.

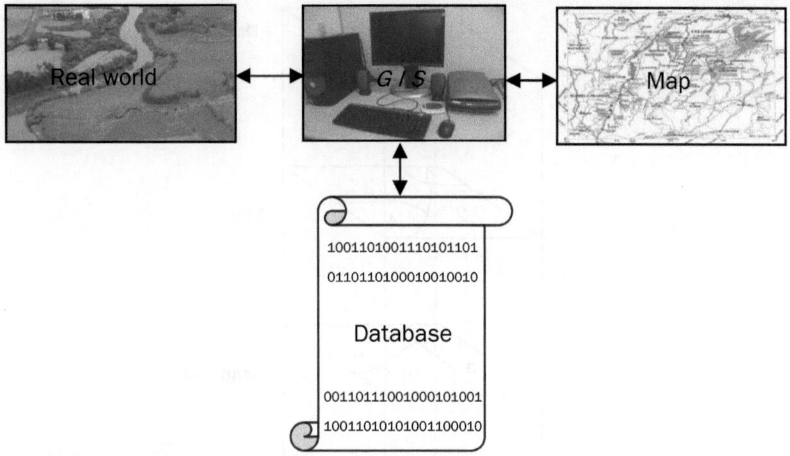

FIGURE 12.6 Block diagram of a GIS.

Use of computers in cartography began in the 1950s and led to the development of the first general purpose mapping software in the 1960s. Development of GISs accelerated in the 1980s and soon organizations began to create spatial databases. Applications of GISs began to multiply and grow from the mid-1990s.

Water resources management has a spatial dimension and hence tools to analyze spatial data are needed. In addition to water resources, GISs are finding increasing applications in natural resources, tourism, governance, transportation, trade, warfare, and so on.

The need for GISs is easy to visualize. Conventionally, maps and their overlays are manually prepared; this process is time consuming and prone to error. If maps for the same area (e.g., land use, soils, river network) are manually drawn, common boundaries may not match in different maps. If a paper map is to be updated, it has to be redrawn. Some features may be missed in annotating a paper map and attribute information (e.g., names) for some of the features may be lost. Big paper maps are difficult to handle and deteriorate with time.

A digital map can be considered to be composed of several layers stacked one over the other (Fig. 12.7), each layer containing the spatial data pertaining to a single category, such as soil, land cover, channel network, and so on. Common boundaries are digitized only once and are used by all layers. A GIS enables the overlaying of a number of map layers and analyzing them; creating new layers thereby may also be possible. An output map of a GIS is automatically updated if the underlying data (or layers) is changed. A GIS stores information in tables which are linked with geographic features.

We will now discuss the geographic data types, data models, and so on.

12.4.1 Geographic Data Types

Geographic or spatial objects have *attributes*, such as location and properties, which indicate where the object is, what it is, and what its relevant properties are. Examples of spatial data are stream networks, well locations, villages, contours, spot elevations, roads, soil, geology, hydrological response units, and so on. The geographic location of a point on (or above) Earth's surface is commonly specified by its latitude, longitude, and altitude; a Cartesian coordinate system or a grid system can also be used. The position of an object can also be given, although not very precisely, with respect to a landmark.

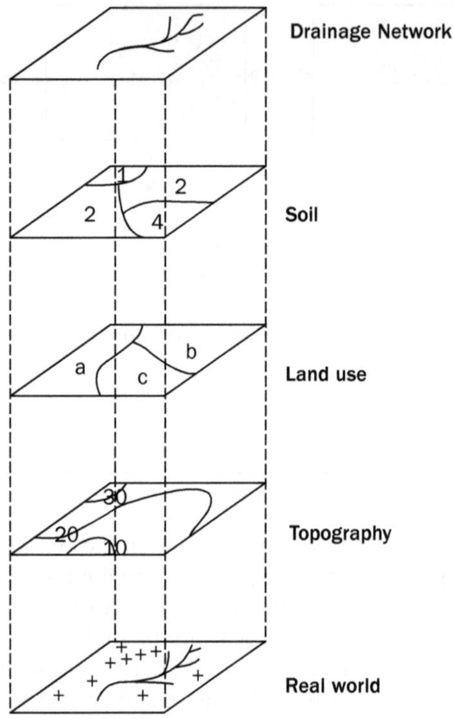

FIGURE 12.7 Categorical approach for handling spatial data.

In the GIS terminology, the real-world things are called *entities* and their digital representations are called *objects*. Entities displayed on a map as spatial objects having zero, one, or two dimensions. Simple objects can be aggregated to create complex spatial objects. A summary of spatial objects in zero, one, and two dimensions is given in Fig. 12.8. In aggregation, the spatial data are grouped at a level of detail that is coarser than the level at which data were collected. For example, census data is collected at the household level but to ensure confidentiality, it is aggregated to geographic units (e.g., districts) during dissemination. The attributes of a spatial object can be the class to which it belongs (e.g., water body, road, or forest); it can be some number associated with the object which also permits mathematical operations. The attributes of geographic entity may change with time.

All the geographic data can be categorized into three types: points, lines, and polygons.

Point Data A point is an object which has zero dimension; its position is specified by its coordinates and it can have no size. Examples of a point are a rain gauge station or a well. For raster data, a point refers to a cell.

Line Data A line is a one-dimensional (1D) object of specified length, for example, a river, a contour, or a railway line. Elevation of a contour line is its attribute; a river will have certain width, depth, and hydraulic properties.

Polygons Polygons are 2D objects with given length, width, and area. Examples are a lake, a watershed, or a country. Boundaries of polygons may be natural features or artificial. Catchments have natural boundaries (ridges), while the boundaries of country are

Number of Dimensions	Geometry	Geometry and Topology	Graphical Representation	Purpose
Zero	Point			Geometric location of point features
		Node		Topological junction
One	Line segment			Direct line between two points
	String		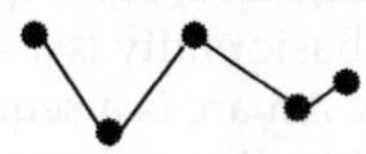	Connected non-branching sequence of line segments
	Arc			Curve defined by a mathematical expression
		Link		Topological connection between nodes
		Chain		Sequence of nonintersecting lines and/or arcs bounded by nodes
	G-ring		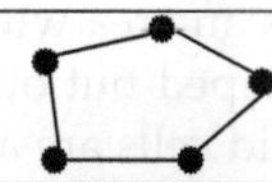	Sequences of nonintersecting strings and/or arcs with closure
		GT-ring	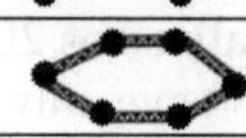	Ring enabled by Chains
Two	Interior area			Area not including its boundary
	G-Polygon		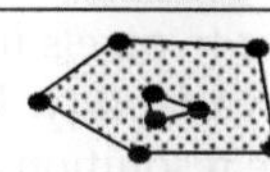	Area bounded by one outer G-ring
		GT-Polygon	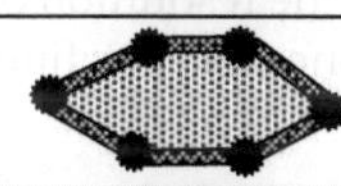	Area bounded by one or more GT-rings
	Pixel			Smallest non-divisible element of a digital image
	Grid cell		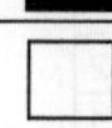	Smallest non-divisible element of a grid

FIGURE 12.8 Summary of simple spatial objects in the SDTS model. (Source: NIST, 1992.)

manmade. Two adjacent polygons share boundaries. A polygon may also be enclosed within another polygon, for example, a lake lying inside a city. Polygons are a common data type in a geographical database in water studies. The most common polygon in hydrological studies is a catchment; attributes associated with a catchment are drainage network, soil types, and land cover.

Note that the data type is scale dependent. For example, on a small-scale map of a city it may be represented as point data but as polygon data on a large-scale map.

12.4.2 GIS Data Structure

In a GIS, spatial data are represented by using two basic models: vector and raster. The actual method of storing the spatial data is known as data structure. Many data structures are used for vector and raster data.

Vector Data In geometry, a vector is a quantity that has a starting point, a magnitude, and a direction. In the vector model, objects (such as a river) are created by connecting two points by a straight line. Vector objects can be assigned attributes, say a canal will have certain water carrying capacity. Advantage of vector data structure is that it provides a compact representation of geographic objects although it may be complex. Generally, vector data structure provides files of small size.

A node is a point where two or more lines join. An arc or a segment is the curve that joins two nodes. To create an area, a region is enclosed by arcs. A polygon is made up of arcs or segments. The basic entity is a segment of straight line the coordinates of whose end points are known. An arc is a sequence of straight lines which begin and end at a node. The starting and ending points of arcs forming an area are the same.

Topology is the mathematical study of the relationships and transformation of geometric computations. In the topological structure, all lines are explicitly linked and spatial relationships between entities are clearly defined. Topology can be efficiently modeled in vector data structure. Popular structures for handling vector data are topological, spaghetti, and TIN.

Raster Data In raster data structure, the map area is divided into a number of cells, each of which represents an area with the same properties. Commonly, these cells are square or rectangular shaped but other shapes, such as hexagonal, are also possible. Square or rectangular grid cells are addressed by the corresponding row and column numbers and can be visualized as 2D arrays. Each raster cell is assumed to represent a homogenous area and can have only one value of the attribute being mapped. Overlays are created if more than one attribute is to be studied (Fig. 12.9). Overlays are seen as stacked layers.

Raster cells are similar to pixels in an RS image. Raster data structures are efficient in modeling high spatial variability. This makes them suitable for the processing of RS data. Depending upon the resolution of data, raster data files may be large. The size of a data file increases exponentially as finer resolution data are used, if the area remains the

Soil

1	1	2	2	2	2	2	2
1	1	2	2	2	2	2	2
3	3	2	2	2	2	2	2
3	3	2	2	2	2	2	2
3	3	3	2	4	4	4	4
3	3	3	3	4	4	4	4
3	3	3	3	3	4	4	4
3	3	3	3	3	3	4	4

Land Use

a	a	b	b	b	b	b	b
a	a	b	b	b	b	b	b
a	a	c	c	b	b	b	b
a	a	c	c	c	b	b	b
a	a	c	c	c	c	b	b
a	a	a	c	c	c	c	c
a	a	a	c	c	c	c	c
a	a	a	a	c	c	c	c

Topography

19	20	20	21	20	20	20	21
20	20	21	20	20	18	20	21
20	21	20	20	18	17	20	20
21	21	20	18	15	16	20	20
21	20	20	15	13	15	18	20
20	20	15	13	11	13	15	20
18	15	10	10	10	13	15	18
15	13	10	8	10	11	13	15

Figure 12.9 Raster representation of soil, land use, and topography categories shown in Fig. 12.7.

same. Both vector and raster data models have some advantages and some limitations. A comparison is shown in Table 12.2.

Thematic maps prepared from the RS data are in raster format. Computed values of some catchment variables (e.g., area) are somewhat scale dependent when raster data are used. For example, the average catchment slope reduces with increase in grid size of the raster map used to determine it.

Vector data are handy, compact, and accurate to describe polygons. Vectors are better suited to represent lines and boundaries and for network analysis. But topographic analysis is difficult in a vector model, because each polygon has a different shape. Overlaying of vector maps is also difficult. When raster data are converted to vectors, lines and boundaries appear as stepped formation.

A good GIS package should have the ability to process both raster and vector data. Most current GIS software have modules to convert raster data into vector data or vice versa.

12.4.3 Geographic Coordinate Systems

A geographic coordinate system is used to specify location of a place on the Earth surface by means of a set of numbers, letters, or symbols. The latitude of a place is the angle defined at the center of Earth between the plane of the equator and the line connecting the point with the center of Earth. While writing the latitudes of places in the Northern Hemisphere, letter N (for north) is suffixed to the angle; the letter S (for south) is suffixed for the locations in the Southern Hemisphere. The latitudes vary between 0° (or the points on equator) and 90° (poles). The longitude of a place is the horizontal angle between the prime meridian and the line connecting the point with Earth's center. Meridians are the imaginary lines connecting the North and the South poles of the Earth. Prime meridian is the reference meridian passing through Greenwich (U.K.). Letter E (for east) is suffixed for the locations lying to the east of the prime meridian and W (for west) for those lying to the west. The longitudes vary between 0 and 180°. Both latitudes and longitudes are expressed in degrees, minutes, and seconds. For example, the longitude of Roorkee (India) is 77° 53′ 01″ E and the latitude is 29° 52′ 02″ N.

Map Projection Although a common belief is that Earth has a spherical shape, its equatorial axis is longer than its polar axis or Earth is bit flattened near the poles. This shape

	Representation or Value in	
Feature	**Vector models**	**Raster models**
Points	Have infinitesimal area	Have finite area
Lines	Have infinitesimal width, curves are smooth	Have finite width and gives jagged or stepped appearance
Data storage space	Require small space	Require large space
Values of catchment variables	Not depend on map scale	Somewhat depends on raster resolution
Data transformation	Data can be transformed from vector to raster and vice versa	
Data structure		Simple and analysis is easy
Overlaying	Somewhat difficult	Easy
Integration with satellite data	Complex	Easy

Table 12.2 Comparison of Raster and Vector Data Models

is close to an ellipsoid or spheroid which better represents Earth's shape (Jain and Singh, 2003). A spherical represents of Earth (e.g., a globe) is not convenient to work; a globe cannot be included in a book. Therefore, it is necessary to use a map projection to display the three-dimensional (3D) Earth on a 2D flat paper. Conceptually, Earth's surface is projected on a cylinder or cone which is "opened" and stretched to conform to the shape of plain paper.

Map projections are of three classes: cylindrical, conical, and azimuthal. The cylindrical projection assumes that a cylindrical-shaped paper is wrapped around the globe. The map globe is opened up and stretched to conform to the cylindrical shape. This cylinder is opened to obtain the map of the world on the rectangular sheet. In the conical projection, the globe is visualized to be wrapped by a cone. When the cone is opened, the fan-shaped map is obtained. Map projections are named based on class, property, the name of the originator, and the modification of Earth's surface. Some commonly used projections are the Mercator projection, Normal Cylindrical Equal Area (Lambert projection), Polyconic projection, and so on. The textbooks by Maling (1992) and Grafarend et al. (2014) discuss coordinate systems in detail.

A number of transformations are employed in map projection which can be mathematically represented as:

$$x = f(\phi,\theta) \tag{12.4}$$

$$y = g(\phi,\theta) \tag{12.5}$$

where ϕ and θ are the latitude and longitude of the place, and x and y are map coordinates.

12.5 GIS-Based Analysis of Spatial Data

The major steps of data analysis by using a GIS are described below.

Data Preparation It involves the identification of data, its source, scale, and geographic coverage. The required data are collected from ground surveys, digitizing existing maps, RS data, or combination. The availability of spatial data is limited in some countries and one may have to start from scratch. In some cases, the requisite data are available only as hard-copy maps. Digitizing a map is labor intensive and requires skilled staff, time, and patience.

Data Input Data are input to GISs in many different ways, namely, by using digitizers, scanners, computer files, and so on. Digitizers are connected with computers and are used to digitize a paper map or to convert it to digital form. Most digitization modules have routines for minimizing checking, and correcting errors. Scanners are used to create an image of a map. After scanning, the digital map is edited to remove errors that might have cropped up.

Output

The final step of GIS analysis is to present the results as maps (most common) or in tabular form. Map design involves design of layout, selection of contents, scale, symbols, legend, and so on. Optional features commonly included in a map are grids, direction arrow, and nameplate (author/creator, date, and other information). These designs depend upon the objective(s) of the study and the requirements of end user. Mode of information dissemination is also important. If the map is to be disseminated by web, the file size should be small. Otherwise, a bigger file can be created and saved on a computer media for subsequent use. If required, results may also be displayed by other tools methods, such as pie charts, histograms, and so on.

12.5.1 Digital Elevation Model

A digital elevation model (DEM) is a dataset of ground elevations at regular spacing in a horizontal plane, most commonly at square grid points. Generally, the grid spacing varies from 10 to 100 m. DEM data can be acquired in many ways: land surveying, RS, Light Detection and Ranging (LiDAR), and so on. A common method of generating DEMs involves digitizing topographic maps produced by land survey.

Lidar is a technique to acquire precise topographic data to make high-resolution maps. It uses a UV, visible, or NIR band to acquire data. Lidar instruments may be fitted to an aircraft, satellite, or a drone to carry out surveying. Its main advantage is high accuracy and fast data collection, although the cost of data is high.

These days DEMs are commonly built by using data that were collected by RS techniques. A powerful RS technique to generate DEM is the interferometric synthetic aperture radar, where two passes of a satellite (such as RADARSAT-1) or a single pass if the satellite is equipped with two antennas, like the Shuttle Radar Topography Mission (SRTM), collect sufficient data to generate a DEM. The SRTM of NASA consisted of a special radar system that flew on-board the space shuttle Endeavour during one of its space missions in 2000. Through this effort, it was possible to capture DEM global data for the zones from 56° S to 60° N. SRTM data are available at 30 and 90 m resolutions. To acquire topographic data, the SRTM payload was outfitted with two radar antennas. Currently, three products are available: (1) One arc-second resolution (~30 m) for all areas between 60° N and 56° S latitudes; (2) three arc-second (~90 m) for all areas between 60° N and 56° S latitudes; and (3) thirty arc-second (~1 km) for all areas between 60° N and 56° S latitudes. Radar data underwent extensive processing and noise filtering before they were released to the public. SRTM DEM uses WGS84 datum and the EGM96 geoid model. SRTM data can be downloaded from the CGIAR-CSI interface: http://srtm.csi.cgiar.org/SELECTION/inputCoord.asp. Note that the SRTM data accuracy decreases drastically in forest areas, since it neither represents the tree canopy nor the ground. Since SRTM data is derived from satellite-based radar, it represents the elevation of the first-reflected surface, often tree tops or buildings. So, the data are not necessarily representative of the ground surface, but the top of whatever is first encountered by the radar. More details about the SRTM mission can be seen at https://www2.jpl.nasa.gov/srtm/mission.htm.

The Ministry of Economy, Trade, and Industry (METI) of Japan and NASA have jointly produced the Advanced Spaceborne Thermal Emission and Reflection Radiometer (ASTER) Global DEM. The first version of the ASTER GDEM was generated by using stereo-pair images collected by the ASTER instrument on-board Terra satellite and was released in June 2009. ASTER GDEM coverage spans from 83° N latitude to 83° S. The improved GDEM V2 has refined production algorithm, providing improved spatial resolution, and increased horizontal and vertical accuracy. The ASTER GDEM V2 data have 30 m accuracy in horizontal, 20 m accuracy in vertical, and can be downloaded from http://gdex.cr.usgs.gov/gdex/.

The DEM generated by using the stereo images of the Cartosat-1 Indian Remote Sensing Satellite can be freely downloaded from the Bhuvan portal of NRSC (http://bhuvan.nrsc.gov.in/). The data are at 1 arc sec (about 30 m) resolution.

The quality of a DEM measures how accurately elevation is measured at each pixel and how accurately the morphology is presented. Several factors govern the quality of DEM and derived products: terrain roughness, sampling density, pixel size, interpolation algorithm, vertical resolution, and terrain analysis algorithm.

Besides the DEMs in the public domain, many national agencies have their own DEMs often with a much higher accuracy than those in the public domain. However, these data may have to be purchased by the prospective user and the cost may be high.

Since water flows on the catchment surface in the downslope direction, the direction of overland and channel flow can be found from an array of elevation data. Slope and flow direction are local properties and GISs permit their correct assessment and representation in the model. Mizukoshi and Aniya (2002) have noted that grid-based DEMS have been extensively used for statistical and quantitative analyses of terrain, because they are suited for calculation and are easy to handle for raster overlay analysis, despite several disadvantages. Figure 12.10 shows how flow directions and a river network can be found by using DEM data.

12.5.2 Analysis of Geographic Data Using a GIS

Some important GIS operations are described here.

Geometric Correction RS images are not maps but these are frequently required to be integrated with maps. Geometric correction is process of transformation of an RS image into a map with a defined scale and projection properties. Need for this correction arises because different maps and RS data can have different coordinate systems and/or projections. These must be reconciled before the different data layers of the same area, for example soil layer and land use layer, can be correctly overlaid and analyzed.

Geometric correction involves at least two layers. Instances of mismatch can arise even when both two layers have the same projection and coordinate systems. Such

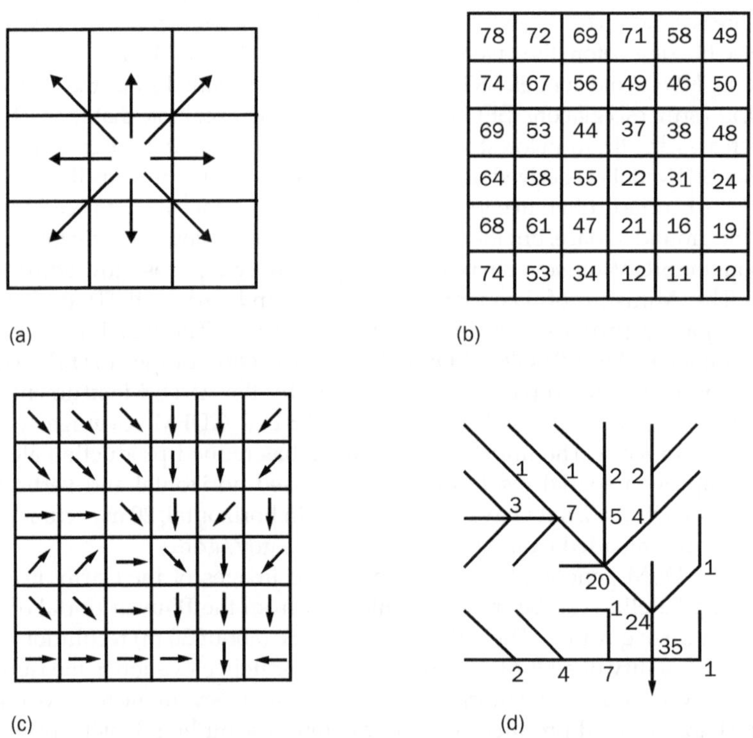

Figure 12.10 Generation of river network from DEM; (a) the eight-direction pour-point model; (b) a grid of terrain elevations; (c) the corresponding flow directions of grids; (d) equivalent river network showing flow accumulation.

mismatches can be resolved by registration by absolute position, registration by relative position, edge matching, etc. Most GISs can transform data among a number of projections. Radiometric correction is a bit similar operation which is performed to account for the degradation in sensor performance, atmospheric conditions, etc.

Data Selection and Query A GIS can be used to pick up data that meets the specified criteria. Data may be selected in different ways: (1) interactively; (2) based on numerical thresholds; or (3) by logical operators. Data can be interactively selected on screen. Selected data can be stored in a new layer. The user can also select all rivers with a particular order or above/below a particular order from a river network. Data can be selected from two or more map layers. For example, a flooded urban area may be selected. The query results can also be saved, for example, as a map.

Mathematical Operations Mathematical operations, for example, addition, multiplication, power, logarithm, and trigonometric operations, can be performed on geographic data and their attributes to derive useful information. Statistical properties of geographical objects (e.g., number, length, area, etc.) can be determined. Summary statistics, for example, land use statistics for catchments, can be generated in a GIS. Logical operations, such as greater than, less than, equal to, and, or, xor, are also frequently performed on spatial data to derive useful inferences.

Query This can be performed by attributes or geometry. In query by attributes, a logical expression works on attributes. For example the query, *land use = agriculture* will select the pixels belonging to agriculture areas. In query by geometry, objects are selected based on geometry to view or analyze their attributes. Conditional operations can be easily performed on geographic data. For example, to generate a flood vulnerability map of residential areas based on return period concept, a simple test is 'IF 25 <return period <=100 AND land use = residential THEN vulnerability = high ELSE vulnerability = low'.

Spatial Aggregation Spatial aggregation is used to generate coarser resolution maps from finer resolution maps. The user inputs the numbers of cells to be aggregated in each *nxn* cluster and the GIS examines successive *nxn* nonoverlapping blocks and assigns a new value to it (Fig. 12.11). This new value could be based on the majority value (the value or class occurring most frequently), average, maximum, and so on. Spatial aggregation reduces database size. Spatial disaggregation is performed to subdivide a cell in fractions, say half, one fourth, and so on.

Buffer Zones In tasks, such as flood plain zoning or highway planning, buffer zones on the sides of the river or highway may have to be created. The buffer zone operation in a GIS generates a buffer zone around spatial objects. If the spatial entity is a point then the buffer would be a circle; if line or arc then it is a polygon. Additional criteria can be applied to control the direction (uphill, downhill), resistance, and barriers to spreading. Typical questions that may arise in hydrologic analysis include determining the number of people living within the flood plain zone (buffer) of river A or valuing all the houses within the flood plain.

Overlay Map overlay is a process in which two or more layers of maps of the same area are digitally stacked over each other. A key advantage of overlay is that it allows to visualize and analyze spatial data of different variables and coming from multiple sources. For example, in a catchment, the areas prone to soil erosion can be identified

F	F	F	A	A	A	A	A
F	F	F	A	A	F	F	A
A	F	F	F	F	F	F	A
A	A	A	A	A	F	F	F
A	A	A	A	A	A	F	F
A	A	A	A	A	A	A	A
F	F	F	A	A	F	A	A
A	F	A	A	F	A	A	A

(a)

F	A	A	A
A	A	F	F
A	A	A	A
F	A	A	A

(b)

F	A	A	A
A	M	F	F
A	A	A	M
F	A	M	A

(c)

A—Agriculture, F—Forest, M—Mixed

Figure 12.11 Spatial aggregation of data of two categories in (a). To resolve tie, majority rule is followed in (b) and a mixed class is created in (c).

by overlaying the maps of topography, slope, soil properties, land use, and rainfall intensity. A mathematical model can be employed to identify the pixels which are erosion prone by analysis of data of corresponding pixels. We assume that these variables impact soil erosion. Map overlay is one of the most useful feature of a GIS.

One can also overlay a vector data layer over a raster data layer. Thus, one can digitize the catchment boundary from a topographic map to create a vector data layer. This can be overlaid over thematic layers, such as soil, land cover, and channel network, to derive desired information by use of mathematical and logical functions. Overlay operation is simple and efficient in raster data. It is bit complex when working with vector data.

Interpolation GISs are used to interpolate data by using spatial interpolation techniques, such as kriging.

A current trend in GIS development is toward an open GIS that will allow the users to acquire data that may be easily and efficiently exchanged and shared between different software systems. Attention is also on interoperability, which will overcome/remove constraints on using specific hardware and software platforms.

12.6 Applications of Remote Sensing and Geographical Information Systems to Water Resources Problems

RS and GISs have become an important source of data as well as tools in hydrologic analysis, modeling, and design. Hydrologists use large volumes of spatial data and RS and GISs are well-oriented toward capturing and managing such data. A GIS is a useful tool for data preparation for hydrologic analysis. DEMs are now continually used for catchment demarcation, and generation of slope maps and channel networks. Using DEMs, catchment features, namely, area, longest flow channel, channel length up to the centroid, average slope, and so on, can be computed in GISs. Interfaces for many popular models, such as the Soil and Water Assessment Tool (SWAT), have been developed in a GIS framework and these are of immense use to prepare input data files for the corresponding models.

Currently, RS data are available at different resolutions from a number of satellites. Hence, the user first identifies which dataset is best suitable for their problems based on the considerations, such as the spatial resolution, the EM bands whose data are available, the frequency at which RS data are available, and the price of data.

The application domains of RS/GISs in the water sector include the measurement of hydrologic variables, such as precipitation, land surface temperature, ET, irrigation water management, snow cover mapping, flood mapping, and hydrological modeling. The GIS helps in managing the map data and display of results. Singh and Fiorentino (1996) have given a good overview of GIS applications in hydrology. Maidment (2002) has provided many examples of the use of ArcGIS in the water sector.

12.7 Summary

New satellites with smaller pixel size and larger number of spectral bands are being launched periodically. Satellites with sensors specific to hydrological applications, such as cryosphere and soil moisture, are also being launched. Besides many other applications, these will provide much needed useful input for water resources studies. This field is bound to see a rapid development. Hydrologists will increasingly apply data produced by next generation of technologies and Earth observation missions for measurements and understanding of hydrological processes. Chen and Wang (2018) have reviewed important and emerging issues in Earth observation big data applications. Many widely used hydrologic models today, such as the SWAT model (Arnold et al., 1996), make extensive use of GISs.

Needless to say, GISs are being employed in a wide variety of applications in diverse fields; the water sector is no exception. The use of GISs in water resources is expected to grow manifold in the future, as the technology penetrates deeper and wider. Technical advancements are resulting in better tools for spatial data collection, database creation, and analysis.

12.8 Questions

12.1 What do you understand by the term remote sensing (RS)? Give a few examples where this technique is used in day-to-day life, in addition to hydrology?

12.2 What do you understand by atmospheric windows?

12.3 What is the difference between active and passive RS?

12.4 What is microwave RS and what are its advantages?

12.5 What do you understand by resolution of RS data? Describe briefly the different types of resolutions of RS data?

12.6 What do you understand by reflectance of RS data? How does it help identify different Earth objects?

12.7 Most RS systems avoid the use of wavelengths in the ultraviolet and blue portions of the EM spectrum. Why?

12.8 What are the best atmospheric conditions and time for RS using the visible portion of EM spectrum?

12.9 Briefly describe digital image processing (DIP).

12.10 Briefly describe two spectral indices that are commonly used in the water sector.

12.11 What are the advantages of digital maps?

12.12 What do you understand by attributes of spatial objects? Explain with examples.

12.13 Define vector and raster data and explain the difference between the two.

12.14 What is a digital elevation model (DEM)? List its use in water resources studies?

12.9 References

Arnold, J. G., J. R. Williams, R. Srinivasan, and K. W. King. *Soil and Water Assessment Tool, User's Manual.* U.S. Department of Agriculture, Agriculture Research Service, Grassland, Soil and Water Research Laboratory, Temple, TX, 1996.

Bastiaanssen, W. G. M. *Remote Sensing in Water Resources Management: The State of the Art.* Colombo, Sri Lanka: International Water Management Institute, 1998.

Chen, Lajiao and Lizhe Wang. Recent advance in earth observation big data for hydrology. *Big Earth Data,* 2(1):86–107,2018. DOI: 10.1080/20964471.2018.1435072.

Gao, Bo-Cai. NDWI—Normalized difference water index for remote sensing of vegetation liquid water from space. *Remote Sensing Environment,* 58:257–266, 1996.

Grafarend, E. W., R. J. You, and R. Syffus. *Map Projections—Cartographic Information Systems.* Berlin: Springer-Verlag, 2014.

Jain, S. K., and V. P. Singh. *Water Resources Systems Planning and Management.* Amsterdam: Elsevier, 2003.

Lillesand, T., R. W. Kiefer, and J. Chipman. *Remote Sensing and Image Interpretation.* New York: John Wiley and Sons, 2008.

Maini, A. K., and V. Agrawal. *Satellite Technology: Principles and Applications.* New York: Wiley, 2014.

McCabe, Matthew F., Matthew Rodell, Douglas E. Alsdorf, Diego G. Miralles, Remko Uijlenhoet, Wolfgang Wagner, Arko Lucieer, Rasmus Houborg, Niko E. C. Verhoest, Trenton E. Franz, Jiancheng Shi, Huilin Gao, and Eric F. Wood. The future of Earth observation in hydrology. *Hydrol. Earth Syst. Sci.,* 21:3879–3914, 2017.

Maidment, D. R. (Ed.). *Arc Hydro: GIS for Water Resources.* ESRI Press, 2002.

Mizukoshi, H. and M. Aniya. Use of contour-based DEMs for deriving and mapping topographic attributes. *Photogrammetric Engineering & Remote Sensing,* 68(1): 83–93, January 2002.

Maling, D. H. *Coordinate Systems and Map Projections.* Oxford: Pergamon Press, 1992.

Singh, V. P., and M. Fiorentino. *Geographical Information Systems in Hydrology.* Dordrecht: Kluwer Academic Publishers, 1996.

PART 4

Analysis and Modeling

CHAPTER 13

Rainfall-Runoff Modeling

Rainfall-runoff relationships are needed for many purposes, such as design of storage and diversion facilities; determination of minimum amounts of water available for agricultural, industrial or municipal use, and power generation; adjustment of long records of runoff for varying rainfall regimes for the study of trends in water yield; and operational planning for storage or diversion structures. This chapter discusses rainfall-runoff models based on their use. The simplest problem that a hydrologist has to address is the computation of peak discharge at a location (e.g., design of a culvert). Next, in addition to peak, the hydrologist may have to estimate the hydrograph volume for designing, for example, a detention basin. Many applications require the estimation of a complete hydrograph. At the next level, complex questions, such as the impact of land use change, climate change, and building of dams and reservoirs, on catchment response, need to be addressed.

13.1 Computation of Hydrograph Volume

The early hydrologists were expected to estimate the volume of flow. This information was needed for tasks, such as the design of detention ponds. Many different approaches could be employed for this purpose, such as the regression method or the Soil Conservation Service–Curve Number (SCS-CN) method. We will now describe the regression method.

13.1.1 Regression Method

Frequently, while modeling environmental systems, the interest is to quantify the relationship between two or more variables. Regression analysis assumes that the variables involved have a cause and effect relationship. If the regression relationship is to be established between rainfall and runoff, we can take rainfall as the cause variable and runoff as the effect. Regression is also used for filling in missing discharge data, validating discharge data series, and predicting future flows. A discharge rating curve is used to transform water levels into discharge series and the commonly used stage-discharge rating curves are of power regression equation type, where for each range of the independent variable (river stage), an equation is established. The regression method has been covered in greater detail in Chap. 19.

Assuming that a linear relationship with a memory length m exists between rainfall P and runoff Q, runoff may be expressed as

$$Q_i = a + b_1 P_i + b_2 P_{i-1} + \dots + b_m P_{i-m+1} + \varepsilon \tag{13.1}$$

where a and b_i = the regression coefficients, and ϵ = the error term. Values of a and bs, which are unknown, are estimated by the method of least-squares which minimizes the sum of squares of errors. Having obtained the regression coefficients a and bs, the values of Q can be obtained by using Eq. (13.1). Of course, there may be additional variables in the equation, such as evapotranspiration and infiltration.

While developing a rainfall-runoff regression relationship, it is important to identify the type of basin, for example, snow-fed or rain-fed or mixed. If the basin is snow and rain fed and snowmelt runoff is a significant part of streamflow, snow water equivalent values are needed in addition to rainfall data. Concurrent rainfall and runoff data should always be plotted first to investigate the type of relationship that applies. One should never blindly apply a particular model.

13.1.2 Simple Linear Regression

The most common model used in hydrology assumes a linear relationship between two variables. Such models, called simple linear regression models, have the following form

$$Q_i = a + b_1 P_i + \varepsilon \tag{13.2}$$

It may be added that the series generated by regression provides a smoothed representation of the true process, having a variance which is smaller than the original. The type of regression equation that is most suitable to describe the relation depends on the physics of the processes driving the variables. It also depends on the range of data under consideration. A nonlinear relation may well be described by a simple linear regression equation within a particular range of the variables in regression. An important tool to judge the regression model is to examine the behavior of residuals which are the differences between observed and computed values of the dependent variable. Ideally, the residuals should have zero mean and zero autocorrelation.

Example 13.1

Annual rainfall and runoff for a catchment for 18 years is given in Table 13.1. Using linear regression, fit a rainfall-runoff relation for the catchment.

Year	Rainfall (x)	Runoff (y)	Year	Rainfall (x)	Runoff (y)
1980	42.39	13.26	1989	36.08	12.89
1981	33.48	7.31	1990	40.89	12.82
1982	47.67	15.17	1991	37.31	11.58
1983	50.24	15.54	1992	37.15	15.17
1984	43.28	14.22	1993	40.38	10.49
1985	52.61	21.24	1994	45.39	18.02
1986	31.06	7.79	1995	33.49	9.01
1987	50.02	17.64	1996	39.28	13.82
1988	47.08	20.91	1997	45.03	17.25

TABLE 13.1 Annual Rainfall (MCM) and Runoff (MCM) Data for a Catchment

Note that to solve real-life problems, regression relations should be established, to the extent feasible, by using a longer data series, for example, about 30 years.

Solution

In any regression analysis, it is always helpful to plot the data for visual appreciation of the relations. Figure 13.1 shows a plot between rainfall and runoff for this example data.

Figure 13.1 shows that the relationship between rainfall and runoff can be considered to be linear. Computations are shown in Table 13.2. After computing averages of the two variables, departures from averages are computed and then the quantities S_{xx}, S_{xy}, S_{yy}.

$$\text{Regression coefficients } b = S_{xy}/S_{xx} = 371.412/681.165 = 0.5453$$

$$a = y_m - b^*x_m = 14.1183 - 0.5453^*41.8239 = -8.6866$$

Hence the regression equation is

$$y = -8.6866 + 0.5453\,x \tag{13.3}$$

Spreadsheet software, such as Microsoft Excel, contain built-in routines to compute regression coefficients. To test the goodness of established relations, values of y estimated from the regression equation are shown in the penultimate column and square of differences is shown in the last column. A small sum of squares and a high value of the coefficient of determination (0.737) show a good relation.

13.1.3 Nonlinear Regression

Many times, a nonlinear relation may explain the data much better than a linear model. Consider that the data follow the following generic relation

$$Q_i = aP_i^bT_i^cY_i^d\varepsilon \tag{13.4}$$

which is not a linear equation. However, it can be linearized by using a logarithmic transformation

$$\ln(Q_i) = \ln(a) + b\ln(P_i) + c\ln(T_i) + d\ln(Y_i) + \varepsilon_T \tag{13.5}$$

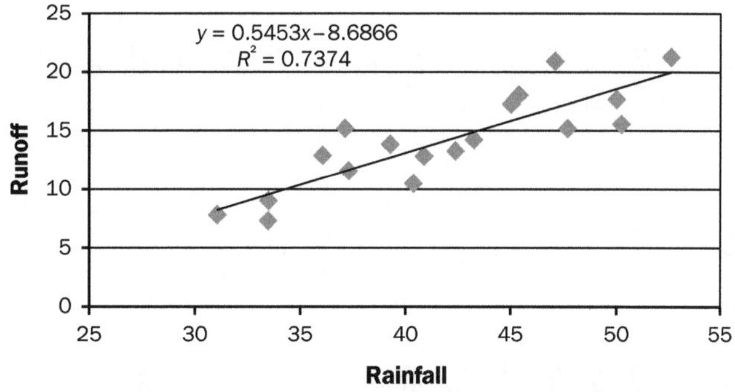

FIGURE 13.1 Plot between annual rainfall and runoff for data of Example 13.1.

Year	Rainfall (x)	Runoff (y)	$(x - x_m)$	$(y - y_m)$	$S_{yy} = (x - x_m)^* (y - y_m)$	$S_{xx} = (x - x_m)$^2	$S_w = (y - y_m)$^2	y estimated by regression (y_{est})	$(y - y_{est})$^2
1980	42.39	13.26	0.566	−0.858	−0.486	0.320	0.737	14.427	1.362
1981	33.48	7.31	−8.344	−6.808	56.808	69.620	46.353	9.569	5.102
1982	47.67	15.17	5.846	1.052	6.148	34.177	1.106	17.306	4.562
1983	50.24	15.54	8.416	1.422	11.965	70.831	2.021	18.707	10.032
1984	43.28	14.22	1.456	0.102	0.148	2.120	0.010	14.912	0.479
1985	52.61	21.24	10.786	7.122	76.815	116.340	50.718	20.000	1.539
1986	31.06	7.79	−10.764	−6.328	68.117	115.861	40.048	8.249	0.211
1987	50.02	17.64	8.196	3.522	28.864	67.176	12.402	18.587	0.897
1988	47.08	20.91	5.256	6.792	35.698	27.627	46.127	16.984	15.411
1989	36.08	12.89	−5.744	−1.228	7.055	32.992	1.509	10.986	3.624
1990	40.89	12.82	−0.934	−1.298	1.212	0.872	1.686	13.609	0.623
1991	37.31	11.58	−4.514	−2.538	11.458	20.375	6.443	11.657	0.006
1992	37.15	15.17	−4.674	1.052	−4.915	21.845	1.106	11.570	12.961
1993	40.38	10.49	−1.444	−3.628	5.239	2.085	13.165	13.331	8.071
1994	45.39	18.02	3.566	3.902	13.914	12.717	15.223	16.063	3.831
1995	33.49	9.01	−8.334	−5.108	42.572	69.454	26.095	9.574	0.318
1996	39.28	13.82	−2.544	−0.298	0.759	6.471	0.089	12.731	1.185
1997	45.03	17.25	3.206	3.132	10.040	10.279	9.807	15.866	1.914
Sum	752.8300	254.1300	0.000	0.000	371.412	681.165	274.645	254.130	72.129
Average	41.8239	14.1183	0.000	0.000	20.634	37.842	15.258	14.118	4.007

TABLE 13.2 Computations for Example 13.1

or

$$Q_{Ti} = a_T + b \times P_{Ti} + c \times T_{Ti} + d \times Y_{Ti} + \varepsilon_T \tag{13.6}$$

where subscript T denotes the transformed variable. Now, the error minimization is carried out on the logarithmic values rather than on the original values. It is important to note that (1) the regression coefficients are estimated such that $\Sigma(Q_{Ti} - \hat{Q}_{Ti})^2$ is a minimum and not the sum of squares of deviations in the original domain, and (2) the error term in the transformed equation is additive, whereas it is multiplicative in the original equation. In some situations, after logarithmic transformation, the data conform more closely to the regression assumptions. In arid regions, the relation between rainfall and runoff is highly nonlinear in general due to large variation in rainfall abstractions due to evaporation. For humid regions, the abstraction is comparatively less and the rainfall-runoff relation is more linear.

13.1.4 Multiple Regression

Often, we wish to model the dependent variable as a function of several other quantities in the same equation. For example, in a large basin, monthly runoff is likely to be dependent on rainfall in the same month and in the previous month(s). Multiple regression is applied in such cases. In multiple regression, the dependent variable is regressed on many independent variables. A multiple regression equation would be

$$Q_i = a + b_1 P_i + b_2 R_{i-1} + \ldots + b_m V_{i-m+1} + \varepsilon \tag{13.7}$$

where a and b_i = the regression coefficients, and ε = the error term. This model is linear in parameters b_j. In hydrologic problem, n observations would be available on Q and on each of the independent variables. Thus n equations can be written, one for each observation, and these are solved for p unknown parameters. Thus n must be equal to or greater than p; in practical problems n should be at least about 4 times p.

Example 13.2

Table 13.3 contains the data of annual flows for the month of August and rainfalls for 2 months: July and August for a watershed for 18 years. Fit simple and multiple regression models and examine the improvement in multiple regression.

Solution

The first four columns of the table contain Year, RF-July, RF-August, and observed flow for August. Following the procedure of Example 13.1, the simple linear regression equation is

$$Q_{Aug} = 358.8786 + 0.4328 \times RF_{Aug} \tag{13.8}$$

The coefficient of determination for this relation is 0.5034 and SSE = 2.58E+07.

Spreadsheet software, such as Microsoft Excel, contain a built-in array formula (LINEST) to compute coefficients of multiple linear regression. We get the following multiple regression equation

$$Q_{Aug} = -2925.83 + 0.5311 \times RF_{Jul} + 0.3668 \times RF_{Aug} \tag{13.9}$$

The coefficient of determination for this relation is 0.834. Also, the sum of square of errors in this case is 8.64E+06 which is much smaller than 2.58E+07 which was

YEAR	RF-JUL (MCM)	RF-AUG	Obs Q Aug	Comp Q LR Q_L	$(Q_{ob} - Q_L)^2$	Comp Q MLR (Q_M)	$(Q_{ob} - Q_M)^2$
1990	5130.04	14561.05	5994.94	6661.21	443917.4	6689.1	481869.2
1991	7781.13	6546.24	2657.92	3192.23	285492.1	3405.1	558333.4
1992	3002.36	12586.63	4895.52	5806.64	830150.0	4860.0	1259.3
1993	8572.75	7532.13	5125.02	3618.94	2268264.5	4219.1	820671.2
1994	5242.03	5799.34	2532.37	2868.95	113287.4	2077.0	207331.4
1995	6311.05	9522.80	3274.52	4480.55	1454510.1	4446.6	1373872.4
1996	6040.00	7285.46	4163.01	3512.18	423580.8	3159.0	1008022.3
1997	1597.33	6922.49	2046.69	3355.08	1711877.9	1336.5	504351.9
1998	8561.71	6889.43	4190.08	3340.77	721329.6	3873.7	100071.8
1999	7153.31	12566.82	6107.45	5798.07	95720.3	6372.2	70105.1
2000	5623.67	10263.08	5145.44	4800.96	118668.0	4587.6	311150.5
2001	4233.30	7108.91	2300.77	3435.77	1288208.4	2402.5	10345.1
2002	12076.88	10472.23	7994.09	4891.48	9626148.6	7066.0	861393.1
2003	6843.64	8068.47	3695.11	3851.08	24328.1	3869.7	30465.3
2004	7819.49	9330.16	4870.40	4397.17	223947.7	4897.7	744.3
2005	9403.82	7424.92	3943.46	3572.54	137577.3	4467.0	274142.5
2006	7040.85	8306.55	3801.73	3954.13	23227.0	4068.4	71133.5
2007	7380.56	9987.30	5895.90	4681.60	1474529.8	5085.7	656480.9
2008	8620.28	4283.79	1501.45	2212.99	506303.3	2511.4	1020067.0
2009	9113.46	5071.52	2670.74	2553.94	13641.6	3110.7	193556.8
2010	1296.93	11168.68	3192.95	5192.92	3999885.7	3481.4	83188.8
2011	7393.84	7785.62	3908.33	3728.66	32281.1	3921.3	167.4
Sum	146238.41	189483.61	89907.88	89907.88	2.58E+07	89907.9	8.64E+06
Average	6647.20	8612.89	4086.72	4086.72			

Table 13.3 Data of Example Problem and Solution

obtained in the case of simple regression. Obviously, we have a much better relation by adding one more dependent variable.

13.1.5 Stepwise Regression

Suppose that we have a big set of potential independent variables for regression and wish to keep the regression equation compact. A commonly used procedure to iteratively build the regression equations is stepwise regression. It consists of creating the regression equation in steps by adding or removing the variables, one at a time. There can be three different strategies in stepwise regression. The forward selection begins with a model without any variable. Iteratively, the variable that contributes the most to the regression equation is added. The procedure stops when the improvement in the model by adding a new variable is not statistically significant. The backward selection procedure starts with all potential variables included in the regression model. Now, iteratively that variable is removed from the model which contributes the least. The procedure stops when all the variables in the model are statistically significant. Sometimes a combination of forward and backward procedures is followed. It needs to be ensured that the final regression equation is rational. Further, there may be certain variable(s) that will be included in the equation irrespective of the error variance reduction they give because the analyst judges them to be of critical importance. Such variable(s) are called forced variable(s).

By following the procedure shown in Example 13.2, a good regression relation can be established when multiple input variables are involved.

The real test of a regression model is its ability to predict the dependent variable for the observations that were not used in computing the coefficients of regression equation. For this purpose, the data is divided into two parts: one part is used to develop the model and the other part to test it.

13.2 Soil Conservation Service–Curve Number Method

In 1954, the United States Soil Conservation Service (SCS, presently known as the Natural Resources Conservation Service, NRCS) developed a procedure to estimate direct runoff from storm rainfall. It is known as the Soil Conservation Service–Curve Number (SCS-CN) method. The SCS-CN method is based on the water balance equation and two fundamental hypotheses. The water balance equation is written as

$$P = I_a + F + Q \tag{13.10}$$

where P is the total rainfall (or maximum potential surface runoff), I_a is the initial abstraction, F is the actual infiltration, and Q is the direct surface runoff. In a storm event, a certain amount of rainfall is initially abstracted as interception, infiltration, and depression storage before runoff begins. These losses are termed as "initial abstraction." Note that the initial abstraction can't exceed rainfall, or $P \geq I_a$. The first hypothesis of the SCS-CN method relates the initial abstraction (I_a) to S

$$I_a = \lambda S \tag{13.11}$$

where S is the maximum potential retention and λ is the initial abstraction coefficient. The second hypothesis equates the ratio of actual direct surface runoff and the potential runoff (which is equal to total rainfall less I_a) to the ratio of actual infiltration (F) and the potential infiltration (which is equal to S)

$$\frac{Q}{P - I_a} = \frac{F}{S} = \frac{(P - Q - I_a)}{S} \tag{13.12}$$

The values of P, Q, and S are in depth dimensions (e.g., mm), while the initial abstraction coefficient (λ) is dimensionless. Substituting Eq. (13.11) in Eq. (13.12) and simplifying yields the expression for Q as

$$Q = \frac{(P - I_a)^2}{P - I_a + S} \tag{13.13}$$

Equation (13.13) is the general form of the popular SCS-CN method and is valid for $P \geq I_a$; $Q = 0$ otherwise. For $\lambda = 0.2$, $I_a = 0.2S$ and we get

$$Q = \frac{(P - 0.2S)^2}{P + 0.8S} \tag{13.14}$$

Equation (13.14) is the widely recognized form of the SCS-CN method. The parameter S of the SCS-CN method depends on soil type, land use, hydrologic condition, and antecedent moisture condition (AMC). Similarly, the initial abstraction coefficient λ is frequently recognized as a regional parameter which depends on geologic, vegetation, and climatic factors. The SCS-CN method assumes λ to be equal to 0.2 for practical applications; this has been frequently questioned for its correctness. Since parameter S can vary in the range of $0 \leq S \leq \infty$, it is mapped onto a dimensionless curve number CN, varying in a more appealing range $0 \leq \text{CN} \leq 100$, as

$$S = \frac{25400}{CN} - 254 \tag{13.15}$$

where S is in mm. Note that CN is nondimensional and has no intrinsic meaning; it is only a convenient procedure to transform S to a 0 to 100 scale. But this is a key parameter of the SCS-CN method which represents the catchment in the model and gives the model its name. The highest numerical value of CN (i.e., 100) yields $S = 0$ which is a condition of zero potential maximum retention, which implies an impermeable watershed. On the other hand, the lowest possible value of CN (which is 0) indicates a situation of highest potential maximum retention ($S = \infty$); this refers to a watershed with infinite abstraction. In practice, S lies in the range (40, 98). All other things remaining the same, higher value of S (or lower CN) implies lesser runoff from a watershed.

Table 13.4 gives CN values for typical land uses. USDA (1986) and Mishra and Singh (2003) have described the SCS-CN method in great details.

13.2.1 Curve Number Estimation

The basic inputs to estimate CN are land cover complex classification, soil group classification, and AMC.

13.2.2 SCS Land Cover Complex Classification

The SCS land cover complex classification consists of three factors: land use, treatment or practice, and hydrologic condition. About 15 different land uses have been identified for estimating curve numbers. Agricultural land uses are often subdivided, based on treatment or practice, for example, contoured or straight row. This separation reflects

Land Use Description	Hydrologic Soil Group			
	A	B	C	D
Cultivated land[1]: Without conservation treatment	72	81	88	91
With conservation treatment	62	71	78	81
Pasture or range land: Poor conditions	68	79	86	89
Good condition	39	61	74	80
Meadow: Good condition	30	58	71	78
Wood or forest land: Thin stand, poor cover, no mulch	45	66	77	83
Good cover[2]	25	55	70	77
Open spaces, lawns, parks, golf courses, cemeteries, etc.				
Good condition: Grass cover on 75% or more of the area	39	61	74	80
Fair condition: Grass cover on 50%–75% of the area	49	69	79	84
Commercial and business areas (85% impervious)	89	92	94	95
Industrial districts (72% impervious)	81	88	91	93
Residential[3]:				
Average lot size — Average % impervious[4]				
1/8 acre (506 m²) or less — 65	77	85	90	92
1/4 acre (1012 m²) — 38	61	75	83	87
1/3 acre (1349 m²) — 30	57	72	81	86
1/2 acre (2023 m²) — 25	54	70	80	85
1 acre (4046 m²) — 20	51	68	79	84
Paved parking lots, roofs, driveways, etc.[5]	98	98	98	98
Streets and roads:				
Paved with curbs and storm sewers[5]	98	98	98	98
Gravel	76	85	89	91
Dirt	72	82	87	89

Source: USDA (1986).

[1] For a more detailed description, please refer to USDA (1986).

[2] Good cover is protected from grazing and liter and brush cover soil.

[3] Curve numbers are computed assuming the runoff from the house and driveway is directed toward the street with a minimum of roof water directed to lawns where additional infiltration could occur.

[4] The remaining previous areas (lawn) are considered to be in good pasture condition for these curve numbers.

[5] In some warmer climates of the country a curve number of 95 may be used.

TABLE 13.4 Runoff Curve Numbers for Selected Agricultural, Suburban, and Urban Land Uses (Antecedent Moisture Condition II, $I_a = 0.2S$)

the differences in runoff potential associated with the variation in land treatment. The hydrologic condition indicates the differences in the level of land management. It has been classified by SCS into three different classes: poor, fair, and good. Not all land uses have been classified in this manner.

Land cover refers to the vegetation in the watershed, agriculture practices or fallow, water bodies, and impervious surfaces (such as roads, buildings, etc.). CN depends on these and also on how agriculture is practiced—row crops, fallow land, terraced fields, and so on. Runoff produced by grassland depends on the hydrologic condition of nature pasture. A good hydrologic condition implies 75% areal coverage by native pasture and light grazing, fair hydrologic condition is for 50%–75% areal coverage and medium grazing, and poor hydrologic condition is less than 50% areal coverage and heavy grazing.

13.2.3 SCS Soil Group Classification

SCS has developed a soil classification system that consists of four groups, identified by the letters A, B, C, and D. Soil characteristics that are associated with each group are as follows:

Group A: Sand, loamy sand, or sandy loam types of soils. These have low runoff potential and high infiltration rates, even when thoroughly wetted. They chiefly consist of deep, well to excessively drained sands or gravels and have a high rate of water transmission.

Group B: Silt loam or loam, having moderate infiltration rate when thoroughly wetted. These consist of moderately deep to deep, moderately well to well drained soils with moderately fine to moderately coarse textures.

Group C: Soils are sandy clay loam. They have low infiltration rates when thoroughly wetted and consist chiefly of soils with a layer that impedes downward movement of water and soils with moderately fine to fine structure.

Group D: Soils are clay loam, silty clay loam, sandy clay, silty clay, or clay. This group has the highest runoff potential. They have very low infiltration rates when thoroughly wetted and consist chiefly of clay soils with a high swelling potential.

13.2.4 Antecedent Soil Moisture Condition

The AMC represents the moisture content of the soil at the beginning of a rainfall event. AMC has a significant effect on rate and volume of runoff produced. Recognizing that it is a significant factor, SCS identified three antecedent soil moisture conditions, which were labelled I, II, and III. The soil conditions for each group are as follows:

I: Soils are dry but not to the wilting point; satisfactory cultivation has taken place.

II: Average condition.

III: Heavy rainfall, or light rainfall and low temperature have occurred within the last 5 days; saturated soil.

The seasonal rainfall limits for each AMC (I, II, III) are described in Table 13.5.

The soil moisture contents in a catchment just before a storm occurs are of utmost importance in determining a hydrograph. According to Hawkins (1975), the errors in CN may have more impact on runoff estimation than the errors of similar magnitude in the estimation of storm rainfall P. Major watershed characteristics, such as soil type, land use/treatment classes, hydrologic soil group, hydrologic conditions, and AMC, play a key role in the accurate estimation of CN. AMC is one of the most important watershed characteristics when determining CN. However, AMC is relatively less

AMC	Total 5-day Antecedent Rainfall (cm)	
	Dormant Season	Growing Season
I	Less than 1.3	Less than 3.6
II	1.3–2.8	3.6–5.3
III	Over 2.8	Over 5.3

TABLE 13.5 Seasonal Rainfall Limits for the Three AMCs

important in extremely wet or dry conditions. AMC mainly depends on catchment conditions. AMC depends on soil hydraulic properties, land cover, land treatment, and runoff generating processes. In hydrologic design, AMC may also be a policy decision rather than a statement of actual soil condition at the site.

The SCS method uses the 5-day antecedent rainfall as antecedent precipitation index (API) for three AMCs: AMC I through AMC III. Hjelmfelt et al. (1982) statistically related the AMC I, AMC II, and AMC III levels to 90%, 10%, and 50% cumulative probability of the exceedance of runoff depth for a given rainfall, respectively.

Despite the widespread use of SCS-CN methodology, the correct estimation of the curve number for the given situation has been a topic of continued discussion among hydrologists and the water resources community.

Conversion of CN from AMC II to AMC I and AMC III Hawkins et al. (2009, p. 106) have reported various formulae for converting CN from one AMC to another AMC. Following Hawkins et al. (1985),

$$S(\mathrm{I}) = 2.281\ S(\mathrm{II}) \text{ and } S(\mathrm{III}) = 0.427\ S(\mathrm{II}) \tag{13.16}$$

where $S(\mathrm{I})$, $S(\mathrm{II})$, and $S(\mathrm{III})$ refer to S corresponding to AMC I, AMC II, and AMC III, respectively.

Hawkins et al. (1985) suggested

$$\mathrm{CN(I)} = \mathrm{CN(II)}/[2.81 - 0.01381\mathrm{CN(II)}];$$
$$\mathrm{CN(III)} = \mathrm{CN(II)}/[0.427 + 0.00573\ \mathrm{CN(II)}] \tag{13.17}$$

Similarly, Chow et al. (1988) proposed

$$\mathrm{CN(I)} = 4.2\ \mathrm{CN}/[10 - 0.058\ \mathrm{CN(II)}];\ \mathrm{CN(III)} = 23\ \mathrm{CN(II)}/[10 + 0.13\ \mathrm{CN(II)}] \tag{13.18}$$

Sobhani (1975) proposed

$$\mathrm{CN(I)} = \mathrm{CN(II)}/[2.334 - 0.01334\ \mathrm{CN(II)}];$$
$$\mathrm{CN(III)} = \mathrm{CN(II)}/[0.4036 - 0.0059\ \mathrm{CN(II)}] \tag{13.19}$$

Example 13.3

A watershed received 47 mm of rain on a given day. It has pasture with heavy grazing and Group B soil. Previous to this event, watershed had received 64 mm rainfall in the past 5 days. Compute the direct runoff by using the SCS method.

Solution

From Table 13.4, for pasture with poor hydrological condition and Group B soil, CN = 79.

For 5-day antecedent rainfall of 61 mm, AMC III condition is considered.

For AMC II CN = 79, AMC III CN = 91

From Eq. (13.15) $S = \dfrac{25{,}400}{\text{CN}} - 254 = \dfrac{25{,}400}{91} - 254 = 25.12$

Hence $Q = \dfrac{(47 - 0.2 \times 25.12)^2}{(47 + 0.8 \times 25.12)} = \dfrac{1761.94}{52.02} = 33.87\,\text{mm}$

Example 13.4

The previous example had only one land use. A real-life catchment will have multiple land uses. Consider that a catchment has the following land uses:

40% cultivated land with conservation treatment

20% open spaces, grass cover on 50%–75% area

20% residential area that is about 38% impervious

10% residential area that is about 65% impervious

10% paved roads with storm sewers

This watershed has Group B soils and received 38 mm rainfall on a given day; the total rainfall in the 5 previous days was 40 mm. Compute runoff by using the SCS method.

Solution

We first compute the weighted CN:

Land use	%	CN	Weighted CN
Cultivated land with cons treatment	40	71	2840
Open space	20	69	1380
Residential area, 38% impervious	20	75	1500
Residential area, 65% impervious	10	85	850
Paved roads with storm sewers	10	98	980
Total	100		7550

Hence weighted CN for the catchment = 7550/100 = 76

For 5-day antecedent rainfall of 40 mm, the AMC II condition is appropriate. Hence CN = 76.

From Eq. (13.15), $S = \dfrac{25{,}400}{\text{CN}} - 254 = \dfrac{25{,}400}{76} - 254 = 80.21$

Hence $S = \dfrac{(38 - 0.2 \times 80.21)^2}{(38 + 0.8 \times 80.21)} = \dfrac{482.154}{102.168} = 4.72\text{ mm}$

13.2.5 Major Strengths and Weaknesses of SCS-CN Methodology

Although the SCS-CN methodology is versatile conceptually and has been widely applied almost all over the world, its success critically depends on how correctly the two major parameters CN and I_a are estimated. These two are indeed sensitive and varying parameters but typically these are assumed constant over space and time. Research has established that even for the same location, the values of these two change with time due to changes in land uses, crop cover, crop growth, and land treatment.

Although the method is attractive for many practicing hydrologists due to its simplicity, it contains some unknowns and inconsistencies. SCS-CN method has been critically reviewed by various researchers who have highlighted its structural inconsistencies (Michel et al., 2005) and uses and limitations (Ponce and Hawkins, 1996; Garen and Moore, 2005). Following Singh et al. (2010), the major strengths and weaknesses of SCS-CN methodology are listed here.

Major Strengths

1. It is a simple, robust, and lumped conceptual model which relies on only one parameter (CN) and it is well suited for ungauged situations.
2. It is responsive to four readily available catchment properties: soil type, land use/treatment, surface condition, and AMC.
3. To estimate direct surface runoff, the method requires only a few basic descriptive inputs that can be easily converted to numerical values (CN).
4. The method works well for agricultural watersheds for which it was originally intended but has been successfully applied for urban or forest watersheds.
5. It is compatible with remote sensing (RS) and geographical information system (GIS) tools in hydrological applications.

Major Weaknesses

1. SCS-CN does not have any explicit expression of time and the impact of spatial and temporal distribution of rainfall on runoff generation is not accounted for.
2. There is a lack of clear guidance on how to accommodate varied AMCs.
3. Fixed initial abstraction coefficient $\lambda = 0.2$ irrespective of variation in catchment properties is a weakness of the technique.
4. The relationship between CN and AMC classes is not continuous and thus the changes in CN are not smooth. Further, the method does not consider effects of spatial scales on CN.
5. Antecedent moisture condition plays important role in runoff generation but the method lacks expression for its assessment.

13.3 Estimation of Peak Discharge

Estimation of peak discharge was one of the important problems that hydrologists were asked to solve. We first discuss a simple but powerful method to solve this problem.

13.3.1 Rational Formula

The rational formula is one of earliest hydrologic methods, which was developed to estimate peak discharge from a catchment. This formula is based on three key assumptions of linearity and proportionality of catchment response: (1) runoff peak is linearly proportional to rainfall intensity; (2) abstractions and losses from rainfall vary linearly and are incorporated in the runoff coefficient; and (3) the duration of storm is equal to or longer than the time of concentration of the catchment. Rational method is applicable to small catchments. Two parameters are used to define a small catchment: area and time of concentration. A catchment with the time of concentration less than 1 hour or area less than 2.5 km^2 can be considered as small. Note that these values are somewhat subjective and there may be different thresholds in different places. In a small catchment, rainfall can be assumed to be uniformly distributed in space and time.

According to the rational formula, peak discharge Q_p (in m^3/s) is:

$$Q_p = 0.278\,CIA \tag{13.20a}$$

$$Q_p = CIA \tag{13.20b}$$

where in Eq. (13.20a), C is the dimensionless runoff coefficient, I is the rainfall intensity (mm/h), and A is the catchment area (km^2). In Eq. (13.20b), Q_p is in cfs, A in acres, and I is in inches/per hour. Runoff coefficient varies from less than 0.1 for forested lands of mild slope to 0.95 for paved areas. Table 13.6 lists common values of the runoff coefficient for various land covers and uses. If the catchment has more than one land use category, the weighted runoff coefficient is calculated and the resulting C value is used in the rational formula.

The rainfall intensity used in the rational formula is the average intensity of rainfall whose duration is equal to the time of concentration for the catchment. It is assumed that the storm covers the whole watershed and is stationary. Note that the rainfall intensity depends on the storm frequency for which the peak flow is being calculated and is obtained from the intensity–duration–frequency (IDF) curves for the catchment. It can be shown that by applying the convolution theorem, the rational formula can also be used to generate the entire flow hydrograph if the rainfall hyetograph and the time of concentration are known.

A higher value for C may be used in the case of design for storms of larger durations. For a storm with duration equal to the time of concentration (T_C), the runoff hydrograph attains peak at T_C and recedes to zero at the time base (T_B). Table 13.7 suggests the time base of hydrograph depending upon the value of C.

The hydrograph as per the rational method or the rational hydrograph is commonly plotted as a triangular hydrograph with a time base equal to 2.67 times T_C. Singh (1988) derived the unit hydrograph for the rational method from which the runoff hydrograph can be computed for any rainfall hyetograph. Figure 13.2 shows a typical rational method hydrograph.

Description of Area	Runoff Coefficient, *C*
Business	
Downtown areas	0.70–0.95
Neighborhood areas	0.50–0.70
Residential	
Single-family areas	0.30–0.50
Multi-units, detached	0.40–0.60
Multi-units, attached	0.60–0.75
Residential, suburban	0.25–0.40
Apartment dwelling areas	0.50–0.70
Industrial	
Light areas	0.50–0.80
Heavy areas	0.60–0.90
Parks	0.10–0.25
Railroad yard areas	0.20–0.35
Unimproved areas	0.10–0.30
Pavement	
Asphalt or concrete	0.70–0.95
Brick	0.70–0.85
Roofs	0.75–0.95
Lawns, heavy soil	
Flat 2%	0.13–0.17
Average, 2%–7%	0.18–0.22
Steep 7% or more	0.25–0.35

Source: ASCE Design and Construction of Urban Stormwater Management Systems, *ASCE Manual and Reports of Engineering Practice, No. 77,* American Society of Civil Engineers (ASCE), New York, 1992.

TABLE 13.6 Typical Values of Runoff Coefficient, *C* (for Storms of up to 10-Year Frequency)

C	T_B/T_C
0.75–1.0	2.00
0.50–0.75	2.33
0.30–0.50	2.50
<0.30	2.67

TABLE 13.7 Time Base in Rational Method

The rational method assumes that the rainfall excess is directly related to and is proportional to the rainfall rate; here C can be assumed to be the proportionality factor. Since this assumption is not strictly valid, particularly for large areas, the use of the rational method is limited to small catchments. ASCE (1992) recommends that the rational method should be used for areas less than 80 ha.

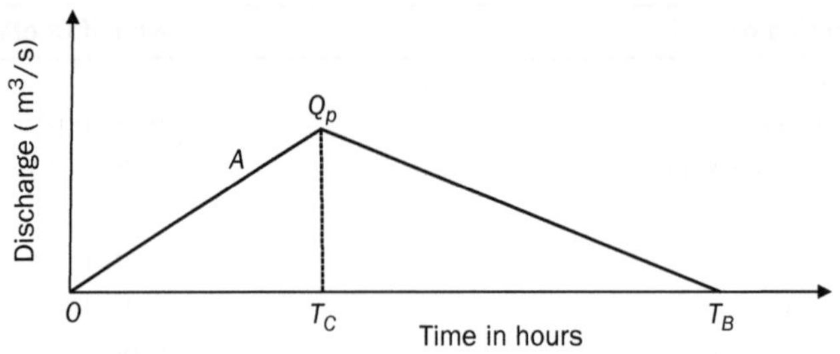

FIGURE 13.2 Rational method hydrograph.

Example 13.5

Calculate the peak runoff rates by using the rational formula for 2- and 20-year frequency storms for a catchment whose area is 10 acres. The intensities for a storm of duration 30 minutes are 2.5 in./hour (2-year storm) and 3.5 in./hour (20-year storm). This area is 80% grass covered (parks) and 20% paved. Assume $T_c = 30$ min.

Solution

For this catchment, we select $C = 0.15$ for vegetated land (parks) and $C = 0.90$ for pavement (see Table 13.6). The composite C value is calculated as follows

$$C = 0.15\left(\frac{80}{100}\right) + 0.90\left(\frac{20}{100}\right) = 0.30$$

Given rainfall intensities for a storm of duration of 30 min

$$I_2 = 2.5 \text{ in./h}$$
$$I_{20} = 3.5 \text{ in./h}$$

Substituting these values in the rational formula:

$$Q = 0.30 \times 10 \times I = 3.0 \times I$$
$$Q_2 = 3.0 \times 2.5 = 7.5 \text{ cfs}$$
$$Q_{20} = 3.0 \times 3.5 = 10.5 \text{ cfs}$$

The rational method is mostly used to estimate the peak flow from small catchments, urban areas, and high drainage systems. This method is applied to estimate peak runoff for storm whose duration is equal to the time of concentration because this storm will produce the largest peak for the catchment. Evidently, for storm durations shorter than the time of concentration, the runoff from the entire catchment will not be able to reach the outlet before the storm is over. Storms lasting longer than the time of concentration and having smaller intensity produce smaller peak flows than the triangular hydrograph.

While longer duration storms produce smaller peaks, they generate greater volumes of flow. Thus in the sizing of detention basins, the storage volume required should be calculated for various storms of different durations so as to determine the critical storm. This will be the storm that requires the largest detention storage. Sometimes a trapezoidal hydrograph is used in such cases.

13.3.2 Peak Discharge–Volume Relation

The relationship between volume and peak of direct runoff is important in a wide range of hydrologic analysis, especially where hydrologic data are scarce. Singh and Aminian (1986) developed a relation between volume and peak of direct runoff by employing data from 134 drainage basins from the United States, Australia, Italy, and Greece. The relation between volume and peak of direct runoff can be expressed as

$$Q_p = \alpha V^b \tag{13.21}$$

Taking the log of both sides, we get a linear equation

$$\log(Q_p) = b + \alpha \log(V) \tag{13.22}$$

in which Q_p is the peak discharge of direct runoff per unit area (cm/h), V is the direct runoff volume per unit area (cm) [it can also be rainfall amount or volume under runoff hydrograph], b is the intercept (cm/h) and α is dimensionless slope. Rogers (1982) proposed the following relation between peak discharge and volume

$$\log(Q_p / V^2) = b + m\log(V) \tag{13.23}$$

Equation (13.22) was calibrated by using data from 134 drainage basins ranging in size from 0.014 to 6075 km²; data for a number of rainfall-runoff events were collected for each basin. Using a standard hydrograph separation procedure, volume and peak of direct runoff were obtained for each event. To eliminate the effect of area, V and Q_p were divided by the drainage area, thereby producing a unit volume and unit peak discharge. The direct runoff peak was further divided by the square of direct-runoff volume, and then transformed logarithmically (base 10). This variable was then related to the logarithm of volume of direct runoff using regression analysis.

The coefficient of determination showed that Eq. (13.23) is a usable relation. The parameter b was found to vary from –2.234 to 0.665, whereas the parameter m varied from –2.315 to –0.699. Using the estimated parameters, Eq. (13.23) was applied to predict peak of direct runoff for a number of additional events on four sample basins. The maximum error was less than 28%, which is satisfactory from a practical point of view.

Basin area A (km²), main stream length L (km), and stream bed slope S (%) were related in various combinations to the parameters b and m of Eq. (13.23). It was found that the basin area alone explains more than 86% of variance in parameter b with

$$b = -0.308 - 0.462\,\mathrm{Log}(A) \tag{13.24}$$

13.4 Rainfall-Runoff Relationship for Different Durations

When we attempt to model the rainfall-runoff relation for different durations, different model forms are obtained, because the dominant processes at different durations are typically different. Here we will discuss models for two different time durations.

13.4.1 Daily Rainfall-Runoff Relationship

A model relating to daily rainfall with daily runoff was developed by Nash and Barsi (1983) and Kachroo et al. (1991). It was assumed that for the year in which the rainfall on each day was exactly the mean for that day (P_d), the corresponding discharges would also agree with their means, Q_d. Hence

$$P_d \rightarrow Q_d \quad (13.25)$$

It was assumed that in any particular year, the departures of rainfall and discharge from these seasonal means are linearly related

$$Q_i - Q_d = f(P_i - P_d) \quad (13.26)$$

The values of P_d and Q_d can be obtained by averaging the rainfall and discharge records for each date d over the period of calibration. The seasonal values of P_d and Q_d are subtracted from the actual values of P_i and Q_i on each day to obtain the departure series, for example, x and y. Thus the time series for x (input) and y (output) whose length is equal to the number of days in the calibration period are obtained. Assuming that a linear relationship with a memory length m exists between the x and y series, y may be expressed as

$$y_i = h_1 x_i + h_2 x_{i-1} \ldots + h_m x_{i-m+1} + \varepsilon_i \quad (13.27)$$

where h_i = the vector of regression coefficients which represent the discrete series of pulse response, and ϵ_i = the error term. The vector of h values is estimated by the method of least-squares. Having obtained the h vector, the y values can be obtained from the regression equation. Finally, the seasonal mean Q_d is added to the y values to get the computed Q values. The catchment memory length m is selected according to the following criteria:

1. It is the value beyond which an increase in m does not significantly improve the model performance.
2. It is the maximum value such that the last ordinate of the pulse response is still above its standard error.
3. The shape of the derived pulse response is physically realistic.

Kothyari et al. (1993) demonstrated the potential of the linear perturbation model (LPM) for simulation of monthly runoff and showed that the LPM is more accurate for monthly than for daily runoff simulation, particularly from medium-sized catchments. They found that the LPM predicted the low flows with high accuracy, and medium and high flows with slightly less accuracy. LPM does not give good results when it is used to generate daily flows on smaller sized catchments not having marked seasonality. However, when LPM is used for monthly runoff generation on the same catchments, results are better produced. LPM has potential for application in generation of monthly runoff data for ungauged catchments.

13.4.2 Monthly Rainfall-Runoff Relationships (Rain-Fed Catchments)

The expression for runoff from large-size catchments, in terms of volume of water, can be given in the following form

$$Q_m = b + b_1 P_m + b_2 P_m \quad (13.28)$$

Another possible form of monthly rainfall-runoff relationship is

$$R_m = a + P_m^b \tag{13.29}$$

13.4.3 Regression Models for Seasonal Rivers

Many places have distinct wet and dry seasons and most of the annual rainfall is received in the wet season. In such cases, rainfall-runoff relationships for the wet season may be developed by using the regression method described previously. In addition, the following form of the relationship may be considered for the wet season

$$Q_{\text{wet}} = a(P_{\text{wet}} - P_0)^b + c \tag{13.30}$$

Here, Q_{wet} = the flow of the wet season, P_{wet} = the rainfall for wet season, and P_0 = the threshold value of rainfall below which no runoff is generated (initial loss).

During dry months, most of the runoff in the stream is due to the baseflow contribution and the contribution of rainfall is small. One alternative is to estimate flows in the dry months as a function of wet season flow and rainfall in the dry season, expressed as

$$Q_{\text{dry}} = a + bQ_{\text{wet}} + cP_{\text{dry}} + \varepsilon \tag{13.31}$$

where Q_{dry} is the flow in the entire dry season. Now, we can distribute the dry season flow in different months in the dry season as a ratio of the average monthly flow for the concerned month to the average dry season flow for that particular site

$$Q_{\text{dry},i} = K_i Q_{\text{dry}} \tag{13.32}$$

$$K_i = Qav_{\text{dry},i} / Qav_{\text{dry}} \tag{13.33}$$

where Qav_{dry} is the average of total flow in the dry season, and $Qav_{\text{dry},i}$ is the average flow for the ith month in the dry season.

13.4.4 Filling in Missing Data

An important application of the regression equation is to fill in missing data. Typically, such techniques are applied to time series of monthly or larger time periods. For instance, consider Example 13.1. The rainfall for the year 1979 was 44.5 MCM, but observed flow data was missing. Using the regression equation, the estimated flow for 1979 is

$$y = -8.6866 + 0.5453 \times 44.5 = 15.579 \text{ MCM} \tag{13.34}$$

13.4.5 Comments on Regression

Typical questions that arise in the use of multiple regression are (1) can a linear model be used and (2) what independent variable should be included? In many cases, the independent variables are not statistically independent but are correlated. One of the first steps in multiple regression analysis is to compute the correlation matrix.

Interpretation of the regression coefficients is difficult when variables retained in a regression equation are highly correlated. At times, the sign of the regression coefficient may be the opposite of what is expected, more so if the variable is highly correlated with another variable. In case of multiple regression, it is common to perform several regressions by using different combinations of the independent variables and the equation that best represents the dependent variable is chosen.

All of the variables retained in a regression should have physical significance and make a significant contribution to the regression, unless there is a strong (theoretical or intuitive) reason for retaining a nonsignificant variable. If two variables are equally significant when used alone, but are highly correlated then, the one that is easier to measure should be used. To avoid "over-fitting" of the regression equation, the number of coefficients estimated should not be more than 25%–35% of the number of observations (HP, 2003).

13.5 Models for Estimating a Hydrograph

As the hydrologists were asked to answer even more complex questions, models that required the estimation of a full hydrograph were developed. Among the techniques that were developed for this purpose, the unit hydrograph method is commonly used and will be discussed in Chap. 14.

13.6 Soil Erosion and Sediment Yield

In a watershed, soil is chiefly eroded due to falling rain and flowing water. Hydrological models used for runoff and sediment yield estimates may be classified as (1) empirical, (2) lumped conceptual, and (3) physically based distributed model.

Wischmeier and Smith (1965) developed the empirical Universal Soil Loss Equation (USLE) to compute soil erosion by rainfall from agricultural lands. USLE assumes that soil erosion mainly depends on the detachment of soil particles by the impact of rainfall energy (called erosivity of the rain) and ability of the soil to resist the detachment of the particles, called the erodibility of the soil. These two factors are functionally related as

$$\text{Soil erosion} = f[(\text{erosivity of rain}) \times (\text{erodibility of soil})] \tag{13.35}$$

USLE estimates soil erosion A (t/ha/year) as

$$A = RKLSCP \tag{13.36}$$

where A is the soil loss per unit area, expressed in the units of K and period selected for R; R is the rainfall erosivity factor; K is the soil erodibility factor; L is the slope length factor; S is the slope steepness factor; C is the crop management factor; and P is the support practice factor. The word "universal" is used probably because the equation considers the five principle factors which influence soil loss: K, R, LS, C, and P. Rainfall erosivity factor (R) is the product of the kinetic energy of raindrops and the maximum intensity of rainfall of duration of 30 minutes. Soil erodibility factor (K) is determined either experimentally or theoretically based on the contents of clay particles (<0.10 mm diameter), sandy particles (0.10–2.0 mm), organic matter (% of humus), and soil texture and soil permeability. The length of overland flow slope (L) is an important control on soil erosion. Larger slope lengths lead to higher concentration of overland flow and also higher flow velocity, leading to higher erosion. The slope gradient factor (S) is the ratio of soil loss from a plot of known slope to soil loss from a unit plot with a slope length of 22.13 m under identical conditions. Wischmeier and Smith (1965) recommended using a combined value of LS, called topographic slope length and gradient factor. Vegetative cover dissipates the impact force of raindrops on the soil surface, protects soil from splash erosion by modifying volume, drop size, coefficient of distribution, impact velocity, and kinetic energy of rainfall. A commonly used method evaluates the intensity of soil erosion under different covers vis-à-vis that from bare soil. Conservation practice factor (P) is the ratio of soil loss from a plot with a specific conservation practice to the corresponding loss from a plot with up and down cultivation under identical conditions.

USLE estimates gross sheet and rill erosion; it does not take into account channel or gully erosion and en route sediment deposition. The ratio of sediment delivered at a location in a watershed to the gross erosion from the upstream drainage area is defined as sediment delivery ratio (SDR). SDR depends mainly on the size of area, density of drainage network, steepness, and catchment characteristics. The USLE can predict average annual sediment yield for large catchments, while computing the delivery ratio at the desired location on the stream. The estimation of SDR requires reliable data on sediment yield at the desired location.

Williams (1975) modified USLE by replacing the rainfall energy factor with the runoff energy factor; the revised model is known as the Modified Universal Soil Loss Equation (MUSLE). MUSLE estimates sediment yield on a storm-to-storm basis, instead of average soil erosion per year as estimated by USLE. Renard et al. (1997) further modified USLE and developed the Revised Universal Soil Loss Equation (RUSLE) which is expressed as

$$A = RKLSCP \tag{13.37}$$

In RUSLE, all factors are the same as USLE except the rainfall erosivity index (R), which is dependent on annual average energy and intensity of rainstorms. RUSLE is best suited for smaller drainage basins of uniform soil types, topography, and crop conditions. The sediment yield is obtained by dividing the total basin into smaller catchments. More details of these models along with examples of application are available in text books and research papers.

In a lumped approach, the average watershed characteristics, that is watershed size, mean slope, drainage density, mean land cover, mean soil type, and so on, are used for the estimation of sediment yield. The United States Department of Agriculture (USDA) with other federal and state agencies developed a process-based lumped model called the Water Erosion Prediction Project (WEPP; Laflen et al., 1991) to predict soil erosion from water and wind.

In a distributed sediment yield model, the watershed is divided into a number of cells (grids). The surface runoff, detachment of sediment particles, and their transport are estimated for each cell with time. The subsequent flow and transported sediments are routed from one cell to another over the catchment.

Systeme Hydrologique Europeen Transport (SHETRAN; Ewen et al., 2000) is a three-dimensional, physically based, coupled surface/subsurface, finite-difference, spatially distributed model which allows a detailed description of the flow and sediment transport in the basin at both temporal and spatial scales. Detailed input data for sediment transport, such as raindrop size distribution, proportion of canopy drainage falling as leaf drip, initial thickness of sediment and channel bed material, sediment concentration and porosities, particle size distribution, and erodibility coefficients, are required for simulation in SHETRAN.

The Kinematic Erosion Simulation (KINEROS) model (Woolhiser et al., 1990) is a physically based event-oriented model. The model describes the processes of interception, infiltration, surface runoff, and erosion in small agricultural and urban watersheds.

Semi-distributed models like the Soil and Water Assessment Tool (SWAT) model (Arnold et al., 2012) are also widely used for sediment yield assessments at the basin scale.

13.7 Snowmelt Runoff Modeling

A significant part of land is covered by snow. Snowmelt is a significant component of the hydrological cycle in many regions of the world. Snow and glacier cover in a drainage basin greatly influence its runoff generation. Snowfall over a watershed is determined by meteorological conditions and snow accumulation depends on topography and elevation (Singh et al., 2011). After snowfall, the snow pack undergoes significant transformation, caused by compaction, action of thermal gradients, and change of the crystal structure resulting from the interactions of ice, liquid water, and water vapor. Modeling of snowmelt runoff generation plays an important role in understanding these processes and their prediction. Here we discuss two approaches for snowmelt runoff modeling.

13.7.1 Temperature Index Method

The temperature index is considered to be the best index of heat transfer processes associated with snowmelt and is widely used in snowmelt estimation. Air temperature measurements are the only data available in many regions from which snowmelt can be computed.

Snowmelt rate is considered to be proportional to air temperature; the proportionality factor is called the melt factor. The air temperature, expressed as degree-days, is used in snowmelt computation as an index of complex energy balance tending to snow melt. A degree-day is the amount of heat in terms of the persistence of temperature for a 24-hour period of 1°C departure from a reference temperature. A commonly used empirical expression relating snowmelt to the temperature index is

$$M = D_f (T_i - T_b) \tag{13.38}$$

where M is the snowmelt (mm) produced in a unit time, D_f is the degree-day or melt factor (mm °C^{-1} day^{-1}), T_i is the index air temperature (°C), and T_b is the base temperature.

It is assumed that D_f and T_i integrate the effects of several of the individual energy-budget components. This assumption can introduce errors in many conditions. To minimize this error, most models adjust D_f and T_i. D_f is adjusted to incorporate knowledge of its relation with measurable spatial and temporal variations in basin and climate. Distributed hydrologic models use a variable known as lapse rate to adjust T_i to take care of the effects of the difference between the elevation of the point where temperature is being measured and the elevation of the zone of application.

Although temperature and other hydrologic conditions vary continuously throughout the day, the daily mean temperature is the most commonly used temperature variable for snow melt. A 0°C base temperature is generally used to compute degree-days assuming that snowmelt mostly results due to the transfer of heat from the air in excess of 0°C. The degree-day method is considered more useful when the daily mean temperature exceeds 2°C. Air temperature is assumed to be a good index of energy available for melt in forested areas also because longwave radiation emitted from canopy and snow surface are estimated by temperature with reasonably high accuracy (Pomeroy et al., 2009).

Most of the temperature indices are based on daily maximum and daily minimum temperatures because of the wide availability and simplicity of use of these measures. The daily maximum temperature has also been extensively used by some investigators as an index of snowmelt. The maximum temperature has been found to be a more accurate index than the daily mean temperature.

Degree-Day Factor The degree-day factor, D_f, is an important parameter and converts the degree-days to snow melt expressed as depth of water. D_f depends on the physical properties of the snowpack. Since these properties change with time, degree-day factor also changes with time. The lower values are seen in the beginning of a melt season and higher toward the end of it. Daily values are highly variable so it is recommended to use long-term data for averaging the degree-day factor. Because D_f changes with season, this aspect should be taken into account while the degree-day approach is used.

A wide range of D_f has been reported in the literature; the range being from 0.7 mm °C^{-1} d^{-1} to 9.2 mm °C^{-1} d^{-1}. The U.S. Army Corps of Engineers reported a mean value of D_f of 1.74 mm °C^{-1} d^{-1} from extensive studies. D_f changes considerably with the exposure of the site to solar radiation. In glacierized basins, the degree-day factor usually exceeds 6.0 mm °C^{-1} d^{-1} toward the end of summer when ice gets exposed. Some distributed models use a different value of D_f in each basin zone, depending on the vegetation characteristics of the zone as well as snowpack density. Other distributed models use different values of D_f for each basin zone and their magnitude is varied through the melt season to account for the effects of variation in day length.

13.7.2 Energy Balance

The energy balance or the heat budget of a snowpack governs the production of melt water. This method maintains an account of incoming energy, outgoing energy, and the change in energy storage for a snowpack for a given period of time. The net energy is expressed in terms of the heat equivalent of snowmelt. The energy balance of a snow surface is significantly affected by the presence of cloud and vegetation cover. The seasonal variability in the energy inputs available for melt general increases as we move toward the poles. The seasonal difference is the minimum at the equator. The energy balance of a snowpack for any time interval can be expressed as

$$H_m = H_{\text{sn}} + H_{\text{ln}} + H_c + H_e + H_g + H_p + H_q \tag{13.39}$$

where H_m = the energy available for snowmelt; H_{sn} = the net shortwave radiation; H_{ln} = the net longwave radiation; H_c = the convective heat flux; H_e = the latent heat flux; H_g = the conduction of heat from the ground; H_p = the heat content of precipitation; and H_q = the change in energy content of the snowpack. The minimum data required to apply Eq. (13.39) are air temperature, incoming solar radiation, vapor pressure, and wind speed. Limited availability of some of these data have prevented wider applications of Eq. (13.39) to snowmelt studies.

Since all the data required to apply the energy balance approach are not frequently available, some models use a modified version of Eq. (13.39). In most of these models, H_{sn}, H_{ln}, H_p, and H_q are computed using measured data and the remaining components are parameterized and fitted, or are assumed to be negligible.

In the energy balance equation, different components of energy are considered in the form of energy flux. This flux is the amount of energy received on a horizontal snow surface of unit area over a unit time. The positive value of H_m will result in melting of snow. The relative magnitude of the various heat transfer processes that are involved in melting of a snowpack depends on time and local conditions. For example, radiation melting dominates when wind is calm. Melting due to sensible heat flux dominates under warm weather and windy conditions. After computing energy flux H_m (kJ/m²d), the snow melt rate can be calculated by

$$M = \frac{H_m}{\rho_w L \beta} \tag{13.40}$$

where M is the depth of melt water (m/day), L is the latent heat of fusion (333.5 kJ/kg), ρ_w is the density of water (1000 kg/m³), and β is the thermal quality of snow (Singh et al., 2011).

13.8 Summary

This chapter covers methods to compute hydrograph volume, peak discharge, and the SCS-CN method. Two related topics—soil erosion and snowmelt modeling—are also briefly covered.

13.9 Questions

13.1 In Table 13.8, annual rainfall and runoff data of a river basin for 18 years (1981–1998) are listed. Apply regression analysis to develop the relationship between rainfall and runoff.

Year	Rainfall (mm)	Runoff (mm)	Year	Rainfall (mm)	Runoff (mm)	Year	Rainfall (mm)	Runoff (mm)
1981	1130	592	1987	1670	872	1993	1650	1240
1982	1280	832	1988	1540	816	1994	1510	1030
1983	1270	768	1989	990	456	1995	1600	1340
1984	1040	488	1990	1190	780	1996	1300	870
1985	1080	472	1991	1520	1090	1997	1490	1060
1986	1150	656	1992	1370	960	1998	1105	513

Table 13.8 Rainfall and Runoff Data (in mm) for the Period 1981–1998

13.2 For the catchment of Question 13.1, the rainfall for the years 2000 and 2001 were 1159 mm and 887 mm. Compute runoff by using the relationship established in Question 13.1.

13.3 A watershed received 68 mm of rain on 1 day. It has open spaces with 60% grass cover and Group A soil. Previous to this event, the watershed had received 34 mm rainfall in the past 5 days. Compute the direct runoff by using the SCS method.

13.4 What are the major strengths and weaknesses of the SCS-CN method?

13.5 If a hypothetical watershed is completely paved, what will be its CN and how much runoff will be generated by the SCS method? What will be the situation if it is highly permeable?

13.6 Consider the following land uses near your office: (1) cultivated land; (2) park; (3) shopping area; (4) paved parking lot. Estimate CNs for each of these. Assume any missing information.

13.7 What do you understand by antecedent soil moisture condition? How does it influence runoff generation from a watershed?

13.8 Calculate the peak runoff rates by using the rational formula for 2- and 25-year frequency storms for a catchment whose area is 20 acres. The intensities for a storm of duration 30 minutes are 2.3 in./h (2-year storm) and 3.8 in./h (25-year storm). This area is 75% grass covered (parks) and 25% paved. Assume $T_c = 60$ min.

13.9 Write a short-note on the USLE and its improved versions.

13.10 References

ASCE Design and Construction of Urban Stormwater Management Systems. *ASCE Manual and Reports of Engineering Practice*, No. 77. New York: American Society of Civil Engineers, 1992.

Arnold, J. G., D. N. Moriasi, P. W. Gassman, K. C. Abbaspour, M. J. White, R. Srinivasan, C. Santhi, et al. SWAT: Model use, calibration, and validation. *Transactions of the ASABE*, 55(4): 1491–1508, 2012.

Chow, V. T., D. R. Maidment, and L. W. Mays. *Applied Hydrology*. New York: McGraw-Hill, 1988.

Ewen, J., G. Parkin, and P. E. O'Connell. SHETRAN: Distributed river basin flow and transport modeling system. *ASCE Journal of Hydrologic Engineering*, 5: 250–258, 2000.

Garen, D. C., and D. S. Moore. Curve number hydrology in water quality modeling: Uses, abuses, and future directions. *Journal of the American Water Resources Association (JAWRA)*, 41(2): 377–388, 2005.

Hawkins, R. H. The importance of accurate curve numbers in the estimation of storm runoff. *Water Resource Bulletin*, 11(5): 887–891, 1975.

Hawkins, R. P., A. T. Hjelmfelt, and A. W. Zevenbergen. Runoff probability, storm depth and curve numbers. *Journal of Irrigation and Drainage Division, ASCE*, 111(4): 330–340, 1985.

Hawkins, R. H., T. J. Ward, D. E. Woodward, and J. A. Van Mullen. *Curve Number Hydrology: State of the Practice*. Reston, VA: ASCE Press, 2009.

Hjelmfelt, A. T., K. A. Kramer, and R. E. Burwell. Curve numbers as random variables, *Proceedings of the International Symposium on Rainfall–Runoff Modeling. Water Resources*, Littleton, CO, 365–373, 1982.

HP. How to do hydrological data validation using regression. Hydrology Project, Govt. of India, New Delhi, 2002.

Kachroo, R. K., M. Bruen, G. C. and Liang. *Workshop on River Flow Forecasting*. Lecture notes, Dept. of Engineering Hydrology, University College, Galway, Ireland, 1991.

Kothyari, U. C., V. Aravamuthan, and V. P. Singh. Monthly runoff generation using the linear perturbation model. *Journal of Hydrology*, 144(1–4): 371–379, 1993.

Laflen, J. M., L. L. Lane, and G. R. Foster. WEPP, a new generation of erosion prediction technology. *Journal of Soil and Water Conservation*, 46(1):34–38, 1991.

Michel, C., A. Vazken, and C. Perrin. Soil conservation service curve number method: How to mend a wrong soil moisture accounting procedure. *Water Resources Research*, V41: W02 011, 1–6, 2005.

Mishra, S. K., and V. P. Singh. *Soil Conservation Service Curve Number (SCS-CN) Methodology*. Dordrecht, Netherlands: Kluwer Academic Publishers, 2003.

Nash, J. E., and B. I. Barsi. A hybrid model for flow forecasting on large catchments. *Journal of Hydrology*, 65(1–3):125–137, 1983.

Ponce, V. M., and R. H. Hawkins. Runoff curve number: Has it reached maturity? *Journal of Hydrologic Engineering*, 1(1): 11–19, 1996.

Pomeroy, John W., Danny Marks, Tim Link, Chad Ellis, Janet Hardy, Aled Rowlands, and Raoul Granger (2009). The impact of coniferous forest temperature on incoming longwave radiation to melting snow. *Hydrological Processes*, 23: 2513–2525, 2009.

Renard, K. G., G. R. Foster, G. A. Weesies, et al. Predicting soil erosion by water: A guide to conservation planning with the Revised Universal Soil Loss Equation (RUSLE). In *USDA Agricultural Handbook No. 703*. Available at http://www.ars.usda.gov/SP2UserFiles/Place/64080530/RUSLE/AH_703.pdf, 1997.

Rogers, W. F. Some characteristics and implications of drainage basin linearity and non-linearity. *Journal of Hydrology*, 55:247–265, 1982.

Singh, V. P. *Hydrologic Systems: Vol. 1. Rainfall-Runoff Modeling*. Englewood Cliffs, NJ: Prentice Hall, 1988.

Singh, V. P., and H. Aminian. An empirical relation between volume and peak of direct runoff. *Water Resources Bulletin*, 22(5): 725–730, 1986.

Singh, P. K., M. L. Gaur, S. K. Mishra, and S. S. Rawat. An updated hydrological review on recent advancements in soil conservation service-curve number technique. *Journal of Water and Climate Change*, IWA 1(2):118–134, 2010.

Singh, V. P., P. Singh, and U. K. Haritashya (eds.). *Encyclopedia of Snow, Ice and Glaciers*. Dordrecht, Netherlands: Springer, 2011.

Sobhani, G. A. *Review of Selected Small Watershed Design Methods for Possible Adoption to Iranian Conditions*. M.S. thesis, Utah State University, Logan, UT, 1975.

USDA. Urban Hydrology for Small Watersheds. United States Department of Agriculture, Natural Resources Conservation Service, Conservation Engineering Division, Technical Release (TR) 55, Washington, DC, June 1986.

Williams, J. R. Sediment yield prediction with universal equation using runoff energy factor. In ARS-S-40: *Present and Prospective Technology for Predicting Sediment Yields and Sources*, 244–252. Washington, DC: Agricultural Research Service, U.S. Department of Agriculture, 1975.

Wischmeier, W. H., and D. D. Smith. *Predicting Rainfall-Erosion Losses from Cropland East of the Rocky Mountains*. Agricultural Handbook No. 282. Washington, DC: U.S. Department of Agriculture, 1965.

Woolhiser, D. A., R. E. Smith, and D. C. Goodrich. *ARS-77: KINEROS, A Kinematic Runoff and Erosion Model: Documentation and User Manual*. Tucson, AZ: USDA-ARS Southwest Watershed Research Center, 1990.

CHAPTER 14

Unit Hydrograph Models

For answering more complex questions regarding the rainfall-runoff relationship, methods that estimate the full hydrograph have been developed. Among the methods that have been developed, the unit hydrograph method is one of the popular methods and is discussed in this chapter in detail.

14.1 Introduction

Sherman (1932) proposed the concept of unit hydrograph (UH) to determine the direct surface runoff (DSRO) hydrograph from the effective rainfall hyetograph (ERH). The word "unit" denotes that the hydrograph has been generated by rainfall excess of one unit for a specified duration and hence, by virtue of continuity, the volume under the hydrograph is also one unit. The UH method has since been widely applied in hydrology for a range of tasks, including design flood estimation and flood forecasting. The UH concept is based on linear system theory which means that the catchment response (outflow hydrograph) is assumed to be a linear function of rainfall. Furthermore, the response to a unit input is independent of the responses to the previous and subsequent inputs. This enables splitting a complex input pattern into a series of individual pulses.

The UH of a watershed can be defined as the DSRO resulting from one unit (typically, 1 cm) of effective rainfall (ER) occurring uniformly over the watershed at a uniform rate during a unit period of time. The UH for a watershed is identified by the unit period. Thus a 1-h UH is the direct runoff hydrograph (DRH) resulting from one unit of ER falling for 1 h; it is the period for which the UH is determined and the UH will change as this period changes. Note that the unit period can be any finite duration up to the time of concentration. Typically, the unit duration is a factor of 24 h. Hence, we normally determine and use 1-h UH, 2-h UH, 3-h UH, 6-h UH, 12-h UH, or 1-day UH.

Thus we can state that

1. A UH is a DSRO hydrograph that would result at a given location on a river from unit rainfall excess occurring in unit time uniformly over the catchment up to that location. Rainfall excess is obtained after deducting losses (abstractions) from the total rainfall.
2. Unit excess rainfall is typically 1 cm or 1 mm but some other value (e.g., 5 mm) can also be selected, depending on the catchment size and the typical rainfall magnitudes.
3. The unit time is arbitrarily selected taking into account the duration of storm and the size of catchment. Short periods of 1 or 2 h are used for small catchments and 3, 4, 6, or even 12 h are used for larger catchments. The excess rainfall is assumed to occur uniformly over the catchment over the unit time period. Time periods which are not multiples of 24 h are not preferred.

Since the UH concept relates to direct or surface runoff only, we separate baseflow from the total runoff before deriving the UH. To obtain the DSRO, several assumptions are made for applying the UH method. Many of these assumptions are not perfectly satisfied, but they must reasonably hold before the UH method can be used. Gross violation of the assumptions may result in erroneous results. The basic assumptions of the UH concepts (Dooge, 1973) are given below:

1. *Fixed duration of runoff:* For a given catchment, the duration of surface runoff is essentially constant for all uniform-intensity storms of the same duration, regardless of the total volume of surface runoff.
2. *Principle of proportionality:* For a given catchment, if two uniform-intensity storms of the same length produce different total volumes of surface runoff, then the rates of surface runoff at corresponding times t, after the beginning of two storms, are in the same proportion to each other as the total volumes of surface runoff.

   DSRO at time t, $q(t)$, is proportional to the total volume of rainfall excess. It means $q_1(t) = k \times P_{E1}$; $q_2(t) = k \times P_{E2}$, where $q_i(t)$ is the ordinate of DSRO at any time t for the ith storm having an excess rainfall volume of P_{Ei}, and k is the proportionality constant. Similarly, the rainfall excess of another event, P_{E3}, over time period T is equal to $P_{E3} \times T$. Therefore at a time t, the following proportionality holds

   $$\frac{q_1(t)}{q_3(t)} = \frac{P_{E1} \times T}{P_{E3} \times T} = \frac{P_{E1}}{P_{E3}} \tag{14.1}$$

   Figure 14.1 shows the surface runoff hydrographs for two different rainfall excess volumes P_{E2} and P_{E1} of 100 mm and 75 mm but for the same unit period. The ratios of P_{E2}:P_{E1} = 100:75 = 4:3.
3. *Principle of superposition:* The time distribution of surface runoff from a given storm period is independent of concurrent runoff from antecedent storm periods. This principle is illustrated in Fig. 14.2.
4. The ER is uniformly distributed throughout the catchment within the specified unit period of time. The assumption requires that the storm selected should be of short duration and the resulting hydrograph should have a single peak. It is also implicit here that the losses from rainfall must be uniform over the entire catchment. Note that as the unit time increases, the validity of this assumption becomes weaker.
5. Since ER is assumed uniform over the catchment, the catchment under consideration should not be very big. Practical considerations suggest that the UH theory should be applied to the catchments of size below 5000 km^2. This theory can be applied to large catchments by subdividing them into smaller catchments.

It is noted that all these propositions are empirical. It is not possible to prove them mathematically. In fact, hydraulic analysis may show that these are not mathematically accurate. Despite such views, the UH theory is an extremely useful practical tool. This theory is particularly adequate in the range of floods experienced on natural watersheds. In addition, this theory is amenable to linear mathematics, whose methods are simple and well understood. Due to these reasons, the UH continues to enjoy popularity.

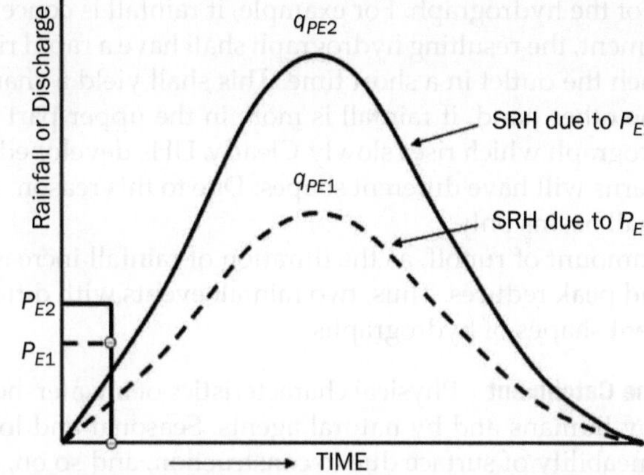

FIGURE 14.1 Surface runoff hydrographs for two different rainfall excess volumes of P_{E1} mm and P_{E2} mm but the same unit period.

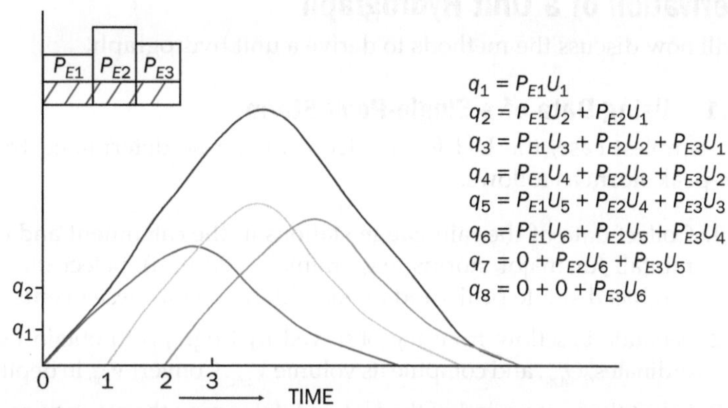

FIGURE 14.2 Principle of superposition: here the hydrograph from a complex storm is calculated by combining the hydrographs from individual storms.

The unit time period of a UH can be determined by using the following criteria: (1) it should be short enough to define the hyetograph and the hydrograph properly, and (2) it should be a simple fraction of 24 h, such as 1 h, 2 h, 3 h, 4 h, and 6 h. Time periods that are not a whole fraction of 24 h, such as 5 h or 7 h, may be avoided. Commonly, a unit period of about one-third to one-quarter of the time of rise of the hydrograph is used.

14.2 Factors Affecting Unit Hydrograph Shape

The factors affecting the shape of the unit hydrograph are the rainfall distribution over the catchment and physiographic characteristics of the catchment, such as shape, slope, vegetation, soil type, and so on.

Rainfall Distribution Rainfall aerial pattern, duration, and intensity patterns greatly affect the shape of the hydrograph. For example, if rainfall is concentrated in the lower part of the catchment, the resulting hydrograph shall have a rapid rise, since most of the rainfall shall reach the outlet in a short time. This shall yield a sharp peak and a rapid recession. On the other hand, if rainfall is more in the upper part of the catchment, it will yield a hydrograph which rises slowly. Clearly, UHs developed from rainfall of different areal patterns will have different shapes. Due to this reason, the UH technique is applicable to small basins only.

For a given amount of runoff, as the duration of rainfall increases, the time base of UH increases and peak reduces. Thus, two rainfall events with different intensities will yield two different shapes of hydrographs.

Physiography of the Catchment Physical characteristics of a watershed change with time due to actions by humans and by natural agents. Seasonal and long-term changes in vegetation, permeability of surface due to construction, and so on, change the physical properties of the watershed. Due to these changes, the catchment response changes and so does the shape of the derived UH.

14.3 Derivation of a Unit Hydrograph

We will now discuss the methods to derive a unit hydrograph.

14.3.1 Using Data of a Single-Peak Storm

In the simplest case, the UH for a catchment can be determined from the data of a single-peak storm as follows:

1. Collect data of the rain gauge stations in the catchment and compute average rainfall for major storms (e.g., rainfall >20 mm). Select the storms that have produced single peaks and extract discharge for these events.
2. Separate baseflow from the observed hydrograph to obtain DSRO hydrograph ordinates, Q_{DR}, and compute its volume V_{DR}. Convert V_{DR} in depth units (e.g., mm).
3. Select the unit period of the UH and determine the average rainfall for this unit period.
4. Determine the loss rate by subtracting the volume of direct surface runoff from the total rainfall to estimate the excess rainfall (P_E).
5. Compute the proportionality factor to obtain the UH ordinates Q_{UH} from Q_{DR} as the ratio of excess rainfall depth (P_E) to V_{DR} or $F = P_E/V_{DR}$.

Example 14.1

For a given P_E of 25 mm, find the proportionality factor (F) to convert the DSRO ordinates to the ordinates of a UH whose unit rainfall is 10 mm ($Q_{UH} = 10$ mm).

Solution

The proportionality factor (F) to convert the DSRO ordinates to UH ordinates is

$$F = \frac{P_E}{Q_{UH}} = \frac{25.0}{10.0} = 2.5$$

Date	July 15	July 16	July 17
Time (h)	0600 1200 1800 2400	0600 1200 1800 2400	0600 1200 1800 2400
DSRO (m^3/s)	10 500 1600 3500	5100 3300 1500 650	250 160 90 20

TABLE 14.1 Ordinates of Surface Runoff Hydrograph for Example 14.2

Date	Time (military time)	Direct Surface Runoff Hydrograph Q_{DR} (m^3/s)	Ordinates of 6-h 100 mm UH, $Q_{UH} = Q_{DR}/F$ (m^3/s)
July 15	0600	10	6.7
	1200	500	334
	1800	1600	1067
	2400	3500	2334
July 16	0600	5100	3400
	1200	3300	2200
	1800	1500	1000
	2400	650	434
July 17	0600	250	167
	1200	160	107
	1800	90	60
	2400	20	13

TABLE 14.2 Computation of 6-h UH

Example 14.2

A storm of 6-h duration with a rainfall excess of 150 mm produces the surface runoff hydrograph shown in Table 14.1.

Find the ordinates of a 6-h UH of a unit volume of 100 mm.

Solution

Since $P_E = 150$ mm and $Q_{UH} = 100$ mm, the proportionality factor $F = 150/100 = 1.50$. The ordinates of UH are computed by dividing the DSRO hydrograph ordinates by 1.5, as shown in Table 14.2.

Convolution After the UH ordinates have been computed, the convolution equation is used to calculate the ordinates of the DSRO (Q_{DR}) for any other event by using the values of excess rainfall, P:

$$Q_{DR,n} = \sum_{m=1}^{n \le M} P_m U_{n-m+1} \quad \text{for } n = 1, 2, ... N \tag{14.2}$$

where n represents the DSRO time index and m represents the rainfall time index, $m = 1, 2, ... M$, and M is the number of PE pulses. The DSRO hydrograph will have $N (= M + n_{UH} - 1)$ nonzero terms, where n_{UH} is the number of UH ordinates. Equation 14.2 is a discrete convolution equation. Expansion of the convolution equation yields a set of equations given in Table 14.3. Figure 14.3 pictorially depicts the concept.

n	Equation
1	$Q_{DR,1} = P_1 U_1$
2	$Q_{DR,2} = P_2 U_1 + P_1 U_2$
3	$Q_{DR,3} = P_3 U_1 + P_2 U_2 + P_1 U_3$
...	...
M	$Q_{DR,M} = P_M U_1 + P_{M\text{-}1} U_2 + \ldots + P_1 U_M$
$M+1$	$Q_{DR,M+1} = 0 + P_M U_2 + \ldots + P_2 U_M + P_1 U_{M+1}$
...	...
$N-1$	$Q_{DR,N-1} = 0 + 0 + P_M U_{N-M} + P_{M-1} U_{N-M-1}$
N	$Q_{DR,N} = 0 + 0 + 0 + P_M U_{N-M+1}$

TABLE 14.3 Convolution Equation in Expanded Form

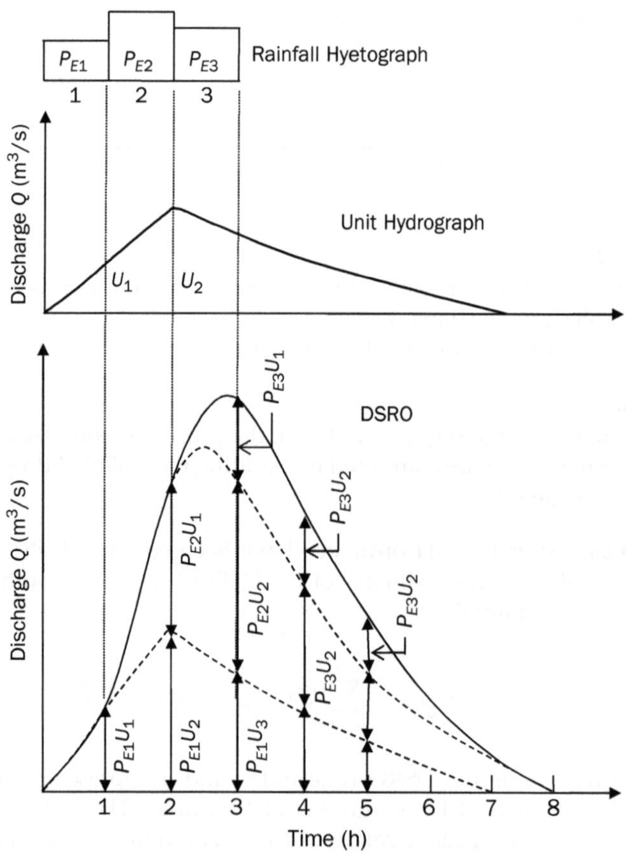

FIGURE 14.3 Unit hydrograph derivation from multiperiod storm data.

Time (h)	1	2	3	4	5	6	7	8
Rainfall (mm)	17	20	11					
UH ordinates (m^3/s)	0	35	150	250	160	90	40	20

TABLE 14.4 Data for Example 14.3

Time (h)	UH Ordinates (m^3/s)	Contribution to DSRO from Excess Rainfall (mm)			DSRO (m^3/s)	Baseflow (m^3/s)	Total Hydrograph (m^3/s)
		8	11	2			
0	0	0			0	15	15
1	35	280	0		280	15	295
2	150	1200	385	0	1585	15	1600
3	250	2000	1650	70	3720	15	3735
4	160	1280	2750	300	4330	15	4345
5	90	720	1760	500	2980	15	2995
6	40	320	990	320	1630	15	1645
7	20	160	440	180	780	15	795
8			220	80	300	15	315
9				40	40	15	55

TABLE 14.5 Solution for Example 14.3

14.3.2 Derivation of Unit Hydrograph from Data of Multiperiod Storm

We will now demonstrate how the DSRO from a catchment can be determined by the use of the convolution equation if the storm rainfall and UH are known.

Example 14.3

For a catchment, a 1-h UH for 1 mm of excess rainfall and the rainfalls for a storm lasting 3 h are given in Table 14.4. If the loss rate is 9 mm/h, determine the DSRO from the catchment.

Solution

Since the loss rate is 9 mm/h, the excess rainfalls for the 3 h will be 8 mm, 11 mm, and 2 mm, respectively. Table 14.5 shows the detailed calculations to compute DSRO and total hydrograph.

We will now show how to determine the UH for a catchment whose total hydrograph is given. The data from Example 14.4 will be used for this purpose.

Example 14.4

For a catchment, the total hydrograph and rainfalls for a storm of 3-h duration are given in Table 14.6. If the loss rate is 9 mm/h, determine the UH for the catchment for 1 mm of excess rainfall and a 1-h unit duration. Baseflow may be assumed at a constant rate of 15 m^3/s.

Time (h)	0	1	2	3	4	5	6	7	8	9
Rainfall (mm)	17	20	11							
Total hydrograph ordinates (m^3/s)	15	295	1600	3735	4345	2995	1645	795	315	55

TABLE 14.6 Data for Example 14.4

Time (h)	0	1	2	3	4	5	6	7	8	9
Excess rainfall (mm)	17 − 9 = 8	20 − 9 = 11	11 − 9 = 2							
DSRO ordinates (m^3/s)	15 − 5 = 10	295 − 15 = 280	1585	3720	4330	2980	1630	780	300	40

TABLE 14.7 Solution for Example 14.4

Solution

We subtract baseflow from the total hydrograph to obtain the DSRO ordinates and determine the excess rainfall by subtracting losses from the total rainfall (Table 14.7).

Now, from Table 14.3, $Q_{DR,1} = P_1 U_1$

Hence, $U_1 = Q_{DR,1}/P_1 = 0/8 = 0$.

Next, $Q_{DR,2} = P_2 U_1 + P_1 U_2$

Hence, $U_2 = (Q_{DR,2} - P_2 U_1)/P_1$

$= (280 - 11 \times 0)/8 = 35$

Next, $Q_{DR,3} = P_3 U_1 + P_2 U_2 + P_1 U_3$

Hence, $U_3 = (Q_{DR,3} - P_3 U_1 - P_2 U_2)/P_1$

$= (1585 - 2 \times 0 - 11 \times 35)/8 = (1585 - 385)/8 = 150$

The remaining ordinates can also be derived in a similar way. All the procedures used for deriving the UHs from multiperiod storms use the principles of proportionality and superposition.

14.4 Instantaneous Unit Hydrograph

Consider that the unit period of excess rainfall is reduced but the amount is kept the same. As a result, the peak flow of the DSRO hydrograph will increase but the base length decreases. If the period of rainfall excess reduces to zero and the rainfall excess depth remains the same, the intensity of the rainfall excess reaches infinity. This corresponds to the instantaneous application of a sheet of water over the entire catchment area. This water drains off the basin through gravity, as shown in the instantaneous unit hydrograph (IUH) in Fig. 14.4. Thus, an IUH is a theoretical concept and represents the UH derived from the unit rainfall excess volume that occurred instantaneously. An IUH represents a characteristic of a catchment and is not affected by the time duration. It is a useful tool in regional UH studies.

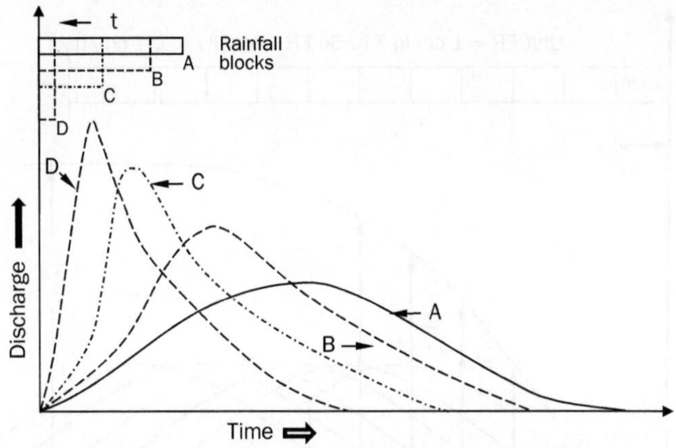

Figure 14.4 Concept of instantaneous UH. Duration of rainfall event B is less than A, C less than B, and D less than C.

The IUH concept assumes that the excess rainfall occurs uniformly over the area for which the IUH is derived. This area can be the entire watershed if it is small (say, less than 250 km^2) or a subwatershed (a portion) therein, if it is large (say, greater than or equal to 2500 km^2). In this manner, the IUH can be applied to large watersheds. The assumption of spatial uniformity of infiltration is implied in the concept of rainfall excess and in turn in the spatial uniformity of antecedent soil moisture and for that matter this assumption also applies to other abstractions, such as evaporation, interception, and depression and detention storage (Singh, 2011). This assumption is only an approximation and can be pesky in some cases.

Figure 14.4 shows a hydrograph produced by a catchment for different rainfall events. Keeping the total rainfall the same, the duration of rainstorm B is less than that of A, duration of storm C is less than B, and so on. It can be seen that the peak of hydrograph B is more than A, peak of C is more than that of B, and so on. The base length of the hydrograph goes on decreasing from A → B → C → D.

The IUH is a single peaked hydrograph with a finite base width and its properties are as follows.

1. $0 \leq u(t) \leq$ a positive value, for $t > 0$
2. $u(t) = 0$ for $t \leq 0$
3. $u(t) \to 0$ as $t \to \infty$
4. $$\int_0^\infty u(t)dt = \text{unit depth over catchment} \tag{14.3}$$

The UH, or more appropriately the IUH, represents the response of a linear time-invariant system, that is, the watershed is assumed to be linear, meaning the principle of superposition and of proportionality apply; and time invariant, meaning the kernel remains independent of time (this is not strictly true for real-life catchments). This implies that the IUH represents the response of the watershed in terms of surface runoff due to an instantaneous excess rainfall of unit volume.

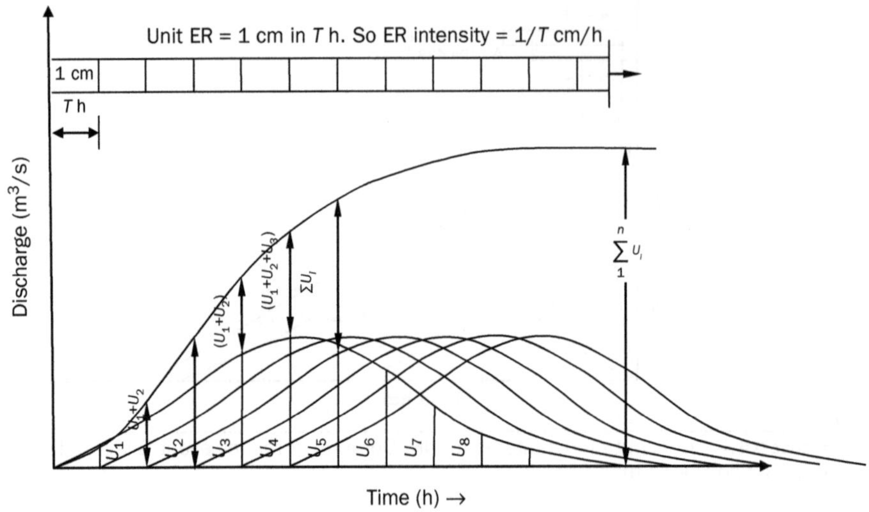

Figure 14.5 S-curve hydrograph.

14.4.1 S-Curve Hydrograph

Recall that for a UH of 1 cm ER and T h duration, the intensity of ER is $1/T$ cm/h. Let the unit excess rainfall of $1/T$ cm/h intensity occur continuously for an indefinite period. The DRH due to this rainfall can be obtained as the summation of these UHs, each lagged by T-h. This hydrograph is called the summation hydrograph or the S-curve hydrograph. The shape of this hydrograph resembles a deformed letter S or integral sign and the ordinates approach the rate of excess rainfall. Thus the S-hydrograph is the DSRO hydrograph that results from excess rainfall of intensity $1/T$ cm/h, occurring for an infinite period. It is obtained as the summation of infinite number of UHs with unit period T h, each lagged by T h, where the T-h UH is the hydrograph that results from $1/T$ cm/h intensity of ER, having occurred for an infinite period (Fig. 14.5).

The S-curve hydrograph may be determined as follows:

1. Plot the UH of unit period T-h at time $t = 0$.
2. Plot a number of such UHs, successively lagged by T-h.
3. Add the ordinates of each UH at a given time, taking the time lag into account.
4. At time t_b (base length of the first UH), the sum of ordinates reaches a constant value.
5. Plot the sum of the ordinates as ordinate and time as abscissa to yield the S-curve for the unit period of T-h. Sometimes, the S-curve has small fluctuations in the region about the constant value. This feature is known as the "hunting effect." To avoid it, a best-fit curve through the points is drawn and used.
6. The maximum ordinate of the S-curve corresponds to the equilibrium discharge, defined as

$$Q_{\max} = \frac{0.2778 A Q_{\mathrm{UH}}}{T} \tag{14.4}$$

Time (h)	1-h, 10-mm Unit Hydrograph (m^3/s)
1	0
2	500
3	1600
4	2400
5	1700
6	1200
7	600
8	150
9	0

TABLE 14.8 Data of Example 14.5

Here A is the catchment area (km^2), Q_{UH} is the unit volume (mm), and T is the time duration of the UH (h). The derivation of an S-curve from a given UH is explained using the following example.

To avoid confusion among different S-curve hydrographs, it is better to label the S-curve hydrographs as S_T-curve, indicating that it has been derived from T-h UHs.

Example 14.5

For a catchment, the ordinates of a 1-h unit hydrograph of 1 cm ER are given in Table 14.8. Determine the ordinates of an S-curve hydrograph.

Solution

Referring to Table 14.9, the computational steps are:

1. Enter the ordinates of the 1-h UH in Column 2.
2. Shift the 1-h UH by 1-h, 2-h, and so on, until the time base of the UH, and enter these values in Columns (3) through (10).
3. Obtain the S-curve ordinates by summing the values of Columns (3) through (10) at each time interval, and obtain the S-curve hydrograph as shown in Column (11).
4. Plot the S-curve obtained at step 3.
5. If the plotted S-curve shows hunting effect near the maximum flow, plot a best fit curve through these points. Figure 14.6 shows the S-curve hydrograph.

A careful view of the S-curve hydrograph computations shows

$$S_t = S_{t-1} + U_t,\ t = 2, 3 \ldots \tag{14.5}$$

where S_t and U_t are the S-curve and UH ordinates at time t and $S_1 = 0$. This equation can be employed to develop a simple algorithm, called the incremental method, to compute an S-curve hydrograph. Here, the S-curve ordinates are obtained by summing the current ordinate of UH with the previous S-curve ordinate. The method is illustrated by the following example.

Time	Original UH (m^3/s)	Shift 1 Steps	Shift 2 Steps	Shift 3 Steps	Shift 4 Steps	Shift 5 Steps	Shift 6 Steps	Shift 7 Steps	Shift 8 Steps	S-Curve
Column 1	Column 2	Column 3	Column 4	Column 5	Column 6	Column 7	Column 8	Column 9	Column 10	Column 11
1	0									0
2	500	0								500
3	1600	500	0							2100
4	2400	1600	500	0						4500
5	1700	2400	1600	500	0					6200
6	1200	1700	2400	1600	500	0				7400
7	600	1200	1700	2400	1600	500	0			8000
8	150	600	1200	1700	2400	1600	500	0		8150
9	0	150	600	1200	1700	2400	1600	500	0	8150
10		0	150	600	1200	1700	2400	1600	500	8150
11			0	150	600	1200	1700	2400	1600	8150
12				0	150	600	1200	1700	2400	8150
13					0	150	600	1200	1700	...
14						0	150	600	1200	...
15							0	150	600	...
16								0	150	...
									0	...

Table 14.9 Solution of Example 14.5

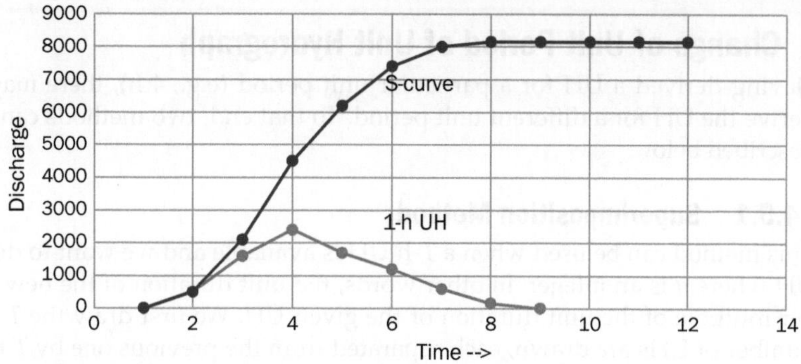

FIGURE 14.6 S-curve hydrograph for Example 14.9.

Time	Original UH (m^3/s)	S-Curve Ordinate at Previous Period (m^3/s)	S-Curve Ordinate (m^3/s) = (Column 2) + (Column 3)
Column 1	**Column 2**	**Column 3**	**Column 4**
1	0	0	0
2	500	0	0 + 500 = 500
3	1600	500	1600 + 500 = 2100
4	2400	2100	2100 + 2400 = 4500
5	1700	4500	4500 + 1700 = 6200
6	1200	6200	6200 + 1200 = 7400
7	600	7400	7400 + 600 = 8000
8	150	8000	8000 + 150 = 8150
9	0	8150	8150 + 0 = 8150
10		8150	8150

TABLE 14.10 Deriving S-Curve Hydrograph by Incremental Method

Example 14.6

Using the UH of Example 14.5, determine the S-curve hydrograph by following the incremental method.

Solution

In Table 14.10, the original UH is tabulated in Columns 1 and 2. For the first time step, the values in Columns (3) and (4) are zero. For time step 2 onward, the ordinate of S-curve equals the S-curve ordinate at previous period (Column 3) + UH ordinate at the current time step.

One can verify that the S-curve hydrograph obtained here is the same as the hydrograph obtained in Example 14.5.

14.5 Change of Unit Period of Unit Hydrograph

Having derived a UH for a particular unit period (e.g., 4 h), there may be a need to derive the UH for a different unit period. To that end, two methods can be applied, as described below.

14.5.1 Superimposition Method

This method can be used when a T-h UH is available and we want to develop an nT-h UH where n is an integer. In other words, the unit duration of the new UH is an integer multiple of the unit duration of the given UH. We first draw the T-h UH. Next, n number of UHs are drawn, each separated from the previous one by T h. If these UHs are combined, we get a direct runoff hydrograph due to n cm of ER (assuming that the input UH is for 1 cm of ER). Hence the ordinates of nT-h UH for 1 cm ER can be obtained by dividing the ordinates of this DRH by n. Figure 14.7 shows the process to obtain a $2T$-h UH from a UH of T-h.

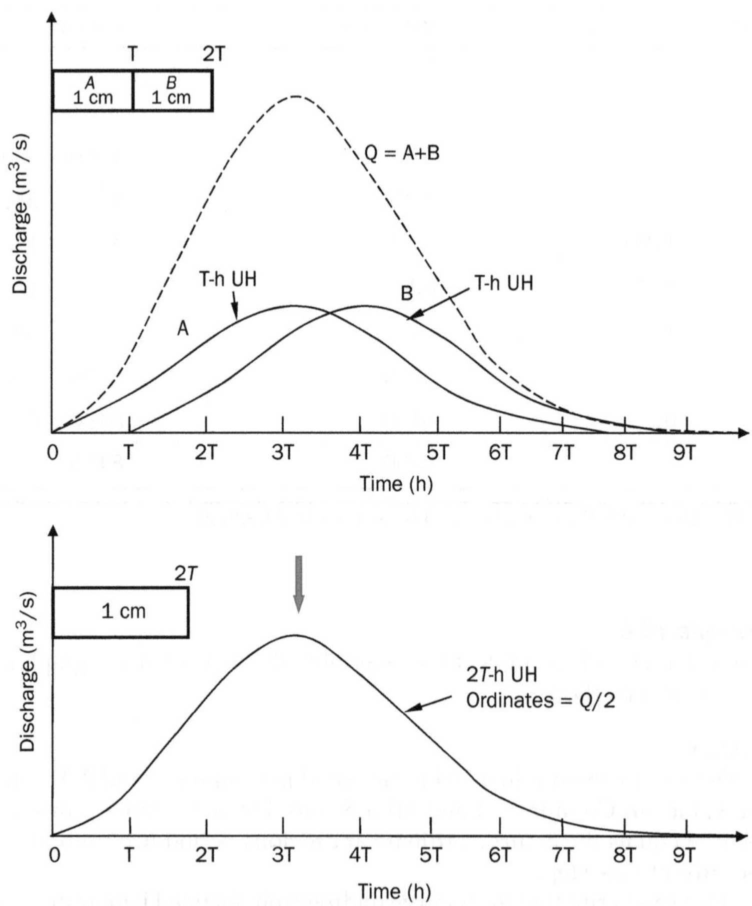

FIGURE 14.7 Change of unit period of unit hydrograph.

Example 14.7

A 2-h UH for a 1 cm ER is given for a catchment in Table 14.11.

It is required to develop a UH for the same catchment for a unit duration of 6-h and 1-cm ER.

Solution

By following the superimposition method, the UH of 6-h duration can be derived from a UH of 2-h duration as follows:

With reference to Table 14.12, the computational steps are:

1. Enter the ordinates of a 2-h UH in Column (2).
2. Lag the 2-h UH by 2-h and enter these values in Column (3).

Time (h)	2-h, 1-cm Unit Hydrograph (m^3/s)
1	0
2	500
3	1100
4	1800
5	1200
6	900
7	500
8	150
9	0

TABLE 14.11 Data of Example 14.7

Time (h)	Original 2-h UH (m^3/s)	2-h UH Ordinates Shifted by 2-h (m^3/s)	2-h UH Ordinates Shifted by 4-h (m^3/s)	Combined Hydrograph	Ordinates of 6-h UH for 1 cm ER = (Column 5)/3
Column 1	Column 2	Column 3	Column 4	Column 5	Column 6
1	0			0	0
2	500	0		500	167
3	1100	500	0	1600	533
4	1800	1100	500	3400	1133
5	1200	1800	1100	4100	1367
6	900	1200	1800	3900	1300
7	500	900	1200	2600	867
8	150	500	900	1550	517
9	0	150	500	650	217
10		0	150	150	50
11			0	0	0

TABLE 14.12 Derivation of a 6-h Unit Hydrograph from a 2-h Unit Hydrograph

3. Lag the 2-h UH by 4-h and enter these values in Column (4).
4. Add the ordinates in Columns (2), (3), and (4). The resulting hydrograph ordinates are shown in Column (5).
5. Divide the ordinates of the hydrograph obtained in step 4 by 3 to obtain a UH of 6-h duration and 1 cm of ER.
6. These ordinates are given in Column (6).

14.5.2 S-Curve Method

It is a more general method than the method of superimposition. After having derived the S-curve of unit (or some other) intensity (e.g., 1 cm/h) from *T*-h UH, the unit hydrograph of *nT*-h (where *n* is any integer) duration can be obtained as follows:

1. Draw the S-curve hydrograph for ER intensity 1/D cm/h,
2. Draw another S-curve hydrograph shifted by *T*-h, as shown in Fig. 14.8.
3. Subtract the S-curve ordinates from the original S-curve.
4. The difference between the two S-curves represents the UH for unit time *T*-h and ER magnitude of $T \times 1/D$ cm.
5. To obtain a UH for an ER of 1-cm unit duration of *T*-h, divide the ordinates obtained in the previous step by *T/D*.

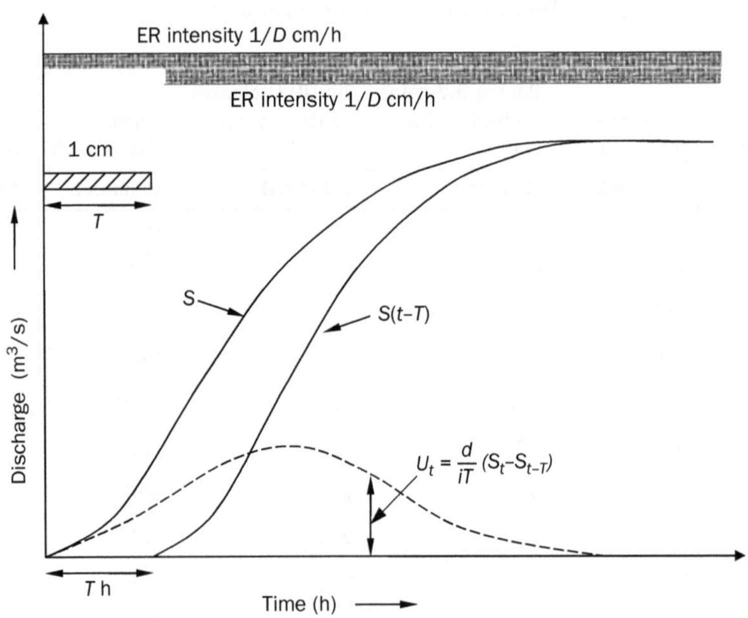

Figure 14.8 Changing UH duration using the S-curve method.

Example 14.8

Table 14.13 gives the ordinates of an S-curve hydrograph, derived from a 1-h UH of 1-cm ER. Develop a 2-h UH by using the S-curve method.

Solution

With reference to Table 14.14, the computational steps follow:

1. Enter the S-curve ordinates in Column (2). This S-curve has been derived from a 1-h unit hydrograph used in Example 14.4 of S-curve derivation.
2. Shift the S-curve by 2 h and enter these values in Column (3).
3. Subtract the two S-curves [Column (2) minus Column (3)] and enter these values in Column (4). This represents the UH of duration 2 h with a unit volume equal to 2 cm.
4. Compute the UH of duration 3 h of unit volume equal to 1 cm by dividing the hydrograph obtained from step 3 by 2. The ordinates of the resulting UH are given in Column (5).

Time (h)	1	2	3	4	5	6	7	8	9
S-curve ordinates (m^3/s)	0	500	2100	4500	6200	7400	8000	8150	8150

TABLE 14.13 Data for Example 14.8

Time	S-Curve (m^3/s)	S-Curve Hydrograph, Shifted by 2 h (m^3/s)	Difference in S-Curve Ordinates (m^3/s) = (Column 2) − (Column 3)	2-h UH (m^3/s) = (Column 4)/2
Column 1	**Column 2**	**Column 3**	**Column 4**	**Column 5**
1	0	–	0	0
2	500	–	500	250
3	2100	0	2100	1050
4	4500	500	4000	2000
5	6200	2100	4100	2050
6	7400	4500	2900	1450
7	8000	6200	1800	900
8	8150	7400	750	375
9	8150	8000	150	75
10	8150	8150	0	0

TABLE 14.14 Change of Unit Period of UH by the S-Curve Method

14.5.3 Relation Between S-Curve and Unit Hydrograph

A simple relationship exists between the S-curve of a catchment and its UH, as shown in Fig. 14.8. The difference in the ordinates of the two S-curves (S and S_{t-T}) represent the ordinates of a hydrograph resulting from the rainfall excess depth of iT cm. Therefore, T h and d mm UH ordinates (U_t) can be derived as

$$U_t = \frac{d}{iT}(S_t - S_{t-T}) \tag{14.6}$$

where S_t is the time (t)-varying S-curve for a continuous rain of i cm/h and S_{t-T} is the same curve shifted by T h to the right. If the unit depth is equal to 1 mm and the intensity of rainfall excess, i, is 1 mm/h, then

$$U_t = \frac{S_t - S_{t-T}}{T} \tag{14.7}$$

14.6 Synthetic Unit Hydrograph

Frequently, measured rainfall and river flow data are not available for the catchment whose UH is to be developed. For such catchments, UHs can be determined from empirical equations which relate salient features of the UH to the watershed characteristics that can be obtained from topographic data. These empirical equations have been derived from analysis of data from a large number of watersheds by using measured rainfall, flow, and topographic data. Since the UHs derived from such empirical equations are not based on actual measured data, these are called synthetic UHs. Note that these equations are applicable to those watersheds whose properties are similar to the watersheds whose data were used to derive the relations. Three of the popular methods are discussed below.

14.6.1 The Snyder Method

Snyder (1938) developed a model to determine the key characteristics of synthetic UH for a catchment from measured topographic data. Snyder derived the underlying formulas from a study of 20 watersheds mainly from the Appalachian Highlands of the eastern United States, whose size varied from about 26 to 25,600 km². The basic parameter of the Snyder method is t_p which is the time to peak (h) or the time from the center of mass of effective rainfall of unit duration to the UH peak and is computed as

$$t_p = C_1 C_t (L \cdot L_c)^{0.3} \tag{14.8}$$

where L is the length of the main stream (km) from the catchment divide to outlet; L_c is the distance (km) from the outlet to a point on the stream which is nearest to the centroid of the watershed; and C_t is a constant which represents catchment storage and swiftness of response. It varies from 1.8 to 2.2; catchments with steeper slopes have lower values.

Steps to derive the UH by the Snyder method are given below.

Step 1: Measure the following information from a topographic map of the catchment:

1. Area of the catchment A (km²).
2. Length of the main channel of the catchment L (km).
3. Length of the main channel from the catchment outlet to a point opposite to the centroid of the catchment L_c (km).

Step 2: Compute the time to peak (t_p) from Eq. (14.7) and t_r, which is the standard duration of the UH adopted by Snyder, expressed as

$$t_r = t_p/5.5 \tag{14.9}$$

Step 3: If a nonstandard duration of the UH is adopted, the modified time to peak (t'_p) is computed by

$$t'_P = t_p + 0.25(t_R - t_r) = 0.25t_R + t_p - (t_p/5.5)/4 = 0.25t_R + \frac{21}{22}t_p \tag{14.10}$$

Step 4: Compute the peak discharge, Q_p (m³/s)

$$Q_P = \frac{2.78C_pA}{t_p} \tag{14.11}$$

Step 5: Compute time base of UH by

$$T_b = 72 + 3t'_p \tag{14.12}$$

Step 5: For ease in drawing and maintaining the correct shape of UH, determine width of the UH at $0.5Q_{PR}$ (W_{50}) and $0.75Q_{PR}$ (W_{75}):

$$W_{50} = 2.14 \Big/ (Q_{PR}/A)^{1.08} \tag{14.13}$$

$$W_{75} = 1.22 \Big/ (Q_{PR}/A)^{1.08} \tag{14.14}$$

Figure 12.9 shows the key elements to draw UH by following the Snyder method.

Example 14.9

For a catchment, the topographic features are: $L = 40$ km, $L_{ca} = 20$ km, and $A = 500$ km². Construct the 2-h UH by the Snyder method, assuming $C_t = 1.25$ and $c_p = 0.58$.

Solution

We compute $t_p = 1.25(40 \times 20)^{0.3} = 9.286$

$$T_r = 9.286/5.5 = 1.689.$$

For 2-h unit duration, $t'_p = \dfrac{2}{4} + \dfrac{21}{22} \times 9.286 = 9.36$

Hence $Q_p = 2.78 \times 0.58 \times 500/9.36 = 86.13$

$$W_{50} = 2.14 \Big/ (86.13/500)^{1.08} = 14.3 \text{ h}$$

$$W_{75} = 1.22 \Big/ (86.13/500)^{1.08} = 8.15 \text{ h}$$

and $t_b = 72 + 3 \times 9.36 = 100$ h

Now a smooth UH can be drawn with the derived parameters.

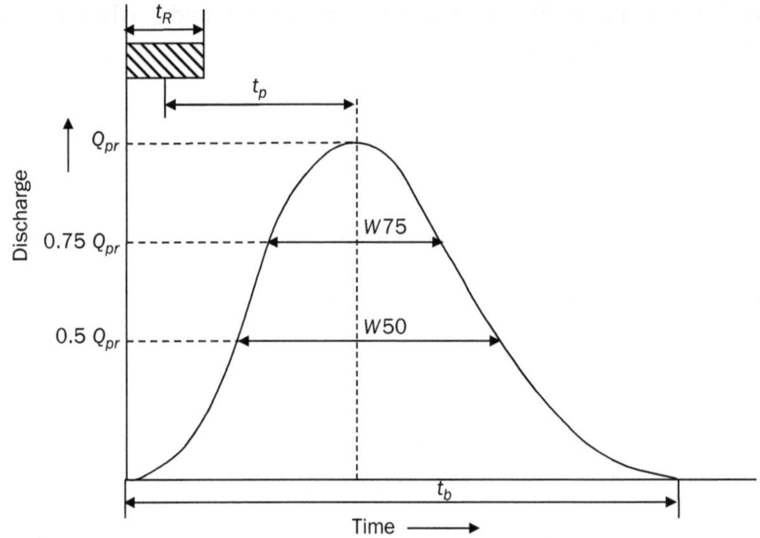

FIGURE 14.9 Key elements of UH by the Snyder method.

14.6.2 Soil Conservation Service Dimensionless Hydrograph

Similar to the Snyder method, the Soil Conservation Service (USDA 1986) of the U.S. Department of Agriculture developed a procedure to construct a synthetic UH based on analysis of data from a large number of catchments in the United States. The SCS method uses dimensionless UH and an average of these was computed. This curvilinear UH can be approximated by a triangular UH having similar characteristics. For comparison, the two dimensionless UHs are shown in Figure 14.10. It is important to note that the triangular unit hydrograph is only used to compute the peak discharge of the curvilinear UH; curvilinear UH is actually used in hydrologic computations. Further, the time base of the triangular UH is only 8/3 of the time to peak, it 5 of the time to peak for the curvilinear UH. The areas under the rising limbs of the two UHs are the same. The steps to construct a UH for a catchment by this method are given here.

Step 1: Compute the basin time lag (t_L) which is the time (h) from the center of mass of rainfall excess to the peak discharge, written as

$$t_L = \frac{L^{0.8}(S+1)^{0.7}}{1900Y^{0.5}} \tag{14.15}$$

where L is the hydraulic length in feet, S is the potential maximum retention, and Y is the slope in percent.

Step 2: Compute the time of concentration

$$t_c = (5/3)\, t_L \tag{14.16}$$

Step 3: Time to peak is the time from the beginning of rainfall to the time of hydrograph peak. Compute the time to peak (t_p) by

$$t_p = 0.67t_c \tag{14.17}$$

Step 4: Time base of the UH is computed by

$$t_b = 2.67t_p \tag{14.18}$$

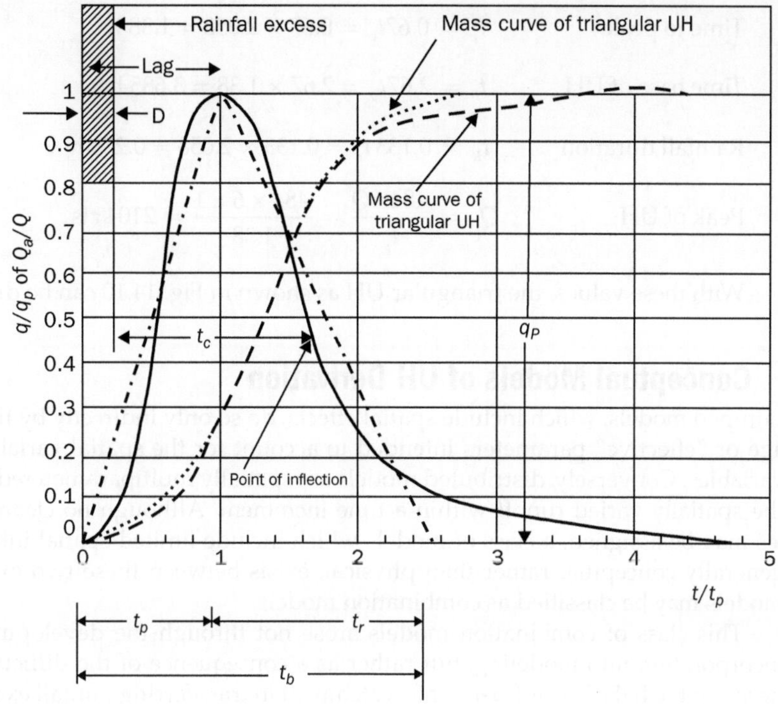

Figure 14.10 SCS triangular hydrograph.

Step 5: Rainfall duration is computed by

$$t_R = 0.133t_c \tag{14.19}$$

Step 6: Compute the peak of UH.

$$q_P = \frac{484AQ}{t_p} \tag{14.20}$$

where Q is the volume, equal to 1 inch for a UH. More details about the method are given in USDA (1986) and USDoT (2002).

Example 14.10

Draw an SCS triangular UH for a catchment whose area is 6.0 m², average slope is 2%, hydraulic length is 2 miles, and maximum potential retention is 1.7 in.

Solution

Given hydraulic length = 2 miles = 10,560 feet.

Basin time lag $t_L = \frac{L^{0.8}(S+1)^{0.7}}{1900Y^{0.5}} = \frac{10{,}560^{0.8}(1.7+1)^{0.7}}{1900 \times 2^{0.5}} = 1.234 \text{ h}$

Time of concentration $t_c = (5/3)\, t_L = (5/3) \times 1.234 = 2.058 \text{ h}$

Time to peak $t_p = 0.67t_c = 0.67 \times 2.058 = 1.38$ h

Time base of UH $t_b = 2.67t_p = 2.67 \times 1.38 = 3.685$ h

Rainfall duration $t_R = 0.133\,t_c = 0.133 \times 2.058 = 0.273$ h

Peak of UH $Q_P = \dfrac{484AQ}{t_p} = \dfrac{484 \times 6 \times 1}{1.38} = 2104$ cfs

With these values, the triangular UH as shown in Fig. 14.10 can be drawn.

14.7 Conceptual Models of UH Derivation

Lumped models, which include spatial effects, do so only indirectly by the use of average or "effective" parameters intended to account for the spatial variability of actual variables. Conversely, distributed models are spatially multiparametered and integrate the spatially varied runoff within a time increment. Although no clean-cut boundaries may be assigned, a class of models which include limited spatial influences, albeit generally conceptual rather than physical, exists between these two extremes. These models may be classified as combination models.

This class of combination models arose not through the development of spatial incorporation into modeling, but rather as a consequence of the difficulty of directly deriving the IUH. The influence of a catchment in transferring rainfall excess into direct runoff was viewed conceptually as equivalent to transferal through a series of reservoirs linked by channels (Huggins and Burney, 1982). In particular, the conceptual elements of linear reservoir and linear channel proved to be valuable blocks in conceptual model development.

Prior to considering specific models and their application, it is helpful to define and elaborate on the linear reservoir and the linear channel.

14.7.1 Linear Reservoir Concept

A linear reservoir is a conceptual reservoir whose storage, S (m^3), is directly proportional to the outflow, Q (m^3/s), or

$$S = kQ \tag{14.21}$$

where the proportionality constant (k) is known as the storage coefficient. By the continuity equation, the difference between inflow I (m^3/s) and outflow is the rate of change in storage with time, written as

$$I - Q = \frac{dS}{dt} = \frac{d(kQ)}{dt} \tag{14.22}$$

Substituting Eq. (14.20) in Eq. (14.21)

$$\frac{dQ}{dt} + \frac{Q}{k} = \frac{I}{k} \tag{14.23}$$

Solution of Eq. (14.22) yields

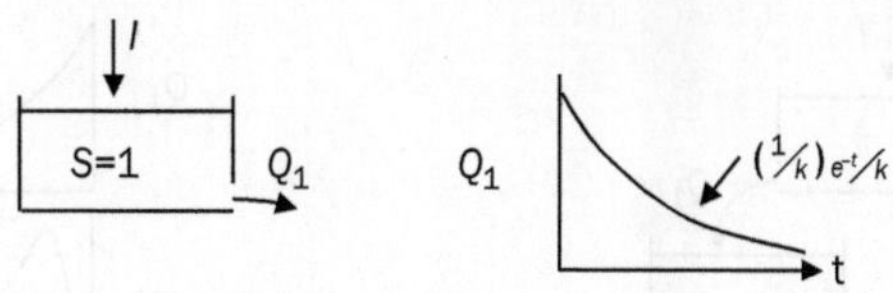

FIGURE 14.11 Schematic diagram of a linear reservoir.

$$Q = I\left\{1 - \exp\left(-\frac{t}{k}\right)\right\} \tag{14.24}$$

In the equilibrium condition when t approaches infinity, the outflow tends to approach the inflow. Suppose the inflow becomes zero at time τ and at that time the outflow is Q_0. Now putting $I = 0$ in Eq. (14.22)

$$\frac{dQ}{d\tau} + \frac{Q}{k} = 0 \tag{14.25}$$

The solution of this equation is

$$Q = Q_0 \exp\left(-\frac{\tau}{k}\right) \tag{14.26}$$

Suppose at time $t = t_0$, the reservoir storage is S_0. Then, for an instantaneous inflow, which fills a reservoir of storage S_0 in an instant, for example, time $t_0 = 0$

$$S_0 = kQ_0 \text{ or } Q_0 = S_0/k \tag{14.27}$$

Combining Eqs. (14.25) and (14.26)

$$Q = \frac{S_0}{K}\exp\left(-\frac{t}{K}\right) \tag{14.28}$$

If the catchment is treated as a linear reservoir that is instantly filled by unit excess precipitation (see Fig. 14.11), the output from such a linear reservoir or IUH for the catchment is

$$u(t) = \frac{1}{k}\exp\left(-\frac{t}{k}\right) \tag{14.29}$$

14.7.2 Nash Model

Nash (1957) proposed that a catchment may be conceptualized to be composed of a series of n linear reservoirs of equal storage coefficient K as shown in Fig. 14.12. The outflow from the first reservoir becomes the input to the second, and so on. The outflow from the first reservoir is an exponential decay curve which is successively modified to yield the familiar hydrograph shape. As seen in Fig. 14.12, the outflow from the first reservoir is $Q_1 = (^1/_k)e^{(-t/k)}$ and the outflow from the nth reservoir or the IUH for the catchment is

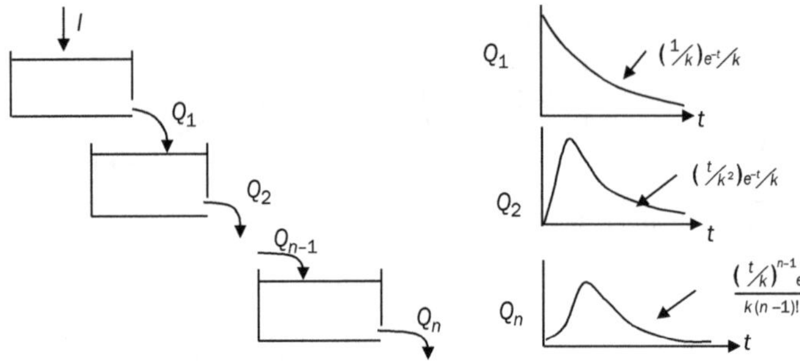

FIGURE 14.12 Cascaded reservoirs of the Nash model. (Adapted from Huggins and Burney, 1982.)

$$Q_n = \frac{1}{K(n-1)!}(t / K)^{n-1}\exp\left(-\frac{t}{K}\right) \tag{14.30}$$

where n may also take on a noninteger value. Catchments with large impervious areas, such as an urban catchment, will have quick response and small values of n and large agricultural catchments have slow response and a large value of n.

The following steps are involved for deriving the UH by the Nash model using the rainfall-runoff data of a particular storm:

1. Obtain mean rainfall values at each computational interval taking the weighted mean of the observed values at different stations.
2. Estimate direct surface runoff separating the baseflow from the discharge hydrograph.
3. Estimate the excess rainfall hyetograph separating the loss from the total rainfall hyetograph.
4. Estimate the first and second moments of effective rainfall hyetograph about the origin.
5. Estimate the first and second moments of direct surface runoff hydrograph about the origin.
6. Solve Eq. (14.30) for the parameters n and K.
7. Estimate the UH of duration T h.

The Nash model has two parameters: n and K. Using the theorem of moments, it can be shown that

$$M_1 = nK \tag{14.31}$$

$$M_2 = nK^2(n+1) \tag{14.32}$$

where M_1 is the first moment of the IUH which equals the difference between first moment of outflow and that of the effective rainfall. Similarly, M_2 is the second moment of IUH which equals the second moment of discharge minus the second moment of effective rainfall. It should also be noted that if T is the watershed lag then $T = nK$, which is the same as the first moment of IUH. For small- to medium-sized watersheds, the value of n is between 3 and 8 and seldom exceeds 20.

Example 14.11
For a watershed subjected to a rainfall-runoff event, it is found that the centroid of the effective rainfall hyetograph is located at 4 h from the origin and the centroid of the surface runoff hydrograph is located at 10 h. Also, the variance of effective rainfall hyetograph is located at 4 h^2 and that of surface runoff hydrograph at 56 h^2. Compute the values of n and K as well as T.

Solution
By definition

$$M_1 = 10\text{ h} - 4\text{ h} = 6\text{ h}$$

$$M_2 = 56\text{ h}^2 - 4\text{ h}^2 = 52\text{ h}^2$$

From Eq. (14.32) $$M_2 = nK^2(n+1) = (nK)^2 + M_1 \times K$$

Or $$52\text{ h}^2 = (6\text{ h})^2 + 6\text{ h} \times K\text{ h}$$

This gives $K = 8/3$ h

$$\text{Then, } n = 6/(8/3) = 2.25$$

$$T = nK = 6\text{ h}$$

In this case, n can be taken as 3.

14.7.3 Dooge Model

Dooge (1959) developed a general UH theory of which most conceptual IUH models, including the Nash model, are special cases. In this theory, a watershed is represented by three elements: time-area concentration (TAC) diagram, linear reservoir, and linear channel. The watershed is drained by a network of linear reservoirs and linear channels, as shown in Fig. 14.13. Consider an isochrone in the watershed. It can be assumed that all points on the isochrone are drained by a single chain of reservoirs and channels. The inflow at any point is proportional to the length of the isochrones ($dA/d\tau$) cutting the main stream at that point. In order to account for the variability of rainfall, rainfall intensity can be averaged along each isochrone so the intensity varies only with distance from the outlet. The general equation of the IUH can be expressed as

$$h(t) = \frac{1}{T}\int_0^{t\leq T} \frac{\delta(t-\tau)}{\prod_{j=1}^{n(\tau)}(1+k_jD)} I(A)\frac{dA}{d\tau}d\tau \tag{14.33}$$

where $dA/d\tau$ is the length of the isochrones, that is the ordinate of the time area diagram, T is the maximum translation time in the watershed, D is the operator d/dt, A is the drainage area, and k is the lag time of the jth reservoir. It may be noted that $I(A)(dA/d\tau)$

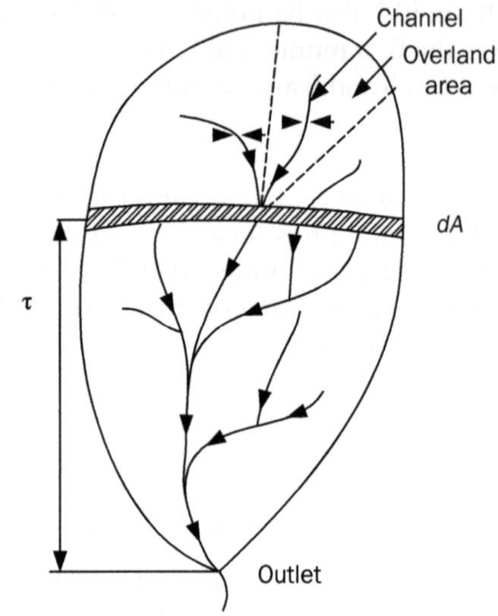

FIGURE 14.13 Watershed drained by a network of linear reservoirs and channels.

can be denoted by the ordinate of the dimensionless time area concentration (TAC) curve, $w(\tau/T)$.

14.7.4 Geomorphological Instantaneous Unit Hydrograph

Rodriguez-Iturbe and Valdes (1980) derived the IUH by coupling empirical laws of geomorphology with the IUH theory. Gupta et al. (1980) simplified this coupling and the IUH so derived is called geomorphological IUH or GIUH. When it rains in a watershed, raindrops fall and start their journey on overland areas and travel through the channel network to reach the outlet . Let a watershed of order W contain channels $c_1, c_2, \ldots, c_W$ and overland regions $r_1, r_2, \ldots, r_W$, where c_j is the channels of order j and r_j the region draining into a channel of order j. The raindrops, depending on their landing, start their travel in the overland region and then go through a string of channels. Their travel follows the following rules dictated by the watershed geomorphology:

Rule 1: The possible path from the overland region r_j is to the channel c_j, $1 \leq j \leq W$.
Rule 2: The possible path from channel c_j is to the channel c_{j+1}, $1 \leq j \leq W$.
Rule 3: The possible path from c_W is the watershed outlet.

These rules define a collection of paths that water follows and the number (N) of paths (R) equals: $N = 2^{W-1}$, in which each path has an overland region and a series of channels as shown below for a third-order watershed shown in Fig. 14.14:

$$\text{Path } R_1\text{: } r_1 \rightarrow c_1 \rightarrow c_2 \rightarrow c_3$$

$$\text{Path } R_2\text{: } r_1 \rightarrow c_1 \rightarrow \quad \rightarrow c_3$$

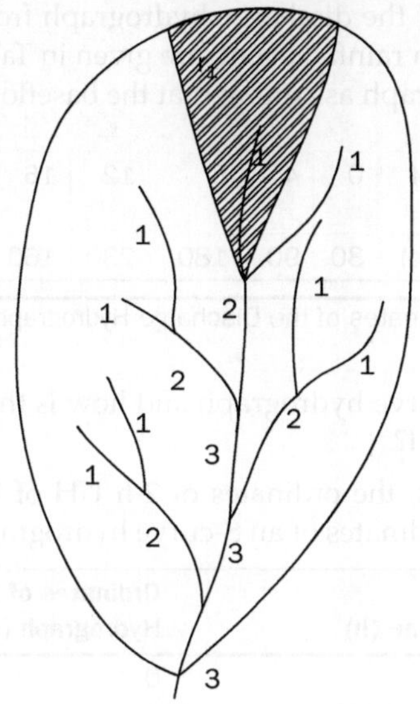

FIGURE 14.14 A third-order watershed.

$$\text{Path } R_3\text{: } r_2 \rightarrow \quad \rightarrow c_2 \rightarrow c_3$$

$$\text{Path } R_4\text{: } r_3 \rightarrow \rightarrow \rightarrow c_3$$

Thus the watershed can be represented by a parallel arrangement of cascades. An overland region can be represented by a linear reservoir and a channel by a linear channel and then the IUH can be determined. Singh (1988) derived the IUH of any cascade with n elements in such an arrangement:

$$h(t) = \sum_{j=1}^{n} \frac{k_j^{\,n-2} \exp(-t/k_j)}{\prod_{i=1, i \neq j}^{n} (k_j - k_i)}, n \geq 2 \tag{14.34}$$

14.8 Summary

This chapter provides a detailed treatment on UH methods, which are widely applied in hydrology. Techniques to derive UHs by conventional methods as well as synthetic UHs have been discussed along with the conceptual models of UHs.

14.9 Questions

14.1 Define a UH and state the assumptions of UH theory.

14.2 List the factors that affect the shape of a UH and briefly describe their impact.

14.3 Write the convolution equation and explain its terms.

14.4 The ordinates of the discharge hydrograph from a catchment, whose area is 500 km^2, due to a 6-h rainfall event, are given in Table 14.15. Derive the ordinates of a 6-h unit hydrograph assuming that the baseflow is zero.

Time from start of rain (h)	0	4	8	12	16	20	24	28	32	36
Discharge (Cumec)	30	90	180	230	160	97	65	43	32	23

TABLE 14.15 Ordinates of the Discharge Hydrograph from a Catchment of 500 km^2

14.5 What is the S-curve hydrograph and how is this concept used to change the unit duration of a UH?

14.6 For a catchment, the ordinates of 2-h UH of 1-cm ER are given in Table 14.16. Compute the ordinates of an S-curve hydrograph.

Time (h)	Ordinates of 2-h, 10-mm Unit Hydrograph (m^3/s)
2	0
4	200
6	1400
8	2800
10	2100
12	1600
14	500
16	175
18	0

TABLE 14.16 Ordinates of 2-h UH of 1-cm ER for a Catchment

14.7 What is a synthetic UH? Describe any one method to derive a UH by synthetic method?

14.8 The topographic features of a catchment are: $L = 50$ km, $L_{ca} = 30$ km, and area = 700 km^2. Construct a 1-h UH by using the Snyder method. Assume $C_t = 1.2$ and $c_p = 0.6$.

14.9 Draw an SCS triangular UH for a catchment whose area is 9 m^2, average slope is 2.5%, hydraulic length is 4 miles, and maximum potential retention is 2 in.

14.10 For a watershed, the centroid of the effective rainfall hyetograph for an event was located at 3 h from the origin and the centroid of the surface runoff hydrograph was at 9 h from the origin. The variance of effective rainfall hyetograph was located at 5 h^2 and that of surface runoff hydrograph at 48 h^2. Compute the values of *n*, *K*.

14.10 References

Dooge, J. C. I. A general theory of the unit hydrograph. *Journal of Geophysical Research*, 64(2): 241–256, 1959.

Dooge, J. C. I. Linear Theory of Hydrologic Systems. Technical Bulletin 1468. U.S. Department of Agriculture, Washington, DC, 1973.

Gupta, V. K., E. Waymire, and C. T. Wang, C.T. Representation of an instantaneous unit hydrograph from geomorphology. *Water Resources Research,* 16(5): 855–862, 1980.

Huggins, L. F. and J. R. Burney. Surface Runoff, Surface runoff, storage and routing. In *Hydrologic Modeling of Small Watersheds*, edited by C. T. Haan, H. P. Johnson, and D. L. Brakensiek. *American Society of Agricultural Engineers*, 1982.

Nash, J. E. The form of the instantaneous unit hydrograph. *Hydrological Sciences Bulletin,* 3: 114–121, 1957.

Rodriguez-Iturbe, I., and J. B. Valdes. The geomorphologic structure of hydrologic response. *Water Resources Research,* 15(6): 1409–1420, 1980.

Sherman, L. K. Streamflow from rainfall by the unit graph method. *Engineering News-Record,* 108: 501–505, 1932.

Singh, V. P. *Hydrologic Systems, Vol. 1: Rainfall-Runoff Modeling*. Englewood Cliffs, NJ: Prentice Hall, 1988.

Singh, V. P. An IUH equation based on entropy theory. *Transactions of the ASABE,* 54(1):1–10, 2011.

Snyder, F. F. Synthetic unit graphs. *Transactions of American Geophysical Union,* 19: 447–454, 1938.

USDA. Urban Hydrology for Small Watersheds. Technical Release (TR) 55, United States Department of Agriculture, Natural Resources Conservation Service, Washington, DC, 1986.

USDoT. Highway Hydrology. Hydraulic Design Series No. 2, Second Edition, U.S. Department of Transportation. Publication No. FHWA-NHI-02-001, National Highway Institute, Arlington, Virginia, 2002.

CHAPTER 15

Flow Routing

Flood movement in a river can be characterized as wave propagation. Flood routing is essentially the determination of the wave propagation, which entails the determination of flow conditions in a river given initial conditions and river morphology (and upstream and/or downstream conditions for a hydraulic method). The wave movement is featured by translation and attenuation which is caused by hydraulic diffusion. This is a nonlinear process that spreads the wave. In any case, the flow volume is always conserved. This chapter deals with basic aspects of flood routing.

15.1 Introduction

Flow routing may be defined as the procedure in which the time and magnitude of water wave at different point(s) on a river are determined from the known upstream conditions.

It is important to introduce two frequently used terms: translation and attenuation of a flood wave. The translation is the difference in time between the occurrence of peak discharges of inflow and outflow hydrographs, and the attenuation is the difference between the inflow and outflow peak discharges. Figure 15.1 explains these terms. The speed of travel (flood wave velocity) and the rate of attenuation depend on the properties of the river in which the flood waves are moving. Having known the peak discharge of the inflow and outflow hydrograph, the attenuation is computed as

$$\text{Attenuation} = \text{Peak of inflow hydrograph} - \text{Peak of outflow hydrograph} \quad (15.1)$$

and translation as

$$\text{Translation} = \text{Time-to-peak of inflow} - \text{Time-to-peak of outflow} \quad (15.2)$$

There are two ways to solve a flood routing problem. First, floods can be routed using diffusion wave approximation that accounts for convection and attenuation or diffusion. Second, floods are routed using the finite difference form of storage or kinematic wave method, which accounts for translation. In order to correct for diffusion, the finite difference form is expanded using the Taylor series to second order and a diffusion wave equation is thus obtained numerically. This diffusion equation is then equated to the actual diffusion equation and parameters of the numerical equation are matched with those of the physical diffusion equation. In this manner, numerical parameters are expressed in terms of hydraulic and geometric parameters.

In the most general classification, flow routing problems can be classified into two groups: (1) stream channel routing and (2) reservoir (or lake) routing. A major difference between these two is that in the reservoir routing, the water surface slope is almost

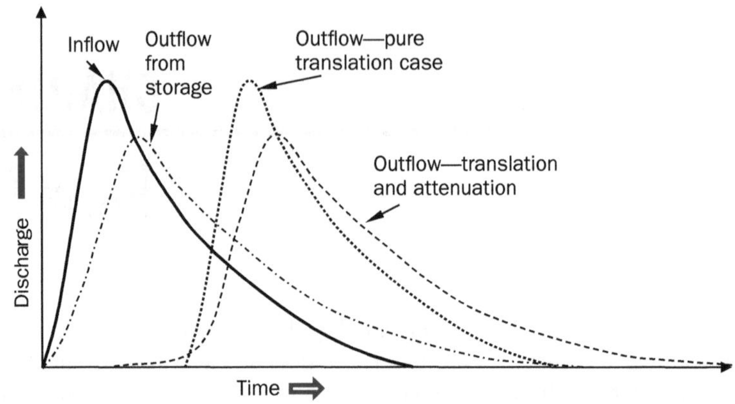

FIGURE 15.1 Inflow hydrograph and outflow hydrographs showing storage effect, translation, and attenuation.

zero, while in stream channel routing, it is nonzero. Several other criteria are used to classify flood routing methods: (1) the equations used to formulate the problem; (2) the overall approach to data collection; and (3) the approach to obtain a solution.

Two broad categories of channel routing methods are: (1) hydrologic methods and (2) hydraulic methods. Hydraulic routing uses two first-order partial differential equations. These equations are the equations of gradually varied unsteady open channel flow and are known as the Saint-Venant equations.

We will first discuss a hydrologic channel routing method: the Muskingum method. This will be followed by a discussion on Saint-Venant equations and their simplified form, then the Muskingum-Cunge method, and finally reservoir routing.

15.2 Muskingum Method of Channel Routing

Nearly 80 years ago, McCarthy (1938) proposed a semiempirical method of flood routing and applied it to the Muskingum River watershed. This method is now popularly called the Muskingum method. Often, it is clubbed as a hydrologic method. The method is still quite popular because it is simple and efficient and its parameters can be estimated easily and appropriately. More details about the method can be seen in Singh (2016).

The Muskingum method is based on the continuity equation which can be expressed as

$$I - O = \frac{dS}{dt} \tag{15.3}$$

Here, I = the inflow, O = the outflow (m^3/s or cumec), S = the storage volume (m^3), K = a time constant or storage coefficient, and X a dimensionless weighting factor. The method involves the concept of wedge and prism storages as shown in Fig. 15.2. The storage volume in a river reach can be correctly related to outflow with a simple linear function only when inflow and outflow are equal. During the advance of a flood wave, however, inflow exceeds outflow, thus producing a wedge of storage, called wedge storage, as shown in Fig. 15.2. The wedge can be related to the difference between the instantaneous values of inflow and outflow.

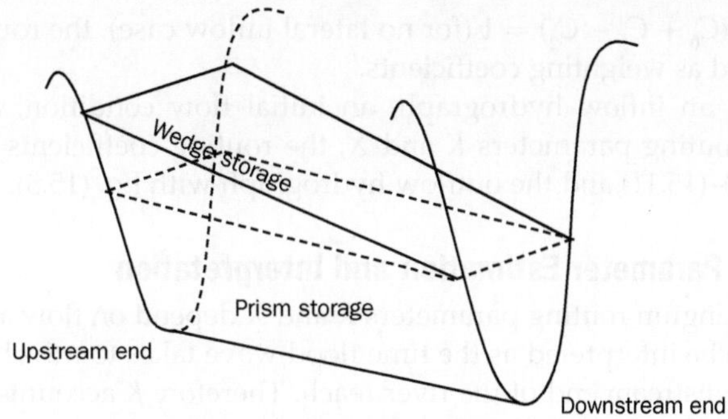

FIGURE 15.2 Prism and wedge storages in a river channel.

The wedge storage is represented by $[KX(I - O)]$ and the prism storage is represented by KO. The total storage is, therefore

$$S = K[XI + (1 - X)O] \tag{15.4}$$

During recession, outflow exceeds inflow, resulting in negative wedge storage.

To derive the Muskingum routing equation, discretization of Eq. (15.3) on the x-t plane yields:

$$\frac{I_1 + I_2}{2} - \frac{O_1 + O_2}{2} = \frac{S_2 - S_1}{\Delta t} \tag{15.5}$$

where subscript 1 denotes the beginning of time interval Δt, and 2 denotes the end of the interval. Equation (15.4) can be written for the beginning and end of the time step as

$$S_1 = K[XI_1 + (1 - X)O_1] \tag{15.6}$$

$$S_2 = K[XI_2 + (1 - X)O_2] \tag{15.7}$$

where O_1 = the outflow discharge at the beginning of time step, O_2 = the outflow discharge at the end of the time step, I_1 = the inflow discharge at the beginning of the time step, and I_2 = the inflow discharge at the end of the time step. Substituting values of S_1 and S_2 in Eq. (15.5) and solving, we get

$$O_2 = C_0 I_2 + C_1 I_1 + C_2 O_1 \tag{15.8}$$

Routing coefficients C_1, C_2 and C_3 are computed as

$$C_0 = \frac{(\Delta t/K) - 2X}{2(1 - X) + (\Delta t/K)} \tag{15.9}$$

$$C_1 = \frac{(\Delta t/K) + 2X}{2(1 - X) + (\Delta t/K)} \tag{15.10}$$

$$C_2 = \frac{2(1 - X) - (\Delta t/K)}{2(1 - X) + (\Delta t/K)} \tag{15.11}$$

Since $(C_0 + C_1 + C_2) = 1$ (for no lateral inflow case), the routing coefficients can be interpreted as weighting coefficients.

Given an inflow hydrograph, an initial flow condition, a chosen time interval Δt, and routing parameters K and X, the routing coefficients can be calculated with Eqs. (15.9)–(15.11) and the outflow hydrograph with Eq. (15.8).

15.2.1 Parameter Estimation and Interpretation

The Muskingum routing parameters K and X depend on flow and channel characteristics; K can be interpreted as the time flood wave takes to travel from the upstream end to the downstream end of the river reach. Therefore K accounts for translation in routing and is a function of flow and channel characteristics.

Referring to Fig. 15.2, the river channel provides storage of water as the flood wave moves from upstream to downstream. Due to storage, the peak flow is reduced and the hydrograph spreads in time. Parameter X accounts for the storage part of the routing. For a given flood hydrograph there is a unique value of X for which the storage in the calculated outflow hydrograph is nearly the same as that of the measured outflow hydrograph.

The Muskingum parameter X can be interpreted as a weighting factor which lies in the range 0.0–0.5. If the value of X is greater than 0.5, it produces hydrograph amplification (i.e., negative diffusion), which is not observed in real-life. When $K = \Delta t$ and $X = 0.0$, Muskingum routing reduces to reservoir routing (discussed later).

Values of Muskingum parameters K and X are computed by using concurrent data of inflow and outflow for the river reach. These parameters K and X can be estimated by (1) graphical method, (2) method of moments, (3) least square, and (4) optimization. The graphical method is popularly used and is described here.

Graphical Method of Parameter Estimation A widely used method for determination of parameters K and X is the graphical method in which the instantaneous S values are plotted on the X axis against the corresponding weighted discharge values $[XI + (1 - X)O]$ on the Y axis. The value of X which gives the narrowest loop between S and $[XI + (1 - X)O]$ is accepted. Note that the values of K and X determined here hold good only for the river reach whose data have been used and strictly for the flood event whose data are used. Use of these parameter values to other river reaches or to other flood events (particularly of very different magnitudes) in the same river reach may be attempted with caution. If data for many flood events are available, each of different magnitude and duration, such data can be used to determine "average" parameters for the river reach.

Example 15.1

Table 15.1 gives inflow and outflow for a river reach for 12 h. Find the Muskingum parameters K and X for the reach.

Solution

The computations are shown in Table 15.2. Here, storage (Col. 4) is calculated by the use of Eq. (15.5). Three trial values of X are chosen and for each of these, the quantity $[XI + (1 - X)O]$ called "weighted flows" is computed and plotted against storage. In each case, a loop curve is obtained.

Time (h)	0	1	2	3	4	5	6	7	8	9	10	11
Inflow (m^3/s)	5	20	50	50	32	22	15	10	7	5	5	5
Outflow (m^3/s)	5	6	12	29	38	35	29	23	17	13	9	7

TABLE 15.1 Inflow and Outflow for a River Reach

Time	Inflow (m^3/s)	Outflow (m^3/s)	Storage ($(m^3/s)/h$)	Weighted Flows (m^3/s)		
				$X = 0.1$	$X = 0.2$	$X = 0.3$
0	5	5	0	–	–	–
1	20	6	14	7.4	8.8	10.2
2	50	12	66	15.8	19.6	23.4
3	50	29	125	31.1	33.2	35.3
4	32	38	140	37.4	36.8	36.2
5	22	35	121	33.7	32.4	31.1
6	15	29	94	27.6	26.2	24.8
7	10	23	67	21.7	20.4	19.1
8	7	17	44	16	15	14
9	5	13	26	12.2	11.4	10.6
10	5	9	14	8.6	8.2	7.8
11	5	7	8	6.8	6.6	6.4

TABLE 15.2 Computations to Determine Muskingum Parameter *X*

We adopt that particular value of *X* for which the loop is the thinnest or close to a straight line. Inspection of Fig. 15.3 shows that for this problem, $X = 0.2$ gives the thinnest loop and so this value is adopted. The slope of the narrowest loop gives the value of *K*. Equation (15.4) can also be written as

$$K = \frac{S}{XI + (1 - X)O}$$

For the line corresponding to $X = 0.2$, slope is $= (110 - 20)/(30 - 10) = 4.5$. Hence *K* for this problem is 4.5 h.

This procedure assumes that the water surface in the river reach is a uniform unbroken profile between upstream and downstream ends of the reach. It is also assumed that *K* and *X* are constant throughout the range of flow in the stage of investigation. If significant departures from these restrictions are present, it may be necessary to work with shorter reaches of the river, or to employ a more sophisticated approach. A rule of thumb is that the Muskingum method will fail if the lag of the channel reach defined by *K* is greater than half the duration of inflow.

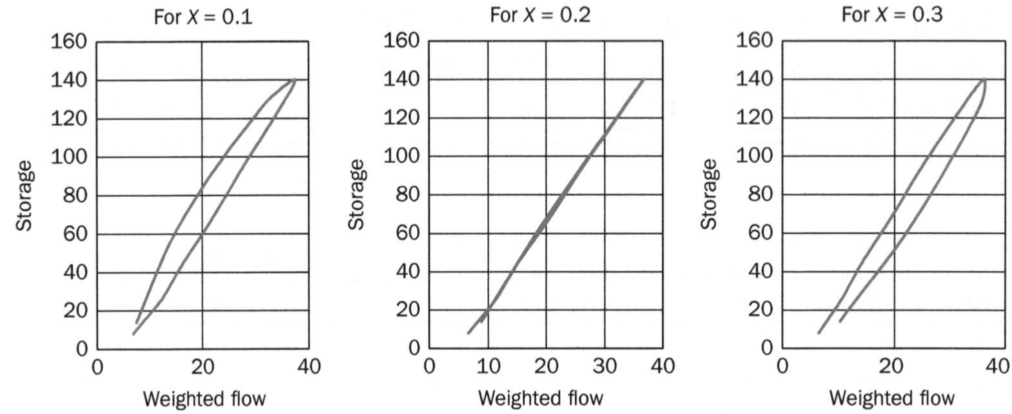

Figure 15.3 Three loop curves obtained in the graphical estimation of parameters *K* and *X*.

Time (h)	0	6	12	18	24	30	36	42	48	54	60	66
Inflow (m^3/s)	5	20	50	75	32	22	15	10	7	5	5	5

Table 15.3 Data for Example 15.2

15.2.2 Routing Procedure

After estimating parameters *K* and *X*, compute routing coefficients C_1, C_2, and C_3. Then, the observed direct inflow hydrograph can be routed through the channel by using Eq. (15.8).

Example 15.2

The inflow hydrograph to a river reach is given in Table 15.3. Route this hydrograph by using the Muskingum method. Adopt $K = 8$ h, $X = 0.15$, and $\Delta t = 6$ h.

Solution

By using the given *K* and *X*, we complete routing coefficients

$$C_0 = \frac{(6/8) - 2 \times 0.15}{2(1 - 0.15) + (6/8)} = \frac{0.45}{2.45} = 0.184$$

$$C_1 = \frac{(6/8) + 2 \times 0.15}{2(1 - 0.15) + (6/8)} = \frac{1.05}{2.45} = 0.428$$

$$C_2 = \frac{2(1 - 0.15) - (6/8)}{2(1 - 0.15) + (6/8)} = \frac{0.95}{2.45} = 0.388$$

It is verified that $C_0 + C_1 + C_2 = 0.184 + 0.428 + 0.388 = 1.0$. Routing computations are shown in Table 15.4. As example, for the second time period

$$\text{Outflow} = 0.184 \times 20 + 0.428 \times 5 + 0.388 \times 5 = 3.68 + 2.14 + 1.94 = 7.76 \text{ m}^3/\text{s}.$$

Time (h)	Inflow (m^3/s)	Partial Flows (m^3/s)			Outflow (m^3/s)
		C_0I_2	C_1I_1	C_2O_1	
0	5	–	–	–	5
6	20	3.68	2.14	1.94	7.76
12	50	9.20	8.56	3.01	20.77
18	75	13.80	21.40	8.06	43.26
24	32	5.89	32.10	16.78	54.77
30	22	4.05	13.70	21.25	39.00
36	15	2.76	9.42	15.13	27.31
42	10	1.84	6.42	10.59	18.85
48	7	1.29	4.28	7.32	12.88
54	5	0.92	3.00	5.00	8.91
60	4	0.74	2.14	3.46	6.33
66	2	0.37	1.71	2.46	4.54

TABLE 15.4 Solution of Example 15.2

15.2.3 Accounting for Lateral Inflow

One of the factors that affect the accuracy of flood routing is the correct accounting of the lateral inflow to the channel. If lateral inflow to the channel is significant but is ignored, then the computed flow at the downstream point would be smaller than the observed flow. The errors propagate as floods are routed down the stream.

Many approaches can be adopted to estimate lateral inflows in the absence of measured data. The volume of lateral inflow for a given storm may be distributed over time equal to the storm duration in a pattern similar to the time distribution of storm rainfall over the intervening catchment of the reach. Lateral inflow can also be estimated as the volume difference between the outflow and inflow hydrographs of the reach and distributed in the pattern of an outflow hydrograph.

15.3 The Saint-Venant Equations

In most situations, river flow has a higher variation in the longitudinal flow direction than in the lateral or vertical direction. Thus one may assume that river flow is typically one-dimensional except in situations when the flood flow covers vast flood plains.

In 1871, Saint Venant proposed the governing equations for conservation of mass and momentum in a river section. These are known as the Saint-Venant equations

$$\frac{\partial Q}{\partial x} + \frac{\partial A}{\partial t} = q_l \quad \text{(Conservation of mass)} \tag{15.12}$$

and

$$\frac{\partial Q}{\partial t} + \frac{\partial}{\partial x}\left(\frac{Q^2}{A}\right) + gA\frac{\partial y}{\partial x} + gA(S_f - S_0) + L = 0 \quad \text{(Conservation of momentum)} \tag{15.13}$$

where Q is the discharge, L is the momentum effect of lateral flow, A is the area of cross section, v is the velocity, S_f is the friction slope, S_0 is the bed slope, y is the depth of flow, x denotes the longitudinal distance along the river, and t is time. Equations (15.12) and (15.13) together are also known as the dynamic wave equations or the full Saint-Venant equations. In Eq. (15.14), the first two terms represent the rate of change of momentum of water, the third term represents the change in the pressure force, the fourth term is due to resistance effects, and the last term is due to lateral inflow. These two equations form a set of nonlinear first-order hyperbolic partial differential equations, describing unsteady flow in open channels. Detailed descriptions of the theory behind these hydrodynamic equations and their derivation have been given by many authors.

The fundamental assumptions underlying the derivation of Saint-Venant equations are (1) flow is one-dimensional; (2) the water surface varies gradually, meaning thereby that the pressure distribution along a vertical is hydrostatic and vertical acceleration is small; (3) the friction losses in unsteady flow are not significantly different from those in steady flow; (4) the velocity distribution along the wetted area does not substantially affect the wave propagation, or the velocity distribution is uniform; (5) the bottom slope of the channel is small so that $\sin\theta \approx 0$, $\tan\theta \approx \theta$, and $\cos\theta \approx 1$; and (6) the channel is straight and prismatic. Attempts to experimentally verify Saint-Venant equations show that the results obtained by numerically solving these equations are sufficiently close to experimental results.

15.3.1 Solution of Saint-Venant Equations

The Saint-Venant equations are nonlinear partial differential equations and their closed form analytical solution is not available. Therefore numerical methods are employed to find a solution.

The momentum equation describes the dynamic behavior of flood wave propagation accounting for the body, surface, and inertial forces of flow at a given time. Solution of the full Saint-Venant equations would give the most accurate results. In many cases of practical applications, all the terms in the momentum equation are not equally important. Furthermore, the local and convective acceleration terms are almost equal but of opposite sign and hence cancel each other. Therefore this equation can be approximated by ignoring some of the terms. This results in simplified equations which are easier to solve and the solution is numerically more stable (of course, there is a trade-off in using simplified version and accuracy). Simplifications of the momentum equation are classified as kinematic wave and diffusive wave models. The attenuation rate of waves is a function of the magnitude of the various forces involved in the motion. Kinematic waves do not attenuate, diffusion waves attenuate at a small to moderate rate, and dynamic waves are subject to very strong attenuation. The equation for these three waves can be depicted as

$$\frac{\partial Q}{\partial t} + \frac{\partial}{\partial x}\left(\frac{Q^2}{A}\right) + gA\frac{\partial y}{\partial x} + gA(S_f - S_0) + Q_s Av = 0 \tag{15.14}$$

Kinematic wave

Diffusive wave

Dynamic wave

The available numerical methods to solve Saint-Venant equations can be classified as (1) method of characteristics; (2) explicit finite difference methods; (3) implicit finite difference methods; and (4) finite element methods. The method of characteristics was developed in the twentieth century and was in use for quite some time. But with the advent of fast computers, numerical methods are used these days. Among these, the implicit finite difference method is most commonly used in practice. We begin with the simplest case, that is kinematic waves.

15.3.2 Implicit Methods of Solving Saint-Venant Equations

Real-life floods have to be routed for a relatively long duration and over large distances. Implicit finite difference methods work best in such cases. In the implicit methods, the x derivatives of the unknown terms of Saint-Venant equations are replaced in terms of finite differences. Since the unknowns implicitly appear in the equations, the name of solution method is known as the implicit method. Here, the channel reach is divided into N computational nodes to yield N–1 rectangular grids between the upstream and downstream boundaries. At each node and at each time, there are two unknowns (flow depth h and discharge Q) and two equations can be written for each grid. Two more equations are obtained from boundary conditions at either end. Thus there are $2N$ unknowns and $2N$ equations.

One of the earliest developed implicit schemes for flood routing was suggested by Preissmann (1961). He adopted a four-point grid in the (x, t) plane and also introduced a weighting factor. Many other modelers have used a four-point or box scheme. This scheme has many advantages over other schemes and has also been used in the DAMBRK model, which is a popular model to simulate the movement of flood waves in the event of failure of a dam. In the four-points implicit (or Preissmann) scheme, the Saint-Venant equations are replaced by two algebraic finite difference equations. The Preissmann type schemes compute both unknown variables at the same computational grid points. These link together flow variables at only two adjacent sections; thus the space intervals may be variable, while the accuracy of approximation is unaffected.

The numerical solution is obtained over a discrete rectangular net of points on the (x, t) plane, as shown in Fig. 15.4. The lines drawn parallel to the x axis represent time and those parallel to the y axis represent channel locations. The spacing of these lines is Δt and Δx, respectively, which need not be constant. The node points (P, R, S, and T) are the intersection points of these lines. The scheme is termed as a four-point scheme, because the continuity and momentum equations are applied to four node points. Let α be any variable, such as Q, y, A, or u. The partial derivatives of α with respect to t and x can be expressed as

$$\frac{\partial \alpha(M)}{\partial t} = \frac{[\alpha(P) + \alpha(T) - \alpha(R) - \alpha(S)]}{2\Delta t} \tag{15.15}$$

$$\frac{\partial \alpha(M)}{\partial x} = \frac{(1-\theta)[\alpha(R) - \alpha(S)]}{\Delta x} + \frac{\theta[\alpha(P) - \alpha(T)]}{\Delta x} \tag{15.16}$$

where θ is a weighting factor $= \Delta t'/\Delta t$.

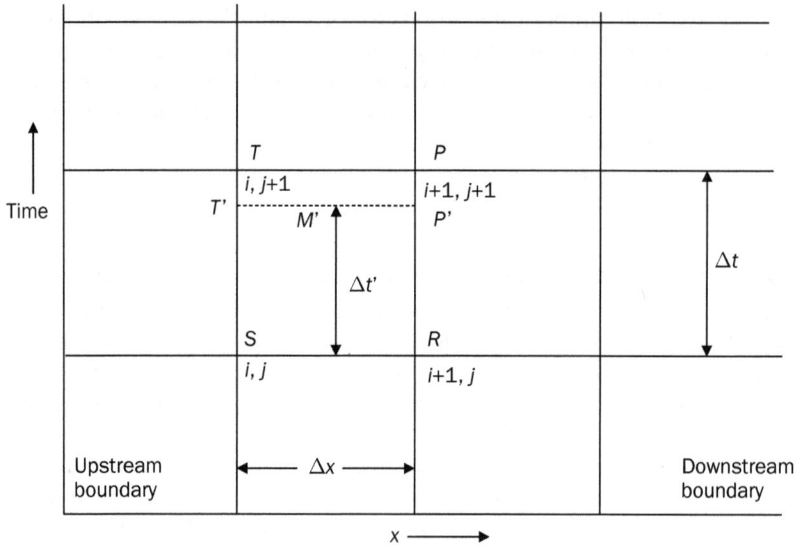

Figure 15.4 Network in (*x*, *t*) plane used in the diffusive scheme.

Factor θ has been introduced to impart artificial damping in the system. It varies from 0 to 1 but $\theta > 0.5$ is necessary to provide unconditional linear stability with respect to time step size. When $\theta = 0$, the scheme reduces to the explicit scheme. The value 0.5 produces the box scheme or centered finite difference scheme. The weighting factor allows flexibility in placing derivative terms. The accuracy of scheme generally decreases as θ approaches unity; the effect becomes more pronounced with the increase in the time step size. However, the resulting equations with $\theta = 1$ are simpler in structure. While the box scheme is stable for slowly varying flow, it may produce numerical oscillations under certain conditions. A value of $\theta = 0.55$ minimizes the loss of accuracy and avoids weak instability. To generalize the approach, the equations are given here with $0.5 \le \theta \le 1$.

Using Eqs. (15.16) and (15.17), the equation of continuity or Eq. (15.8) can be expressed in the finite difference form as

$$\begin{aligned}&\left[(-\theta)\left(Q_{i+1}^{j} - Q_{i}^{j}\right) + \theta\left(Q_{i+1}^{j+1} - Q_{i}^{j+1}\right)\right] / \left(x_{i+1} - x_{i}\right)\\&+ 0.5\left[\left(A_{i+1}^{j+1} + A_{i}^{j+1}\right) - \left(A_{i+1}^{j} + A_{i}^{j}\right)\right] / \left(t^{j+1} - t^{j}\right) - Q_{s}(x,t) = 0\end{aligned} \tag{15.17}$$

Here, subscripts refer to the longitudinal location and superscripts refer to the time step. Simplifying Eq. (15.17)

$$\begin{aligned}&\left[(1-\theta)\left(Q_{i+1}^{j} - Q_{i}^{j}\right) + \theta\left(Q_{i+1}^{j+1} - Q_{i}^{j+1}\right)\right]\\&+ 0.5\frac{\Delta x}{\Delta t}\left[\left(A_{i+1}^{j+1} + A_{i}^{j+1} - A_{i+1}^{j} - A_{i}^{j}\right)\right] - Q_{s}(x,t)\Delta x = 0\end{aligned} \tag{15.18}$$

Similarly, the finite difference form of momentum Eq. (15.9) is

$$\begin{aligned}
&0.5\left[\left(Q_{i+1}^{j+1}+Q_i^{j+1}\right)-\left(Q_{i+1}^{j}-Q_i^{j}\right)\right]/\left(t^{j+1}-t^{j}\right)\\
&+\left[(1-\theta)\left\{\left(Q_{i+1}^{j}\right)^2/A_{i+1}^{j}-\left(Q_i^{j}\right)^2/A_i^{j}\right\}+\theta\left\{\left(Q_{i+1}^{j+1}\right)^2/A_{i+1}^{j+1}-\left(Q_i^{j+1}\right)^2/A_i^{j+1}\right\}\right]/\\
&\left(x_{i+1}-x_i\right)-0.5g\left[\theta\left(A_{i+1}^{j+1}+A_i^{j+1}\right)+(1-\theta)\left(A_{i+1}^{j}+A_i^{j}\right)\right]\\
&\left[\left\{(1-\theta)\left(y_{i+1}^{j}-y_i^{j}\right)+\theta\left(y_{i+1}^{j+1}-y_i^{j+1}\right)\right\}/\left(x_{i+1}-x_i\right)\right.\\
&\left.+0.5\left\{(1-\theta)\left(S_{f_{i+1}}^{j}+S_{f_i}^{j}\right)+\theta\left(S_{f\ i+1}^{j+1}+S_{f\ i}^{j+1}\right)\right\}-2S_o\right]\\
&+Q_s(x,t)V(x,t)=0
\end{aligned} \tag{15.19}$$

or

$$\begin{aligned}
&0.5\left[Q_{i+1}^{j+1}+Q_i^{j+1}-Q_{i+1}^{j}-Q_i^{j}\right]+\Delta t/\Delta x\left[(1-\theta)\left\{\left(Q_{i+1}^{j}\right)^2/A_{i+1}^{j}-\left(Q_i^{j}\right)^2/A_i^{j}\right\}\right.\\
&\left.+\theta\left\{\left(Q_{i+1}^{j+1}\right)^2/A_{i+1}^{j+1}-\left(Q_i^{j+1}\right)^2/A_i^{j+1}\right\}\right]\\
&-0.5g\Delta t\left[\theta\left(A_{i+1}^{j+1}+A_i^{j+1}\right)+(1-\theta)\left(A_{i+1}^{j}+A_i^{j}\right)\right]\left[\left\{(1-\theta)\left(y_{i+1}^{j}-y_i^{j}\right)+\theta\left(y_{i+1}^{j+1}-y_i^{j+1}\right)\right\}\right.\\
&\left.+0.5\left\{(1-\theta)\left(S_{f_{i+1}}^{j}+S_{f_i}^{j}\right)+\theta\left(S_{f\ i+1}^{j+1}+S_{f\ i}^{j+1}\right)\right\}-2S_o\right]+Q_s(x,t)V(x,t)=0
\end{aligned} \tag{15.20}$$

Equations (15.18) and (15.20) constitute a system of two nonlinear algebraic equations in four unknowns. All the variables with superscript j are known and those with superscript $(j+1)$ are unknown [i.e., the values of discharge and depth at grid points $(i, j+1)$ and $(i+1, j+1)$]. However, all the unknowns are not independent. It is noted that Δx and Δt need not be constant and Δx can be varied at any x_i and Δt can be varied at any time t^j. If a channel has N sections or $(N-1)$ grids, then in all 2N unknowns need to be evaluated at any state of iteration.

Equations (15.18) and (15.20) can be written for any grid. Since two unknowns are common to two adjacent grids, the application of equations gives $2(N-1)$ equations. Hence two additional equations are needed which are provided by the boundary conditions. The upstream boundary can be flow depth or discharge as a function of time and the downstream boundary condition can also be flow depth or discharge as a function of time or the rating curve. Special treatment is required if a structure, such as a weir, is present in the river, since the Saint-Venant equations are not valid at such locations. In such situations, the momentum equation is replaced by empirical relations for the structure, for example, the weir formula. If the flow depth at the upstream boundary is known as a function of time, then

$$y_1^{j+1}-y_N'(t^{j+1})=0 \tag{15.21}$$

where $y_N'(t^{j+1})$ is the known depth of flow at the upstream boundary at time t^{j+1}. Likewise, at the downstream boundary, the depth may be known as a function of time

$$y_N^{j+1}-y_N''(t^{j+1})=0 \tag{15.22}$$

If discharge is known as a function of time, then

$$Q_N^{j+1} - Q_N''(t^{j+1}) = 0 \tag{15.23}$$

Again, Eq. (15.19) or (15.20) can be used as a second supplementary equation. In general form

$$F_N(y_{N,}Q_N) = 0 \tag{15.24}$$

Putting together all the equations, 2*N* simultaneous equations for 2*N* unknowns are obtained. These are assembled here:

$$F_0(y_{1,}Q_1) = 0$$

$$F_1(y_1, Q_1, y_2, Q_2) = 0$$

$$G_1(y_1, Q_1, y_2, Q_2) = 0$$

$$F_i(y_i, Q_i, y_{i+1}, Q_{i+1}) = 0$$

$$G_i(y_i, Q_i, y_{i+1}, Q_{i+1}) = 0$$

$$F_N(y_{N,}Q_N) = 0 \tag{15.25}$$

Here F_i denotes equations obtained by the application of Eq. (15.18) and G_i denotes those obtained from Eq. (15.20). Equation (15.25) forms a system of 2*N* equations in 2*N* unknowns. Iterative techniques, such as the Newton-Raphson method, can be used to solve the equations. Every iterative solution of the system leads to a better approximation of unknowns. The process can be stopped when the required accuracy is obtained. The computations can be then be advanced to the next time step. As for initial approximation, the value of unknowns at the previous time step can be taken as a starting value. It has been reported that instead of linear interpolation, a parabolic interpolation gives faster convergence.

Although many implicit schemes are unconditionally linearly stable, instability could occur if the time step size is large and the *x*-derivative terms are not sufficiently advanced toward the future time line when modeling rapidly varying transients. Also, too large a time step size should not be used for the sake of accuracy. In general, Δt depends on the Δx size, the shape of the hydrograph, the type of implicit scheme chosen, and the nonlinearities introduced due to irregularities of the cross section, and so on.

Fread (1993) discusses use of extended Saint-Venant equations for flood modeling.

15.3.3 Kinematic Wave Modeling

The term kinematic involves consideration of such macroscopic properties of a flow phenomenon as velocity, discharge, concentration, time of travel, and so on. It does not involve consideration of forces and mass. Thus the connotation of kinematic is that it is phenomenological and less physical but its representation may be derived from or related to dynamical considerations involving forces, mass, momentum, and energy (Singh, 2002). In kinematics, not only are we concerned with the position of a point at a certain time but also with the displacement accomplished and the distance traversed in a certain time interval. Note that the introduction of mass does not necessarily make some motions strictly dynamical; they may remain strongly and functionally kinematical.

The kinematic wave (KW) theory is comprised of three components: (1) theory of kinematic continuous waves; (2) theory of shock waves; and (3) theory of formation of shock waves out of continuous waves. Kinematic waves exist if it can be assumed with sufficient accuracy that there is a functional relationship between flux, concentration, and position. Thus the KW theory is mathematically expressed by the law of conservation of mass through the continuity equation and a flux-concentration relation. Following Singh (2002), the coupling of these two equations leads to a first-order partial differential equation having only one system of wave characteristics. In other words, the wave property is derived from the continuity equation alone. Since the governing equation (or continuity equation) is of the first-order only, the kinematic waves possess only one system of characteristics. This means that KWs travel only in the downstream direction and that the theory of KW cannot consider those waves that travel in the upstream direction, for example, the case of backwater flow. In the absence of any lateral inflow and/or outflow, the flux along the characteristics is constant and the KWs are nondispersive and nondiffusive. However, the waves are dispersive and diffusive when there is lateral inflow and/or outflow, as for example in the case of rainfall runoff. The mathematical description of KWs assumes the absence of pressure and inertial terms in the momentum equation.

According to the principle of conservation of mass, the difference between inflow and outflow in a control volume in the given time interval would be balanced by a change in volume

$$(Q_2 - Q_1)\Delta t + (A_2 - A_1)\Delta x = 0 \tag{15.26}$$

where Q is the flow, A is cross-section area, Δt is time interval, and Δx is the distance between upstream and downstream sections. Dividing by $\Delta t \Delta x$, Eq. (15.26) can be written in the differential form as

$$\frac{\partial Q}{\partial x} + \frac{\partial A}{\partial t} = 0 \tag{15.27}$$

Manning's equation is used to compute discharge in a river:

$$Q = \frac{1}{n} A R^{2/3} S_f^{1/2} \tag{15.28}$$

where n is Manning's roughness coefficient, $R = A/P$ is the hydraulic radius (m), P is the wetted perimeter, and S_f is the friction slope. Assuming that n, S_f, and P are constant (P is assumed to be independent of A for wide channels)

$$Q = \alpha A^{\beta} = \frac{1}{nP^{5/3}} S_f^{1/2} P^{5/3} \tag{15.29}$$

Differentiating Eq. (15.30)

$$\frac{dQ}{dA} = \beta \alpha A^{\beta-1} = \beta \frac{Q}{A} = \beta V \tag{15.30}$$

where V is the mean (cross-sectional) flow velocity. Multiplying Eq. (15.27) by $\frac{dQ}{dA}$ and simplifying

$$\frac{\partial Q}{\partial t} + \frac{dQ}{dA}.\frac{\partial Q}{\partial x} = 0 \tag{15.31}$$

or

$$\frac{\partial Q}{\partial t} + \beta V.\frac{\partial Q}{\partial x} = 0 \tag{15.32}$$

where θ is a weighting factor $= \Delta t' / \Delta t$.

In Eq. (15.32), inertia and flow depth gradient have been ignored and the waves described by this equation are known as KWs. These waves do not attenuate and move with celerity βV. The KW equation can be solved by a finite difference method. Taking centralized difference (Fig. 15.5), we get

$$\frac{1}{\Delta t}\left(\frac{Q_{j+1}^{n+1} + Q_j^{n+1}}{2} - \frac{Q_{j+1}^{n} + Q_n^{j}}{2}\right) + \frac{\beta V}{\Delta x}\left(\frac{Q_{j+1}^{n} + Q_{j+1}^{n+1}}{2} - \frac{Q_j^{n} + Q_j^{n+1}}{2}\right) \tag{15.33}$$

which can be simplified to

$$Q_{j+1}^{n+1} = C_0 Q_j^{n+1} + C_1 Q_j^{n} + C_2 Q_{j+1}^{n} \tag{15.34}$$

Courant number is the ratio of βV (physical wave celerity) to $\Delta x / \Delta t$ (grid wave celerity). Hence Courant number $C = \beta V \frac{\Delta t}{\Delta x}$. By using this number, the routing coefficients are

$$C_0 = \frac{C-1}{C+1},\ C_1 = 1, \ \text{and}\ C_2 = \frac{1-C}{1+C} \tag{15.35}$$

Example 15.3

Route the following wave through a river by using KW theory. Consider $\Delta t = 2$ h, $\Delta x = 1000$ m, $V = 1.5$ m/s, and $\beta = 5/3$.

Time (h)	0	1	2	3	4	5	6	7	8
Flow Q_1 (m^3/s)	0	25	40	65	90	70	50	25	0

Solution

We first compute

$$C = \beta V \frac{\Delta t}{\Delta x} = \left(\frac{5}{3}\right)(1.5)\left(\frac{7200}{1000}\right) = 2.4\ \text{m/s}$$

$$C_0 = \frac{2.4-1}{2.4+1} = 0.412,\ C_1 = 1, \text{and}\ C_2 = \frac{1-2.4}{1+2.4} = -0.412$$

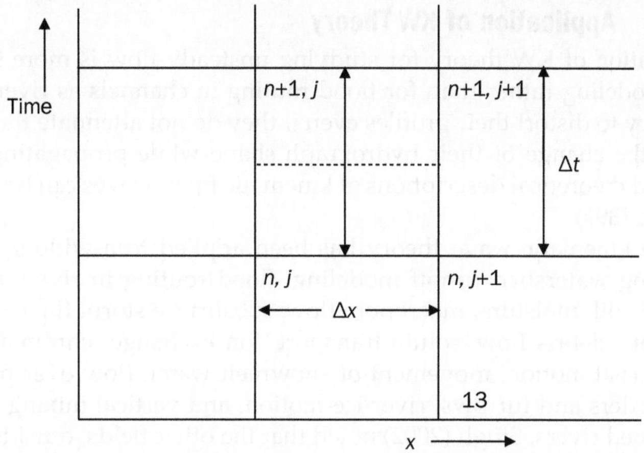

FIGURE 15.5 Discretization in space and time for use in the kinematic wave scheme.

Time	Inflow (m³/s)	Partial Flows (m³/s)			Outflow (m³/s)
		$C_0Q_j^{n+1}$	$C_1Q_j^n$	$C_2Q_{j+1}^n$	
0	0	–	–	–	0
2	25	10.30	0.00	0.00	10.30
4	40	16.48	25.00	–4.24	37.24
6	65	26.78	40.00	–15.34	51.44
8	90	37.08	65.00	–21.19	80.89
10	70	28.84	90.00	–33.33	85.51
12	50	20.60	70.00	–35.23	55.37
14	25	10.30	50.00	–22.81	37.49
16	0	0.00	25.00	–15.45	9.55
18	0	0.00	0.00	–3.94	-3.94
20	0	0.00	0.00	1.62	1.62

TABLE 15.5 Solution of Example 15.3

It is verified that $C_0 + C_1 + C_2 = 1.0$

Routing computation are shown in Table 15.5.

It can be seen that flood hydrograph is translated corresponding to Δt time. We note that $\beta V = \dfrac{5 \times 1.5}{3} = 2.5; \dfrac{\Delta x}{\Delta t} = \dfrac{1000}{7200} = \dfrac{5}{36}$. Since βV is not equal to $\dfrac{\Delta x}{\Delta t}$, there is a small dispersion due to numerical approximation. Also, note negative outflow can occur near the fag end of the hydrograph due to errors in the numerical scheme that has second-order accuracy.

15.3.4 Application of KW Theory

Application of KW theory for studying unsteady flow is more suitable for overland flow modeling rather than for flood routing in channels as river flood waves have a tendency to distort their profiles even if they do not attenuate their peak flows, resulting in the change of their hydrograph shape while propagating down the channel. Detailed theoretical descriptions of kinematic flood waves can be found in Singh (1996, pp. 830, 1399).

The kinematic wave theory has been applied to a wide spectrum of problems, including watershed runoff modeling, flood routing in rivers and channels, movement of soil moisture, macropore flow, subsurface storm flow, erosion and sediment transport, debris flow, solute transport, ion exchange, chromatography, sedimentation, glacial motion, movement of snowmelt water, flow over porous beds in irrigation borders and furrows, river-ice motion, and vertical mixing of coarse particles in gravel-bed rivers. Singh (2002) noted that the other fields, outside of the realm of geophysics, applying the KW theory, include traffic flow, movement of agricultural grains in a bin, and blood circulation. One can argue that within certain limitations, which may vary with the problem, the hydrologic nature can indeed be approximated quite closely by KW theory. Outside of these limitations, the theory provides only a crude approximation and the full hydrodynamic wave theory or one of its higher-order variants may be needed to mimic nature.

15.3.5 Diffusion Waves

Diffusion, meaning "spreading out," is the movement of water particles from a region of high concentration to a region of low concentration. Diffusion takes place due to flow depth gradient, $\partial h/\partial x$, which is present in unsteady open channel flow. Inclusion of the diffusion term yields a better description of flood wave movement in an open channel compared to KW but it also substantially increases the difficulty of obtaining the solution.

The KW equation was desired by considering the flow to be uniform. If the flow is considered nonuniform, or friction slope is replaced by water surface slope, and Manning's equation is replaced by the following

$$Q = \frac{1}{n} A R^{2/3} \left(S_0 - \frac{dy}{dx} \right)^{1/2} \tag{15.36}$$

where water surface slope is $\left(S_0 - \dfrac{dy}{dx} \right)$; the term $\dfrac{dy}{dx}$ accounts for the diffusion process that occurs in natural river flow. The continuity Eqs. (15.8) and (15.36) together describe diffusive waves. These equations can be combined and expressed as

$$\frac{\partial Q}{\partial t} + \left[\frac{\partial Q}{\partial A} \right] \frac{\partial Q}{\partial x} = \left[\frac{Q_0}{2TS_0} \right] \frac{\partial^2 Q}{\partial x^2} \tag{15.37}$$

Here, T is the top width. The left side of Eq. (15.37) is the same as the KW equation. The second-order partial difference term of Eq. (15.37) accounts for physical diffusion in a river. The coefficient inside [ ] on right-hand side is the hydraulic diffusivity. It is noted that when slope is higher, S_0 is large and hydraulic diffusivity is small. For flat slopes, hydraulic diffusivity will be large. A wave will be termed a diffusion wave if it satisfies the inequality (Ponce, 1989)

$$t_r S_0 \left[\frac{g}{d_0}\right]^{0.5} \geq M \tag{15.38}$$

where t_r is the time of rise of inflow hydrograph, d_0 is the average depth of flow, and g is the acceleration due to gravity. Ponce (1989) recommended $M = 15$.

15.4 Muskingum-Cunge Method

Cunge (1969) showed that the Muskingum method is essentially a linear KW solution and the flood wave attenuation shown by the calculation is due to the numerical diffusion of the scheme. Cunge (1969) developed a physically based alternative to the Muskingum method, popularly known as the Muskingum-Cunge method (MCM).

We rewrite the kinematic wave equation as

$$\frac{\partial Q}{\partial t} + c\frac{\partial Q}{\partial x} = 0 \tag{15.39}$$

in which c is the kinematic wave celerity, $c = \beta v$; $\beta = 5/3$ for wide channels. Cunge (1969) discretized Eq. (15.39) on the x-t plane (Fig. 15.6) such that the spatial derivative was centered and the temporal derivative was off-centered by means of a weighting factor X. The resulting equation is given as

$$\frac{X\left(Q_j^{n+1} - Q_j^n\right) + (1 - X)\left(Q_{j+1}^{n+1} - Q_{j+1}^n\right)}{\Delta t} + c\frac{\left(Q_{j+1}^n - Q_j^n\right) + \left(Q_{j+1}^{n+1} - Q_j^{n+1}\right)}{2\Delta x} = 0 \tag{15.40}$$

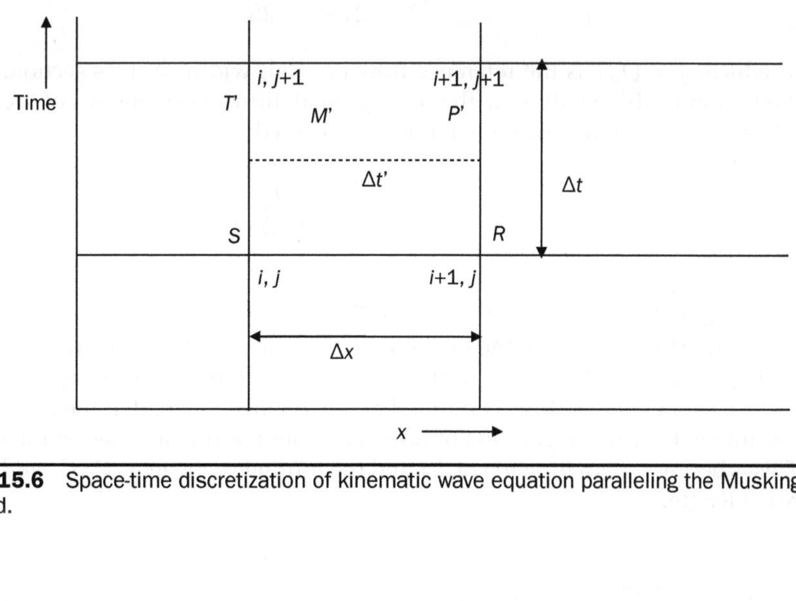

FIGURE 15.6 Space-time discretization of kinematic wave equation paralleling the Muskingum method.

Solving Eq. (15.39) for the unknown discharge leads to the following equation

$$Q_{j+1}^{n+1} = C_0 Q_j^{n+1} + C_1 Q_j^n + C_2 Q_{j+1}^n \tag{15.41}$$

where the routing coefficients are given as

$$C_0 = \frac{c(\Delta t/\Delta x) - 2X}{2(1-X) + c(\Delta t/\Delta x)} \tag{15.42}$$

$$C_1 = \frac{c(\Delta t/\Delta x) + 2X}{2(1-X) + c(\Delta t/\Delta x)} \tag{15.43}$$

$$C_2 = \frac{2(1-X) + c(\Delta t/\Delta x)}{2(1-X) + c(\Delta t/\Delta x)} \tag{15.44}$$

Defining

$$K = \frac{\Delta x}{c} \tag{15.45}$$

the equations to compute C_0, C_1, and C_2 in Muskingum and MCM become identical. Equation (15.45) suggests that K is the time taken for a water mass moving with the kinematic celerity c to cover the reach length Δx. In reality, c varies with discharge. The numerical diffusion coefficient of the Muskingum method can be derived as (Ponce, 1989)

$$\nu_n = c\Delta x\left(\frac{1}{2} - X\right) \tag{15.46}$$

The hydraulic diffusivity (v_h) is a characteristic of flow and channel and defined as

$$v_n = \frac{Q_o}{2TS_o} = \frac{q_o}{2S_o} \tag{15.47}$$

in which, $q_o = Q_o/T$ is the reference flow per unit width of cross section. If we match the hydraulic diffusivity v_h with the numerical diffusion coefficient of the Muskingum scheme, the following expression for X is obtained:

$$X = \frac{1}{2}\left(1 - \frac{q_o}{S_o c\Delta x}\right) \tag{15.48}$$

When X is calculated by Eq. (15.48), the Muskingum method is referred to as the Muskingum-Cunge method (NERC, 1975).

In the Muskingum-Cunge method, there is the grid independence of the calculated outflow hydrograph. If numerical dispersion is minimized (keeping Courant number C close to one), the outflow computed at the downstream end of the river reach will be essentially the same regardless of how many subreaches are used in the computation. This is because X is a function of Δx and the routing coefficients C_0, C_1, and C_2 vary with reach length.

An improved version of the MCM was presented by Ponce (1989). The Courant number, C, is defined as the ratio of wave celerity (c) to grid celerity $\Delta x/\Delta t$, or

$$C = c\frac{\Delta t}{\Delta x} \tag{15.49}$$

The grid diffusivity is defined as the numerical diffusivity for $X = 0$. From Eq. (15.46), the grid diffusivity is

$$v_g = c\Delta x/2 \tag{15.50}$$

The cell Reynolds number is the ratio of hydraulic diffusivity to grid diffusivity. This leads to

$$D = \frac{q_o}{S_o c\Delta x} \tag{15.51}$$

in which D = the cell Reynolds number. Therefore from Eqs. (15.48) and (15.51)

$$X = 0.5(1 - D) \tag{15.52}$$

Substitution of Eqs. (15.49) and (15.52) into Eqs. (15.42)–(15.44) leads to the routing coefficients expressed in terms of the Courant and cell Reynolds numbers

$$C_0 = \frac{-1 + C + D}{1 + C + D} \tag{15.53}$$

$$C_1 = \frac{1 + C - D}{1 + C + D} \tag{15.54}$$

$$C_2 = \frac{1 - C + D}{1 + C + D} \tag{15.55}$$

Thus C and D are the two routing parameters required to be estimated for the MCM.

15.4.1 Estimation of Muskingum-Cunge Parameters

Parameter C can be estimated by Eq. (15.48). It requires an estimate of wave celerity (c) in addition to grid size (Δx, Δt). The wave celerity can be calculated with either

$$c = \beta v \tag{15.56}$$

or

$$c = \frac{1}{T}\frac{dQ}{dY} \tag{15.57}$$

where v is the average flow velocity; T is the top width; and β is an exponent in the discharge-area rating equation. Parameter β is a function of cross-sectional properties.

While estimating c, if a stage-discharge rating and cross-section geometry are available, Eq. (15.57) is preferred over Eq. (15.56) because it directly accounts for the cross-sectional shape. In the absence of these data, Eq. (15.56) can be used to estimate the celerity of wave. The velocity term in Eq. (15.56) can be taken as the velocity at reference flow which can be taken as the average flow.

The cell Reynolds number (D) can be calculated by using the reach length (Δx), reference discharge per unit width q_o, kinematic wave celerity (c), and bottom slope (S_o) in Eq. (15.51).

15.4.2 Flood Routing by the Muskingum-Cunge Method

The steps involved in flood routing through a channel reach using MCM are as follows

(1) Estimate parameter C (Courant number) by using the expression $C = c\dfrac{\Delta t}{\Delta x}$.

The wave celerity c is computed using the procedure described earlier. The temporal and spatial resolutions (Δt and Δx) should be chosen such that the routing co-efficient C_o should not be negative. Value of Courant number (C) should be close to one so as to minimize numerical dispersion.

(2) Estimate parameter D (cell Reynolds number) as $D = \dfrac{q_o}{S_o c \Delta x}$.

One should ensure that $-1 + C + D > 0$ which is the practical criterion to avoid the negative values of C_0 in the Muskingum-Cunge routing.

(3) Estimate routing coefficients C_0, C_1, and C_2.

(4) Route the inflow hydrograph (Q) using the following equation in order to have the outflow hydrograph (Q_j+_1):

$$Q_{j+1}^{n+1} = C_0 Q_j^{n+1} + C_1 Q_j^n + C_2 Q_{j+1}^n \qquad (15.58)$$

(5) If the channel is divided into subreaches, steps (1)–(4) should be repeated for all the subreaches considering the outflow from the first subreach as inflow to the second subreach, and so on.

Example 15.4

Use the Muskingum-Cunge method to route a flood wave with the following flood and channel features: peak flow $Q_p = 750$ m³/s, channel bottom slope $S_o = 0.0009$, flow area at peak discharge $A_p = 300$ m², top width at peak discharge $T_p = 50$ m, rating exponent $\beta = 1.7$, reach length $\Delta x = 12.0$ km, and time interval $\Delta t = 1$ h.

Time (h)	0	1	2	3	4	5	6	7	8	9
Flow (m³/s)	0	150	300	450	600	750	550	350	150	0

Solution

The mean velocity (based on the peak discharge)

$$v = Q_p / A_p = 750/300 = 2.5 \text{ m/s}$$

Wave celerity $c = \beta v$. Here $\beta = 1.7$.

Hence $c = 1.7 \times 2.5 = 4.25$ m/s

Courant number $C = c\,\Delta t/\Delta x = 4.25 \times 1 \times 3600 \times 10^{-3}/12 = 1.275$

Cell Reynolds number $D = q_o/S_o\, c\Delta x$

Time (h)	Inflow	Partial Flows		Outflow	
	Q_j^{n+1} (m³/s)	$C_0Q_j^{n+1}$ (m³/s)	$C_1Q_j^n$ (m³/s)	$C_2Q_{j+1}^n$ (m³/s)	Q_{j+1}^{n+1} (m³/s)
(1)	(2)	(3)	(4)	(5)	(6) = (3) + (4) + (5)
0	0	–	–	–	0
1	200	18.2	0	0	18.2
2	400	36.4	163.6	1.66	201.66
3	600	54.6	327.2	18.35	400.15
4	800	72.8	490.8	36.41	600.01
5	1000	91.0	654.4	54.60	800.00
6	800	72.8	818.0	72.80	963.60
7	600	54.6	654.4	87.69	796.69
8	400	36.4	490.8	72.50	599.70
9	200	18.2	327.2	54.57	399.97
10	0	0	163.6	36.40	200.00
11	0	0	0	18.20	18.20
12	0	0	0	1.66	1.66
13	0	0	0	0.16	0.16

TABLE 15.6 Channel Routing by the Muskingum-Cunge Method

where q_o = the flow per unit width (based on the peak discharge)
$= Q_p/T_p = 1000/100 = 10$ m²/s

Hence $D = 10/(0.000868 \times 4 \times 14.4 \times 10^3) = 10/0.868 \times 57.6 = 0.2$

Routing coefficients:

$$C_0 = (-1+C+D)/(1+C+D) = (-1+1+0.2)/(1+1+0.2) = 0.091$$

$$C_1 = (1+C-D)/(1+C+D) = (1+1-0.2)/(1+1+0.2) = 0.818$$

$$C_2 = (1-C+D)/(1+C+D) = (1-1+0.2)/(1+1+0.2) = 0.091$$

Check $\quad C_0 + C_1 + C_2 = 0.091 + 0.818 + 0.091 = 1.000$

(v) Compute the outflow hydrograph using the following routing equation:

$$Q_{j+1}^{n+1} = C_0Q_j^{n+1} + C_1Q_j^n + C_2Q_{j+1}^n$$

$$= 0.091\, Q_j^{n+1} + 0.818\, Q_j^n + 0.091\, Q_{j+1}^n$$

The routing calculations are shown in Table 15.6.

15.4.3 Advantages and Limitations of the Muskingum-Cunge Method

The Muskingum-Cunge method works best when the numerical dispersion is minimized. Since the method is limited to diffusion waves and does not take into account the substantial loops in stage-discharge rating (i.e., dynamic waves), its application is restricted to the routing of flows in natural streams whose backwater effects are without significance and to unsteady flows that classify under the diffusion wave.

1. MCM is a physically based alternative to the Muskingum method. In the Muskingum method, parameters are determined by using streamflow data but MCM parameters are calculated by the use of flow and channel physical characteristics, such as rating curves, cross-sectional data, and channel slope. Hence routing parameters can be computed without time-consuming calibration. Further, channel routing can be carried out in ungauged rivers with good accuracy.
2. With the variable parameter feature, nonlinear properties of flood waves can be described within the context of the Muskingum formulation (Perumal et al., 2001).
3. The Muskingum-Cunge method does not consider a looped rating curve. This method does not take into account strong nonuniformity in flows or unsteady behavior of flows that display significant loops in the rating relationship, for example, the dynamic waves. MCM is a suitable method for flow routing in natural streams which do not display significant backwater effects and for unsteady flows.
4. The Muskingum method is based on the storage concept and its parameters K and X are reach averages. However, the Muskingum-Cunge method is kinematic in nature and its parameters C and D are based on the values evaluated at channel cross sections; these are not average values for the reach.

15.5 Muskingum Method and Diffusion Wave Approximation

Koussis (1978, 1980) has shown that the Muskingum method can give rise to the diffusion wave approximation. To that end, Eq. (15.3) can be rewritten with discharge denoted by Q as

$$\frac{dS}{dt} = Q(x,t) - Q(x + \Delta x,t) \tag{15.59}$$

It is noted that $Q(x, t)$ denotes inflow and $Q(x + \Delta x, t)$ denotes outflow. Similarly, Eq. (15.4) with $K = k$ can be written as

$$S = kwQ(x,t) + k(1 - w)Q(x + \Delta x,t) \tag{15.60}$$

Expanding $Q(x + \Delta x,t)$ by the Taylor series up to second order, we get

$$Q(x + \Delta x,t) = Q(x,t) + \Delta x\frac{\partial Q}{\partial x} + \frac{(\Delta x)^2}{2}\frac{\partial^2 Q}{\partial x^2} \tag{15.61}$$

Substituting Eq. (15.61) in Eq. (15.60), the result is

$$S = kwQ(x,t) + k(1 - w)\left[Q(x,t) + \Delta x\frac{\partial Q}{\partial x} + \frac{(\Delta x)^2}{2}\frac{\partial^2 Q}{\partial x^2}\right] \tag{15.62}$$

Differentiating Eq. (15.62) and substituting in Eq. (15.59), we obtain

$$kw\frac{\partial Q}{\partial t} + k(1-w)\left[\frac{\partial Q}{\partial t} + \Delta x\frac{\partial^2 Q}{\partial x \partial t} + \frac{(\Delta x)^2}{2}\frac{\partial}{\partial t}\left(\frac{\partial^2 Q}{\partial x^2}\right)\right] = Q(x,t) - Q(x+\Delta x,t) \tag{15.63}$$

Inserting Eq. (15.61) in Eq. (15.63), ignoring the third-order term, we obtain

$$\frac{\partial Q}{\partial t} + \frac{\Delta x}{k}\frac{\partial Q}{\partial x} = \left[(1-w)c\Delta x - \frac{(\Delta x)^2}{2k}\right]\frac{\partial^2 Q}{\partial x^2} \tag{15.64}$$

Equation (15.64) is a diffusion equation and was derived by Koussis (1978). It may be recalled that $\partial^2 Q/\partial x \partial t = -c\partial^2 Q/\partial x^2$, $\partial^2 Q/\partial t^2 = c^2 \partial^2 Q/\partial x^2$ and recalling that celerity from pure convection equation can be expressed as $c = -\Delta x/\Delta t$.

Now, by linearizing the Saint-Venant equations about a reference flow (characterized by discharge, velocity, flow depth, flow cross section, and surface width), the convective-diffusion equation can be expressed following Lighthill and Whitham (1955) and Cunge (1969) as

$$\frac{\partial Q}{\partial t} + c_1\frac{\partial Q}{\partial x} = D\frac{\partial^2 Q}{\partial x^2} \tag{15.65}$$

in which c_1 is the celerity of discharge, and D is the diffusion coefficient. Equations (15.64) and (15.65) are of the same form. Now recall the definition of kinematic celerity c as

$$c = \left(\frac{dQ}{dA}\right)\Big|_{x=\text{constant}} = \frac{1}{B_0}\left(\frac{dQ}{dy}\right)_{x=\text{constant}} \tag{15.66}$$

which defines the slope of a rating curve. The diffusion coefficient D is expressed as

$$D = \frac{Q_0}{2B_0 S_0} \tag{15.67}$$

where Q_0 is the reference discharge used for linearization, B_0 is the surface width of channel flow section corresponding to the reference discharge, and S_0 is bed slope. The diffusion coefficient D is also expressed as

$$D = \frac{Q_0}{2B_0 S_0}(1 - \beta^2 F_0^2) \tag{15.68}$$

where F_0^2 is the Froude number defined as

$$F_0^2 = \left(\frac{Q_0^2}{gA_0^2}\right)^{1/2} \tag{15.69}$$

in which A_0 is the referenced flow cross-sectional area, and β is a function of flow equation, that is equal to two-thirds for Manning's equation. Often, $\beta^2 F_0^2 << 1$.

Equating Eq. (15.64) to Eq. (15.65), we obtain

$$k = \frac{\Delta x}{c} = \frac{\Delta x}{\left(\dfrac{dQ}{dA}\right)_{x=\text{constant}}}; \quad w = \frac{1}{2} - \frac{wQ_0}{2B_0S_0(\Delta x)^2} \tag{15.70}$$

Equation (15.70) states that the storage coefficient k can be interpreted as the propagation of discharge, or travel time or residence time, as indicated before. It also shows that the weighting factor w is less than 0.5. It also shows that for the Kalinin-Miljukov method, the reach length $\Delta x = L$ can be expressed (with $w = 0$) as

$$L = \frac{Q_0}{B_0S_0c} \tag{15.71}$$

For flow routing in a river, the number of sections will need to be determined, which will depend on the nonuniformity of river geometry. Koussis (1978) used average discharge and corresponding values of $B_{0,}$ k, and slope of the water surface for determining the weighting factor for routing flow in the Rhine River.

15.5.1 Kinematic Wave Approximation and Muskingum Method

The kinematic wave (KW) approximation for flow routing corresponds to advection and can be expressed in the form of the Muskingum method presented above. For flow in a river reach, conservation of mass can be expressed as

$$\frac{\partial A}{\partial t} + \frac{\partial Q}{\partial x} = 0 \tag{15.72}$$

where A is the flow cross-sectional area, and Q is the discharge. The KW hypothesis can be written for a wide prismatic channel as

$$Q = \alpha B y^{m-1}; \quad \alpha = C_r\sqrt{S_f} \tag{15.73}$$

where y is the flow depth, B is the channel width, m is exponent, S_f is the friction slope approximated by the bed slope $S_{0,}$ C_r is a resistant coefficient and equals Chezy's coefficient if exponent m is equal to half and equals 1 divided by Manning's roughness factor if m equals two-thirds (in the SI system of units, otherwise 1.49 divided by Manning's roughness factor in the FPS system).

With the introduction of celerity, Eq. (15.72) can be written as

$$\frac{\partial Q}{\partial t} + \left(\frac{dQ}{dA}\right)_{x=\text{constant}} \frac{\partial Q}{\partial x} = 0 \tag{15.74}$$

For simplicity, it is assumed that the flow is uniform so the celerity becomes kinematic celerity (a single valued function) which then yields a relation with mean cross-sectional velocity as

$$c = (m + 1)u \tag{15.75}$$

Equation (15.74) can be written as

$$\frac{\partial Q}{\partial t} + c\frac{\partial Q}{\partial x} = 0 \tag{15.76}$$

Equation (15.76) can also be expressed in terms of flow depth as

$$\frac{\partial y}{\partial t} + c\frac{\partial y}{\partial x} = 0 \tag{15.77}$$

Equations (15.76) and (15.77) are purely convective equations.

For purposes of generalization, let a variable, denoted as f, represent, either flow depth y, flow discharge Q, or flow velocity. Then, the KW approximation can be expressed as

$$\frac{\partial f}{\partial t} + a\frac{\partial f}{\partial x} = 0 \tag{15.78}$$

where a has the connotation of celerity.

In Eq. (15.78), the space derivative is discretized over Δx and the time derivative is approximated by a weighted average (with a weighting factor w). We then obtain

$$w\left(\frac{\partial f}{\partial t}\right)_j + (1-w)\left(\frac{\partial f}{\partial t}\right)_{j+1} + a\frac{f_{j+1} - f_j}{\Delta x} = 0 \tag{15.79}$$

where j denotes the space discretization index, such that $x_j = j.\Delta x$. Let

$$\frac{\partial f}{\partial t} \approx \frac{f_{i+1} - f_i}{\Delta t}; \ \Delta t = t_{i+1} - t_i \tag{15.80}$$

Centering the spatial derivative in Δt, we obtain

$$w\frac{f(x_j,t_{i+1}) - f(x_j,t_i)}{\Delta t} + (1-w)\frac{f(x_{j+1},t_{n+1}) - f(x_{j+1},t_i)}{\Delta t} + a\frac{f(x_{j+1},t_i) - f(x_j,t_i)}{2\Delta x}$$
$$+ a\frac{f(x_{j+1},t_{i+1}) - f(x_i,t_{i+1})}{2\Delta x} = 0 \tag{15.81}$$

Rearranging terms of Eq. (15.81), we obtain

$$f(x_{j+1},t_{i+1}) = a_1 f(x_j, t_i) + a_2 f(x_j,t_{i+1}) + a_3 f(x_{j+1},t_i) \tag{15.82}$$

where

$$a_1 = \frac{2w\Delta x + a\Delta t}{2\Delta x(1-w) + a\Delta t}; \quad a_2 = \frac{a\Delta t - 2w\Delta x}{2\Delta x(1-w) + a\Delta t};$$
$$a_3 = \frac{2\Delta x - a\Delta t - 2w}{2\Delta x(1-w) + a\Delta t}; \quad \sum_{i=1}^{3} a_i = 1 \tag{15.83}$$

Equation (15.82) is analogous to the Muskingum method. Coefficients a_i, $I = 1, 2,$ and 3, are the coefficients and depend on the weighting factor w.

15.5.2 Another Kinematic Approximation Through Numerical Integration

Koussis (1983) proposed a numerical integration method for solving linear Eqs. (15.76), (15.77), and (15.78) which exploits the spatial discretization error of a finite difference scheme developed for the numerical integration of the pure advection equation to approximate the solution of the AD equation. Koussis (1983) termed this error as numerical diffusion/dispersion. It is composed of a fixed error term that originates from the discretization of the first spatial derivative, and a variable error term that results from the spatial off-centering when evaluating the temporal derivative. The AD equation is approximated by matching the numerical diffusion/dispersion coefficient with the physical one. Of course, the same transport velocity is used.

To illustrate, consider Eq. (15.78) where f may represent either flow depth y, flow discharge Q, or velocity. Its numerical approximation is given by Eq. (15.79), which can be rearranged with unknown terms on the left side and known terms on the right side as

$$\left(\frac{\partial f}{\partial t}\right)_{j+1} + \frac{a}{\Delta x(1-w)} f_{j+1} = -\frac{w}{1-w}\left(\frac{\partial f}{\partial t}\right)_j + \frac{a}{\Delta x(1-w)} f_j \tag{15.84}$$

To solve Eq. (15.84), let the initial condition be defined as

$$f_{j+1} = f_j = f_\infty = f(x, t = 0) \tag{15.85}$$

Now the solution of Eq. (15.36) can be expressed using the integrating factor method as

$$f_{j+1}(t) = \frac{f_\infty}{1-w}\exp\left[-\frac{wt}{\Delta x(1-w)}\right] + \frac{a}{\Delta x(1-w)^2}\int_0^t f_j(\tau)\exp\left[-\frac{a(t-\tau)}{\Delta x(1-w)}\right]d\tau - \frac{w}{1-w} f_j(t) \tag{15.86}$$

Let $f_j(t)$ vary linearly over Δt and i be the time discretization index such that $t_i = i.\Delta t$. Then, the integral in Eq. (15.86) can be evaluated and the finite difference solution can be written as

$$f_{j+1}^{i+1} = C_1 f_j^i + C_2 f_j^{i+1} + C_3 f_{j+1}^i; \quad \sum_{i=1}^{3} C_i = 1 \tag{15.87}$$

in which

$$C_1 = \frac{1-\omega}{N_c} - \omega; \quad C_2 = 1 - \frac{1-\omega}{N_c};$$
$$C_3 = \omega = \exp\left[-\frac{a\Delta t}{\Delta x(1-w)}\right] = \exp\left[-\frac{N_c}{1-w}\right]; \quad N_c = \frac{a\Delta t}{\Delta x} \tag{15.88}$$

The quantity N_c is regarded as the Courant number.

A Taylor series expansion of Eq. (15.86) up to the second order leads to the linear convection diffusion equation (Koussis, 1980) as

$$\frac{\partial Q}{\partial t} + c\frac{\partial Q}{\partial x} = D\frac{\partial^2 Q}{\partial x^2_x} \tag{15.89}$$

in which c is the celerity and D is the diffusion coefficient expressed as

$$D = \left[\frac{2N_c}{1 - \exp\left(-\dfrac{N_c}{1-w}\right)} - 1 - N_c\right]\frac{c\Delta x}{2} \tag{15.90}$$

Since D is based on hydraulic considerations and is hence physically based, Eq. (15.83) can be used to determine the weighting factor w of the Muskingum method as

$$w = 1 - \frac{N_c}{\ln\left(\dfrac{1+M+N_c}{1+M-N_c}\right)} \tag{15.91}$$

in which

$$M = \frac{Qk}{BS_0(\Delta x)^2} \tag{15.92}$$

where k is the average propagation time $\Delta x/c$.

It may be noted that a finite-difference solution similar to Eq. (15.84) can be derived by approximating the derivatives in Eq. (15.78) by finite differences weighted by factor w for time derivative and p for space derivative as follows:

$$w\left(\frac{\partial f}{\partial t}\right)_j + (1-w)\left(\frac{\partial f}{\partial t}\right)_{j+1} + pa\left(\frac{\partial f}{\partial x}\right)_i + a(1-p)\left(\frac{\partial f}{\partial x}\right)_{i+1} = 0 \tag{15.93}$$

Equation (15.93) can be expressed in finite difference form as

$$w\frac{f(x_j,t_{i+1}) - f(x_j,t_i)}{\Delta t} + (1-w)\frac{f(x_{j+1},t_{i+1}) - f(x_{j+1},t_i)}{\Delta t} + ap\frac{f(x_{j+1},t_i) - f(x_j,t_i)}{\Delta x}$$
$$+ a(1-p)\frac{f(x_{j+1},t_{i+1}) - f(x_i,t_{i+1})}{\Delta x} = 0 \tag{15.94}$$

Collecting the unknown term on the left side and known terms on the right side, we obtain

$$f(x_{j+1},t_{i+1})\left[\frac{(1-w)}{\Delta t} + \frac{a(1-p)}{\Delta x}\right] = f(x_j,t_i)\left[\frac{w}{\Delta t} - \frac{ap}{\Delta x}\right]$$
$$+ f(x_j,t_{i+1})\left[-\frac{w}{\Delta t} - \frac{a(1-p)}{\Delta x}\right] + f(x_{j+1},t_i)\left[\frac{a(1-w)}{\Delta t} - \frac{ap}{\Delta x}\right] \tag{15.95}$$

Equation (15.95) takes on the same form as Eq. (15.87) with coefficients expressed as

$$C_1 = \frac{w + pN_c}{1 + N_c - (w + pN_c)}; \quad C_2 = \frac{N_c - (w - pN_c)}{1 + N_c - (w + pN_c)}; \quad C_2 = \frac{1 - (w + pN_c)}{1 + N_c - (w + pN_c)} \tag{15.96}$$

The sum of coefficients in Eq. (15.96) sums to one.

15.5.3 Diffusion Wave Approximation

Assuming that convective and local (inertial) acceleration terms as negligible, the momentum equation can be written for a wide prismatic channel as

$$\frac{\partial y}{\partial x} = S_0 - S_f; \; Q = C_r B y^{m+1} \left(S_0 - \frac{\partial y}{\partial x} \right)^{1/2} \tag{15.97}$$

where y is the flow depth, B is the channel width, S_0 is the bed slope, S_f is the friction slope, R is the hydraulic radius, C_r is a resistant coefficient, and m is an exponent equal to half for Chezy's equation and two-thirds for Manning's equation. Expanding the square root term as a binomial series and truncating it, Eq. (15.45) can be simplified as

$$Q = C_r B y^{m+1} S_o \left(1 - \frac{\partial y / \partial x}{2S_0} \right) \tag{15.98}$$

Equations (15.71) and (15.98) can be combined to produce a nonlinear diffusion equation as

$$\frac{\partial y}{\partial t} + c\frac{\partial y}{\partial x} = \frac{yc}{2(m+1)S_0} \frac{\partial^2 y}{\partial x^2} \tag{15.99}$$

Equation (15.99) can be linearized by fixing the coefficients at some reference value flow depth, say y_R

$$\frac{\partial y}{\partial t} + c(y_R)\frac{\partial y}{\partial x} = \frac{y_R c(y_R)}{2(m+1)S_0} \frac{\partial^2 y}{\partial x^2} \tag{15.100}$$

where

$$D_H(y_R) = \frac{y_R c(y_R)}{2(m+1)S_0} \tag{15.101}$$

can be regarded as hydraulic diffusivity or dispersivity.

One can also derive analogous expressions for Q. This requires differentiating Eq. (15.70) spatially and differentiating Eq. (15.98) temporally, eliminating the mixed derivative of depth, and combining

$$\frac{\partial Q}{\partial t} + c\frac{\partial Q}{\partial x} = \frac{Q}{2BS_0} \frac{\partial^2 Q}{\partial x^2} \tag{15.102}$$

Fixing the coefficients at the values at a reference discharge, a linear form of Eq. (15.102) is obtained as

$$\frac{\partial Q}{\partial t} + c(Q_R)\frac{\partial Q}{\partial x} = \frac{Q_R}{2BS_0}\frac{\partial^2 Q}{\partial x^2} \tag{15.103}$$

where

$$D_H(Q_R) = \frac{Q_R}{2BS_0} \tag{15.104}$$

can be regarded as hydraulic diffusivity.

15.6 Reservoir Routing

The passage of flood hydrograph through a reservoir is an unsteady flow phenomenon. The primary equation used here is the equation of continuity which states that the difference between the inflow and outflow is equal to the rate of change of storage, that is

$$I - Q = dS/dt \tag{15.105}$$

where I = the inflow, Q = the outflow, S = the storage, and t = time. Over a small time interval Δt, the difference between total inflow volume and total outflow volume in a reach is equal to the change in storage in that reach. Hence Eq. (15.105) can be written as

$$I_m\Delta t - Q_m\Delta t = \Delta S \tag{15.106}$$

where I_m, Q_m, and ΔS denote the average inflow, average outflow, and change in storage during time period Δt, respectively.

A schematic representation of reservoir routing is given in Fig. 15.7.

15.6.1 Reservoir Routing Techniques

The outflow from a reservoir can be either controlled or uncontrolled. The controlled reservoirs have spillways and other outlets with gates which are operated to release

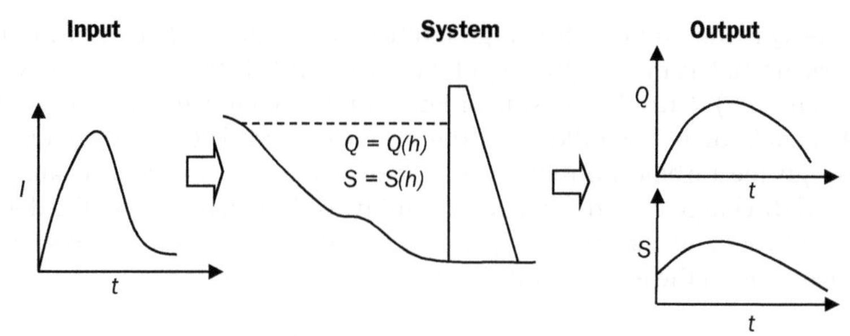

FIGURE 15.7 Schematic representation of reservoir routing. *I:* Inflow, *Q:* Outflow, *S:* Storage, *h:* elevation, *t:* time.

water. Outlets of uncontrolled reservoirs are not controlled by gate operation. For reservoir routing computations, the relationship between reservoir elevation, storage, and discharge is required. This relationship depends on topography of the site near the dam and the characteristics of the outlets and is typically available in tabular or graphical form.

Using the basic Eq. (15.105), several methods to route a flood wave through a reservoir have been developed, namely the Modified Puls Method, Mass Curve Method, Wisler Brater Method, Goodrich Method, Steinberg Method, and Coefficient Method. A brief description of the first two methods that are most widely used is given next.

15.6.2 Modified Puls Method

This method is also known as the Storage-Indication method. The basic law used in the Modified Puls method in the continuity equation states: *For a reservoir, the difference between average inflow and outflow is equal to the rate of change in storage.* Assuming $I_m = (I_1 + I_2)/2$, $Q_m = (Q_1 + Q_2)/2$, and $\Delta S = S_2 - S_1$, Eq. (15.106) is written as

$$(I_1 + I_2)\Delta t/2 - (Q_1 + Q_2)\Delta t/2 = S_2 - S_1 \tag{15.107}$$

where suffixes 1 and 2 denote the beginning and end of time interval Δt and Q may incorporate controlled as well as uncontrolled releases. Here the time interval Δt must be sufficiently small so that the inflow and outflow hydrographs can be assumed to be linear in any time interval. Furthermore, Δt must be smaller than the time of transit of flood wave through the reservoir. Separating the known quantities from the unknown ones and rearranging

$$(I_1 + I_2) + (2S_1/\Delta t - Q_1) = (2S_2/\Delta t + Q_2) \tag{15.108}$$

Here, the known quantities are I_1 (inflow at time 1), I_2 (inflow at time 2), Q_1 (outflow at time 1), and S_1 (storage in the reservoir at time 1), and the unknown quantities are S_2 and Q_2. Since one equation with two unknowns cannot be solved, one must have another relation that relates storage, S, and outflow, Q. As the major outflow from the reservoir takes place through the spillway, the discharge passing through the spillway can be conveniently related with the water surface elevation in the reservoir which, in turn, can be related to the reservoir storage. Such a relationship is invariably available or can be developed for any reservoir. Also, it can be computed from the following relation

$$Q = C_d\, L\, H^{1.5} \tag{15.109}$$

where Q is the outflow discharge (cumec); C_d is the coefficient of discharge (=1.70 in metric unit); L is the length of spillway (m); and H is the depth of flow above the spillway crest (m). Thus the left side of Eq. (15.108) contains the known terms and the right side is unknown. The inflow hydrograph is known. The discharge Q, which may pass through the turbines, outlet works, or over the spillway is also known. The uncontrolled discharge goes freely over the spillway. It depends upon the depth of flow over the spillway and the spillway geometry. Further, the depth of flow over the spillway depends upon the level of water in the reservoir. Therefore

$$S = S\,(Y)$$

$$Q = Q\,(Y) \tag{15.110}$$

where Y represents the water surface elevation. The right side of Eq. (15.101) can be written as

$$2S/\Delta t + Q = f(Y) \tag{15.111}$$

Adding the crest elevation to the depth of flow, the elevation for which storage in the reservoir is known can be computed. Therefore one can develop a relation between storage and outflow. This storage outflow relation is used to develop the storage indication $[(2S/\Delta t) + Q]$ versus outflow relation. To develop this relation, it is necessary to select a time interval such that the resulting linearization of the inflow hydrograph remains a close approximation of the actual nonlinear (continuous time varying) shape of the hydrograph. For smoothly rising hydrographs, a minimum value of $t_p/\Delta t = 5$ was recommended by Jain and Singh (2003), in which t_p is the time to peak of the inflow hydrograph. In practice, a computer aided calculation would normally use a much greater ratio, such as, 10–20.

To utilize Eq. (15.108), the elevation storage and elevation-discharge relationship must be known. Before routing, the curves of $(2S/\Delta t \pm Q)$ versus Q are constructed.

The routing computations are performed as follows. At the start of flood routing, the initial storage and outflow discharge are known. In Eq. (15.108) all the terms on the left-hand side are known at the beginning of time step Δt. Hence the value of $(S_2 + Q_2\Delta t/2)$ at the end of the time step is calculated by Eq. (15.108). Since the relations $S = S(h)$ and $Q = Q(h)$ are known, $(S_2 + Q_2\Delta t/2)$ will enable one to determine the reservoir elevation and hence the discharge at the end of the time step. This procedure is repeated to cover the full inflow hydrograph.

Example 15.5

The spillway of a dam is ogee crested whose crest is at an elevation of 101 m. Initially, the reservoir level is at an elevation of 102 m and the initial outflow is 27 cumec. Route the hydrograph given in Table 15.7 through the reservoir. Take time interval $\Delta t = 1$ h.

Route the flood through this reservoir.

Solution

The calculations for the Modified Puls method are shown in Table 15.8 in which

- Column 1 shows water surface elevation, from 101 m to 107 m.
- Column 2 shows the head above spillway crest.
- Column 3 shows the outflow, calculated by Eq. (15.110).
- Column 4 shows the storage volume in cubic meters above the spillway crest elevation.
- Column 5 shows storage volume in cumec/h, computed by dividing the values in Column 4 by $\Delta t \times 3600$.
- Column 6 shows the storage indication quantities $[(2S/\Delta t + Q]$, in m^3/s.

Time (h)	1	2	3	4	5	6	7	8
Inflow (m^3/s)	27	40	60	100	160	240	180	130
Time (h)	9	10	11	12	13	14	15	16
Inflow (m^3/s)	90	60	40	35	30	27	27	17

TABLE 15.7 Inflow Hydrograph for Example 15.5

(1)	(2)	(3)	(4)	(5)	(6)
Elevation (m)	Head (m)	Outflow (m^3/s)	Storage (10^6 m^3)	Storage (m^3/s)$-h$	[(2S/Δt)+Q] (m^3/s)
101	0	0	0.00	0.00	0.00
102	1	17.00	1.50	416.67	860.33
103	2	48.08	2.50	694.44	1446.97
104	3	88.33	3.50	972.22	2042.77
105	4	136.00	4.50	1250.00	2666.00
106	5	190.07	5.50	1527.78	3275.63
107	6	249.85	6.50	1805.56	3910.96

TABLE 15.8 Reservoir Water Elevation versus Outflow, Storage Table

(1)	(2)	(3)	(4)	(5)	(6)	(7)
Time (h)	Inflow (m^3/s)	Elevation (m)	Storage (10^6 m^3)	[2S/Δt) − Q] (m^3/s)	[2S/Δt) + Q] (m^3/s)	Outflow (m^3/s)
0	27	102.00	1.5	806.33	860.33	27.00
1	40	102.02	1.52	817.96	873.33	27.69
2	60	102.10	1.6	857.85	917.96	30.05
3	100	102.27	1.77	947.16	1017.85	35.35
4	160	102.59	2.09	1116.41	1207.16	45.37
5	240	103.12	2.62	1390.87	1516.41	62.77
6	180	103.61	3.11	1645.54	1810.87	82.66
7	130	103.85	3.35	1770.67	1955.54	92.44
8	90	103.91	3.41	1801.05	1990.67	94.81
Remaining lines deleted.						

TABLE 15.9 Computations and Results of Example 15.5

The routing is summarized in Table 15.9 in which Column 1 shows time; Column 2 shows the inflow hydrograph; Column 3 shows elevation, and so on.

Here, the initial outflow is 27 m^3/s; the initial storage indication value is 860.33 m^3/s (it corresponds to the initial elevation of 102 m in Table 15.8). The recursive procedure should continue until the outflow has substantially declined. Sample calculations are shown in Table 15.9.

15.6.3 Mass Curve Method

This is one of the most versatile methods of reservoir routing. It has different versions that include (1) direct, (2) trial and error, and (3) graphical. Here the trial and error version is discussed in detail. For solution by the trial and error method, Eq. (15.106) can be rewritten as

$$M_2 - (V_1 + Q_m \Delta t) = S_2 \tag{15.112}$$

where M is the accumulated mass inflow, and V is the accumulated mass outflow.

A storage-discharge relation and the mass curve of inflow should be plotted before obtaining the trial and error solution. Necessary adjustments are made to show zero storage at the beginning elevation and, correspondingly, spillway discharge is obtained. Now, the following steps are involved in the trial and error solution:

1. A time is chosen and Δt is computed. Mass inflow is also computed.
2. Mass outflow is assumed. As a guideline, it is a function of accumulated mass inflow.
3. Reservoir storage is computed by deducting mass outflow from mass inflow.
4. The instantaneous and average spillway discharges are calculated.
5. Outflow for the time period Δt is computed by multiplying Δt with average discharge. Then, mass outflow is computed.
6. Now, the computed mass outflow is compared with the assumed mass outflow. If the two values agree within an acceptable degree of accuracy, then routing is complete. If this agreement is not acceptable, then another mass outflow is assumed and the above procedure is repeated.

Example 15.6

Route the hydrograph of Example 15.5 through the reservoir by using the Mass Curve method. All the data remains the same.

Solution

The calculations for the Mass Curve method by following the suggested methodology are shown in Table 15.10, which is easy to follow.

Comparison of the results with the previous example shows that the results are comparable.

(1)	(2)	(3)	(4)	(5)	(6)	(7)
Time (h)	Inflow Cumec	Cumulative Inflow (MCum)	Average Outflow (Cumec)	Cumulative Outflow (MCum)	Reservoir Storage (MCum)	Reservoir Elevation (m)
0	27	0.097	27.00	0.097	1.5	102
1	40	0.241	27.71	0.196	1.546	102.046
2	60	0.457	30.15	0.300	1.657	102.157
3	100	0.817	35.64	0.418	1.899	102.399
4	160	1.393	46.06	0.565	2.328	102.828
5	240	2.257	64.53	0.764	2.993	103.493
6	180	2.905	85.52	1.034	3.371	103.871
7	130	3.373	95.97	1.361	3.512	104.012
8	90	3.697	98.30	1.711	3.486	103.986
Remaining lines deleted.						

TABLE 15.10 Computations and Results of Example 15.6

15.6.4 Reservoir Routing with Controlled Outflow

Most of the big dams have spillways with gates that can be raised or lowered to control the outflow from the reservoir. The dam may also have undersluices to control the outflow for purposes such as irrigation, water supply, and so on. The release of water from storage through the spillway gates and undersluices depends on the state of the reservoir, magnitude of demands, their priorities, and the operation policy for the reservoir.

In gated dams, the reservoir outflow can be either (1) controlled, (2) uncontrolled, and (3) a combination of these two. The continuity equation can be written as

$$(I_1 + I_2)/2 - (Q_1 + Q_2)/2 - Q_c = (S_2 - S_1)/\Delta t \quad (15.113)$$

where Q_c is the mean controlled outflow from the reservoir during the time interval Δt. Rearranging the terms, Eq. (15.113) can be written as

$$2S_2/\Delta t + Q_2 = I_1 + I_2 + 2S_1/\Delta t - Q_1 - 2Q_c \quad (15.114)$$

When the controlled outflow Q_c is known, the solution can be obtained as explained above. The solution of Eq. (15.114) is simple if the entire outflow is controlled. The spillway rating chart can be used to determine the outflow if the reservoir elevation and the gate opening are known.

15.6.5 Major Applications of Storage Routing

Although the simplicity of storage routing is well known and it is easy to apply, the variety of problems which can be solved by this routing method gives it an important practical meaning. Some of the major applications are given below.

Determination of Reservoir Capacity for Flood Control Routing of different flood waves through a reservoir gives useful information for economic and risk analyses of flood control storage regulation. It also helps in regulation of the storage capacity reserved for flood control.

Establishing Capacities of Outlet Structures Routing of flood waves entering the reservoirs permits the determination of maximum levels in reservoirs for given reservoir features, operation policy, and initial conditions. The larger is the overall outlet capacities to release water, the lower will be the maximum level of water attained by the reservoir for a given incoming flood and the given reservoir level at the beginning of this flood. Storage routing helps determine the reservoir levels for various flood hydrograph and discharge capacity curves of outlets. Repeated routing of flood hydrographs of different return periods will provide the necessary information on the optimal design of outlet structures. The bottom outlets and the intermediate outlets are then economically analyzed and decisions made on their type and size by using the information produced by all the relevant storage routing cases.

Sizing of Spillway Capacities The storage routing provides information for the design of spillways, such as the type of spillway, number of spillway openings, the width and height of spillway openings with their discharge capacities at various reservoir levels, and so on.

Computation of Probabilities of Highest Water Levels Some fixed level reservoirs pose two problems of importance: (1) what probabilities of occurrence (return periods) should be assigned to the highest historic flooding levels and inundation consequences, and (2) what are the probabilities of reaching or exceeding still higher levels than historically observed? By storage routing of well-defined inflow hydrographs for initial reservoir levels, regulating policies, and similar information, it then becomes feasible to resolve these two practical problems.

Rate of Change of Reservoir Levels The stability of banks along the reservoir shores is an old but important problem for planning new reservoirs and the defense of shores of existing reservoirs. Wave erosion is another problem. Landslides into reservoirs are often crucial in many respects. Any evaluation of dangers of landslides should be based on the rate of change of reservoir level besides the analysis of all the other factors that trigger landslides, especially during the highest reservoir levels. To obtain an insight of how fast levels may decrease after high for a given time, in the form of the rates of change of levels in these decreasing phases, may be determined by storage routing of flood inflow discharge hydrograph for given initial and boundary conditions.

Effects of Reservoirs on Downstream Floods Downstream flood conditions can be well estimated only if both the forced or uncontrolled reservoir releases are combined with intermediate inflows between a reservoir on the river and the locations of flood problems. Proper flood storage routing through reservoirs will provide reliable information, with the corresponding probability statements, of what will the flood conditions downstream of that reservoir be.

Computation of Dam Breach Outflows To determine the hydrologic impacts in the downstream areas of a dam due to a dam breach, the outflow hydrograph from the dam is needed. To compute this hydrograph, storage routing is carried out.

15.6.6 General Observations on Reservoir Routing

For all flood routing problems, it is important that the computational time interval Δt is properly selected. This time interval should be neither too big nor too small. If a big time interval is chosen which exceeds the travel time of the wave through the reservoir, then the crest segment of outflow containing the peak discharge might pass through the reservoir between two time steps and may not be accounted for in the computations. However, if Δt is too small, then the routing computations will take longer time. Further, it is assumed that the inflow hydrograph is approximately linear during the period Δt. Jain and Singh (2003) suggested that Δt should be at least one-third to one-half of the travel time through the reservoir. Note that one may use a variable Δt, keeping it small when there are large changes in inflows and large when changes are small. However, this will make the computations complex.

The routing computations by following the trial and error procedure of the Mass Curve method are simple. The Modified Puls method has two noteworthy drawbacks. First, it assumes that the outflow from the reservoir begins at the same time as the inflow enters it; it implies that the inflow almost instantaneously passes through the reservoir regardless of its length. Second, it is difficult to choose an appropriate Δt in some cases since negative outflow may be observed during recession whenever $\Delta t > 2S_2/Q_2$ or $Q_2/2 > S_2/\Delta t$. The former drawback is not a serious one if the ratio of T_t/T_m is less than or equal to half, where T_m denotes the time to peak of inflow hydrograph and T_t denotes the travel time. T_t is defined as L/u, with L being the length of the reach and u being average steady-state velocity. The latter weakness can be overcome by plotting discharge versus $[(2S/\Delta t) + Q]$ curve on a log-log paper and comparing the plot with the line of equal values. If the plotted values lie above the line of equal values, the drawn figure must be abandoned and a new value of Δt must be selected. Occurrence of negative outflow can be mostly avoided by adopting Δt less than T_t.

15.7 Summary

Flow routing is an important component of hydrologic analysis and design for hydro-infrastructure, disaster management, dam-break flood simulation, urban planning, and so on. The chapter presents a detailed description of flow routing. With this background, the readers may want to pursue developments in flood forecasting, river rejuvenation, and hosts of other topics.

15.8 Questions

15.1 What do you understand by flood routing? Explain the terms attenuation and translation?

15.2 Explain the difference between prism and wedge storage?

15.3 What is the physical meaning of Muskingum routing parameters K and X?

15.4 Inflow and outflow for a river reach for 11 h is given below. Find the Muskingum parameters K and X for the reach.

Time (h)	0	1	2	3	4	5	6	7	8	9	10
Inflow (m^3/s)	15	25	45	70	93	82	63	44	32	18	10
Outflow (m^3/s)	15	22	37	50	63	73	72	58	46	34	20

15.5 Observed values of inflow and outflow hydrographs at the upstream and downstream ends of a river are given below.

Time (h)	0	3	6	9	12	15	18	21	24	27	30
Inflow (m^3/s)	20	60	180	210	160	110	68	40	24	14	8
Outflow (m^3/s)	20	25	67	163	179	155	112	76	49	28	10

Determine the values of K and X for the Muskingum method of flood routing.

15.6 The ordinates of flood hydrograph entering in a river are given below. Assume $K = 12$ h and $X = 0.25$. Route the hydrograph through the river and compute the outflow hydrograph.

Time (h)	0	3	6	9	12	15	18	21	24	27	30
Inflow (m^3/s)	15	60	160	234	180	120	78	57	37	28	18

15.7 Crest of the spillway of a dam is ogee shaped with crest elevation at 200 m and discharge over the crest is given by $Q = 25 \times H^{1.5}$. A flood, whose ordinates are given below, hits the initial reservoir level of 201 m. The initial outflow is 25 cumec.

Time (h)	1	2	3	4	5	6	7	8	9
Inflow (m^3/s)	48	56	70	93	124	176	231	202	180
Time (h)	10	11	12	13	14	15	16	17	18
Inflow (m^3/s)	150	132	115	101	85	70	61	52	50

Route the hydrograph through the reservoir. Take time interval $\Delta t = 1$ h.

15.8 Write the Saint-Venant equations and explain what are kinematic, diffusion, and dynamic wave models?

15.9 Route the following wave through a river by using kinematic wave theory. Consider $\Delta t = 2$ h, $\Delta x = 1500$ m, $V = 1.8$ m/s, and $\beta = 5/3$.

Time (h)	0	1	2	3	4	5	6	7	8	9	10	11	12	13
Flow Q_1 (m^3/s)	0	40	89	136	86	60	47	58	73	61	48	37	26	18

15.9 References

Cunge, J. A. On the subject of a flood wave propagation computation method (Muskingum method). *Journal of Hydraulic Research*, 7(2): 205–230, 1969.

Fread, D. L. Flow routing. In *Handbook of Hydrology*. New York: McGraw Hill Book Company, 1993.

Koussis, A. D. Theoretical estimators of flood routing parameters. *Journal of the Hydraulics Division, ASCE*, 104(HY1): 109–115, 1978.

Koussis, A. D. Comparison of Muskingum method difference schemes. *Journal of the Hydraulics Division, ASCE*, 106(HY5): 925–929, 1980.

Jain, S. K., and V. P. Singh. *Water Resources Systems Planning and Management*. Amsterdam: Elsevier, 2003.

Lighthill, M. J., and G. B. Whitham. On kinematic waves:1. Flood movement in long rivers. *Proceedings of the Royal Society of London, England, Series A*, 22g:281–316, 1955.

McCarthy, G. T. The unit hydrograph and flood routing. Manuscript presented at a *Conference of the North Atlantic Division, U.S. Army Corps of Engineers*, June 24, 1938 (unpublished).

Natural Environment Research Council (NERC). *Flood Studies Report, Vol.5: Flood Routing*. London: NERC, 1975.

Perumal, M., E. O'Connell, and K. G. Ranga Raju. Field Applications of a Variable-Parameter Muskingum Method. *Journal of Hydrologic Engineering, ASCE*, 6(3): 196–207, 2001.

Ponce, V. M. Stream channel routing. In *Engineering Hydrology: Principles and Practices*. Englewood Cliffs, NJ: Prentice Hall, 1989.

Preissmann, A. Propagation of translator waves in channels and rivers. In *Proceedings of First Congress of French Association for Computations*, Grenoble, France, 433–442, 1961.

Singh, V. P. *Kinematic-Wave Modeling in Water Resources: Surface-Water Hydrology*. New York: John Wiley & Sons, Inc., 1996.

Singh, V. P. Is hydrology kinematic? *Hydrological Processes*, 16:667–716. DOI: 10.1002/hyp.306, 2002.

Singh, V. P. (Ed.). *Handbook of Applied Hydrology*. 2nd ed. New York: McGraw-Hill, 2016.

CHAPTER 16

Pollutant Modeling

Pollution modeling is fundamental to developing strategies for environmental protection. This chapter discusses the rudiments of pollutant transport process, presents mathematical elements of model building, and derives simple pollution models.

16.1 Elements of Transport Process

Mechanisms of transport depend on the type of pollutant. A pollutant can be completely dissolved in water and cannot be visualized by the naked eye and separated from water. Such a pollutant is called a solute. There can also be a pollutant which, when carried by water, mixes with water but does not become part of water, such as sand. Such a pollutant is normally a physical pollutant. A pollutant can be conservative or nonconservative. A conservative pollutant conserves its mass, that is, a pollutant introduced in a flow system does not either increase or decrease in mass. An example of a conservative pollutant is sediment. Most physical pollutants are conservative. On the other hand, a nonconservative pollutant may undergo biochemical reactions and hence may increase or decrease in mass in time during its transport, as for example, agricultural pesticides. Further, certain pollutants interact with solids and hence there is solid-pollutant interaction.

Pollutant transport mechanisms may thus be divided into three groups. The first group includes mechanisms related to flow of water, such as advection dispersion, and miscibility. The second group includes mechanisms related to the pollutant itself and its concentration, such as diffusion, chemical reactions, decay, solubilization, precipitation, volatilization, and particle transport. The third group includes mechanisms related to solid-pollutant interaction, such as adsorption and desorption. Each of these mechanisms is now briefly discussed.

16.1.1 Advection

Advection refers to the mechanism where the pollutant is transported by the bulk motion of water. It is the dominant mechanism of pollutant transport in rivers, bayous, canals, channels, estuaries, and lakes. If a pollutant is nonreactive, it is transported by flow in a river at a rate which is equal to the average flow velocity. Some pollutants travel faster than the average velocity and some travel slower. In porous media, the average velocity will be affected by porosity. In groundwater, the average velocity will equal the Darcy flux divided by the medium porosity.

For pollutant modeling, let C denote the concentration of pollutant defined as the mass of solute per unit volume of solution or flow (ML^{-3}). If u is the average velocity of flow in a channel carrying the pollutant, then the flux of pollutant, called advective flux ($ML^{-2}T^{-1}$), F_a, is the product of u and C, that is Cu.

$$F_a = Cu \tag{16.1}$$

If the pollutant is transported in a porous medium, such as soil, whose porosity is n, then the average flow velocity is $u = q/n$ where q is the Darcy flux.

16.1.2 Dispersion and Diffusion

In a river, flow velocity varies longitudinally as well as transversely. This also applies to the pollutant and its concentration. Hence the pollutant spreads during its movement from the path defined by advection, as shown in Fig. 16.1, and this spreading is called dispersion or sometimes hydrodynamic dispersion. The spread occurs both longitudinally and transversely and is irreversible and its effect is dilution. In surface water systems, longitudinal dispersion is dominant.

Hydrodynamic dispersion comprises mechanical dispersion and molecular diffusion or simply diffusion. Mechanical dispersion, also called hydraulic dispersion, is caused by advection of the fluid. At a small scale, dispersion in a channel is caused by turbulence, eddy formation, vorticity, and viscous sublayer. In groundwater flow, the bulk flow velocity varies due to heterogeneity and anisotropy, manifested by different pore sizes along flow paths, different pore channels, tortuosity, branching, and interfingering of pore channels.

Now, the dispersive flux, F, defined as the mass of pollutant per unit area ($ML^{-2}T^{-1}$) per unit time, can be expressed in one-dimensional flow using Fick's law which expresses the dispersive flux as linearly proportional to the gradient of average concentration. Thus F can be written as

$$F = -D\frac{\partial C}{\partial x} \tag{16.2}$$

where C is the solute concentration which is the mass of solute per unit volume of water (or flow) (ML^{-3}), and D is the coefficient of (mechanical) dispersion (L^2/T).

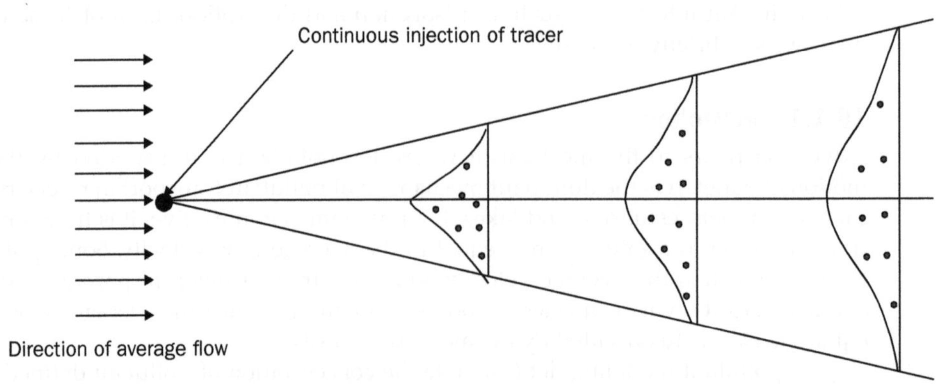

FIGURE 16.1 Pollutant dispersion.

The coefficient of dispersion in porous media depends on the characteristics of porous media, average flow velocity, and molecular diffusion, and can be expressed as

$$D = \alpha u \tag{16.3}$$

in which α is the characteristic property of the porous medium and is referred to as the dynamic dispersivity, longitudinal dispersivity, or simply dispersivity (L).

Molecular diffusion occurs due to the concentration gradient and may occur in the absence of movement of fluid. Consider putting a sugar cube in a cup of coffee in which the water is still. Immediately, the sugar starts moving down and within seconds it gets mixed with the water. The movement of sugar downward from the top where it is placed in the cup occurs because of the difference in sugar concentration which is higher upward and lower downward. This movement due to the concentration gradient is called diffusion. The movement will continue until the sugar is uniformly mixed and the gradient tends to zero. The diffusion is also called ionic or self-diffusion, because a pollutant may move under the influence of thermal kinetic energy. It causes mixing of pollutants and may be important at low velocities. Due to the concentration gradient across the flow path, diffusion contributes to the growth of transversal dispersion.

Now, let J denote the molecular diffusive flux. Using Fick's law, J can be written as

$$J = -D_d \frac{\partial C}{\partial x} \tag{16.4}$$

in which D_d is the coefficient of molecular diffusion or simply diffusion coefficient (L^2/T) and is approximately 10^{-5} cm/s in dilute systems. Equation (16.4) has a form similar to Eq. (16.2). Hence the dispersive and diffusive fluxes can be summed to obtain the hydrodynamic dispersion as

$$F + J = -(D + D_d)\frac{\partial c}{\partial x} = -D_h \frac{\partial c}{\partial x} \tag{16.5}$$

where D_h, which combines the coefficients of mechanical dispersion and molecular dispersion, is called the coefficient of hydrodynamic dispersion or simply dispersivity (L^2/T).

16.1.3 Adsorption and Desorption

Some pollutants, such as agricultural pesticides, react with soil through adsorption, are washed away by runoff, and infiltrate into the soil. Adsorption results in binding with sediment particles and hence partitioning of pollutant (or solute) into the immobile soil surface phase and the mobile solution phase. There occurs an exchange of ions between ions in the solution and ions on the surface of soil particles. The binding of pollutant on the soil particle surface is through electrical attraction and chemisorption. The attraction comprises intermolecular forces of attraction between soil particles and adsorbed solute particles. Chemisorption is caused by the chemical interaction between soil particles and adsorbed pollutant. Adsorption increases the pollutant mass on the soil at the fluid-soil interface. Desorption is the reverse of adsorption and entails dislodgement of pollutant from soil particles. Both adsorption of pollutant to and desorption of pollutant from soil particles depend on the physical and chemical characteristics of soil and pollutant, temperature, flow characteristics, and other pollutants in the flow.

The amount of pollutant attached to the soil particles is estimated using what is called an adsorption isotherm, denoted by s, which is a relation between the adsorbed concentration (s) and the solution concentration C under isothermal conditions. Different adsorbate-adsorbent pairs exhibit different types of isotherm. For a single type ion transport, an isotherm can be either an equilibrium type or a nonequilibrium type. The equilibrium isotherm implies that the quantities on the soil particle surface and the surrounding solution are continuously at equilibrium, but the equilibrium is not achieved in the nonequilibrium isotherm. Singh (1997, p. 830) has provided a table of equilibrium and nonequilibrium isotherms. The simplest equilibrium isotherm is of linear form expressed as

$$s = kC \tag{16.6}$$

in which k is referred to as the partitioning coefficient. Equation (16.6) states that the adsorption is instantaneous and reversible. Parameter k, ($k = s/C$), specifies at every instant the mass of solute per unit mass of soil particles per unit concentration of the solute in the liquid phase.

Equation (16.6) is modified in an unsaturated porous medium where water occupies only part of the void space and this part defines the degree of saturation denoted as θ. This means that only a portion of the soil particle surface is in contact with the liquid and the amount of this portion depends on θ. Thus the linear equilibrium isotherm is modified as

$$s = \theta kC \tag{16.6a}$$

An accurate partitioning of solute between solid and liquid phases is needed for transport modeling in soil and groundwater. Thus the choice of an isotherm is critical.

16.1.4 Growth and Decay

A pollutant can grow or decay in mass in the liquid phase and/or solid phase. The factors that cause loss or gain of the pollutant mass include chemicals in the liquid, biodegradation, radioactive decay, and growth due to bacterial activities.

In pollutant modeling, the gain of pollutant is described as a source and the loss as a sink. The growth is often expressed by a zeroth-order production reaction and the decay by a first-order decay reaction. Let ρ_s denote the bulk density of solid (ML^{-3}), ρ_w denote the bulk density of liquid (ML^{-3}), r_w rate constant for the first-order decay in the liquid phase (T^{-1}), r_s the rate constant for the first-order decay in the solid phase (T^{-1}), p_w the rate constant for zero-order production in the liquid phase (T^{-1}), and p_s the rate constant for zero-order production in the solid phase (T^{-1}). The loss or gain of solute mass in unsaturated flow can be expressed as

$$L = r_s s + r_w \theta C - p_s \rho_s - p_w \rho_w \theta \tag{16.7}$$

where L is in mass change per unit volume per unit time. If θ equals unity, Eq. (16.7) will apply to streams. In streams, however, biochemical oxygen demand (BOD) is often assumed to follow a first-order kinetics.

16.1.5 Volatilization

The process by which liquids and solids vaporize and escape to the atmosphere is called volatilization. The vaporization of a pollutant from the surface of a liquid or solid entails escape from the solid-liquid interface depending on the vapor pressure of the pollutant at the given temperature, diffusion through the surface boundary, and dispersion by advection and turbulent dispersion.

16.1.6 Dissolution-Precipitation

Most of the constituents found in groundwater result from the dissolution of minerals with which water comes in contact. The dissolution or precipitation affects the water quality and depends on temperature, pressure, pH, and concentration. An example is the dissolution of calcium carbonate which follows the following reaction

$$CaCO_3 + H_2CO_3 \Leftrightarrow C_a^{+2} + 2HCO_3 \qquad (16.8)$$

16.2 Transformation of Pollutants in the Environment

The movement of pollutants occurs in surface water, groundwater, and between the air-water, water-soil, and soil-air interfaces. There are important transformations that occur in the environment that affect the concentration and fate of pollutants therein. These transformations are exemplified by bacterial conversion, natural decay, hydrolysis reactions, photochemical reactions, and oxidation-reduction reactions.

Bacterial conversion is one of the most important transformations in surface water and groundwater. Commonly found examples of bacterial conversion are carbonaceous BOD (CBOD) and nitrogenous BOD (NBOD). Decay of pollutants is inherent in nature. For example, bacteria have finite life and certain organic compounds decay because of photo-oxidation. There are reactions between pollutants and water. Such hydrolysis reactions convert organic matter that occurs in wastewater, bottom deposits of rivers and lakes, and groundwater. Photochemical reactions are caused by solar radiation. A variety of organic compounds are decomposed by radiation in the ultraviolet and visible range. These contaminants must typically be present near the water surface. Oxidation-reduction reactions may mediate other reactions with contaminants that may be present in such reactions, where one pollutant may be oxidized and the other may be reduced by donating or accepting electrons.

16.3 Elements of Pollutant Modeling

Development of a pollutant transport model depends on the governing equation, and initial and boundary conditions (ICs and BCs). The fundamental governing equation is the conservation of mass which states that the net rate of change of pollutant mass within an elemental volume is equal to the flux of pollutant out of the element minus the flux of pollutant into the element and the loss or gain of pollutant mass as a result of reactions. The physical processes that control the movement of flux in and out of the elemental volume are advection due to flow and hydrodynamic dispersion due to mechanical mixing and molecular mixing. The loss or gain of pollutant mass is the result of chemical or biochemical reactions or radioactive decay. Mathematical expressions of mass conservation and ICs and BCs depend on the flow medium in which the pollutant moves. Therefore it is more convenient to write them for the medium under consideration.

The mass balance equation for a pollutant entails the amount of pollutant into or out of control volume through advection, dispersion, and diffusion; pollutant leaving the liquid phase through liquid-solid interface due to ion exchange and adsorption;

pollutant added to or subtracted from the liquid phase due to chemical interactions among species inside water or by various decay processes; addition of pollutant through the injection of other means; and the time rate of change of pollutant mass in the liquid and solid phases. When the pollutant mass balance equation is coupled with the hydrodynamic dispersion, the resulting equation is the advection-dispersion equation, which is a parabolic partial differential equation whose solution will require two boundary conditions: one at the upstream boundary and the other at the downstream boundary.

For pollutant transport, mathematical models can be developed in two ways. First, flow transporting the pollutant can be considered steady and uniform so the model needs to be developed for pollutant transport only. Here also, depending on the situation, pollutant concentration can be considered unsteady but uniform, steady but nonuniform, or unsteady and nonuniform. The second way, which is more realistic, is to consider both flow and pollutant transport simultaneously. Again, simplifications can be achieved by assuming steadiness or uniformity in flow and pollutant transport. We will focus on the second way in the discussion here. Irrespective of the way adopted for pollutant transport modeling, one can derive solutions of transport equations using one of the two schemes. First, the advection-dispersion (AD) transport equation can be solved numerically or analytically after linearization. Note this AD equation will have parameters, like coefficient of hydrodynamic dispersion, which can be expressed in terms of measurable hydraulic and geometric parameters. Second, the transport equation can be reduced to an advection equation (like kinematic) by eliminating the dispersion term, expressing the equation in finite difference form using weighting factors, expanding the difference terms by the Taylor series up to second order, obtaining the advection-dispersion (AD) type equation, equating this AD equation with the original AD equation, expressing the weighting factors in terms of hydraulic and geometric parameters, and then solving the numerically obtained AD equation, which is quite simple to do. This scheme of solution will be followed in this chapter. It may be emphasized that this scheme will hold for both flow of water and pollutant transport.

16.4 Pollutant Transport in Urban Watersheds

The amendments of the Federal Water Pollution Control Act in 1972 emphasized control of the pollution of water resources from nonpoint sources. In order to improve water quality, the 1976 Clean Water Act mandated industries and municipalities to control pollutant discharges. In 1978, the U.S. Environmental Protection Agency (EPA) established the 5-year Nationwide Urban Runoff Program to better understand urban runoff. In urban watersheds, pollutants are generated from road pavement material, automobiles, sweepings of sidewalks, litter, droppings from animals and birds, house clean-up activities, acid precipitations, garbage, and so on. Such pollutants accumulate during dry periods on impervious surfaces, are washed off when it rains, and are discharged into the sewer system and eventually into the receiving stream. Thus stormwater in urban areas is a major source of pollution in the receiving water bodies.

Urban watersheds normally have mixed land use, comprising impervious surfaces, pavements and roads, lawns, parks, houses, schools, hospitals, and office buildings. Depending on the location and the degree of urbanization, a watershed may be predominantly impervious or pervious or equally pervious and impervious. Further, pollutants can mix in water instantaneously or over a finite time period. Over pervious surfaces, there can be a mixing zone. There can be all kinds of pollutants and all kinds of pollutant sources. However, for economy of space, it is assumed that pollutants considered in the discussion here are conservative and are lumped together to form one pollutant mass. The urban surface is primarily impervious. Urban runoff can be adequately modeled using one-dimensional kinematic wave theory. For simplicity, we assume that rainfall is steady, uniform, and of finite duration. Therefore transport of pollutant on a plane by runoff involves flow of water, pollutant washoff, and transport or flow of pollutant. The transport of pollutant is by advection only, thus neglecting dispersion and biochemical reactions.

The urban pollutant transport problem is cast as a couple problem of flow and pollutant transport. For flow on a plane, the kinematic wave equation can be written on a unit width basis as

$$\text{Mass conservation:} \quad \frac{\partial A}{\partial t} + \frac{\partial Q}{\partial x} = I - f \tag{16.9}$$

and kinematic wave depth discharge relation

$$Q = uh = \alpha h^n; \quad u = \alpha h^{n-1} \tag{16.10}$$

where A is the flow cross-sectional area $= Bh$, h is the depth of flow, Q is the discharge per unit width, u is the mean cross-sectional flow velocity, I is rainfall intensity, f is the rate of infiltration, α is the resistance parameter, and n is the exponent in the depth-discharge relation indicating if the flow is laminar, transient, or turbulent. Combining Eqs. (16.9) and (16.10), the result is

$$\frac{\partial h}{\partial t} + \alpha n h^{n-1}\frac{\partial h}{\partial x} = I - f \tag{16.11a}$$

Equation (16.11) can also be expressed as

$$\frac{\partial h}{\partial t} + c\frac{\partial h}{\partial x} = I - f;\ c = \left(\frac{dQ}{dh}\right)_{x=\text{constant}} = \alpha n h^{n-1} \tag{16.11b}$$

where c is wave celerity.

The finite period rainfall can be expressed as

$$I(x,t) = I > 0, 0 < t \leq T;\ I(x,t) = 0,\ t \geq T, T > 0 \tag{16.12}$$

where T is the duration of rainfall.

Now we will consider the one-dimensional pollutant mass conservation equation for both instantaneous as well as finite period pollutant mixing. For instantaneous mixing, the mass conservation ion can be expressed as

$$\frac{\partial (Ch)}{\partial t} + \frac{\partial (Ch)}{\partial x} = w\delta(t) \tag{16.13}$$

and for finite-period mixing as

$$\frac{\partial (Ch)}{\partial t} + \frac{\partial (Ch)}{\partial x} + \frac{\partial w}{\partial t} = C_r I;\ \frac{dw}{dt} = -Kh \tag{16.14}$$

in which C is the pollutant concentration defined as the mass of pollutant per unit volume of water, w is the mass of pollutant per unit area A of the plane, $K = kSw$, k is the washoff rate constant ($L^{-1}T^{-1}$), S is the bed slope, $\delta(t)$ is the instantaneous unit flux (1/T), and C_r is the mass of pollutant per unit volume of rainwater, that is, concentration in rainwater and is assumed constant. The pollutant discharge Q_s can be expressed as

$$Q_s = CQ \tag{16.15}$$

Thus the problem of pollutant transport is formulated by Eqs. (16.9)–(16.12) or Eqs. (16.9)–(16.11), and (16.13). To solve these equations, the following initial and boundary conditions (IC & BC) can be specified

$$h(x,t) = 0, 0 \leq x \leq L; h(0,t) = 0, 0 \leq t \leq T \tag{16.16}$$

where L is the length of the plane. The rate of infiltration depends on the depth of flow as

$$f(x,t) > 0 \text{ if } h(x,t) > 0; f(x,t) = 0 \text{ if } h(x,t) = 0 \tag{16.17}$$

It is assumed that $I(x, t) > f(x, t)$, $0 \leq t \leq T$, $0 \leq x \leq L$; and that f is constant. For pollutant transport, the IC and BC can be written as

$$C(x,0) = C_0(x) > 0,\ x > 0; \quad C(0,t) = C_I(t) > 0,\ t > 0 \tag{16.18a}$$

The initial condition for w can be expressed as

$$w(x,0) = w_0;\ w(0,t) = w_0 \tag{16.18b}$$

where w_0 denotes the mass of pollutant per unit surface area accumulated before the rain event.

It may be noted when considering Eq. (16.16) that at zero flow or depth a pollutant already exists on the plane. Thus the pollutant concentration would amount to infinity and it does not have an a priori definite value. Since pollutant concentration is a physical quantity, it must be bounded in order to be physically meaningful and the bounding condition must be imposed on the solution. Using the kinematic wave theory, Singh (2002a, 2002b) derived analytical solutions for pollutant transport on pervious and impervious planes with instantaneous and finite period mixing. Analytical solutions were based on the method of characteristics. However, analytical solutions are tractable for only simple cases and numerical solutions are the only resort for real-world cases. Here we will present simple numerical solutions which will apply in most practical cases.

16.5 Pollutant Transport in Agricultural Runoff

Pollution in agricultural watersheds is mostly nonpoint source. Agricultural runoff, containing pesticides, weedicides, and chemical fertilizers, has been identified as the largest polluter of water in many U.S. states. In particular, if there is heavy rainfall after the application of pesticides, then surface runoff may transport significant amounts of these pesticides. Hence there has been a lot of public concern about the migration and fate of agricultural chemicals in the aquatic environment and the health of soil.

In agricultural watersheds, soil-surface applied or soil-incorporated chemicals are transferred from soil to surface runoff due to rainfall. The release of chemicals from soil to surface runoff depends on rainfall-soil-infiltration-surface runoff interaction. The partitioning of pollutant between surface runoff and infiltration depends on this interaction as well as soil-chemical sorption characteristics. Some chemicals are strongly adsorbed to soil particles; these move with sediment in surface runoff. The chemicals that are weakly adsorbed to soil particles can infiltrate into the soil. The depth of soil that interacts with surface runoff is relatively small. Often it is less than 2 cm and defines the mixing zone. On the soil surface, mixing is due to turbulence in surface runoff and impact of raindrops. In the mixing zone, mixing is mechanical, caused by raindrop impact, boundary turbulence due to overlying flow, and chemical sorption characteristics. Pollutants from the soil surface and mixing zone are transported by surface runoff.

For pollutant transport over infiltrating surfaces, the governing equation can be expressed by Eq. (16.1) and then combining it with Eq. (16.13) for finite period mixing as

$$h\frac{\partial C}{\partial t} + \alpha h^{n-1}\frac{\partial C}{\partial x} + C(I - f) - Kh = C_r I - Cf \tag{16.19}$$

The rate of infiltration depends on the depth of flow as

$$f(x,t) > 0 \quad \text{if} \quad h(x,t) > 0;\ f(x,t) = 0 \quad \text{if} \quad h(x,t) = 0 \tag{16.20}$$

It is assumed that $I(x, t) > f(x, t)$, $0 \le t \le T$, $0 \le x \le L$.

If pollutant transport also considers transport from the mixing zone, then the governing equations can be expressed as follows:

For surface runoff

$$\frac{\partial (Ch)}{\partial t} + \frac{\partial (CQ)}{\partial x} = K(C_m - C) - fC + C_r I \tag{16.21a}$$

For mixing zone

$$[\theta d - d\rho_s K_d(1-\theta)]\frac{dC_m}{dt} = (f + K_m)C - (f + K_m)C_m \tag{16.21b}$$

where C is the pollutant concentration in surface runoff, C_m is the pollutant concentration in the mixing zone, C_r is the pollutant concentration in the rainwater, ρ_s is the soil density, θ is the soil porosity, d is the depth of the mixing zone, S_p is the pollutant source or sink, K_d is the partition coefficient, and K_m is the film transport coefficient. Concentration C_m can be simplified as

$$C_m = C_{m0} - aI - a_1 Q \tag{16.22}$$

in which C_{m0} is the initial value of C_m, and a and a_1 are parameters.

16.6 Pollutant Transport in Streams

The source of pollutants in the case of streams is usually a point source, such as an accidental spill, industrial discharge, or sewage waste, although nonpoint source of pollution can also occur along with lateral inflow from land surfaces. The flow in the river will carry the pollutant where the primary interest is in modeling the pollutant concentration. From a hydraulic view point, modeling pollutant transport involves modeling of flow and modeling of pollutant. Pollutant transport in streams, rivers, and channels is usually modeled using the advection-dispersion equation (ADE) which, for simplicity, will be expressed as one-dimensional. Koussis (1983b) developed a unified treatment for flow and pollutant routing, which will be followed in the discussion here.

Equations for flow in a river reach have been presented in Chap. 15 on flood routing and will therefore not be repeated here. For a river reach, subjected to distributed inflow, the conservation of mass can be expressed as

$$\frac{\partial A}{\partial t} + \frac{\partial Q}{\partial x} = Q_x \tag{16.23}$$

where A is the flow cross-sectional area, Q is the discharge, and Q_x is the lateral inflow and is known. With the introduction of celerity Eq. (16.23) can be written as

$$\left(\frac{dA}{dQ}\right)\Bigg|_{x=\text{constant}} \frac{\partial Q}{\partial t} + \frac{\partial Q}{\partial x} = Q_x \quad c = \left(\frac{dQ}{dA}\right)\Bigg|_{x=\text{constant}} \tag{16.24}$$

where c is celerity. Equation (16.24) can be written as

$$\frac{\partial Q}{\partial t} + c\frac{\partial Q}{\partial x} = cQ_x \tag{16.25}$$

Equation (16.25) is a convection (kinematic equation) equation.

A linear diffusion wave equation can be expressed as

$$\frac{\partial Q}{\partial t} + c(Q_R)\frac{\partial Q}{\partial x} = \frac{Q_R}{2BS_0}\frac{\partial^2 Q}{\partial x^2} + c(Q_R)Q_x \tag{16.26}$$

where

$$D_H(Q_R) = \frac{Q_R}{2BS_0} \tag{16.27}$$

can be regarded as hydraulic diffusivity, and Q_R is a reference discharge, B is the channel width, and S_0 is the bed slope. Celerity and width are computed at the reference discharge. It may be recalled that

$$Q = C_r B y^{m+1} (S_0)^{1/2} \tag{16.28}$$

where y is the flow depth, B is the channel width, C_r is a resistant coefficient, and m is an exponent equal to half for Chezy's equation and two-thirds for Manning's equation. In terms of flow depth, the linear diffusion wave equation can be written as

$$\frac{\partial y}{\partial t} + c(y_R)\frac{\partial y}{\partial x} = \frac{y_R c(y_R)}{2(m+1)S_0}\frac{\partial^2 y}{\partial x^2} + \frac{Q_x}{B} \tag{16.29}$$

where

$$D_H(y_R) = \frac{y_R c(y_R)}{2(m+1)S_0} \tag{16.30}$$

can be regarded as hydraulic diffusivity or dispersivity, and y_R is a reference value flow depth.

For pollutant transport, the mass conservation equation can be expressed as

$$\frac{\partial C}{\partial t} + \frac{\partial Q_C}{\partial x} = Q_{Cx} \tag{16.31}$$

where Q_C is the pollutant flux, and Q_{Cx} is the lateral inflow of pollutant. Assuming that pollutant dispersion/diffusion follows Fick's law, flow is uniform, and time averaging is used, the linear advection-dispersion (AD) equation is obtained as

$$\frac{\partial C}{\partial t} + u\frac{\partial C}{\partial x} = D_L\frac{\partial^2 C}{\partial x^2} + Q_{Cx} \tag{16.32}$$

in which D_L is the longitudinal dispersion coefficient accounting for both longitudinal hydraulic dispersion and molecular diffusion.

For unsteady nonuniform flow, Eq. (16.32) can be modified as

$$\frac{\partial C}{\partial t} + u\frac{\partial C}{\partial x} = \frac{1}{A}\frac{\partial}{\partial x}\left(AD_L\frac{\partial C}{\partial x}\right) + Q_{Cx} - \frac{Q_x}{A}C \tag{16.33}$$

The last term on the right-hand side of Eq. (16.33) may be small by comparison with other terms. Therefore this equation becomes

$$\frac{\partial C}{\partial t} + u\frac{\partial C}{\partial x} = \frac{1}{A}\frac{\partial}{\partial x}\left(AD_L\frac{\partial C}{\partial x}\right) + Q_{Cx} \tag{16.34}$$

Example 16.1

It is assumed that the pollutant concentration varies in time but not in space and that the lateral pollutant inflow is constant in time. Derive the pollutant concentration as a function of time.

Solution

Under the given assumption, Eq. (16.34) can be written as

$$\frac{\partial C}{\partial t} = Q_{cx}$$

Using the initial condition $C = C_0$ at $t = 0$, the solution of this equation becomes

$$C(t) = C_0 + Q_{cx}t$$

This equation shows that pollutant concentration varies linearly with time and the rate of change is defined by the constant pollutant inflow.

16.7 Pollutant Transport in Unsaturated Flow

The one-dimensional advection-dispersion equation with zero-order production and first-order decay can be expressed as

$$R\frac{\partial C}{\partial t} + u\frac{\partial C}{\partial x} = \frac{\partial^2 C}{\partial x^2} - \gamma C - rC \tag{16.35}$$

where $u = q/\theta$ is the pore velocity, R is the retardation factor,

$$R = 1 + \rho_s \frac{K}{\theta} \tag{16.36}$$

and r and γ are constants given as

$$r = r_w + r_s \rho_s \frac{K}{\theta} \tag{16.37}$$

$$\gamma = \gamma_w + \gamma_s \frac{\rho_s}{\theta} \tag{16.38}$$

Example 16.2

It is assumed that the pollutant concentration in the unsaturated zone varies in time only. Determine the pollutant concentration as a function of time. The initial concentration is defined as C_0.

Solution

Under the assumed conditions, Eq. (16.35) becomes

$$R\frac{\partial C}{\partial t} = -C(\gamma - r)$$

The solution of this equation, subject to the initial condition, becomes

$$C(t) = C_0 \exp\left[-\frac{(\gamma + r)}{R}t\right]$$

This equation shows that the concentration declines exponentially from the initial concentration.

16.8 Pollutant Transport in Groundwater

The one-dimensional AD equation describing pollutant transport in groundwater flow can be expressed as

$$\frac{\partial C}{\partial t} + u\frac{\partial C}{\partial x} = D_H \frac{\partial^2 C}{\partial x^2} \tag{16.39}$$

where C is the pollutant or solute concentration, D_H is the dispersion coefficient describing the dispersive ability of the pollutant, and u is the groundwater velocity at any position x and time t. The ratio of D_H and u is referred to as the dispersivity of the porous medium.

Example 16.3
It is assumed that the pollutant concentration in groundwater varies in time only. Derive the pollutant concentration as a function of time, subject to the initial condition C at $t = 0$ is C_0.

Solution
Equation (16.39), under the given assumption, becomes

$$\frac{\partial C}{\partial t} = 0$$

The solution of this equation becomes

$$C(t) = C_0$$

This equation shows that the pollutant concentration in groundwater remains constant.

16.9 Numerical Solution of Convective (or Kinematic Wave) Approximation

For purposes of generalization, let a variable, denoted as f, represent, either flow discharge Q or pollutant concentration C. Considering, for simplicity, lateral source and sink terms as zero, the pure advection equation [Eqs. (16.25) and (16.35)] can be expressed as

$$\frac{\partial f}{\partial t} + a\frac{\partial f}{\partial x} = 0 \tag{16.40}$$

In Eq. (16.40), the space derivative is discretized over Δx and the time derivative is approximated by a weighted average (with a weighting factor w). We then obtain

$$w\left(\frac{\partial f}{\partial t}\right)_j + (1 - w)\left(\frac{\partial f}{\partial t}\right)_{j+1} + a\frac{f_{j+1} - f_j}{\Delta x} = 0 \tag{16.41}$$

where j denotes the space discretization index, such that $x_j = j.\Delta x$. Let

$$\frac{\partial f}{\partial t} \approx \frac{f_{i+1} - f_i}{\Delta t};\ \Delta t = t_{i+1} - t_i \tag{16.42}$$

Centering the spatial derivative in Δt, we obtain

$$w\frac{f(x_j,t_{i+1}) - f(x_j,t_i)}{\Delta t} + (1 - w)\frac{f(x_{j+1},t_{n+1}) - f(x_{j+1},t_i)}{\Delta t} + c\frac{f(x_{j+1},t_i) - f(x_j,t_i)}{2\Delta x}$$
$$+ c\frac{f(x_{j+1},t_{i+1}) - f(x_i,t_{i+1})}{2\Delta x} = 0 \tag{16.43}$$

Here $c = a$. Rearranging terms of Eq. (16.43), we obtain

$$f(x_{j+1}, t_{i+1}) = a_1 f(x_j, t_i) + a_2 f(x_j, t_{i+1}) + a_3 f(x_{j+1}, t_i); \quad \sum_{i=1}^{3} a_i = 1 \tag{16.44}$$

where

$$a_1 = \frac{2w\Delta x + c\Delta t}{2\Delta x(1-w) + c\Delta t}; \quad a_2 = \frac{c\Delta t - 2w\Delta x}{2\Delta x(1-w) + c\Delta t}; \quad a_3 = \frac{2\Delta x - c\Delta t - 2w}{2\Delta x(1-w) + c\Delta t} \tag{16.45}$$

Equation (16.44) is analogous to the Muskingum method of flood routing. Coefficients a_i, $I = 1, 2,$ and 3, are the coefficients and depend on the weighting factor w.

16.10 Another Numerical Solution of Convective (Kinematic Wave) Approximation

Equation (16.40) can be approximated by finite differences weighted by factor w for the time derivative and p for the space derivative as follows

$$w\left(\frac{\partial f}{\partial t}\right)_j + (1-w)\left(\frac{\partial f}{\partial t}\right)_{j+1} + pa\left(\frac{\partial f}{\partial x}\right)_i + a(1-p)\left(\frac{\partial f}{\partial x}\right)_{i+1} = 0 \tag{16.46}$$

Equation (16.46) can be expressed in finite difference form as

$$w\frac{f(x_j, t_{i+1}) - f(x_j, t_i)}{\Delta t} + (1-w)\frac{f(x_{j+1}, t_{i+1}) - f(x_{j+1}, t_i)}{\Delta t} + ap\frac{f(x_{j+1}, t_i) - f(x_j, t_i)}{\Delta x}$$
$$+ a(1-p)\frac{f(x_{j+1}, t_{i+1}) - f(x_i, t_{i+1})}{\Delta x} = 0 \tag{16.47}$$

Collecting the unknown term on the left side and known terms on the right side, we obtain

$$f(x_{j+1}, t_{i+1})\left[\frac{(1-w)}{\Delta t} + \frac{a(1-p)}{\Delta x}\right] = f(x_j, t_i)\left[\frac{w}{\Delta t} - \frac{ap}{\Delta x}\right]$$
$$+ f(x_j, t_{i+1})\left[-\frac{w}{\Delta t} - \frac{a(1-p)}{\Delta x}\right] + f(x_{j+1}, t_i)\left[\frac{a(1-w)}{\Delta t} - \frac{ap}{\Delta x}\right] \tag{16.48}$$

Equation (16.48) takes on the same form as Eq. (16.44) with coefficients expressed as

$$C_1 = \frac{w + pN_c}{1 + N_c - (w + pN_c)}; \quad C_2 = \frac{N_c - (w - pN_c)}{1 + N_c - (w + pN_c)}; \quad C_2 = \frac{1 - (w + pN_c)}{1 + N_c - (w + pN_c)} \tag{16.49}$$

The sum of coefficients in Eq. (16.49) equals one. Koussis (1983a, 1983b) called this as a fractional scheme.

A Taylor series expansion of $f(t_{j+1}, t_{i+1})$ about $f(x_j, t_i)$ would yield an approximation of the AD equation with the numerical dispersion D_N, written as

$$D_N = \frac{a\Delta x}{2}[(1-2w) + a(1-2p)] \tag{16.50}$$

If D_N is greater than zero then the concentration wave would attenuate and if D_N is less than zero then the wave would amplify, indicating instability. If $D_N = 0$, then the wave would be advected without dispersion. Thus the performance of the numerical scheme will depend on the values of a, Δx, Δt, and weighting factors. For the correspondence between numerical and physical dispersion, $w < 0.5$ and $p < 0.5$ for all a.

16.11 Solution by Numerical Integration

Equation (16.41) can be rearranged with unknown terms on the left side and known terms on the right side as

$$\left(\frac{\partial f}{\partial t}\right)_{j+1} + \frac{a}{\Delta x(1-w)} f_{j+1} = -\frac{w}{1-w}\left(\frac{\partial f}{\partial t}\right)_j + \frac{a}{\Delta x(1-w)} f_j \tag{16.51a}$$

To solve Eq. (16.51a), let the initial condition be defined as

$$f_{j+1} = f_j = f_\infty = f(x, t = 0) \tag{16.51b}$$

Following Koussis (1983a, 1983b), the solution of Eq. (16.51a) can be expressed using the integrating factor method as

$$f_{j+1}(t) = \frac{f_\infty}{1-w}\exp\left[-\frac{wt}{\Delta x(1-w)}\right] + \frac{a}{\Delta x(1-w)^2}\int_0^t f_j(\tau)\exp\left[-\frac{a(t-\tau)}{\Delta x(1-w)}\right]d\tau - \frac{w}{1-w}f_j(t) \tag{16.52}$$

Let $f_j(t)$ vary linearly over Δt and i be the time discretization index such that $t_i = i.\Delta t$. Then, the integral in Eq. (16.52) can be evaluated and the finite difference solution can be written as

$$f_{j+1}^{i+1} = C_1 f_j^i + C_2 f_j^{i+1} + C_3 f_{j+1}^i \tag{16.53}$$

in which

$$C_1 = \frac{1-\omega}{N_c} - \omega \tag{16.54}$$

$$C_2 = 1 - \frac{1-\omega}{N_c} \tag{16.55}$$

$$C_3 = \omega = \exp\left[-\frac{a\Delta t}{\Delta x(1-w)}\right] = \exp\left[-\frac{N_c}{1-w}\right] \tag{16.56}$$

$$N_c = \frac{a\Delta t}{\Delta x} \tag{16.57}$$

The quantity N_c is regarded as the Courant number. The sum of coefficients equals one. Equation (16.53) is termed as exponential scheme (Koussis, 1983a, 1983b). Here $w < 0.5$ for all a to assure correspondence between numerical and physical dispersion.

The AD equation can be obtained from the advection (kinematic) wave equation by expanding $f_{j+1}^{i+1} = f((j+1)\Delta x;(i+1)\Delta t)$ about (x_j, t_i) by the Taylor series which leads to the numerical dispersion coefficient as

$$D_N = \frac{a\Delta x}{2}\left[\frac{2N_c}{1-\omega} - 1 - N_c\right] \tag{16.58}$$

Like Eq. (16.48), the performance of the numerical scheme will depend on the values of a, Δx, Δt, and weighting factors and its behavior will be influenced by D_N as before. However, Koussis (1983a, 1983b) reasoned that a numerical integration method would be preferable. This method exploits the spatial discretization error of a finite difference scheme developed for the numerical integration of the pure advection transport equation to approximate the solution of the AD equation. This error is called numerical diffusion/dispersion. It consists of a fixed error term originating from the discretization of the first spatial derivative, and a variable error term that results from the spatial off-centering when evaluating the temporal derivative. The AD equation is approximated by matching the numerical diffusion/dispersion coefficient with the physical. Of course, the transport velocity used is the same.

For characterizing advection-dispersion phenomenon, the Peclet number P_N is often used

$$P_N = \frac{al}{D_l} \tag{16.59}$$

where D_l is the longitudinal dispersion, $a = u$ is velocity, and l is the characteristic length. For the grid or cell Peclet number, l is replaced by Δx. If $P_N \leq 1$ the flow is dispersion dominated and if $P_N \geq 0.5$ the flow is advection dominated (Leonard, 1980). The ratio u/D_1 typically varies from 0.1/m for small streams to 0.001/m for large rivers (Fischer et al., 1979).

16.12 Modeling of Dissolved Oxygen in Streams

When waste is discharged into a stream, it impacts the dissolved oxygen (DO) of the streamflow. Since water quality standards in the United States and some other countries are expressed in terms of DO, the DO concentration is of particular concern. A waste discharge is often characterized by the BOD. It may therefore be desirable to compute BOD and DO along a stream as well as in time. We also note that BOD affects DO but not vice versa. Although advection plays a dominant role in the transport of DO and BOD, the effect of dispersion is not negligible. One can, however, accurately model the DO and BOD concentration using the scheme alluded to earlier, as proposed by Koussis (1983a, 1983b) and Koussis et al. (1990). For modeling, a few simplifying assumptions are made. First, a one-dimensional approximation will suffice. Second, longitudinal mixing due to turbulent and molecular diffusion is not significant. Third, BOD degradation follows a first- order kinetic reaction. Fourth, reaeration follows Henry's law which states that the rate of reaeration is proportional to the DO deficit ($C_s - C$), C_s is the DO value at saturation, and C is the DO concentration. With these assumptions, transport of BOD and its impact on DO can be mathematically expressed.

16.12.1 Model Formulation

For modeling using the indicated scheme, the mass balance equations for BOD, denoted by C_B, and DO denoted by C, can be written, respectively, as

$$\frac{\partial C_B}{\partial t} + u(x)\frac{\partial C_B}{\partial x} = \frac{1}{A}\frac{\partial}{\partial x}\left(D_L A\frac{\partial C_B}{\partial x}\right) - k_1 C_B \tag{16.60}$$

$$\frac{\partial C}{\partial t} + u(x)\frac{\partial C}{\partial x} = \frac{1}{A}\frac{\partial}{\partial x}\left(D_L A\frac{\partial C}{\partial x}\right) - k_1 C_B + k_2(C_s - C) \tag{16.61}$$

in which C_B is the mean BOD concentration, C is the mean DO concentration, A is the flow cross-sectional area, u is the mean cross-sectional flow velocity, D_L is the coefficient of longitudinal dispersion, k_1 is the rate of oxidation of BOD in first-order kinetics, k_2 is the coefficient of aeration, x is the distance along the channel, and t is time. It may be convenient to express Eq. (16.61) in terms of the DO deficit $C_d = C_s - C$ as

$$\frac{\partial C_d}{\partial t} + u(x)\frac{\partial C_d}{\partial x} = \frac{1}{A}\frac{\partial}{\partial x}\left(D_L A\frac{\partial C_d}{\partial x}\right) + k_1 C_B - k_2 C_d \tag{16.62}$$

First, Eq. (16.60) can be solved for BOD and then Eq. (16.61) can be solved for DO wherein C_B can be considered known. For solving Eqs. (16.60) and (16.61), the coefficients can be considered constant over grid segment. Hence $u(x)$ can be represented as U and the second-order term on the right side can be approximated by multiplying D_L with the second-order spatial derivative of concentration.

Considering constant coefficients and variable transformations as $C_B = C_b \exp(-k_1 t)$ which removes the BOD decay term in Eq. (16.60) and $C = C_D \exp(-k_2 t)$ which removes the source term from Eq. (16.61). Thus Eq. (16.60) becomes

$$\frac{\partial C_b}{\partial t} + U\frac{\partial C_b}{\partial x} = D_L\frac{\partial^2 C_b}{\partial x^2} \tag{16.63}$$

$$\frac{\partial C_D}{\partial t} + U\frac{\partial C_D}{\partial x} = D_L\frac{\partial^2 C_D}{\partial x^2} + k_1 C_b \exp[-(k_1 - k_2 t)] \tag{16.64}$$

16.12.2 Numerical Solution

To develop the numerical solution, it is assumed that advection is the dominant pollutant transport mechanism and dispersion is secondary whose effect is corrected for afterward. With the dispersion term deleted, Eq. (16.63) can be expressed in finite form with spatial weighting factor w for the time derivative and the arithmetic average of difference for the spatial derivative at time t and $t + \Delta t$ as

$$\begin{aligned} w\frac{C_b(i,j+1) - C_b(i,j)}{\Delta t} &+ (1-w)\frac{C_b(i+1,j+1) - C_b(i+1,j)}{\Delta x} \\ &+ \frac{U}{2\Delta x}[C_b(i+1,j) - C_b(i,j) + C_b(i+1,j+1) - C_b(i,j+1)] \end{aligned} \tag{16.65}$$

Note that t is denoted by index j and x by index i and 1 implies one grid interval or size. Collecting the unknown term on the left side and known terms on the right side, one obtains

$$C_b(i+1,j+1) = C_1C_b(i,j) + C_2C_b(i,j+1) + C_3C_b(i+1,j) \tag{16.66}$$

where

$$C_1 = \frac{\left(w + \dfrac{U\Delta t}{2\Delta x}\right)}{Den};\ C_2 = \frac{\left[\dfrac{U\Delta t}{(2-w)\Delta x}\right]}{Den};\ C_3 = \frac{\left[1 - w - \dfrac{U\Delta t}{2\Delta x}\right]}{Den};\ Den = 1 - w - \frac{U\Delta t}{\Delta x} \tag{16.67}$$

Quantity *Den* is the Courant number.

Equation (16.66) is first-order approximation of the advection equation. In a similar manner, the first-order approximation for DO can be expressed as

$$C_D(i+1,j+1) = C_1C_D(i,j) + C_2C_D(i,j+1) + C_3C_D(i+1,j) + C_4C_b(i,j) \tag{16.68}$$

in which

$$C_4 = \frac{2(k_1\Delta t)\exp[-(k_1 - k_2)j\Delta t]}{Den} \tag{16.69}$$

Equation (16.68) can be expanded in the Taylor series up to a second-order derivative around point (i, j) as

$$C_D(i+1,j) = C_D(i,j) + \Delta x\frac{\partial C_D}{\partial x} + \frac{(\Delta x)^2}{2}\frac{\partial^2 C_D}{\partial x^2} \tag{16.70}$$

$$C_D(i,j+1) = C_D(i,j) + \Delta t\frac{\partial C_D}{\partial t} + \frac{(\Delta t)^2}{2}\frac{\partial^2 C_D}{\partial t^2} \tag{16.71}$$

$$\begin{aligned} C_D(i+1,j+1) = C_D(i,j) &+ \Delta x\frac{\partial C_D}{\partial x} + \Delta t\frac{\partial C_D}{\partial t} + \frac{(\Delta x)^2}{2}\frac{\partial^2 C_D}{\partial x^2} \\ &+ \frac{(\Delta t)^2}{2}\frac{\partial^2 C_D}{\partial t^2} + \Delta x\Delta t\frac{\partial^2 C_D}{\partial x\partial t} \end{aligned} \tag{16.72}$$

Now the second-order time derivatives and mixed derivatives are transformed with the use of advection transport equation for C_b. Hence

$$\begin{aligned} \frac{\partial}{\partial t}\left(\frac{\partial C_D}{\partial t}\right) = U^2\frac{\partial^2 C_D}{\partial x^2} - k_1\exp[-(k_1-k_2)t]\left[U\frac{\partial C_D}{\partial x} - \frac{\partial C_D}{\partial t}\right) \\ - k_1(k_1-k_2)C_D\exp[-(k_1-k_2)t] \end{aligned} \tag{16.73}$$

$$\frac{\partial}{\partial x}\left(\frac{\partial C_D}{\partial t}\right) = -U\frac{\partial^2 C_D}{\partial x^2} + k_1\exp[-(k_1-k_2)t]\frac{\partial C_D}{\partial x} \tag{16.74}$$

Inserting Eqs. (16.70) to (16.74) in Eq. (16.68), we obtain the following approximate AD equation

$$\frac{\partial C_D}{\partial t} + U\frac{\partial C_D}{\partial x} = \left(\frac{U\Delta x}{2}\right)(1 - 2w)\frac{\partial^2 C_D}{\partial x^2} + k_1 C_b \exp[-(k_1 - k_2)t](1 + \text{error}) \quad (16.75)$$

Neglecting the last term on the right side, Eq. (16.75) can be written as

$$\frac{\partial C_D}{\partial t} + U\frac{\partial C_D}{\partial x} = D_N\frac{\partial^2 C_D}{\partial x^2};\ D_N = \left(\frac{U\Delta x}{2}\right)(1 - 2w) \quad (16.76)$$

Koussis et al. (1990) have reasoned that the dispersion term arises numerically and is hence referred to as numerical diffusion. In order to account for physical dispersion, the numerical diffusion term is equated with the actual dispersion term and hence this approach is called the matched artificial dispersion (MAD) approach. This approach results in the value of w as

$$w = 0.5 - \frac{D_L}{U\Delta x} = 0.5 - \frac{1}{(U\Delta x/D_L)} \quad (16.77)$$

where quantity $U\Delta x/D_L$ is the Peclet number. In a similar manner, the BOD solutions can be derived.

16.13 Summary

With growing pollution in water bodies worldwide, hydrologists are expected to provide cost-effective and sustainable solutions to clean these water bodies. This chapter has discussed basics of pollutant modeling. These days, versatile models are available to assist the hydrologists in this task. As the coverage of water quality monitoring is improving in terms of network density and variables, significant rise in applications of models is expected.

16.14 Questions

16.1 Fischer et al. (1979) provided an analytical solution for a semi-infinite stream with a uniform initial concentration C_i that receives a constant concentration C_0 input for $t \geq 0$ at a cross section located at the origin ($x = 0$). The input concentration is a step function. The solution is

$$\frac{C(x,t) - C_i}{C_0 - C_i} = \frac{1}{2}\left[erfc\left(\frac{x - ut}{\sqrt{4D_i t}}\right) + erfc\left(\frac{x + ut}{\sqrt{4D_i t}}\right)\exp\left(\frac{ux}{D_i}\right)\right]$$

in which *erfc* is the complimentary error function defined as

$$erfc(z) = 1 - erf(z) = 1 - \frac{2}{\sqrt{\pi}}\int_0^z \exp(-k^2)dk$$

The concentration can be normalized by using $C_i = 0$ and $C_0 = 1$. Consider $D_i = 40\ \text{m}^2/\text{s}$ and $u = 1$ m/s. With these values, Koussis (1983a, 1983b) computed the values of concentration at $x = 100$ m as

Time, s	C (analytical)	Time, s	C (analytical)
10	0.0013	180	0.8671
20	0.0392	240	0.9300
30	0.1250	300	0.9615
60	0.4093	360	0.9782
90	0.6063	420	0.9873
120	0.7316		

Compute the concentration using the fractional scheme given by Eq. (16.43). Compute the values of weighting factors a_i, $i = 1, 2,$ and 3. Use different time intervals, for example, 10, 15, 20 seconds. In the beginning, negative values of concentration may be encountered. Compare the computed values with the analytical values and comment on the goodness of computed values.

16.2 Compute the numerical dispersion coefficient for solution in Question 16.1 and then compute the Peclet number.

16.3 Solve Question 16.1 using the exponential scheme.

16.4 Compute the numerical dispersion coefficient for the solution in Question 16.3 and then compute the Peclet number.

16.5 Compare the analytical solution with the two numerical solutions obtained in Question 16.1 and Question 16.3.

16.6 Compare the values of numerical dispersion coefficients and Peclet numbers obtained in Question 16.2 and Question 16.4 with analytical values.

16.7 Compute concentration at $x = 200, 400, 600, 800,$ and 1000 m for the values in Question 16.1 using the fractional scheme. Comment on how concentration varies with distance and how the role of advection changes.

16.8 Compute concentration at $x = 200, 400, 600, 800,$ and 1000 m for the values in Question 16.1 using the exponential scheme. Comment on how concentration varies with distance and how the role of advection changes.

16.9 Fischer (1967) experimentally measured pollutographs at two stations 11 m apart in a 40 m long rectangular flume. The mean velocity was 0.362 m/s and the dispersion coefficient determined was $D_l = 0.0236$ m^2/s. Compute concentration using the fractional scheme and compare with laboratory measurements and Fischer's values. Also, compute the numerical dispersion coefficient (Fischer's measurements can be obtained from his publication).

16.10 Compute concentration using the exponential scheme for the data in Question 16.9 and compare with laboratory measurements and Fischer's values. Also, compute the numerical dispersion coefficient.

16.11 Fischer (1968) reported the concentration data collected by Godfrey and Frederick in Copper Creek, Virginia. The mean advective velocity was 1.91 m/s and the dispersion coefficient was $D_l = 230$ ft^2/s and the reach length between sections 4 and 6 was 5600 ft. Compute the concentration using the fractional scheme and compare with field observations. Also compute the dispersion coefficient (Fischer's data can be obtained from his publication).

16.12 Compute the concentration using the exponential fractional scheme for the data in Question 16.11 and compare with field observations. Also compute the dispersion coefficient.

16.15 References

Fischer, H. B. The mechanics of dispersion in natural streams. *Journal of the Hydraulics Division, ASCE,* 93(HY6):187–214, 1967.

Fischer, H. B. Dispersion predictions in natural streams. *Journal of the Sanitary Engineering Division, ASCE,* 94(SA5):927–943, 1968.

Fischer, H. B., E. J. List, R. C. Y. Koh, J. Imberger, and N. H. Brooks. *Mixing in Inland and Coastal Waters.* New York: Academic Press, 1979.

Koussis, A. D. Pollution routing in streams. *Journal of Hydraulic Engineering,* 109(12): 1636–1651, 1983a.

Koussis, A. D. Unified theory for flood and pollution routing. *Journal of Hydraulic Engineering,* 109(12):1652–1664, 1983b.

Koussis, A. D., P. Kokitkar, and A. Mehta, A. Modeling DO concentration in streams with dispersion. *Journal of Environmental Engineering,* 116(3):601–614, 1990.

Leonard, B. P. A survey of finite differences of opinion on numerical middling of the incomprehensible defective confusion equation. In *Finite Element Methods for Convection Dominated Flows,* edited by T. J. R. Hughes, AMD-Vol. 34. New York: American Society of Mechanical Engineers, 1980.

Singh, V. P. *Kinematic Wave Modeling in Water Resources: Environmental Hydrology.* New York: John Wiley, 1997.

Singh, V. P. Kinematic wave solutions for pollutant transport by runoff over an impervious plane, with instantaneous or finite period mixing. *Hydrological Processes,* 16: 1831–1861, 2002a.

Singh, V. P. Kinematic wave solutions for pollutant transport by runoff over an infiltrating plane with finite-period mixing and mixing zone. *Hydrological Processes,* 16: 2441–2477, 2002b.

CHAPTER 17

Environmental Flows

It is well known that rivers provide numerous benefits to humans, such as water for domestic needs, agriculture, energy (hydropower, cooling of thermal stations), food (e.g., fish), waterways, fertile sediments, and ecosystems. Rivers also have a central place in many social, cultural, and religious activities in different civilizations. Rivers are perceived differently by different groups. Hydrologists see rivers as channels to transport water and sediment, energy specialists see them as a means of energy generation, hydraulicians see them as a means for navigation and flood control, tourism managers see them as a means for recreation, and land planners see them as an essential part of the landscape. Rivers provide water to farmers to irrigate crops, and for religious leaders, river water has a spiritual value that is central to many cultures worldwide. But river water is not always a boon; rivers can also be hazardous when they are in flood, and support disease vectors, such as mosquitoes when their water is grossly polluted.

Rising demands and increasing withdrawal of water for various needs have caused large reductions in flow in vast segments of rivers. Among other things, this has led to severe degradation of river ecosystems. The health of a river's ecosystem depends on many factors: complete flow regime in the river, sediment properties and transport by water, physical structure of the channel and riparian zone, quality of water, channel management works (such as dredging, macrophyte removal), level of exploitation (e.g., fishing), and the presence of physical barriers (e.g., dams) impacting river connectivity. Storage dams or diversions can have a profound impact on many of these features. Although a myriad of factors influence the ecology of aquatic ecosystems (e.g., temperature, water quality, and turbidity), the flow regime is considered as the primary driving force (Richter et al., 1997).

17.1 Ecohydrology

It is now being increasingly realized that most water-related problems can be best addressed by following a multidisciplinary approach. Yet the interactions between hydrologists and professionals from other disciplines, such as geologists, biologists, sociologists, and economists, are not strong enough, leading to, at times, suboptimal solutions. This calls for a new paradigm of problem-solving by establishing interfaces between hydrology and other disciplines.

Rodriguez-Iturbe (2000) defined ecohydrology as the science which seeks to describe the hydrological mechanisms that underlie ecological patterns and processes. Here, the attention was focused on the processes in which soil moisture is the key link between climate fluctuations and vegetation dynamics. Hydroecology was defined by Dunbar and Acreman (2001) as "the linkage of knowledge from hydrological, hydraulic, geomorphological and biological/ecological sciences to predict the response of freshwater biota and ecosystems to variation of abiotic factors over a range of spatial and temporal scales." According to this definition, a task of hydroecologist is to study changes in aquatic ecosystems from various disciplines and develop ability to predict how the biota and ecosystems will be impacted due to alterations to a river's flow regime, geomorphological, and other reasons.

To be precise, the terms ecohydrology and hydroecology refer to the integration of two disciplines: ecology and hydrology. The prefix "eco" in ecohydrology implies the study of hydrology wearing the lens of ecology. Similarly, "hydro" in hydroecology emphasizes that a student of ecology wants to understand the topic from a hydrologic perspective. In earlier times, ecohydrology concepts were focused on "wet ecosystems"—wetlands and aquatic ecosystems. Depending on the perspective, an ecohydrologist may be interested to study the constituents of aquatic habitat such a flow regime and its variation, habitat conditions, water temperature, chemical composition, etc. He/She may also be interested in biological aspects such as growth of plants including water weeds in open water bodies, algae growth, eutrophication, wetland dynamics, presence of nutrients and pesticides in return flow from agricultural fields, riparian vegetation, and so on. Plants are an important component of biosphere and actively participate in the hydrological cycle. Many water fluxes, such as interception, stemflow, throughfall, transpiration, evapotranspiration, soil moisture, and water quality, have direct links with vegetation and its growth. We also note here that there are a number of other definitions of eco-hydrology and this term does not mean the same thing to everyone.

Ecohydrology can also be understood as the interface between ecology and hydrology. Ecology can be defined as the continuum of environment and biota, wherein the environmental continuum is comprised of air, water, and soil in which water occupies the central position. Clearly then, water occupies the central position in the ecological continuum. Hydrology being the science of water occupies the key position in ecology. Thus, ecohydrology encompasses those hydrological processes that are directly connected to ecological processes. According to Newman et al. (2006), ecohydrology seeks to elucidate (1) how hydrological processes influence the distribution, structure, function, and dynamics of biological communities and (2) how feedbacks from biological communities affect the water cycle.

It is easy to appreciate that several hydrological processes impact human life and ecosystems. It is indeed hard to identify a subarea of hydrology which does not (directly or indirectly) impact the biosphere and, in turn, gets impacted by it. Ecohydrology may provide environment-friendly solutions to many problems related to water abundance, scarcity, and pollution, and such solutions may be more sustainable. Leonardo da Vinci (1452–1519), the famous scientist, engineer, painter, and architect, had stated "Water is the driving force of all nature."

17.1.1 River Health

Just like humans, the health of a river denotes its ability to maintain its structure and function; ability to support local biota, including human society, and to maintain key processes, such as sediment transport, nutrient cycling, and energy exchange. Much like a living being, a healthy river should be able to recover from shocks with minor injury marks.

A healthy river can support a diverse range of habitats, animals, and plants. River health can be described in terms of its biodiversity and ecological functioning, compared to its preintervention status. These days, due to various instream, riparian and (degrading) catchment changes, flow regulation, water diversion, and influx of pollutants, biodiversity, and ecological functioning of most rivers have been degraded and are functioning with less biological activity compared to their predevelopment condition. Rivers are also under stresses due to deterioration of water quality and biodiversity, and nutrient enrichment.

An ecologically healthy river need not be in pristine condition. In fact, except for headwaters areas, very few rivers are in pristine condition these days. A balance between the human use and river health needs to be struck so that neither of these suffers unduly.

17.1.2 River Health Assessment Tools

A number of tools are available to assess the health of a river. Most of these assume that the river was healthy in virgin condition, that is, before any human intervention was introduced. A tool, known as the Indicators of Hydrologic Alteration (IHA), was developed by Nature Conservancy (Richter et al., 1997) for hydrologic analysis with ecological applications. It computes about 67 ecologically relevant statistics from daily hydrologic data and calculates 33 IHA parameters and 34 EFC (environmental flow components), including major parameters like high flow threshold, low flow threshold, high flow start-rate threshold, high flow end-rate threshold, small flood minimum peak flow, large flood minimum peak flow, and extreme low flow threshold.

A tool called "Flow Health" (Gippel et al., 2012) compares the values of hydrological attributes of a river with the values in reference condition and computes a "flow health score" for the river based on the magnitude and frequency of flows. The reference condition is a period of time when the river did not have (or had small) human interventions (Fig. 17.1). This tool assumes that the ecosystem will be restored to a greater extent when the flow magnitude and frequencies are made healthy.

Flow health compares the flow time series data of test year with a reference flow series. The virgin flow is assumed to be the healthy river flow and taken as the reference.

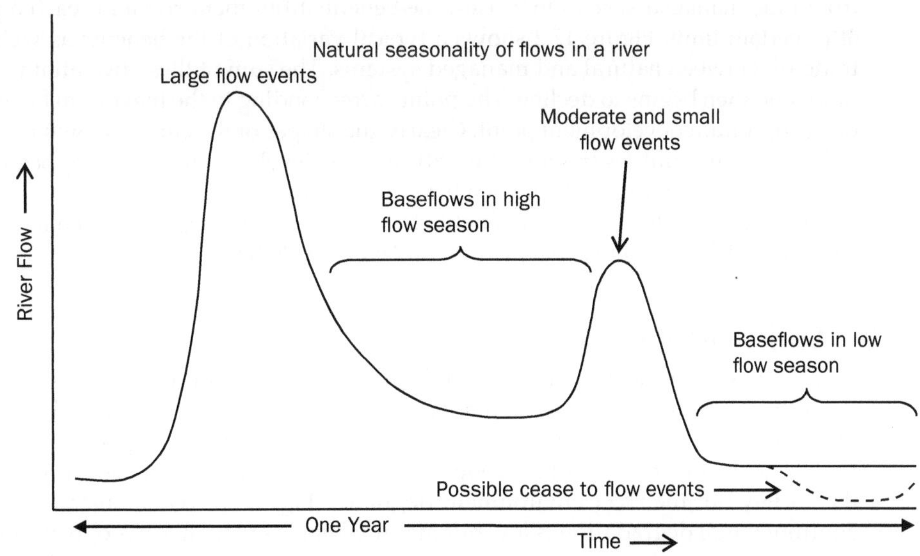

Figure 17.1 Ecologically main relevant flow components. (Source: Gippel et al., 2012.)

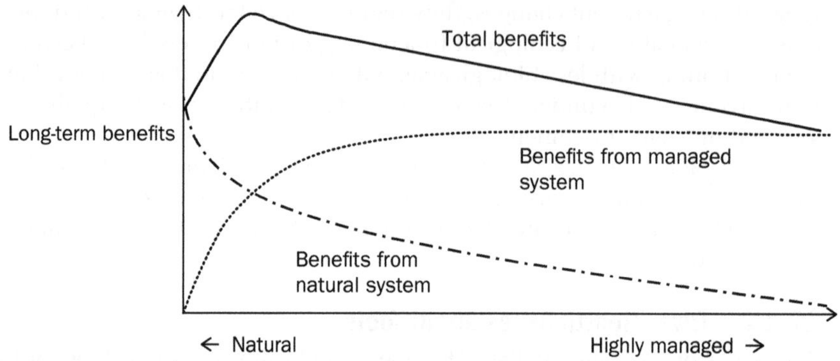

Figure 17.2 Trade-offs in river regulation showing typical variation of benefits from natural and managed systems. (Source: After Jain and Acreman, 2017.)

The tool assigns a score in such a way that the flow, which is more or less the same as that of the virgin condition, will have a flow health score close to 1, while the flow which deviates considerably from the virgin condition will be assigned a value close to zero. The scoring system assumes that flow reductions are more detrimental to river health than flow increases and occasional increased flows in the high flow season are not detrimental to river health.

17.2 Trade-Off Between Development and Conservation

There is always a trade-off between development (using water to meet societal needs) and conserving the water bodies. As the natural systems are progressively modified, the benefits from ecosystem services gradually decline. However, as the degree of control in the managed system increases, the benefits from them rise and reach a plateau after certain limit. Figure 17.2 shows a typical variation of the benefits as well as the trade-off between natural and managed systems. The sum of these two attains a maximum and then begins to decline. The point corresponding to the maximum benefits can be the optimum development point. Clearly, the shapes of the curves shown in Fig. 17.2 will vary with countries/regions. Identification of development space is a sociopolitical exercise involving trade-off between benefits and costs.

It is essential that costs and benefits to society of allocating water to maintain ecosystems and for social use are quantified (Acreman, 2001).

17.3 Environmental Flows

Flow is considered as the master variable that governs the condition of a river because it exerts great impact on aquatic habitat, river morphology, biotic life, river connectivity (upstream-downstream and with floodplains), and water quality. As water is abstracted from a river (and consequently flow reduces), there are economic gains to the society but the river environment degrades (Fig. 17.3). When abstraction is low, environmental degradation is low but marginal gains are high. As progressively more water is abstracted from a river, economic gains to the society increase, first rapidly

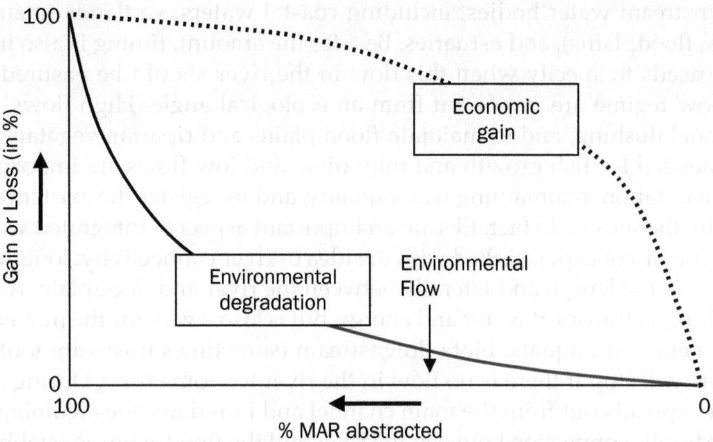

Figure 17.3 Typical pattern of environmental degradation and economic gain as a function of flow abstracted from a river. MAR = mean annual runoff.

and then slowly but environmental degradation also starts, first slowly and then at rapid rates.

In the 1970s, the hydrologists and ecologists realized that the health of a river ecosystem significantly deteriorates if flow falls below a certain minimum value. Gradually the concept of minimum flow in the rivers came into practice with the hope that as long as discharge in a river exceeds a certain value, the ecosystem of the river will function satisfactorily. However, detailed studies on the different components of river ecosystems carried out subsequently led to new understanding. It was realized that ensuring minimum flow alone is not sufficient. All elements of a flow regime of a river, including high, medium, and low flows, are important (Poff et al., 1997). Therefore, any changes in the flow regime will have some (adverse) influence on the river ecosystem. Hence, if the natural ecosystem in a river is to be maintained in a healthy condition, the modified flow regime should closely follow the natural flow regime. However, it is not always possible to let the rivers flow undisturbed and most rivers are managed to different degrees to meet the needs of the society. Some societal needs, such as bathing in the river, do not require that water be removed from the river. However, many needs, such as irrigation, require removal of water from the river. For some uses (e.g., hydropower generation or cooling of a thermal power plant), water is diverted from the river but this water returns back to the river after serving the intended use.

The concept of environmental flows has been developed to check the adverse impact of large withdrawals on the river ecosystem and the uses to which the river is put. The underlying concept says that the entire ecosystem should remain in healthy condition and its integrity should be ensured to sustain human well-being.

17.3.1 Definition of Environmental Flows

Environmental flows (EFs) are the flows characterized by their quantity, timing, duration, frequency, and quality required to sustain freshwater, estuarine, and near shore ecosystems and the human livelihoods and well-being that depend on them (Acreman and Ferguson, 2010). EFs are necessary to maintain the health and biodiversity of downstream water bodies, including coastal waters, wetlands (mangroves, sea grass beds, floodplains), and estuaries. Besides the amount, timing is also important and one also needs to specify when this flow in the river should be ensured. All components of flow regime are important from an ecological angle. High flows are important for channel flushing, and to maintain flood plains and riparian vegetation; medium flows are needed for fish growth and migration; and low flows are important to avoid river fragmentation, maintaining water quality, and to register the existence of a river and its use by the society. In fact, EFs are an important aspect of integrated water management.

The EF concept is linked with the idea of river connectivity: longitudinally throughout its entire length and laterally between the river and floodplain. A river is not only a medium to transport water and energy, but is also a way for the movement of nutrients, sediments, and aquatic biota downstream (sometimes upstream too). All these movements will stop if there is no flow in the river for some time. During high flows, floodwater spreads out from the main channel and inundates the adjoining floodplains, and a hydraulic connection between the river and the floodplains is established. This allows the exchange of food, nutrients, and carbon between the river and flood plains. Thus, EFs are a mechanism to ensure river continuum in the longitudinal direction by having a certain amount of flows and lateral connectivity by flood pulses.

Environmental flow requirement (EFR) represents a compromise between water resources development and maintenance of a river in ecologically acceptable conditions (Fig. 17.3). A balance is needed so that development is carried out but the river remains healthy. Realizing the importance of EFR, many countries have now formulated policies and laws to ensure allocation of water to river ecosystems. After allocating water to meet the basic human needs, environmental needs typically get the highest priority.

The natural flow paradigm (Poff et al., 1997) marked a significant point in the conceptual development of EFs. It considers that the natural flow regime of a river, described by six components (magnitude, timing, frequency, duration, rate of change, and overall variability of flow), is key to the integrity of ecosystem of a river. Large modifications of the natural flow regime can adversely impact riverine, riparian, and floodplain species and processes. Beyond certain hydrological changes, significant (and unacceptable) ecological alteration occurs. Scientists understand and agree that if the ecosystem of a river is to be kept healthy, the EFs in the rivers will have to be provided so as to closely follow the natural flow regime. However, the society needs to manage rivers to meet its needs. As stated earlier, some societal needs, such as municipal water supply and/or irrigation, require that water is removed from the river. The needs where water is consumed are called consumptive needs. Some uses, such as bathing, do not require that water be diverted from the river. Such uses are called instream uses. In some uses (hydropower generation or cooling of a thermal power plant), almost the entire diverted water is returned to the river after serving the desired purpose. These types of uses are termed as nonconsumptive uses.

EFR for a particular location on a river depends on many factors, including size of rivers, natural state, desired state, sensitivity of river ecosystem, choice of the society, and what are the uses of river water. Therefore, before the EFR are determined, it is helpful to set broader objectives and clearly state the condition in which the society desires its river to be. For example, for some rivers, EFR are set to achieve specific predefined ecological, economic, or social objectives. This is called objective-based flow setting (Acreman and Dunbar, 2004). As an example, the Water Framework Directive (WFD) of the European Union requires member states to achieve "Good Status" in all surfaces and groundwaters. Good status of a river is a combination of good chemical status and good ecological status.

17.3.2 Typical Stages of an Environmental Flow Assessment

A typical environmental flow assessment (EFA) process can be divided into six stages.

Stage 1: Define the issue: what is the key reason for changes in a river flow regime and why the river is perceived to be degraded or changed?

Stage 2: Define the scope and objective of setting EFs with participation of stakeholders. It is necessary that all stakeholders agree on the type of river desired.

Stage 3: Decide the locations where EFs are to be estimated. Typical locations are reaches based on the classification of river zones, or on morphology, or on biological patterns, or at strategic points where the river is seriously degraded and is to be restored.

Stage 4: Collect required hydrological and ecological data. EFAs require data from a range of domains. Typical data requirements are:

- Hydrological data: time series of water depth and river flow, channel cross-sectional properties, and sediment transport dynamics.
- Ecological data: such as fish species/abundance/life history, invertebrate diversity and abundance, composition and distribution patterns of riparian vegetation
- Economic data: viz. economic value of fisheries or recreational uses of the river
- Social data: such as details of river-based festivals and water needs at those times

Stage 5: Data analysis and consultations: Dependencies between flow events, channel hydraulics, and ecological/social components are established. Flow needs of different species and ecosystem components during different months/seasons is checked through consultation with the literature, experts, and stakeholders familiar with the river. Based on these, EFs are computed.

Stage 6: Implement, monitor, and feedback. Computed EFs are implemented and the health of the river is monitored. The outcome is used to review the EFA process. Many times, all the data listed in Stage 4 may not be available and field measurements may be needed to collect the missing data.

17.3.3 Objectives and Factors Governing Environmental Flows

Flows of many rivers have been overallocated to off-stream uses. Consequently, downstream ecosystems are highly degraded, since little water is left for river ecosystems; often the quality of water left in the river is poor. In an objective-based EF setting (Acreman and Dunbar, 2004), EFs are set to achieve specified ecological, economic, or social objectives. Member states of the European Union are required to meet at least good ecological status in all water bodies.

Ideally, the EF objectives are set by involving a range of stakeholders. In some cases, different objectives are set for different rivers in the region/country and for different stretches of the same river. In estimating EFs, a common practice is to refer to the river conditions at some earlier time as a reference state. The question is how far back to go? Most rivers have changed substantially over the past 50 years and it will be very difficult to restore them to the conditions of the 1950s. While looking back, one should not go so far back that the previous river state is not relevant.

17.3.4 Locations and Reasons for Estimation of Environmental Flows

Environmental flows are most frequently computed at hydro projects. Storage or diversion of large quantities of water by a project adversely affects a river's ecosystem and hence an EFA is carried out to ensure good health of the river. In addition, an EFA may frequently be needed at places of social/cultural gatherings, near a major city or industry.

EFs may be estimated for many reasons. Most common of these are the desire to rejuvenate a river to an improved ecological condition or prevent further degradation of a river. If a new water resources development project is expected, it will be necessary to carry out EFA downstream of the project to ensure that the river ecosystems remain in good ecological condition.

17.4 Methodologies to Assess Environmental Flow Requirements

Over the past 40 years, there has been a rapid growth in the development of methods to estimate EF for a given river stretch. The methods that have been developed range from simple, low-confidence, desktop approaches, to resource-intensive, high-confidence approaches. The choice of a particular method depends on the type of issue (abstraction, dam, run-of-river scheme), management objective, expertise, time and funds available, and the regulatory framework in which EFA is carried out.

The comprehensive methods require large amounts of hydrological, geomorphological, and ecological data. These also use inputs from multidisciplinary teams and consultation among experts. Typically, such studies take many months, sometimes years, to complete. A key constraint in undertaking comprehensive EF assessments in many countries is the lack of data from all the disciplines and for the locations where such data are needed.

Approaches developed to estimate EF requirements have been categorized differently by different experts. Based on the evolution of methods, a broad classification in three groups is hydrological methods, hydrology hydraulics methods, and hydrobiology methods. A brief description of the methods based on this classification follows.

17.4.1 Hydrological Methods

Hydrology forms the bedrock of water resources management; flow regime is recognized as the primary driving force that influences the ecology of aquatic systems. Hence the initial estimates of EFs were based on hydrological indices. Generally, long series of river flow data are available for many places and indices based on flow data can be calculated. Streamflow data offers very useful and appropriate indicators of river ecosystem integrity.

Although numerous factors influence the ecology of aquatic ecosystems (e.g., water temperature, water quality, and sediment being transported by the river), the flow regime is the primary driving force that influences their ecological structure and functions (Poff et al., 1997; Richter et al., 1997). Many abiotic characteristics of riverine ecosystems, such as the size of suspended and bed-load sediment being moved by river, condition of aquatic habitat, stability channel bed dissolved oxygen content of water, and water temperature, have been found to vary with flow conditions. Further, channel and floodplain morphology is shaped by fluvial processes which are driven by streamflow; high-flow conditions play an important role.

In many countries, biological data are scarce and their temporal and spatial coverage is limited. Usually, the coverage of streamflow data is much better. In such cases, the hydrological tools can be employed to obtain at least the preliminary estimate of EFR.

In the United States, Tennant (1976) analyzed data from a number of river gauging sites in the Midwestern states and attempted to separate the entire range of mean annual flow (MAF) at a site into several ecologically relevant ranges. Tennant specified percentages of the MAF that provide different quality habitat for fish. For example, 10% of MAF is necessary for survival of fish and represents poor quality habitat, 30% of MAF is needed for moderate habitat, and 60% for excellent habitat. Although Tenant developed these indices for certain climatic region in North America, these have been used in other countries. But the results may not be correct because the periods used to separate spawning from rearing that Tennant originally used may not be valid for other areas and the flow required to maintain rivers in different categories may be different. Tennant's original dataset came from low gradient (<1% slope) streams and hence it may not be applicable to high gradient (>1% slope) streams, even in the United States. Modifications of Tennant's method have been suggested. Tessman (1980) considered natural variations in flow on a monthly basis to determine the flow thresholds and the resulting approach is called the Modified Tennant Method or Tessman Method.

Freshwater fishing law in France requires that the residual flows in the bypassed section of a river must be a minimum of one-fortieth of the mean flow for existing schemes and one-tenth for new schemes (Souchon and Keith, 2001). In regulating abstraction from the rivers in the United Kingdom, an index of natural low flow has been employed to define the EF. In the United Kingdom, a flow-based index, Q_{95} (flow which is equaled or exceeded 95% of the time) is often used to define EFs (Acreman and Dunbar, 2004). The figure of Q_{95} was chosen purely on hydrological grounds. In other cases, indices of rarer events (such as mean annual minimum flow) have been used. In Greece, at least one-third of the average summer flow of the river shall be reserved for EFs. In Spain, residual flows are to be kept at 10% of interannual average flow. In South Africa, a hydrological index [i.e., coefficient of variation of flows (CV) divided by the proportion of total flow that is baseflow (BFI) or (CV/BFI)] is used to assess the EF needs (Hughes et al., 2014).

EFR may also be determined, based on the habitat requirements for keystone species. For example, habitat requirements of Golden Mahseer (estimated by Wildlife Institute of India and others), which is found in the Upper Ganga basin in India, are given in Table 17.1 (based on the data provided by the Wildlife Institute of India and Central Water Commission). Hence, EF can be decided such that the desired flow depth is available at different times.

	Adults	Juveniles	Spawning	Incubation and Larval Development
Depth	Deep (>1 m) 0.5–1.5 m	Shallow (<0.75–1.5 m)	Shallow to high (0.5–2.0 m)	Shallow to high (0.3–2.00 m)
Velocity	Medium to high (0.5–1.5 m/s)	Low to medium (0.1–1.5 m/s)	Low to medium (0.1–1.0 m/s)	Low to medium (0.1–0.5 m/s)
Habitat	Riffles, pools, glides	Pools, backwater pools closer to the banks and run habitats	Low gradient riffles, backwater pools, secondary channel	Backwater pools and secondary channels
Breeding period	March–April; October–December			
Migration timing	March–April; October–December			

TABLE 17.1 Habitat Requirements for Golden Mahseer

Example 17.1

Golden Mahseer are found in a given river which is 40 m wide, has a trapezoidal-shaped cross section, and has side slopes of 1:1. The river slope is 0.003 and Manning's roughness is 0.04. Compute the flow required to maintain a depth of 1 m, which is considered adequate for this fish.

Solution

For river width $B = 40$ m, depth of flow $d = 1$ m, and trapezoidal cross section

Wetted perimeter $P = 42.82$ m
Flow area $A = 40 \times 1 + 1 \times 1 = 41\ \text{m}^2$
Hydraulic radius $R = A/P = 41/42.82 = 0.9575$
Applying Manning's formula

$$Q = \frac{1}{n} A R^{2/3} S^{1/2} = \frac{1}{0.04} 41 \times 0.9575^{2/3} 0.003^{1/2} = 54.5\ \text{m}^3/\text{s}$$

Hence the river should always have 54.5 cumec flow to meet the requirement of Golden Mahseer. With this minimum flow, the EF hydrograph can be reconstructed by mimicking natural flow variation.

17.4.2 Range of Variability Approach

In the Range of Variability Approach (RVA) developed by Richter et al. (1997), different aspects of flow variability are expressed through 32 indices. These indices are estimated from virgin daily flow and are calculated on an annual basis. They reflect different aspects of flow variability and there is a one-to-one relationship between hydrological and ecological variables. Smakhtin and Anputhas (2006) noted that the number of indices used in RVA is too large and either many of these are likely to be correlated or there is little difference between their values. Redundancy analysis can overcome this problem (Olden et al., 2011).

An important question here is to find the acceptable range of variation of the indices. For each parameter, a threshold of one standard deviation from the mean has been suggested by various researchers as a default arbitrary limit for setting EF targets in the absence of other supporting ecological information.

Indices based purely on hydrological data do not take account site-specific conditions. Therefore these are particularly appropriate for low controversy situations but tend to be precautionary rather than refined.

17.4.3 Hydrology- and Hydraulics-Based Methods

A refinement of hydrologic methods is the use of hydrologic and hydraulic (HH) data to estimate EFs. The HH methods use the relationship between river flow and channel hydraulic characteristics such as water depth, wetted perimeter, and flow velocity to calculate EF. Since the HH methods make use of river channel properties, they are able to consider landscape characteristics, such as slope, sediment movement, and geology. To estimate EF, cross-section data are collected at a site on the river where the maintenance of environmental flow is critical or where hydraulic habitat in the river is most responsive to flow reduction. The relationship between hydraulic habitat and discharge Q is developed by plotting the relevant hydraulic variable (e.g., depth, velocity, cover) against discharge. This relationship is used to estimate the EF. An adequate number of channel cross sections should be considered to ensure that the longitudinal hydraulic variability of stream reaches is accounted for. A breakpoint is identified on the Q-habitat response curve and is interpreted as a threshold below which habitat quality is significantly degraded. EFR is set as the discharge Q corresponding to the breakpoint.

17.4.4 Hydraulic Rating Methods

Among the environmental conditions required by an animal or plant living in a river, the most important and visible are the physical features that are affected when the flow regime changes. EFR for rivers whose flows have the same properties (e.g., FDC) but are located in different ecological regions will be different. Methods based on only hydrologic data are unable to relate to the properties of flow regime with the response of aquatic species. Hydraulic Rating Methods (HRMs) were essentially developed to overcome this lacuna. HRMs use data coming from hydrologic, hydraulic, and ecological domain and also elicit opinion of experts from these fields. HRMs use indices of habitat conditions (morphological and ecological data) in the river to compute EF.

Hydraulic characteristics of a river which can be easily measured are water depth, velocity, wetted perimeter, and so on. The wetted perimeter of a cross section is defined as the length of the channel bottom and sides which are in direct contact with water. The wetted perimeter depends upon flow depth and cross-section properties and is a useful surrogate for habitat factors of the target biota. Hence it is a commonly applied hydraulic rating parameter. Wetted-perimeter-based methods use the relationship between wetted perimeter and hydro-ecological variables such as river flow, cross-section properties, and habitat quality to calculate EF. Data of habitat response versus discharge are plotted and breakpoints are identified. It is assumed that if the flow falls below the flow at breakpoints, the habitat quality will significantly and sharply degrade. EFR is set as the discharge that produces an acceptable reduction in the habitat attribute.

17.4.5 Hydrobiology Methods

Depending on the location of the study area, EFs are typically computed by determining their suitability for a particular or target species. It is assumed that the integrity of the riverine ecosystem is maintained if flow depth and velocity are maintained at levels that are appropriate for the most sensitive species. However, a broader view is that EFs should suit not only the most critical but all aspects of river ecosystem. Therefore the approach to determine EF should consider all aspects of river ecosystem; in other words, it should be "holistic."

Hydrobiology (also called holistic) methods use hydrological, hydraulic, and biological data and habitat simulation models. The features of ecosystem that are typically considered in holistic assessment include hydraulic habitat, channel morphology, geomorphology, water quality, riparian and aquatic vegetation, macroinvertebrates, and fish and others that depend upon the river ecosystem. Each of these components can be evaluated by using field and desktop techniques (Tharme, 1996, 2003; Arthington, 2012). Based on these evaluations, the flow requirements are systematically incorporated in EF assessments. A number of holistic methodologies have been developed and applied to real-life situations.

Holistic methods are not specific methods; these are frameworks that combine hydrological, hydraulic, and habitat simulation models and explicitly adopt a comprehensive ecosystem based approach to EF determination.

17.4.6 Building Block Methodology

The Building Block Method (BBM), developed in South Africa, is based on the premise that the survival and growth of riverine communities depends on the elements (high, medium, low) of flows which can be termed as building blocks of the flow regime (King and Louw, 1998). Hence it is important to preserve the key elements of natural flow regime, including variation to keep the river ecosystem in healthy condition. When combined together, these building blocks create a flow regime that facilitates the maintenance of the river in a prespecified condition.

One of the most popular and versatile method in this category is the BBM. Building blocks are basically different components of river flow, such as low flows, medium flows, and high flows. When these blocks are combined, they yield an altered flow regime that can maintain the river in the desired condition. The flow blocks comprise low flows as well as higher flows required for channel maintenance, breeding stimuli, fish movement, and for connectivity with floodplains (Fig. 17.4). Each flow component may vary with season, and the flow regime may be defined for "normal years" and "drought years." The flow needs in normal years were called "maintenance requirements" by Acreman et al. (2009) and are divided between high and low components. Table 17.2 lists the building blocks suggested by Acreman et al. (2009).

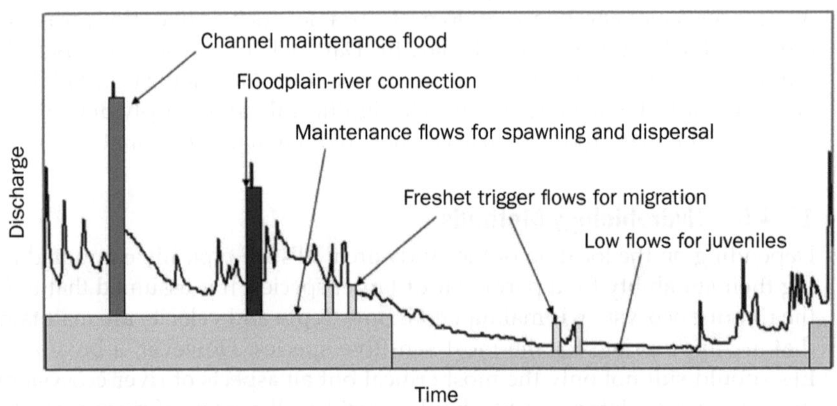

FIGURE 17.4 Typical building blocks making up an environmental flow regime. (Adapted from presentation by Prof. Mike Acreman.)

Building Block	Purpose
Low flows	Habitat for juveniles and prevention of invasive species
Maintenance flows	Stimulate species migration, spawning, and dispersal
Freshets	Stimulate species migration, spawning, and dispersal
Small floods	Sort river sediments, connect river and floodplain habitats
Large floods	Remove undesired species, maintain channel structure and evolution

TABLE 17.2 List of Building Blocks Making up an Environmental Flow Regime

The BBM makes use of the knowledge of experts who may come from different domains: hydrology, geology, geomorphology, environmental science, aquatic ecology, and sociology. A series of structured stages is followed by the experts who assess available information, such as hydraulic geometry and its relationship to species requirements. Based on this information and professional judgment, experts arrive at a consensus on the building blocks of the EF regime.

17.4.7 Ecological Limits of Hydrologic Alteration

The Ecological Limits of Hydrologic Alteration (ELOHA) framework was developed by a group of river scientists (Arthington et al., 2006; Poff et al., 2010) as a synthesis of a number of techniques and EF methods that can support comprehensive regional EF management. It essentially consists of four steps: (1) build a hydrologic foundation of streamflow data; (2) classify natural river types, and define how flows have been altered within each river type; (3) determine flow-alteration—ecological response relationships associated with each river type; and (4) implement water management policy to achieve river condition goals defined through social processes involving stakeholders.

The hydrologic foundation consists of two databases of flow time-series that represent simulated baseline conditions (minimally altered flow regime) and developed conditions (altered flow regimes) throughout the study region during a common time period. ELOHA assumes that as flow alteration increases from baseline conditions, it leads to increasing ecological change; the relationships may vary. The hydrologic component provides baseline and current data as well as a comparison between the two to highlight what features of the river have changed. River classification compares variation in observed characteristics with the river types that are similar in terms of hydrologic and other environmental features (Kennard et al., 2010).

By assigning the river or its reaches to a particular category and using data obtained from a set of rivers of that type within the region, relationships between ecological metrics and flow alteration can be developed (Poff et al., 2010).

Relationships between altered flow and ecological characteristics of a river are important in EF assessment. Identification of these relationships is an important part of the ELOHA framework also. These relationships vary with locations. How the river ecology responds to changes in flow regime is something that also depends on the natural flow regime and geomorphic settings of the river. ELOHA recognizes that in addition to ecological impacts, socioeconomic and environmental consequences of altered flow regime are also important and should be considered while framing the policies for the use and protection of water resources of a region. The decisions to exploit these resources are taken by governments and stakeholders, based on their trade-off between development and environmental conservation.

17.4.8 Habitat Simulation Methods

As stated earlier, difficulties arise in relating changes in the flow regime to the response of species and communities dependent on rivers. Hence the next level of improvement in the methodology to estimate EF is to understand how the suitability of physical habitat of indicator species changes with changes in river flows. The resulting methods which are based on simulating the response of species have been called Habitat Simulation Methods (HSMs). To apply HSMs, detailed hydrological, hydraulic, ecological, and biological data are collected. According to Tharme (2003), these techniques attempt to assess EFRs on the basis of detailed analyses of the quantity and suitability of instream physical habitat available to target species under different flow regimes.

Results of integrated analysis of hydrological, hydraulic, and biological data provide the necessary support.

Fish are commonly taken as the indicator species of river habitat. Productivity of fish in a river depends on four major factors: flow regime, structure of physical habitat (channel properties and geomorphology, substrate structure, and riparian vegetation), water temperature and quality (organic and chemical constituents), and inputs from the catchment (sediments, organic matter, and nutrients). Production and population of fish in a river reach depend upon the way the related components interact.

A popular model known as the Physical Habitat Simulation Model (PHABSIM) was developed by Bovee (1986). The PHABSIM model has been designed to predict the micro-habitat (depth, velocities, channel indices) conditions in rivers as a function of streamflow, and the relative suitability of those conditions to aquatic life (http://www.fort.usgs.gov/products/software/phabsim/). PHABSIM deals with the physical habitat results of various flow regimes. This model housed in the Instream Flow Incremental Methodology (IFIM) is the preeminent modeling platform of this type.

Application of PHABSIM to a river reach has two components: hydraulic simulation and habitat simulation. Results of hydraulic simulation and data of physical structure of the stream and streamflow are input to habitat simulation where the behavior of physical habitat is simulated and "weighted usable area" is calculated. PHABSIM is implemented in four major steps: (1) simulate water surface elevations; (2) simulate flow velocities; (3) simulate the physical habitat versus streamflow relationship; and (4) simulate the physical habitat when combinations of flows are involved. The relationship between physical habitat and streamflow can be used to understand the trade-off between the values of instream and out-of-stream water uses. These trade-offs are of use in developing acceptable decisions. This model represents the current state-of-art in HRMs.

17.4.9 Methodology for Environmental Flow Assessment

An important decision in EFA is the choice of methodology which depends upon the availability of data, technical expertise, finances, time, purpose of assessment, and so on. Surveys reported in hydrologic literature have found that hydrology-based methods were globally followed in the highest number of cases (30%) closely followed by habitat simulation methods. If the applications pertaining to developing countries only are studied, the proportion of hydrology-based methods will be still higher. The main reason behind this situation is the paucity of data pertaining to ecology and biotic components. In many countries, such data are rarely collected and the collected data are not documented and easily accessible. This is a big handicap in any EF assessment. Further, multidisciplinary teams and huge resources are needed to apply HRMs and holistic methods and, at times, it becomes difficult to arrange requisite resources.

17.4.10 Sediment Aspects

EFs are concerned not only with water; sediment movement is also important, which is hindered by projects. Run-of-river projects trap little sediments due to continuous sluicing and flushing of sediments. However, storage projects or the projects with high (e.g., >15 m) dams trap large quantities of sediment. Hence some flood pulses may be released during the high flow period, to the extent feasible, to ensure that the sediment is transported in downstream river reaches.

17.5 Implementation of Environmental Flows

Implementation of EFs is a challenge for many reasons. Most countries have different organizations, and regulatory agencies have responsibility for different components of environmental protection, for example, rivers, forests, wetlands, wildlife, and aquatic life. The importance of these components for river ecosystems typically changes from river source to sea. It is not easy to have coordination among the various agencies. Another challenge is that all the water of a river may already be allocated to various sectors, such as irrigation, drinking, hydropower, and very little is left to be allocated. Thus to give water for ecological needs, supplies of some of the existing uses have to be curtailed, which may result in economic loss and would invite resistance from the concerned sectors. This makes implementation of EFs a political challenge. It will also require other actions, such as improvement in irrigation efficiency, recycle of water, and so on.

EF norms can be comparatively easily applied on new projects. However, many of the existing projects which were not planned and designed as per EF norms may face difficulty in meeting the EF requirements easily. For ensuring good river health, requisite EFs should be released by all projects. However, there might be technological or economic constraints in bringing the existing projects to desired EF norms. Hydropower projects will generate less energy if more water is released to the river and less water is diverted to a powerhouse. Some economic incentive may be considered to compensate for the loss of revenue and encourage the project owner to release EF.

Usually, it is difficult to estimate the benefits from EFs, such as improvements in the environment, healthy rivers, and higher biodiversity. Hence the application of traditional economic analysis poses challenges. Some models have been developed that can produce ecological outcomes of basin-wide planning, but their use is not widespread.

Operational management of a project to ensure EFs involves setting rules and taking day-to-day decisions about how much water to divert/release/abstract at a given time. Some existing dams may not have gates or sluices at desired locations to release water from storage to meet the EF requirements. In such cases, reservoir operation has to be modified to release EFs. For example, in hydropower projects where the powerhouse is at the toe of dam and there is no other way to release EFs, at least one machine may be run continuously. When designing the outlets of a new dam, it is desirable to incorporate a range of water release capabilities (sluices at different elevations and sizes) so that the full array of dam-operating objectives can be accommodated and there is enough flexibility for the future. Releases should be made to coincide with water availability within the catchment downstream.

Since experience and expertise in implementing EFs and their consequences (LeQuesne et al., 2010) is limited at present, adaptive management is recommended. The basic idea is "learn by doing." This involves estimating and implementing EFs, monitoring the ecological outcomes for the ecosystem, and then revising the EF decision (Fig. 17.5). Sharing results with all stakeholders and with the global community of EF practitioners will help the practice of EF management to grow.

In recent times, allocation of water for the environment has been mainstreamed in many countries across the world. For example, the water law of South Africa recognizes that water for the environment should be accorded the highest priority along with that for basic human needs. However, the expertise and experience vary widely among the countries and there are frequent difficulties in arranging for requisite expertise to incorporate environmental goals amidst the competing demands from other sectors.

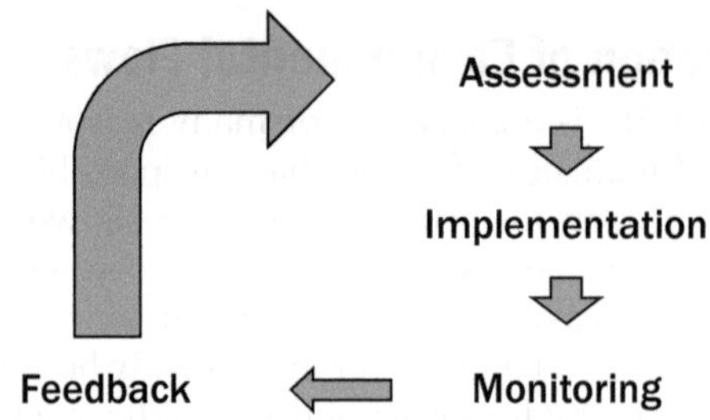

FIGURE 17.5 Mechanism for implementing environmental flows.

The key to operationalize EFs is a mechanism that explicitly links EFs to (socio)-economic values by specifically focusing on ecosystem services. Existing tools can be applied to determine the economic valuation of ecosystem services until a better approach is developed.

17.6 Summary

Water from many rivers is being used for a long time to meet societal needs. It is not feasible to undo the developments and go back to the conditions existing, for example, 75 years ago. Hence an objective of water resources management should be to develop flow regimes that are feasible and provide desired benefits, including setting aside some rivers and reaches for biodiversity protection. The science of EFs has advanced considerably in the last 25 years from little knowledge and awareness to a focus on individual aquatic species and thence to a broader concern about ecosystem protection and restoration (Arthington, 2012; Acreman et al., 2014). At the same time, there have been considerable advances in basic scientific understanding and the development of EFA techniques. However, many of these advances in knowledge are limited to regions (mostly in developed countries) where the scientific studies have been undertaken. The same kind of understanding of ecological responses is not present in many other areas where EFAs are being applied and the knowledge is not directly transportable.

Frequently, there is a disconnect between water, environment, and policy-making groups which hinders environmentally sensitive and sustainable development. Appropriate mechanisms need to be developed to bridge the gap.

17.7 Questions

17.1 Collect five pairs of interdisciplinary terms, such as, ecohydrology and hydroecology, and list the disciplines they cover.

17.2 Collect data for a nearby river and identify the causes and changes in its flows in recent times. What alternate development strategies could have been implemented to save the river yet meet the water needs of the society?

17.3 List the factors that control the trade-off between benefits from the managed systems and those from the natural systems.

17.4 Define environmental flows (EFs) and explain how these help in keeping a river in an acceptable state.

17.5 List important elements of natural flow design of a river and what role they play in keeping a river in a healthy state.

17.6 What do you understand by river health? How do you find out if a river is healthy or not?

17.7 Dolphins are found in a given river which is 50 m wide. Cross-sectional area is of trapezoidal shape, side slopes are 1:1. The river slope is 0.004 and Manning's roughness is 0.03. Compute the flow required to maintain a depth of 0.95 m, which is considered adequate for dolphins.

17.8 List the indicators of hydrologic alterations. Which indicators are more important?

17.9 Why is there a disconnect between water, environment, and policy-making groups? What can be done to eliminate the disconnect or minimize it?

17.10 What are the indices embedded in the range of variability approach?

17.11 What are the difficulties typically faced in implementing EFs and how these can be overcome?

17.8 References

Acreman, M. C. (ed.). Hydro-ecology: Linking Hydrology and Aquatic Ecology. (Proc. Birmingham Workshop, July 1999). *IAHS Publ* 266, 2001.

Acreman, M. C., and A. J. D. Ferguson. Environmental flows and the European Water Framework Directive. *Freshwater Biology*, 55:32–48, 2010. doi:10.1111/j.1365-2427.2009.02181.x.

Acreman, M., and M. J. Dunbar. Defining environmental river flow requirements: A review. *Hydrology and Earth System Sciences*, 8(5):861–876, 2004.

Arthington, A. H. *Environmental flows. Saving rivers in the Third Millennium*. University of California Press, Berkeley, 406, 2012.

Bovee, K. D. Development and evaluation of habitat suitability criteria for use in Instream flow incremental methodology. *U.S. Fish and Wildlife Service Biological Report*, 86(7): 235, 1986.

Dunbar, M. J., and M. C. Acreman. Applied hydro-ecological science for the twenty-first century. In *Hydro-ecology: Linking Hydrology and Aquatic Ecology* (Proc. Birmingham Workshop, July 1999) edited by M. C. Acreman, 3–20. IAHS Pub. No. 266, 2001.

Gippel, C., N. Marsh, and T. Grice. Flow Health: Software to Assess Deviation of River Flows from Reference and to Design a Monthly Environmental Flow Regime. Technical Manual and User Guide Version 2.0. International Water Centre, Brisbane Fluvial Systems Pty Ltd, Brisbane, Australia, 2012.

Hughes, D. A., and F. Munster, F. *Hydrological Information and Techniques to Support the Determination of the Water Quantity Component of the Ecological Reserve. Report TT 137/00*. Pretoria, SA: Water Research Commission, 2000.

Hughes, D. A., A. Y. Desai, A. L. Birkhead, and D. Louw. A new approach to rapid, desk-top-level, environmental flow assessments for rivers in South Africa. *Hydrological Sciences Journal*, 59(3–4):673–687, 2014.

Kennard, M. J., B. J. Pusey, J. D, Olden, S. J. Mackay, J. L. Stein, and N. Marsh. Classification of natural flow regimes in Australia to support environmental flow management. *Freshwater Biology*, 55:171–193, 2010.

King, J. M., and D. Louw. Instream flow assessments for regulated rivers in South Africa using the building block methodology. *Aquatic Ecosystem Health and Restoration*, 1: 109–124, 1998.

King, J. M., and R. E. Tharme. *Assessment of the Instream Flow Incremental Methodologies for South Africa Water Research Commission, Report No. 295/1/94*, 590. Pretoria, SA: Water Research Commission, 1994

King, J., C. Brown, H. Sabet. A scenario-based holistic approach to environmental flow assessments for rivers. *River Research and Applications*, 19:619–639, 2003.

Krchnak, K., B. Richter, and G. Thomas. *Integrating Environmental Flows into Hydropower Dam Planning, Design, and Operations. Water Working Note No. 22*. Washington, DC: World Bank, 2009.

Le Quesne, T., E. Kendy, and D. Weston. *The Implementation Challenge: Taking stock of government policies to protect and restore environmental flows*. WWF-UK and The Nature Conservancy, 2010.

Newman, B. D., B. P. Wilcox, S. R. Archer, D. D. Breshears, C. N. Dahm, C. J. Duffy, N. G. McDowell, et al. Ecohydrology of water-limited environments: A scientific vision. *Water Resources Research*, 42(6):W06032 (1–15), 2006. doi:10.1029/2005WR004141.

Norris, R. H., and M. C. Thoms. What is river health? *Freshwater Biology*, 41:197–209, 1999.

Olden, J. D., M. J. Kennard, and B. J. Pusey. A framework for hydrologic classification with a review of methodologies and applications in ecohydrology. *Ecohydrology*, 5:503–518, 2011.

Poff, N. L., et al. The ecological limits of hydrologic alteration (ELOHA): a new framework for developing regional environmental flow standards. *Freshwater Biology*, 55: 147–170, 2010. doi:10.1111/j.1365-2427.2009.02204.x

Poff, N. L., J. D. Allan, M. B. Bain, J. R. Karr, K. L. Prestegaard, B. D. Richter, R. E. Sparks, et al. How much water does a river need? *Freshwater Biology*, 37(1):231–249, 1997.

Richter, B. D., J. V. Baumgartner, R. Wigington, and D. P. Braun. How much water does a river need? *Freshwater Biology*, 37:231–249, 1997.

Rodriguez-Iturbe, I. Ecohydrology: A hydrologic perspective of climate-soil-vegetation dynamics. *Water Resource Research*, 36(1):3–9, 2000.

Smakhtin, V. U., and M. Anputhas. *An Assessment of Environmental Flow Requirements of Indian River Basins. Research Report 107*. Colombo: International Water Management Institute, 2006.

Souchon, Y., and P. Keith. Freshwater fish habitat: Science, management and conservation in France. *Aquatic Ecosystem Health Management*, 4:401–412, 2001.

Tennant, D. L. Instream flow regimes for fish, wildlife, recreation and related environment resources. *Fisheries*, 1(4):6–10, 1976.

Tessmann, S. *Environmental assessment. Technical Appendix E in: Reconnaissance elements - Western Dakotas region of South Dakota study*. Water Resources Institute, South Dakota State University, Brookings, South Dakota, 1980.

Tharme, R. E. *Review of International Methodologies for the Quantification of the Instream Flow Requirements of Rivers. Water Law Review Final Report for Policy Development, for the Department of Water Affairs and Forestry*. Pretoria, SA: Freshwater Research Unit, University of Cape Town, 1996.

Tharme, R. E. *A Global Perspective on Environmental Flow Assessment: Emerging Trends in the Development and Application of Environmental Flow Methodologies for Rivers*. Cape Town, SA: Freshwater Research Institute, University of Cape Town, 2003.

PART 5

Hydrologic Design

CHAPTER 18

Statistical Analysis of Hydrological Data

Hydrologic variables often exhibit statistical characteristics, for they are not always deterministic and are either fully random or partly random. Measurements of these variables are therefore analyzed statistically. Commonly used statistical descriptors are moments and probability distributions. The probability distributions that are used in hydrology contain parameters that are estimated using standard methods. The estimated parameters are often biased so the properties of these estimates are determined. This chapter provides a snapshot of more commonly used statistical concepts and methods.

18.1 Basic Concepts

All possible data related with a process constitute its population. However, in general, we are able to collect only limited data for a population and this is called a sample. If the sample size is big enough, it can be assumed that the properties of the sample closely match that of the population.

Hydrologists routinely use large volumes of data of a number of variables. Hydrological data often have some typical characteristics as listed below:

1. Lower limit of most hydrometeorological data is zero. A common exception is temperature data.
2. Outliers are frequently seen in the data. Outliers are observations whose value is considerably higher or lower than most of the remaining data.
3. Most water resources data do not follow the normal distribution, although it is common to assume in the analysis that the data are normally distributed.
4. Most hydrological data are positively skewed or skewed to the right, but some are skewed to the left.
5. Many regions of the world have distinct wet or dry, and hot or cold seasons. Consequently, many hydrologic time series have distinct seasonality.
6. Many hydrologic variables, such as streamflow, have high (positive) correlation, particularly when observations are taken at close time intervals.

A hydrologic process may be characterized as deterministic or stochastic. The behavior of a deterministic process can be completely described by using the physical laws. However, a stochastic process cannot be described with certainty, since the process mostly evolves randomly. A variable whose value at a given time is not a function of the previous values is known as a random variable. Discrete random variables can only take on a finite set of values, for example, the number of days it rains in Copenhagen. A continuous random variable can take on any value, for example, the discharge of Nile river.

The observed pattern or variability of many hydrologic variables cannot be fully explained by applying the laws of physics. This is because we do not have correct or full understanding of the underlying processes, all the needed data are not available, and the process has considerable randomness. Hence we need to employ statistical analysis to model the behavior of such variables.

We now briefly discuss the basic concepts of probability.

18.1.1 Probability

Probability is the chance or likelihood that some event will take place. When a coin is tossed, we see either head or tail and both are equally likely outcomes. If a dice is thrown n times and a particular face is seen s times then the ratio s/n is the probability that the outcome is tail.

The notation $P[A]$ denotes the probability that the event A will take place. $P[A]$ can take on a value between zero and unity

$$0 \leq P[A] \leq 1 \tag{18.1}$$

If the event A is certain to occur, then $P[A] = 1$ or if it does not occur with certainty then $P[A] = 0$.

18.1.2 Properties of a Random Variable

Let X denote a random variable and x be a particular value of X. The cumulative distribution function (CDF) of X, $F_X(x)$, is the probability that X is less than or equal to x

$$F_X(x) = P(X \leq x) \tag{18.2}$$

The probability distribution function (PDF) represents the relative probability that the variable X takes on different values. PDF is the derivative of the CDF

$$f_X(x) = d\,\{F_X(x)\}/dx \tag{18.3}$$

Figure 18.1 shows the PDF and CDF of a random variable. Conventionally, the CDF is denoted by capital letter F and the PDF by lower case letter f.

We now describe some properties of $f(x)$ and $F(x)$ for continuous and discrete cases.

1. The probability density function of a random variable can take only positive values

$$f(x) \geq 0, \qquad -\infty < x < \infty \tag{18.4}$$

2. The sum of probabilities of all possible outcomes is equal to 1, that is, the area under the PDF is unity, written as

$$\int_{-\infty}^{\infty} f(x)dx = 1 \tag{18.5}$$

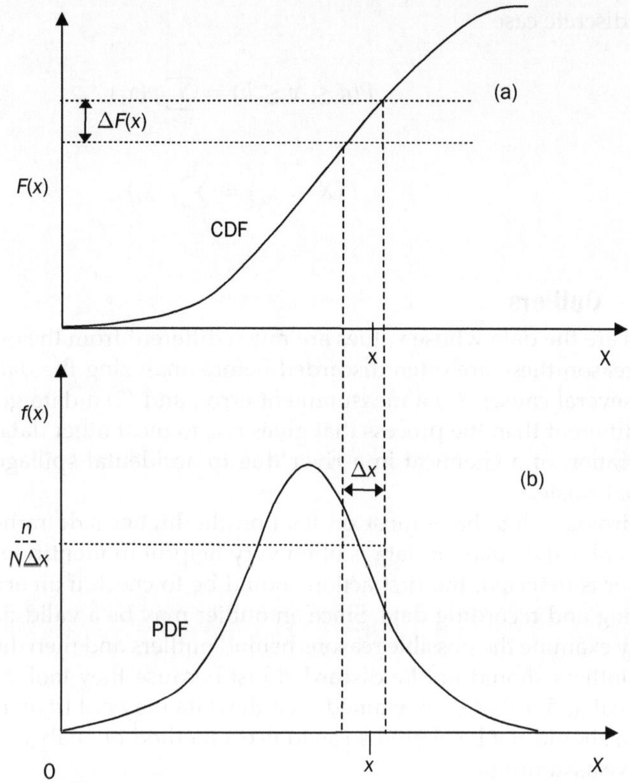

FIGURE 18.1 PDF and CDF of a random variable.

For discrete random variables, we can write

$$\sum_i f(x_i)\Delta x = 1 \tag{18.6}$$

where $f(x_i)\Delta x$ is the probability of $X = x_i$ in the sample space if the sample contains finite and $f(x_i)$ of observations can be replaced by $p(x_i)$. The probability of "X assuming a value $\leq x$" is equal to the area to the left of x under PDF curve:

$$P(X \leq x) = F(X \leq x) = F(x) = \int_{-\infty}^{x} f(x)dx \tag{18.7}$$

3. $$P(a < X \leq b) = P(X \leq b) - P(X \leq a) = \int_{-\infty}^{b} f(x)dx - \int_{-\infty}^{a} f(x)dx$$

$$= \int_{a}^{b} f(x)dx, \quad for\ a < b \tag{18.8}$$

For discrete case

$$P(a \le x \le b) = \sum_{x_i \ge a}^{x_i \le b} p(x_i) \tag{18.9}$$

$$P(X \le x_k) = \sum_{i=1}^{k} p(x_i) \tag{18.10}$$

18.1.3 Outliers

Outliers are the data whose values are much different from the other data and because of this reason these are often discarded before analyzing the data. Outliers can occur due to several causes: (1) a measurement error, and (2) a data generated by a process that is different than the process that gives rise to most other data. For example, a high concentration of a chemical in a river due to accidental spillage rather than regular industrial waste.

Hydrologic data have most outliers on the higher side rather than on the lower side. Graphical display of data is often very helpful in identifying outliers. Whenever an outlier is detected, the first action should be to check if an error has been made in measuring and recording data. Since an outlier may be a valid data, it is important to carefully examine the possible reasons behind outliers and then decide to keep or delete them. Outliers should not be discarded just because they look to be suspicious and/or unusual and if these are retained then the data may not fit an assumed distribution. The data should not be altered so as to use a method of analysis that requires a rather restrictive assumption.

18.2 Key Statistical Measures of Data

Four measures that characterize a probability distribution represented by its PDF, $f(x)$, are (1) the central tendency, (2) the spread, (3) the asymmetry, and (4) the peakedness. These measures are estimated from the sample data of the random variable.

18.2.1 Central Tendency

Several measures of central tendency can be computed.

(1) Arithmetic Mean: The arithmetic mean is the first moment of the probability distribution, μ, about the origin and can be expressed as

$$\mu = \int_0^\infty x f(x) dx \tag{18.11}$$

The lower limit of the integral can also be minus infinity. Here m denotes the mean of population. For a sample consisting of n observation, $x_1, x_2 \dots x_n$, the arithmetic mean of the sample can be written as

$$\overline{x} = \frac{1}{n}\sum_{i=1}^{n} x_i \tag{18.12}$$

where $\bar{x}$ is the arithmetic mean of the sample. Equation (18.12) is a discrete form of Eq. (18.11) where the continuous distribution $f(x)$ is decomposed into n class intervals with $f(x_i)$ equal to $1/n$.

(2) Mode: It is the peak value of the PDF or the value which occurs most frequently. A dataset may have more than one peak.

(3) Median: It is the middle value when the data are ranked according to the magnitude. The median divides the distribution in two equal parts or it is the 50th percentile.

It is seen that while computing the mean, the result is heavily influenced by those values which are quite far from it. Since median is not much influenced by a single value of data, it is a resistant measure, whereas arithmetic mean is not a resistant measure.

Example 18.1

Let the data in a sample be 5, 12, 18, 10, 21, 14, and 20. Compute the mean and median of the sample.

Solution

Arithmetic mean $= (5 + 12 + 18 + 10 + 21 + 14 + 20)/7$
$= 100/7 = 14.3$

Arranging the data in increasing order, the series is 5, 10, 12, 14, 18, 20, 21.
So the median is 14.

Example 18.2

In Example 18.1, replace the data value 21 by 60 and compute the arithmetic mean and median.

Solution

The new series is 5, 12, 18, 10, 21, 14, and 60.

Arithmetic mean $= (5 + 12 + 18 + 10 + 21 + 14 + 60)/7$
$= 140/7 = 20.$

Arranging the new series in increasing order, the series is 5, 10, 12, **14**, 18, 20, 60.
The median is 14.

Note that due to the change in one value, the mean changes significantly but in this case, the median remains the same. Due to this reason, median is called a resistant measure, whereas arithmetic mean is not.

Example 18.3

Compute the median of the series: 5, 12, 18, 10, 21, 14, 20, 24.

Solution

This series has an even number (8) of data values. If the series has an even number of observations, median is the average of two middle values.

Arranging the series in increasing order: 5, 10, 12, 14, 18, 20, 21, 24.
The median is $= (14 + 18)/2 = 16.$

18.2.2 Measure of Dispersion or Spread

Variance represents the scatter of the data about the mean and is computed by

$$s^2 = \frac{1}{n}\sum_{i=1}^{n}(x_i - \overline{x}) \tag{18.13}$$

A small value of variance implies that the data are close to the mean. For small sample sizes, the denominator n in Eq. (18.13) is replaced by $(n - 1)$ to correct for the bias due to size.

Standard deviation (SD) is the square root of the variance

$$s = \left[\frac{1}{n}\sum_{i=1}^{n}(x_i - \overline{x})^2\right]^{0.5} \tag{18.14}$$

Greek letter σ is used to denote the standard deviation of population. As for the variance, the denominator n in Eq. (18.14) is replaced by $(n - 1)$ when computing the SD.

(iii) Coefficient of Variation (CV) is a dimensionless quantity and is obtained by dividing the SD by the mean

$$C_V = s/\overline{x} \tag{18.15}$$

C_v is not defined when the mean is zero. CV is useful to compare the spread of different populations. Given many samples of data, the one with smallest C_v will have the least dispersion of the values around the mean. It is also useful when doing reliability analysis.

Example 18.4

Average annual flows at a river gauging site are given in the table below. Compute the mean, variance, standard deviation, and the coefficient of variation of the flows. This example is drawn from Jain (2011).

Year	**1970**	**1971**	**1972**	**1973**	**1974**	**1975**	**1976**	**1977**	**1978**	**1979**	**1980**
Flow (cumec)	195.5	145.4	148.1	324.7	205.6	302.9	210.3	194.4	71.2	126.8	216.0
Year	1981	1982	1983	1984	1985	1986	1987	1988	1989	1990	1991
Flow (cumec)	136.4	403.9	145.3	161.6	112.5	110.0	90.2	129.8	80.5	136.3	243.3

Solution

We have a total of 22 values. The mean of the flows can be computed as

$$\text{Mean} = (195.5 + 145.4 + 148.1 + \ldots. 80.5 + 136.3 + 243.3)/22 = 176.8 \text{ cumec.}$$

The variance can be computed by

$$\text{Variance } s^2 = [(195.5 - 176.8)^2 + (145.4 - 176.8)^2 + (148.1 - 176.8)^2 + \ldots$$
$$+ (136.3 - 176.8)^2 + (243.3 - 176.8)^2)]/22 = 6926.46 \text{ cumec}^2$$

$$\text{SD} \quad s = (6926.46)^{0.5} = 83.22 \text{ cumec}$$

$$\text{CV} = 83.22/176.8 = 0.47.$$

If we use $(n - 1)$ in place of n in Eq. (18.14)

$$\text{SD} \quad s = (152382.12/21)^{0.5} = 85.18 \text{ cumec}$$

18.2.3 Measures of Symmetry

Commonly, hydrologic data are not symmetrically distributed around the mean or median. If the data have a longer tail to the right of the mean than to the left, then the data are said to be positively skewed and vice versa for negatively skewed (see Fig. 18.2).

Coefficient of skewness (C_s) is a nondimensional measure of the asymmetry of the sample data. It is computed by

$$C_S = \frac{n\sum_{i=1}^{n}(x_i - \bar{x})^3}{(n-1)(n-2)s^3} \tag{18.16}$$

The coefficient of skewness, C_s, is small for symmetrical frequency distributions, while C_s for asymmetrical frequency distributions is either positive or negative. For the normal distribution, $C_s = 0$. The symmetrical and skewed distributions are shown in Fig. 18.2.

Example 18.5

Compute the coefficient of skewness of the data of Example 18.1.

Solution

The coefficient of skewness can be computed by Eq. (18.20)

$$C_s = [22/(21{*}20{*}83.22^3)]{*}[(195.5 - 176.8)^3 + (145.4 - 176.8)^3 + \ldots$$

$$+ (136.3 - 176.8)^3 + (243.3 - 176.8)^3)] = 1.238$$

A positive value of C_S implies that the PDF of the data has a heavy tail on the right-hand side.

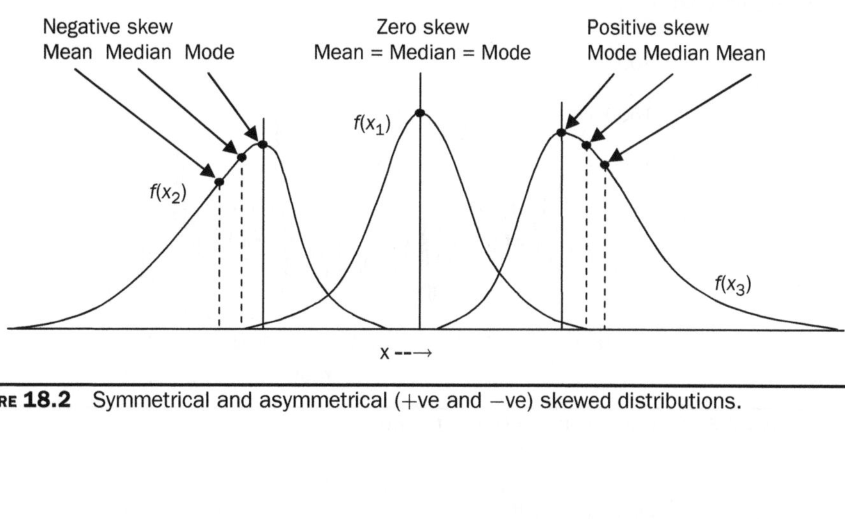

FIGURE 18.2 Symmetrical and asymmetrical (+ve and –ve) skewed distributions.

18.3 Graphical Presentation of Data

Graphical presentation helps in better visualization of the behavior and pattern of the data. Graphically, data can be presented by various means, for example, histograms, box plots, and so on. In a histogram, continuous data are divided into discrete categories. To present the data by histograms, the range of data is determined which is the difference between the highest and lowest value. This range is divided into a number of class intervals and the number of values falling in each interval is noted. The appearance of a histogram critically depends upon the class interval size. If the class interval is large, many details will be lost. If the interval is small, histogram may be too big.

A guideline is that the number of class intervals be between 5 and 20. Also, the number of class intervals not exceed one-fourth to one-half the standard deviation.

To prepare the histogram, the following steps are followed:

(1) Arrange the data in increasing or decreasing order of magnitude and compute range.

(2) Determine the number of class intervals (NC). Size of each class, ΔX, is found by dividing the range by NC.

(3) Divide the ordered data into NC intervals and count the number of values in each class.

(4) Determine the absolute frequency n_j which is the number of observations that fall in the jth class interval, $j = 1, \ldots, \text{NC}$.

(5) Compute the relative frequencies of various classes as n_j/n, $j = 1,\ldots, \text{NC}$. Here, n is the number of observations.

(6) Compute the cumulative relative frequencies F_j, $j = 1, \ldots, \text{NC}$.

(7) Plot the relative frequencies as well as cumulative relative frequencies as ordinate and group interval as abscissa.

Example 18.6

The annual flow of the Sabarmati River at Dharoi is plotted in Fig. 18.3 for the period 1868–1965. Plot the histogram and the cumulative histogram. This example is drawn from Jain (2011).

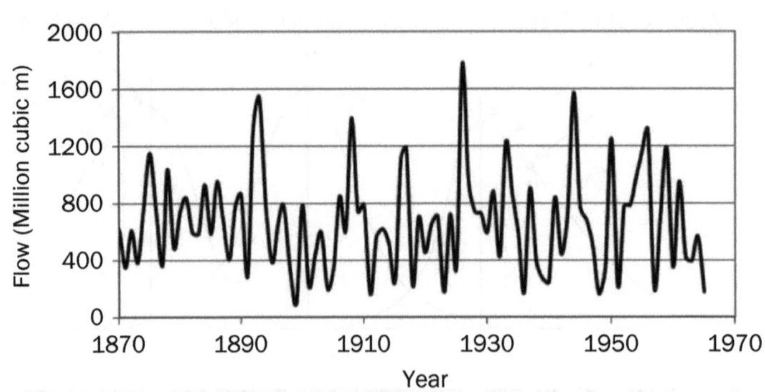

FIGURE 18.3 Plot of the annual flow of the river.

Solution

After examining the data, the class interval was chosen as 100 MCM (million cubic meters), giving 17 classes. The table below shows the mid-values of classes in which the data has been divided, the frequency of values in each class, and the cumulative frequencies.

Mid-value of Class (MCM)	Frequency	Cumulative Frequency	Mid-value of Class (MCM)	Frequency	Cumulative Frequency
150	6	6	1050	1	85
250	9	15	1150	5	90
350	11	26	1250	2	92
450	9	35	1350	3	95
550	9	44	1450	0	95
650	9	53	1550	2	97
750	19	72	1650	0	97
850	6	78	1750	1	98
950	6	84			

The histogram of the given data is plotted in Fig. 18.4.

For the same data, the resultant histogram if the class interval is 200 MCM is shown in Fig. 18.4b. It can be seen that the two figures give different visual impressions, although the underlying data is the same. Of course, many other histograms are possible by adopting different class intervals. The cumulative histogram from the same data is shown in Fig. 18.5.

18.3.1 Box Plots

Box plots are a useful graphical method to display summary information of a dataset and for screening the data for possible outliers. The upper and lower limits of the box

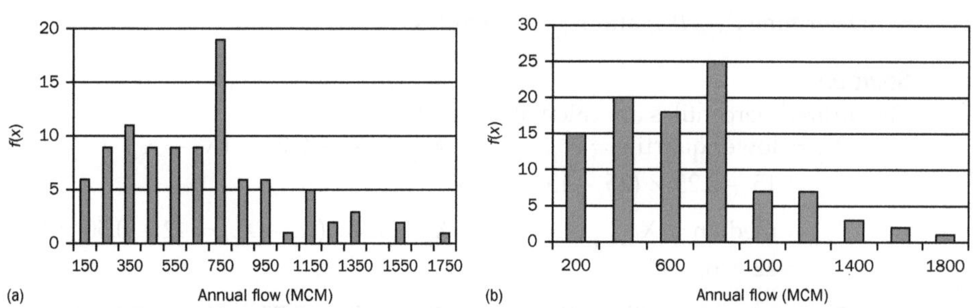

FIGURE 18.4 Histogram of the annual river flows; (a) class interval = 100 MCM, (b) class interval = 200 MCM.

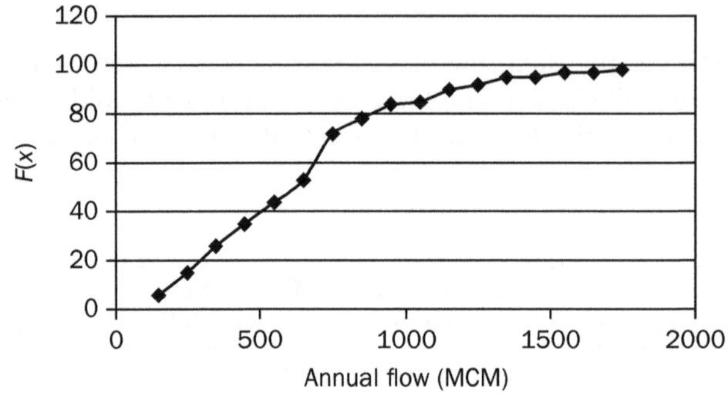

FIGURE 18.5 Cumulative histogram of annual flows of the river.

are typically defined by using quartiles. Three quartiles which are 25th, 50th, and 75th percentiles of the data are used and are computed by

$$P_j = X_{(n+1)\times j} \tag{18.17}$$

where $j = 0.25$, 0.50, or 0.75 and n is the sample size. In case values of $(n + 1) \times j$ are noninteger, a linear interpolation is carried out.

Example 18.7

A sample of data of size 11 has the following values:

$$5, 9, 15, 25, 30, 42, 54, 63, 76, 81, 87.$$

Determine $P_{0.25}$, $P_{0.5}$, and $P_{0.75}$ percentiles.

Solution

The desired percentiles are calculated as follows:

$$P_{0.25} = \text{lower quartile} = X_{(11+1)\times 0.25} = X_3 = 15$$
$$P_{0.50} = \text{median} = X_{(11+1)\times 0.50} = X_6 = 42$$
$$P_{0.75} = \text{upper quartile} = X_{(11+1)\times 0.75} = X_9 = 76$$

Example 18.8

A sample of data of size 12 has the following values:

$$5, 9, 15, 25, 30, 42, 54, 63, 76, 81, 87, 87.$$

Determine $P_{0.25}$, $P_{0.5}$, and $P_{0.75}$ percentiles.

Solution

The desired percentiles are calculated as follows:

$$P_{0.25} = \text{lower quartile} = X_{(12+1)\times 0.25} = X_{3.25} = X_3 + 0.25 \times (X_4 - X_3)$$
$$= 15 + 0.25 \times (25 - 15) = 17.5$$

$$P_{0.50} = \text{median} = X_{(12+1)\times 0.50} = X_{6.5} = X_6 + 0.50\times(X_7 - X_6) = 42 + 0.50 \times (54 - 42)$$
$$= 42 + 6 = 48$$

$$P_{0.75} = \text{upper quartile} = X_{(12+1)\times 0.75} = X_{9.75} = X_9 + 0.75 \times (X_{10} - X_9)$$
$$= 76 + 0.75 \times (81 - 76) = 79.75$$

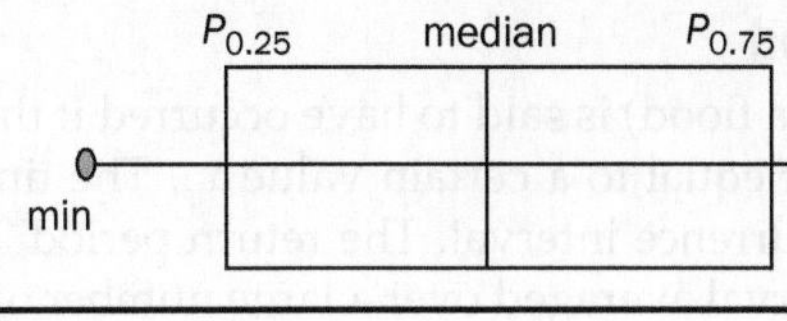

FIGURE 18.6 A box plot.

A box and whisker plot is also called a box plot. It displays the summary of a dataset by using five numbers: the minimum, first quartile, median, third quartile, and the maximum. To prepare a box plot, we first draw a box from the first quartile to the third quartile (see Fig. 18.6). Next, a vertical line inside the box is drawn at the median. Two whiskers are drawn on either side, one for the minimum and the other for maximum.

18.3.2 Quantiles

Quantiles divide a sample into equal-sized, adjacent, subgroups. The median is also a quantile which divides a probability distribution into two equal halves. Quartiles are also quantiles as they divide the distribution into four equal parts. Percentiles are quantiles that divide a distribution into 100 equal parts. The p-th quantile or 100p percentile xp is the quantity which has cumulative probability p

$$F_X(x_p) = p \tag{18.18}$$

In a quantile plot, quantiles or cumulative frequency is plotted. Such plots are sample approximations of the CDF of a continuous random variable (Helsel and Hirsch, 1992). These plots are useful, since no class intervals are required and the complete data are displayed.

One way to compare two datasets is by plotting quantiles of one set against the quantiles of the other. This plot is called a quantile-quantile plot. If the two datasets have the same distribution, the quantiles of the two would fall on a straight line and the median, the quartiles, the 10th and 90th percentiles, and so on, of the two datasets would be equal.

18.3.3 Quartiles

Quartiles are the values that divide a dataset in four equal parts or quarters. Let the water level data at a gauging site (in m) be

$$1.4, 0.9, 1.5, 1.7, 3.2, 2.8, 2.6, 1.8, 2.2, \text{and } 2.4.$$

Arranging the data in increasing order, we have

$$0.9, 1.4, \mathbf{1.5}, 1.7, 1.8, 2.2, 2.4, \mathbf{2.6}, 2.9, 3.2.$$

Hence the lower quartile $= 1.5$, middle quartile (mean) $= 1.8 + 2.2 = 2.0$, and the upper quartile $= 2.6$.

Here, the middle quartile falls between two numbers and hence their average is taken.

The interquartile range (IQR) is the difference between the upper and lower quartiles. In this case, IQR $= 2.6 - 1.5 = 0.9$.

18.3.4 Return Period

An extreme event (e.g., a flood) is said to have occurred if the random variable X (flood peak) is greater than or equal to a certain value x_T. The time interval between events $X \geq x_T$ is called the recurrence interval. The return period T of the event $X \geq x_T$ is the value of recurrence interval averaged over a large number of occurrences. Thus, a 1000-year return period flood would be, on average, equaled or exceeded once in 1000 years. The probability of occurrence of this extreme event is the inverse of the return period

$$P(X \geq x_T) = \frac{1}{T} \tag{18.19}$$

Hence, for a 1000-year flood, probability of occurrence in a given year is 1/1000 = 0.001. Considering the occurrences to be independent variables, the probability of non-exceedances

$$P(X < x_T) = 1 - \frac{1}{T} \tag{18.20}$$

The probability of non-exceedance in n-years

$$P(X < x_T) = \left(1 - \frac{1}{T}\right)^n \tag{18.21}$$

The probability of this flood exceeding at least once in n-years

$$P(X \geq x_T) = 1 - \left(1 - \frac{1}{T}\right)^n \tag{18.22}$$

Example 18.9

The design life of a dam is 100 years. Determine the probability of occurrence of a 1000-year return period flood during the design life of the dam.

Solution

The risk of a 1000-year flood occurring 100 years is

$$P(X \geq x_{1000}) = 1 - \left(1 - \frac{1}{1000}\right)^{100} = 0.095$$

18.4 Probability Distributions

A probability distribution (PD) gives either the probability of occurrence of each value of a discrete random variable, or the probability of a continuous variable falling within a particular interval. The probability distribution also shows the range of values that a random variable can take on.

A probability distribution also indicates how the values of a random variable are spread in a dataset—whether they are symmetrically placed, clustered together, spread, and so on.

Probability distributions commonly used in water resources studies are the Normal, Log Normal, Extreme Value type-1 (Gumbel or EV1), Gamma, Pearson Type III, Log Pearson Type III, Weibull, and Generalized Extreme Value (GEV) distributions. A brief description of the commonly used PDs is given next.

18.4.1 Normal or Gaussian Distribution

The origin of the normal distribution lies in the central limit theorem. Let x_i, $i = 1, 2, \ldots, n$, be a sequence of random variables independently and identically distributed with mean μ and standard deviation σ. As n becomes large, distribution of n such random variables $Y = \sum_{i=1}^{n} x_i$ tends to the normal distribution with mean $n\mu$ and standard deviation $\sqrt{n}\sigma$. This theorem holds good irrespective of the probability distribution of x. The normal distribution assumes that when a random variable results from the sum of many influencing factors, then it can be expressed as the sum of these factors.

The normal distribution has a symmetrical bell-shaped probability density function. The two parameters of this distribution are mean μ and standard deviations. Its PDF is

$$f(x) = \frac{1}{\sigma\sqrt{2\pi}}\exp\left[-\frac{(x-\mu)^2}{2\sigma^2}\right] - \infty < x < \infty \tag{18.23}$$

and the CDF is

$$F(x) = \frac{1}{\sigma\sqrt{2\pi}}\int_{-\infty}^{x}\exp\left[-\frac{(x-\mu)^2}{2\sigma^2}\right]dx \tag{18.24}$$

The reduced variate of the normal distribution is defined as $Z = (x - \mu)/\sigma$. The reduced variate has mean = 0, standard deviation $\sigma_z = 1$, and coefficient of skewness = 0. Equation (18.25) can be written as

$$F(z) = \frac{1}{\sqrt{2\pi}}\int_{-\infty}^{x}\exp\left[-z^2/2\right]dz \tag{18.25}$$

Abramowitz and Stegun (1965) have numerically approximated Eq. (18.25) as

$$F(z) = 1 - f(u)(a_1q + a_2q^2 + a_3q^3 + a_4q^4 + a_5q^5) + \varepsilon(z) \tag{18.26}$$

where

$$q = 1/(1 + kz),\ k = 0.2316419,\ 0 \le z \le \infty;\ a_1 = 0.319381530,$$
$$a_2 = -0.356563782,\ a_3 = 1.781477937,\ a_4 = -1.821255978,\ a_5 = 1.330274429 \tag{18.27}$$

$\varepsilon(z)$ is the error in Eq. (18.26) which is less than 7.5×10^{-8}. Note that $F(-z) = 1 - F(z)$.

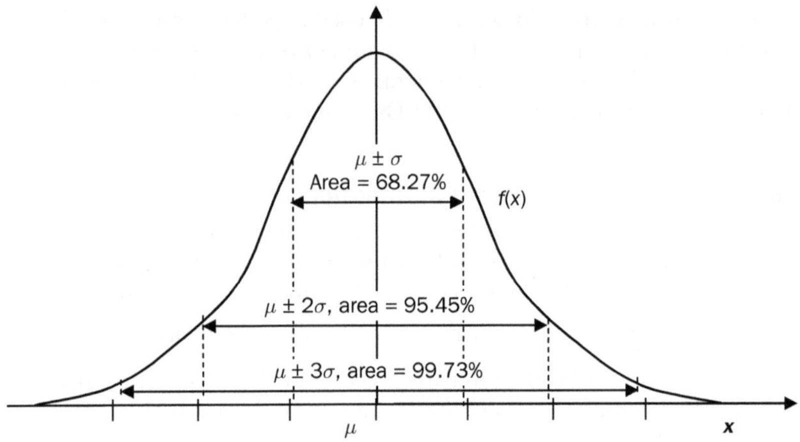

FIGURE 18.7 Normal distribution and the area for three ranges of the standard variate.

Figure 18.7 shows the normal distribution. It also shows the percent area under the curve for three values of the reduced variate. It can be seen that the area under the curve within $\mu \pm \sigma$ is 68.27%, within $\mu \pm 2\sigma$ is 95.45, and it is 99.73 within $\mu \pm 3\sigma$. This means that almost the whole area is covered in the zone $\mu \pm 3\sigma$.

Normal distribution is the most widely used probability distribution in hydrology. This distribution is also employed in many tasks such as the analysis of variance, hypothesis testing, synthetic generation of random numbers, and so on. Many hydrologic variables are not normally distributed, but after transformation, such as logarithmic transformation, power transformation, and so on, they may follow approximately the normal distribution.

Example 18.10

Assuming that the data of the Sabarmati River follows the normal distribution, find the parameters of the distribution and plot it.

Solution

For the data of the Sabarmati River, the mean and SD are as follows:

Mean of the data $\bar{x} = 665.37$ million cubic m.
Standard deviation $s = 346.9$ million cubic m.
Coefficient of variation $C_V = 346.9/665.37 = 0.521$.
Coefficient of skewness $C_s = 0.76$ (positively skewed).
Kurtosis $C_k = 3.65$.

Figure 18.8 shows the plot of the probability distribution of the Sabarmati River data.

For the Sabarmati River data, the coefficient of skewness C_s is 0.76, implying that the data is positively skewed. This is easily verified by Fig. 18.8.

18.4.2 Log-Normal Distribution

If logarithm of a variable is normally distributed then that variable will follow the log-normal distribution. Hence if $X = \ln(Y)$ is normally distributed then Y is log-normally distributed. If a random variable represents a process that is the result of multiplicative

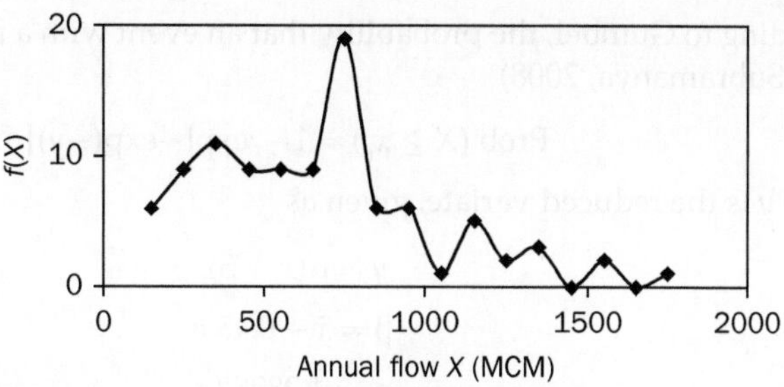

Figure 18.8 Probability distribution of the Sabarmati River data.

effects of many processes, each of which is positive, then that random variable can be expressed as the sum of logarithms of these processes. It is likely that the logarithm of this random variable will follow normal distribution. Note that the log-normal distribution is not useful to model the variables that can have negative values. This distribution has been applied to model rainfall amounts, size distributions of aerosol particles, and so on.

The PDF of the log-normal distribution is

$$f(x) = \frac{1}{x\sigma_y\sqrt{2\pi}}\exp\left[-\frac{(\ln x - \mu_y)^2}{2\sigma_y^2}\right] \quad x > 0 \tag{18.28}$$

The log-normal distribution has two parameters μ_y and σ_y, which can be estimated by transforming all x_i's to y_i's by

$$y_i = \ln x_i \tag{18.29}$$

18.4.3 Extreme Value Type 1 Distribution

Consider a series having a large number of observations (N) of a random variable. Let the series of riverflow data be subdivided into n subsamples of size m each; hence $N = nm$. Each of the subseries will have a maximum and a minimum. These may correspond to, for example, floods and droughts. Gumbel (1958) showed that the n largest values of subsamples asymptotically follow an extreme value type 1 (EV1) distribution. Named after him, the Gumbel or double negative exponential distribution is widely used in hydrologic studies for frequency analysis of variables such floods, maximum rainfall, and so on. The PDF and CDF of EV1 distribution are as follows

$$f(x) = \alpha \exp\{-\alpha(x - \beta) - \exp[-\alpha(x - \beta)]\} \; -\infty < x < \infty; \quad -\infty < \beta < \infty; \alpha > 0$$

$$F(x) = \exp\{-\exp[-\alpha(x - \beta)]\} \tag{18.30}$$

where α and β are scale and location parameters, respectively. The estimates of parameters using the method of moments (discussed later) are

$$\hat{\alpha} = \frac{1.283}{s}; \quad \hat{\beta} = \overline{x} - 0.45s \tag{18.31}$$

This distribution has a constant skewness (approximately 1.14).

According to Gumbel, the probability that an event with a magnitude larger than x_0 occurs is (Subramanya, 2008)

$$\text{Prob}\,(X \geq x_0) = 1 - \exp[-\exp(-y)] \tag{18.32}$$

where y is the reduced variate, given as

$$y = \alpha(x - \beta),$$
$$\beta = \bar{x} - 0.45\,\sigma_x$$
$$\alpha = 1.2825/\sigma_x \tag{18.33}$$

Substituting the values of a and α

$$y = 1.285\,(x - \bar{x})/\sigma_x + 0.577 \tag{18.34}$$

The expression for the reduced variate y for return period T is

$$y = -\left[0.834 + 2.303\log\left(\log\frac{T}{T-1}\right)\right] \tag{18.35}$$

Now we can compute variate x with return period T using Chow's (1954) frequency factor method as

$$X_T = \bar{x} + k\sigma_x \tag{18.36}$$

where $$k = (y_T - 0.577)/1.2825 \tag{18.37}$$

Equations (18.36) and (18.37) assume that the statistics have been computed by using a long data series. If the record length is finite, the following equation may be used

$$x_T = \bar{x} + k_n\sigma_{n-1} \tag{18.38}$$

where σ_{n-1} is the standard deviation of the sample of size n. The frequency factor for use with sample of size n is given as

$$k_n = (y_T - \bar{y}_n)/s_n \tag{18.39}$$

where $\bar{y}_n$ and s_n are the reduced mean and the reduced standard deviations which are functions of n. The values of these can be read from standard tables. Note that as $n \to \infty$, $\bar{y}_n \to 0.577$ and $s_n \to 1.2825$.

Example 18.11

From the flow data of a river, the mean and standard deviation were computed and these were 600 MCM and 350 MCM, respectively. Compute the parameters of EV1 distribution.

Solution

Given that the mean and standard deviation of the data are 600 MCM and 350 MCM, respectively. Therefore the estimates of the parameters by the method of moments are

$$\alpha = 1.2825/350 = 0.0037$$

and $$\beta = 600 - 0.45 \times 350 = 442.5$$

Example 18.12

From the annual maximum flood data of a river for 30 years, the mean and standard deviation were computed as 5200 m^3/s and 1600 m^3/s. Compute the peak of the flood whose return period is 100 years by using the Gumbel or EV1 distribution.

Solution

From standard tables, for $n = 30$ years

$$y_n = 0.5362,\ s_n = 1.1124.$$

$$\text{Hence}\quad y_T = -\left[0.834 + 2.303\log\left[\log\frac{100}{99}\right]\right] = 4.601$$

$$k_{100} = (4.601 - 0.5362)/1.1124 = 3.654$$

$$x_{100} = 5200 + 3.654 \times 1600 = 11046\ \text{m}^3/\text{s}$$

18.4.4 Gamma Distribution

The gamma distribution is one of the most commonly used distributions in hydrology. Its PDF and CDF can be expressed, respectively, as

$$f(x) = \frac{1}{a\Gamma(b)}\left(\frac{x}{a}\right)^{b-1}\exp\left(-\frac{x}{a}\right) \tag{18.40}$$

$$F(x) = \frac{1}{a\Gamma(b)}\int_0^x\left(\frac{x}{a}\right)^{b-1}\exp\left(-\frac{x}{a}\right)dx \tag{18.41}$$

If $y = x/a$ then the CDF can be expressed as

$$F(y) = \frac{1}{\Gamma(b)}\int_0^x (y)^{b-1}\exp(-y)dy \tag{18.42}$$

Abramowitz and Stegun (1965) have shown that $F(y)$ can be expressed as

$$F(y) = F(\chi^2|\upsilon) \tag{18.43}$$

where $F(\chi^2|\upsilon)$ is the chi-square distribution with $\upsilon = 2b$ degrees of freedom and $\chi^2 = 2y$.

One way to compute F for a given value of x is with the use of the expression given by Kendall and Stuart (1964)

$$u = \left[\left(\frac{\chi^2}{\upsilon}\right)^{1/2} + \frac{2}{9\upsilon} - 1\right]\left(\frac{9\upsilon}{2}\right)^{1/2} \tag{18.44}$$

which is approximately normally distributed with zero mean and unit variance when $\upsilon > 20$. Thus, $y = x/a$ is computed first, then $\chi^2 = 2y$, and then u from the above equation. Then, $F(x)$ is computed from the normal distribution expression.

18.4.5 Log Pearson Type III Distribution

The Pearson type III distribution is a three-parameter gamma distribution. Like lognormal distribution, the log Pearson type III distribution (LP3) is obtained by the logarithmic transformation of the variable that follows the Pearson type III distribution. If $Y = \log X$ follows the Pearson type III distribution then X follows the LP3 distribution. The LP3 distribution has been identified as the standard distribution for flood frequency analysis in the United States. Use of LP3 distribution for flood frequency analysis was recommended by the U.S. Water Resources Council. LP3 is a three-parameter distribution whose PDF is given as

$$f(x) = \frac{1}{a\Gamma(b)}\left(\frac{x-c}{a}\right)^{b-1} \exp\left(-\frac{x-c}{a}\right) \tag{18.45}$$

where a, b, and c are the scale, shape, and location parameters, respectively, and G(b) is a gamma function. Parameters a, b, and c of LP3 are related to mean, standard deviation, and coefficient of skewness as (method of moment estimates)

$$a = \sigma/\sqrt{b} \tag{18.46a}$$

$$b = (2/C_s)^2 \tag{18.46b}$$

$$c = \mu - \sigma\sqrt{b} \tag{18.46c}$$

If $c = 0$, LP3 becomes a two-parameter log gamma distribution. To determine the magnitude of flood for a return period T-year by using the LP3 distribution, the frequency factor, K_T, is computed by (Chow et al., 1988)

$$K_T = z + (z^2 - 1)k + (z^3 - 6z)k^2/3 + (z^2 - 1)k^3 + zk^4 + k^5/3 \tag{18.47}$$

where $k = C_s/6$. To compute z for a given return period T, first the exceedance probability p is determined as $p = 1/T$. Now, we compute a variable w as

$$w = \sqrt{\ln(1/p^2)} \quad 0 < p \le 0.5 \tag{18.47a}$$

Next, z is calculated by the following equation given by Abramowitz and Stegun (1965)

$$z = w - \frac{2.515517 + 0.802853w + 0.010328w^2}{1 + 1.432788w + 0.189269w^2 + 0.001308w^3} \tag{18.48}$$

when $p > 0.5$, p in Eq. (18.47a) is replaced by $(1 - p)$ and the negative sign is put before z is computed Eq. (18.48). Now, following the frequency factor method, the magnitude of T-year return period flood is computed by

$$y_T = \overline{y} + K_T s_y \tag{18.49}$$

Example 18.13

For the data of Example 18.11, find the parameters of the Pearson type III distribution. The parameters are mean = 600 MCM, standard deviation = 350 MCM, and $C_s = 0.76$.

Solution

The estimates of parameters using the method of moments are

$$b = (2/C_s)^2 = (2/0.76)^2 = 6.93 \text{ MCM}$$

$$a = 350/\sqrt{6.93} = 132.95 \text{ MCM}$$

$$c = 600 - 350 \times \sqrt{6.93} = -321.37$$

Example 18.14

Logarithms of the annual flood peak data of a river gave the mean as 3.6, SD as 0.41, and $C_s = -0.06$. Find the 50-year return period flood by using the LP3 distribution.

Solution

First we find the value of K_{50} by the following equation

$$K_{50} = z + (z^2 - 1)k + (z^3 - 6z)k^2/3 + (z^2 - 1)k^3 + zk^4 + k^5/3$$

Here $k = C_s/6 = -0.06/6 = -0.01$. For $T = 50$, $p = 1/50 = 0.02$.

$$w = \sqrt{\ln(1/0.02^2)} = 2.797$$

From Eq. (18.48)

$$z = 2.797 - \frac{2.515517 + 0.802853 * 2.797 + 0.010328 * 2.797^2}{1 + 1.432788 * 2.797 + 0.189269 * 2.797^2 + 0.001308 * 2.797^3} = 2.054$$

Now K_{50} is calculated as

$$K_{50} = 2.054 + (2.054^2 - 1)*(-0.01) + (2.054^3 - 6*2.054)* (-0.01)^2/3$$

$$+ (2.054^2 - 1)* (-0.01)^3 + 2.054*(-0.01)^4 + (-0.01)^5/3 = 2.022$$

Hence $y_{50} = 3.6 + 2.022*0.41 = 4.429$

So, the 50-year flood $x_{50} = (10)^{4.429} = 26859$.

18.4.6 Discrete Probability Distributions

Discrete probability distributions are generally applied only to those random events where only two mutually exclusive outcomes are possible, for example, success or failure. Further, the successive trials are assumed independent and the probability of success does not change with the trials. To find the probability of occurrence of an event r times in n successive years, the Binomial or Poisson distribution can be used.

18.4.7 Binomial Distribution

Binomial distribution arises in Bernoulli processes where, in any trial, a given event may or may not occur. The probability of occurrence of the event remains fixed from trial to trial (called Bernoulli trials). Binomial distribution is usually observed while dealing with complementary events. For example, when tossing a coin, the probability that head appears remains the same in each trial. Likewise, the occurrence of wet and dry days during a given time interval is a complementary event. In such cases, the probability that the event takes place r times in n successive years is computed by

$$P_{r,n} = n\,C_r\,P^r q^{n-r} = \frac{n!}{r!(n-r)!} p^r q^{n-r} \tag{18.50}$$

where $P_{r,n}$ is the probability that a random event of a given magnitude and exceedance probability P occurs r times in n successive years; q is the probability that the event does not occur. The probability that the event does not occur at all in n successive years is

$$P_{o,n} = q^n = (1 - p)^n \tag{18.51}$$

The probability that an event takes place at least once in n successive years is

$$P_1 = 1 - q^n = 1 - (1 - p)^n \tag{18.52}$$

Example 18.15

Data of maximum one-day rainfall depth at a station was analyzed and it was found that a depth of 275 mm had a return period of 50 years. Determine the probability that one-day rainfall depth equal to or greater than 275 mm occurs (1) once in 20 successive years, and (2) twice in 20 successive years.

Solution

Here, $P = 1/50 = 0.02$.

(1) In the first case, $n = 20$, $r = 1$. We apply Eq. (18.50)

$$P_{1,20} = \frac{20!}{19!\ 1!} * (0.02) * (0.98)^{19} = 0.272.$$

(2) In the second case, $n = 20$, $r = 2$. Therefore

$$P_{2,20} = \frac{20!}{18!\ 2!} * (0.02^2) * (0.98)^{18} = 0.0528$$

Example 18.16

What is the probability that a 5-year flood will not occur at all in a 10-year period?

Solution

Here, $p = 1/5 = 0.2$, $n = 10$, and $r = 0$. Hence the probability is

$$P_{0,10} = \frac{10!}{0!\ 10!} * 0.2^0 * (0.8)^{10} = 0.1074$$

18.5 Point and Interval Estimates

The estimates of statistical properties of a dataset, such as mean or standard deviation, which have a single value are called point estimates. A single value does not give any idea of the reliability of these estimates. Consider two datasets A and B having the same number of data (e.g., 50) and the same mean of 100. But data in A lie closely near the mean, while data in B are spread out. The point estimate of mean for A is more reliable than for B because data of set B have more spread and when we state that the mean of two datasets is 100, we are relatively more confident in case of A. However, when only single value is reported, this difference is not revealed.

In contrast to point estimates, interval estimates are a range of values or intervals. In addition, the probability that the true population value lies in this interval is stated. Obviously, the intervals will be wider for the datasets that have greater variability. Interval estimates with stated probability that the interval contains the true population value are also called confidence intervals.

The probability that the given interval includes the true value is called the confidence level of the interval. The width of the confidence interval depends on the variability in the data, the size of sample, and the confidence level. As the confidence level increases, the width of the confidence interval increases. Thus all things remaining the same, the width of 95% confidence interval will be more than that of the 90% confidence interval. The probability that this interval will not cover the true value is known as the alpha (α) level. So, $\alpha = 1 -$ confidence level.

18.5.1 Confidence Intervals for the Mean

As sample sizes get larger, the distribution of sample mean is closely approximated by a normal distribution. Hence, for large sample sizes, the variation of the mean is described by symmetric intervals.

Symmetric confidence intervals for the mean can be computed as

$$\bar{x} - t_{(\alpha/2,n-1)}\sqrt{s^2/n} \le \mu \le \bar{x} + t_{(\alpha/2,n-1)}\sqrt{s^2/n} \qquad (18.53)$$

where $t_{(\alpha/2,n-1)}$ is a value of Student's *t*-distribution with $(n - 1)$ degrees of freedom and probability of exceedance of $\alpha/2$.

Example 18.17

The mean of a sample (of size = 25) is = 84 and SD is 42.

(1) Determine 95% confidence interval for the mean.

(2) Determine the confidence interval if the SD reduces by half, or SD = 21.

Solution

(1) For 95% confidence interval, $\alpha = 1 - 0.95 = 0.05$. Hence

$$84 - t_{(0.025,25-1)}\sqrt{42^2/25} \le \mu \le 84 + t_{(0.025,25-1)}\sqrt{42^2/25}$$

$$84 - 2.064 \times 8.4 \le \mu \le 84 + 2.064 \times 8.4$$

$$66.66 \le \mu \le 101.34$$

Thus there is a 95% probability that the population mean lies in the interval between 66.66 and 101.34.

(2) If SD is 21, the confidence interval is

$$84 - 2.064\sqrt{21^2/25} \le \mu \le 84 + 2.064\sqrt{21^2/25}$$

$$84 - 2.064 \times 4.2 \le \mu \le 84 + 2.064 \times 4.2$$

$$75.33 \le \mu \le 92.67$$

Note that the confidence interval is reduced due to the reduction in SD.

18.5.2 Standard Error

Standard error (SE) is a measure of the variability of the estimated value of, say, quantile and is a function of sample size, parameters of the distribution, and method of parameter estimation. It measures the error due to sample size but not due to the improper distribution used. This means that the SE of estimate of quantile will be different for different methods of parameter estimation.

The SE of mean or the standard deviation of the mean is used to estimate the expected change in mean if the mean is estimated by drawing different samples from the population. The SE of the mean s_M is computed by

$$s_M = s/\sqrt{n} \tag{18.55}$$

where s is the SD of the dataset and N is the sample size. It can be seen that s_M decreases as n increases. If the random component in the process is large, the SE-mean will be higher and vice versa. The SE-mean is extensively used to calculate the margin of error which is used to calculate confidence intervals.

18.6 Parameter Estimation Methods

Many methods have been developed to estimate parameters of hydrologic models (Jain and Singh, 2003). Some methods commonly used in hydrologic analysis include (1) method of moments; (2) method of maximum likelihood; (3) method of least squares; and (4) method of *L*-moments. These methods are briefly discussed here.

18.6.1 Method of Moments

This method is commonly employed to estimate parameters of probability distributions. This method is based on the assumption that if the distribution parameters have been correctly estimated, the moments of the probability density function (PDF) will be equal to the corresponding moments of the sample data.

Let X be a continuous random variable and $f(x)$ be its function. The rth moment of $f(x)$ about an arbitrary point "a" is denoted as $M_r^a(f)$. The rth moment of the function $f(x)$ can be expressed as

$$M_r^a(f) = \int_{-\infty}^{\infty} (x-a)^r f(x)dx \tag{18.56}$$

Figure 18.9 explains the various terms used in Eq. (18.56). Consider the zeroth moment, or $r = 0$. For this case, Eq. (18.56) yields

$$M_0^a = \int_{-\infty}^{\infty} (x-a)^0 f(x)dx = \int_{-\infty}^{\infty} f(x)dx = 1 \tag{18.57}$$

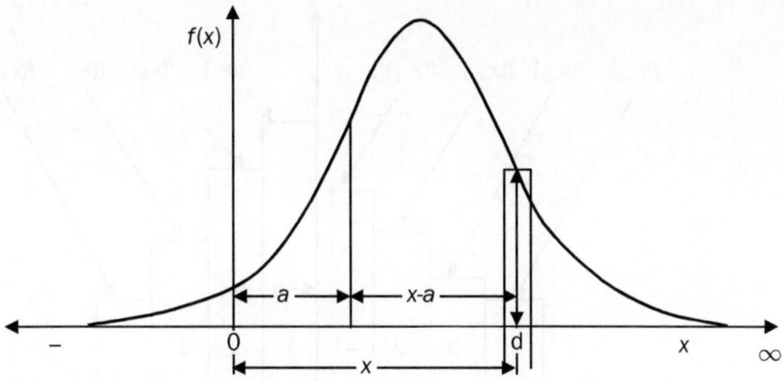

FIGURE 18.9 Concept of moment of a function $f(x)$ about an arbitrary point

Thus the zero-order moment gives the area under the curve $f(x)$ subject to $-\infty < x < \infty$. This area is unity for a probability distribution. If $r = 1$, Eq. (18.54) yields

$$M_1^a = \int_{-\infty}^{\infty} (x - a)^1 f(x)dx = \mu - a \tag{18.58}$$

where μ is the mean. In the case of the moment around the origin, $a = 0$, and this moment equals the mean. The rth moment about the mean ($a = \mu$) is

$$M_r^\mu = \int_{-\infty}^{\infty} (x - \mu)^r f(x)dx \tag{18.59}$$

If the superscript is dropped when the moment is taken about the origin, the various moments can be written as: M_0 = area, M_1 = mean, $M_2{}^\mu$ = variance, $M_3{}^\mu$ = measure of skewness of the function, and measure $M_4{}^\mu$ = kurtosis.

18.6.2 Method of Moments for Discrete Systems

For a discrete function, f_j, $j = -\infty,\ldots, -1,0,1,\ldots, \infty$, the rth moment about any arbitrary point can be expressed in a similar way as for continuous case. The rth moment about the origin is

$$M_r = \sum_{m=-\infty}^{\infty} m^r f_m \tag{18.60}$$

Figure 18.10 (redrawn from Jain, 2011) shows how the moment of a discrete function is computed.

Example 18.18

The frequency table of annual flows of the Sabarmati River is given below. Find the mean and variance of the data by using the method of moments. This example is drawn from Jain (2011).

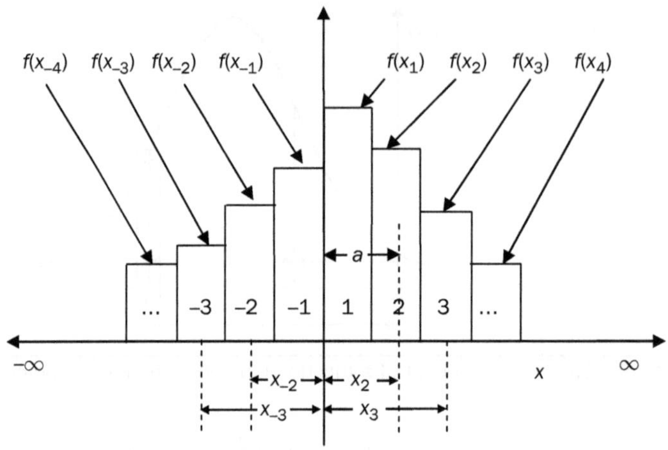

FIGURE 18.10 Concept of moment of a discrete function about an arbitrary point "a."

Discharge Range	Frequency	Discharge Range	Frequency
100–200	6	200–300	9
300–400	11	400–500	9
500–600	9	600–700	9
700–800	19	800–900	6
900–1000	6	1000–1100	1
1100–1200	5	1200–1300	2
1300–1400	3	1400–1500	0
1500–1600	2	1600–1700	0
1700–1800	1		

Solution

The sum of frequencies for all discharges = (6 + 11 + 9 + 19 + ...) = 98. For the first discharge range (100–200), the average value = 150; for the second range (200–300), the average value = 250, and so on. The first moment of the data is computed as

$$M_1 = (150 \times 6 + 250 \times 9 + 350 \times 11 + \ldots + 1750 \times 1)/98 = 664.29 \text{ cumec.}$$

Hence, the mean (x_m) of annual flows is 664.29 cumec.

The second moment about the mean gives the variance of data.

$$M_2^{\mu} = \sum_{i=1}^{N} \frac{(x_i - x_m)^2 f(x_i)}{N}$$

Second moment

$$= [\,(150 - 664)^{2}*6 + (250 - 664)^{2}*9 + \ldots + (1750 - 664)^{2}*1]/98$$

$$= 120{,}000 \text{ cumec}^2.$$

Hence, the standard deviation (s) = $(120{,}000)^{0.5}$ = 346.61 cumec.

The coefficient of variation $c_v = s/x_m = 346.61/664.29 = 0.52$.

18.6.3 Method of Maximum Likelihood

The maximum likelihood (ML) method is a powerful and widely used method to estimate parameters of a frequency distribution. The parameter estimates obtained from the ML method are unbiased, have minimum variance, and are normally distributed.

Let $f(x; a_1, a_2, \ldots a_m)$ be the PDF of random variable X and the problem is to estimate the parameters a_i, $I = 1, 2, \ldots, m$. Let a random sample of data $x_1, x_2, \ldots x_n$ be drawn from this PDF. Hence, the joint PDF will be given by

$$f(x_i, x_2, \ldots x_n; a_1, a_2, \ldots a_m) = \underset{i=1}{\overset{n}{\pi}} f(x_i; a_1, a_2, \ldots a_m) \quad (18.61)$$

In a similar manner, the probability of obtaining the random sample $x_1, x_2, \ldots x_n$ from the population of random variable X is proportional to the product of individual probability densities or the joint PDF. This joint PDF is termed as the likelihood function, L, and is given by

$$L = \underset{i=1}{\overset{n}{\pi}} f(x_1; a_1, a_2, \ldots a_m) \quad (18.62)$$

where a_i, $i = 1, 2, \ldots m$, are the unknown distribution parameters which are to be estimated.

In the ML method, the unknown parameters are determined by maximizing the likelihood that the given sample would be obtained if n random observations were drawn from $f(x; a_1, a_2, \ldots a_m)$. Since the maximum of L and of $\ln(L)$ occur at the same values of a_i, $i = 1, 2, \ldots m$, the ML function can also be expressed as

$$\ln L = L^* = \ln \underset{i=1}{\overset{n}{\pi}} f(x_1; a_1, a_2, \ldots a_m) = \sum_{i=1}^{n} \ln f(x_1; a_1, a_2, \ldots a_m) \quad (18.63)$$

In the ML method, frequently $\ln(L)$ is maximized, since many times finding the maximum of the logarithm of function L is easier than finding the maximum of L. The values of parameters obtained by following this method are known as ML estimators.

The method finds the point where the maximum of ML function lies by taking partial differences of L or $\ln(L)$ with respect to each parameter and equating these differences to zero. By doing so, as many equations are obtained as the unknown parameters. Let there be m unknown parameters:

$$\begin{aligned} \frac{\partial L(a_1, a_2, \ldots a_m)}{\partial a_1} &= 0 \\ \frac{\partial L(a_1, a_2, \ldots a_m)}{\partial a_2} &= 0 \\ \frac{\partial L(a_1, a_2, \ldots a_m)}{\partial a_m} &= 0 \end{aligned} \quad (18.64)$$

These m equations are solved to get the values of m unknown parameters.

Example 18.19

Use the ML method to compute the parameter a of the exponential distribution for the data of the Sabarmati River, given in Example 18.11.

Solution

The probability density function of one-parameter exponential distribution is given by

$$f_X(x) = a\,[exp(-ax)]$$

The likelihood function is given by

$$L(\alpha) = \prod_{i=1}^{n} \alpha \exp(-\alpha x_i) = \alpha^n \exp\left(-\alpha \sum_{i=1}^{n} x_i\right)$$

Hence, the log-likelihood function is

$$\ln L(\alpha) = n\ln(\alpha) - \alpha\left(\sum_{i=1}^{n} x_i\right)$$

where n is the sample size. Differentiating Eq. (18.59) with respect to

$$\frac{d\ln L(\alpha)}{d\alpha} = \frac{n}{\alpha} - \sum_{i=1}^{n} x_i = 0$$

or

$$\alpha = n/\left(\sum_{i=1}^{n} x_i\right) = \frac{1}{\bar{x}}$$

In Example 18.11, the mean of the data was 600 cumec. Hence the estimate of α is

$$\alpha = 1/600 = 1.67 \times 10^{-3}\ \text{cumec}^{-1}.$$

18.6.4 Method of Least Squares

The method of least squares (MOLS) is frequently used to estimate parameters in the field of hydrology. Natale and Todini (1974) developed constrained MOLS for linear models.

Consider that parameters a_i, $i = 1, 2, \ldots m$, of a function $Y = f(X; a_1, a_2, \ldots a_m)$, are to be estimated. In MOLS, the sum of squares of deviations, D, between observed and computed values of all data is minimized to determine the parameters. This sum is computed by

$$D = \sum_{i=1}^{n} d_1^2 = \sum_{i=1}^{n} [y_0(i) - y_c(i)]^2 = \sum_{i=1}^{n} [y_0(i) - f(x; a_1, a_2, \ldots a_m)]^2 \tag{18.65}$$

where $y_0(i)$ and $y_c(i)$ are the ith observed and computed values of Y, respectively, and $n > m$ is the number of observations. The expression for D is partially differentiated with respect to each parameter and each differential is equated to zero

$$\frac{\partial \sum_{i=1}^{n} [y_0(i) - f(x_i; a_1, a_2, \ldots a_m)]^2}{\partial a_1} = 0 \quad j = 1, 2, \ldots m \tag{18.66}$$

The resulting equations are called the normal equations. These m equations are solved to estimate m parameters.

18.6.5 Method of L-Moments

Linear moments or L-moments were developed by Hosking (1986). These moments give a descriptive summary of the location, scale, and shape of a probability distribution. The L-moments are similar to ordinary moments. These are expressed as linear combinations of order statistics, which gives them this name. L-moments can also be expressed by linear combinations of probability weighted moments.

Hosking (1986) mentioned several important advantages of L-moments over ordinary moments. They have less bias than ordinary moments because they are linear combinations of ranked observations. For example, variance (second moment) and skewness (third moment) involve squaring and raising the observations to the third power, respectively. Due to this, the observations far from the mean get greater weight. As a result, ordinary moments have substantial bias and variance.

Let the data in a sample of size n be arranged in ascending order to yield the series $x_{1:n} \leq x_{2:n} \leq x_{3:n} \leq \ldots x_{n:n}$. The unbiased estimator of probability weighted moments is

$$b_r = \frac{1}{n}\sum_{j=r+1}^{n}\frac{(j-1)(j-2)\ldots(j-r)}{(n-1)(n-2)\ldots(n-r)}x_{j:n} \tag{18.67}$$

In particular

$$b_0 = \frac{1}{n}\sum_{j=1}^{n}x_{j:n} \tag{18.68}$$

$$b_1 = \frac{1}{n}\sum_{j=2}^{n}\frac{(j-1)}{(n-1)}x_{j:n} \tag{18.69}$$

$$b_2 = \frac{1}{n}\sum_{j=3}^{n}\frac{(j-1)(j-2)}{(n-1)(n-2)}x_{j:n} \tag{18.70}$$

Now, the sample L-moments are defined by (Hosking and Wallis, 1997)

$$l_1 = b_0 \tag{18.71}$$

$$l_2 = 2b_1 - b_0 \tag{18.72}$$

$$l_3 = 6b_2 - 6b_1 + b_0 \tag{18.73}$$

$$l_4 = 20b_3 - 30b_2 + 12b_1 - b_0 \tag{18.74}$$

The sample L-moment ratios are defined by

$$t_r = l_r/l_2 \tag{18.75}$$

and the sample $L - CV$ is

$$t = l_2/l_1 \tag{18.76}$$

Similar to the method of moments, the L-moments method can also be employed to compute parameters of probability distributions by expressing these parameters as functions of L-moments. Hosking and Wallis (1997) have given expressions of L-moments for many probability distributions.

Example 18.20

The annual maximum flood (AMF) data (in cumec) of a river for 22 years is given in the table below. Assuming that the data follows GEV distribution, compute the parameters of the distribution by L-moments and the quantile for exceedance probability of 0.5.

SN	1	2	3	4	5	6	7	8	9	10	11
AMF	1297	1597	3002	4233	5020	5242	5624	6040	6311	6844	7041
SN	12	13	14	15	16	17	18	19	20	21	22
AMF	7153	7381	7494	7819	7980	8562	8573	8620	9113	9404	13077

Solution

From the given data, the following are computed

$$l_1 = 6701,\quad l_2 = 1492.21 \quad l_3 = 129.259 \quad l_4 = 326.766$$

$$t_r = 0.222 \quad t_3 = -0.0866 \quad t_4 = -0.219$$

Parameters of the GEV distribution are calculated by using the formulas provided by Hosking and Wallis (1997) as

$$k \approx 7.859 \times c + 2.955 \times c^2 \quad \text{where } c = \frac{2}{3 + t_3} - \frac{\log(2)}{\log(3)}$$

For the given data, $c = 0.056$, giving $k = 0.446$

$$\text{Parameter } \alpha = \frac{l_2 k}{(1 - 2^{-k})\Gamma(1 + k)} = \frac{1492.21 \times 0.446}{(1 - 2^{-0.446})\Gamma(1 + 0.446)} = 2825.389$$

$$\xi = l_1 - \frac{\alpha\{1 - \Gamma(1 + k)\}}{k} = 6701 - \frac{2825.389\{1 - \Gamma(1 + 0.446)\}}{0.446} = 5976.842$$

Hence the GEV quantile for $F = 0.5$ is

$$x(F = 0.5) = \xi + \alpha\frac{1 - (-\ln F)^k}{k} = 5976.842 + 2825.389\frac{1 - (-\ln(0.5))^{0.446}}{0.446}$$
$$= 6932.217 \text{ cumec}$$

18.6.6 Bias in Parameter Estimation

In estimating parameters of a probability distribution by using the sample data, errors may be introduced due to several causes: the sample data may have errors, the assumption of the parameter estimation method may not be valid, and errors may arise due to truncation and round-off. These may introduce errors in the estimates of parameters. A number of statistical measures are employed to evaluate the estimates of parameters. Bias is one of them.

Let a_c be the estimate of parameter a. This estimate a_c is called an unbiased estimate of a if $E(a_c) = a$ where $E(a_c)$ is the expected value of a. Often, a_c may have certain bias $b(a)$, so that

$$E(a_c) = a + b\,(a) \tag{18.77}$$

If the estimate is unbiased, $b(a) = 0$. Note that even if $b(a) = 0$, an individual a_c may not be close to a. Unbiasedness means that the average of many independent estimates of a will be equal to a.

18.7 Summary

Beginning with a discussion of descriptors of data, this chapter discusses commonly used continuous and discrete probability distributions. Methods for estimating distribution parameters are then presented. Also discussed is the bias in parameter estimates. The discussion in the chapter should suffice for a preliminary statistical analysis of data often encountered in hydrology.

18.8 Questions

18.1 Write five important characteristics of hydrologic data?

18.2 Probability distribution formula (PDF) of a random variable can be obtained from the cumulative distribution function (CDF) by

I. Integrating CDF II. Differentiating CDF

III. By inverting CDF IV. PDF cannot be obtained from CDF

18.3 If a probability distribution has positive skewness then

I. Mean of the data equals median II. Mean is greater than median

III. Mean is smaller than median IV. Any of these conditions may exist

18.4 Annual rainfall for a city for 10 years from 1980 to 1989 is given below.

Year	Annual Rainfall (mm)
1980	840
1981	940
1982	1120
1983	985
1984	1070
1985	901
1986	1187
1987	884
1988	993
1989	1065

Compute the median, coefficient of variation, and the coefficient of skewness of the data.

18.5 When the sample size is large, parameters estimated by the maximum likelihood method are

I. Biased II. Unbiased

III. Biased irrespective of sample size IV. None of the above

18.6 A sample of data of size 13 has the following values:

105, 119, 125, 158, 230, 342, 354, 403, 476, 481, 527, 556, 599.

Determine $P_{0.25}$, $P_{0.5}$, $P_{0.75}$ percentiles.

18.7 The average value of flow in a river is 312 MCM and the standard deviation of this data is 219 MCM. Assuming that the data follow EV1 distribution, find the parameters of the distribution.

18.8 The mean annual flood of a river is 32,000 cumec and the standard deviation of flood peaks is 6000 cumec. Assuming that the peaks follow the Gumbel distribution, what is the probability of a flood of magnitude of 45,000 cumec occurring in the river in the next 5 years?

18.9 The mean of a sample (of size = 50) is = 41.3 and SD is 22.5. (1) Determine 95% confidence interval for the mean, (2) determine 90% confidence interval for the mean, and (3) compare the two results.

18.10 Determine 95% confidence interval for the mean for a sample (of size = 100) whose mean is = 41.3 and SD is 22.5. Compare the results with results of previous problem.

18.11 In a statistical test, a significance level of 0.05 implies that

I. The decision of a statistical test may be in error 5 times.

II. The decision of a statistical test may be in error 1 time out of 20.

III. The results are significant only if the test is carried out 5 times.

IV. None of the above.

18.12 Write a short note on hypothesis testing, clearly bringing out the need for it and briefly describing two commonly used tests in the field of hydrology.

18.9 References

Abramowitz, M., and I. A. Stegun. *Handbook of mathematical functions with formulas, graphs, and mathematical tables*. Dover Books on Advanced Mathematics, New York, 1965.

Chow, V. T. The log-probability law and its engineering applications. *Proceedings, American Society of Civil Engineers*, 80:1–25, 1954.

Chow, V. T., D. R. Maidment, and L. W. Mays. *Applied Hydrology*. New York: McGraw-Hill Book Company, 1988.

Haan, C. T. *Statistical Methods in Hydrology*. Ames, IA: Iowa State Press, 2002.

Helsel, D. R. and R. M. Hirsch. *Statistical Methods in Water Resources*. Amsterdam, The Netherlands: Elsevier, 1992.

Hosking, J. R. M., and J. R. Wallis. *Regional Frequency Analysis: An Approach Based on L-Moments*. Cambridge: Cambridge University Press, 1997.

Jain, S. K. Hydrological Measurements and Analysis of Data. Web-based course under National Program on Technology Enhanced Learning (NPTEL). http://nptel.ac.in/courses/105107129/, 2011.

Jain, S. K. and V. P. Singh, *Water Resources Systems Planning and Management*. Amsterdam: Elsevier, 2003.

Kendall, M. G., and A. Stuart. *The Advanced Theory of Statistics*. Oxford University Press, Oxford, England, 1978.

McCuen, R. H. *Hydrologic Analysis and Design*. Eaglewood Cliffs, NJ: Prentice Hall, 1989.

Natale L., and E. Todini. *A Constrained Parameter Estimation Technique for Linear Models in Hydrology. Publ. No. 13*. Pavia, Italy: Institute of Hydraulics, University of Pavia, 1974.

Sturges, H. A. The choice of a class interval. *Journal of American Statistical Association*, 21:65–66, 1926.

Subramanya, K. *Engineering Hydrology*. Tata McGraw Hill, New Delhi, 2008.

Yevjevich, V. *Probability and Statistics in Hydrology.* Fort Collins, CO: Water Resources Publications, 1972.

CHAPTER 19

Correlation, Regression, Hypothesis, and Trend Analysis

In continuation with the previous chapter, this chapter discusses some more statistical techniques which students and practitioners of hydrology need to master. These techniques have also been described in textbooks, such as Davis (2015) and Haan (2002).

19.1 Covariance and Correlation

Frequently, there is a cause-and-effect relation among hydrologic variables due to the physics of the underlying processes. Consequently, changes in the value of the cause variables lead to changes in the effect variable. Such variables are said to be correlated. Correlation among the variables can help with hydrologic analysis and interpretations. For that, we have to find out the strength of association. Linear association between hydrologic variables is expressed by covariance and correlation.

Consider that we have N pairs of observations, $(x_1, y_1), (x_2, y_2), \ldots, (x_N, y_N)$, of two hydrologic variables X and Y. The (sample) covariance between the variables can be computed by

$$C_{XY} = \frac{1}{N-1}\sum_{i=1}^{N}(x_i - m_X)(y_i - m_Y) \tag{19.1}$$

where m_X and m_Y are the mean values of X and Y, respectively.

Covariance is a measure of "linear dependence" between two variables and shows the degree to which the variables change together. When the two variables X and Y are the same, the covariance is termed variance. Since covariance depends on the magnitude of the data, values of covariance among variables having different magnitude or scales cannot be compared. This issue is addressed through dimensionless correlation coefficient which is obtained by normalizing the covariance by the product of the standard deviations of the variables. Thus, correlation coefficient r_{XY} is obtained by dividing the covariance by the standard deviations of X and Y

$$r_{XY} = \frac{C_{XY}}{s_X s_Y} = \frac{\dfrac{1}{N-1}\displaystyle\sum_{i=1}^{N}(x_i - m_X)(y_i - m_y)}{s_X s_Y} \tag{19.2}$$

where s_x, s_Y = the standard deviations of X and Y, respectively.

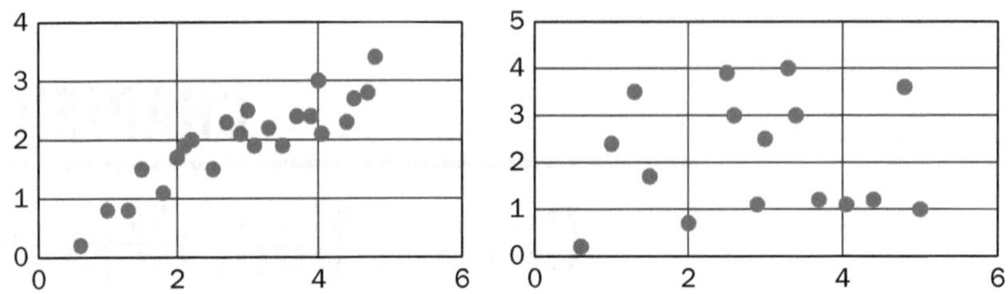

FIGURE 19.1 Scatter plots showing very high (left) and very low (right) correlation among the variables.

The correlation coefficient computed by Eq. (19.2) measures linear association between the variables and does not account for nonlinear association. Value of r_{XY} varies between −1 (perfect negative correlation) and +1 (perfect positive correlation); $r_{XY} = 0$ means there is no linear association between X and Y.

A set of variables may be related with each other due to several reasons. If one variable drives the other, they will be correlated, for example, rainfall and runoff for a catchment. Two variables that share the same cause may also be correlated. For example, river discharge and concentration of suspended sediment that are transported in association with flow. Note that high magnitude of correlation coefficient per se is not an indicator of a causal relationship between two variables.

Example 19.1

The table below contains annual rainfall and runoff data for a watershed. Compute correlation between rainfall and runoff.

Year	1980	1981	1982	1983	1984	1985	1986	1987	1988
Rainfall	42.39	33.48	47.67	50.24	43.28	52.61	31.06	48.02	47.08
Runoff	13.26	6.31	12.17	15.54	10.22	20.24	7.79	15.64	22.91
Year	**1989**	**1990**	**1991**	**1992**	**1993**	**1994**	**1995**	**1996**	**1997**
Rainfall	36.08	40.89	34.31	37.15	40.38	45.39	33.49	39.28	42.43
Runoff	12.89	11.02	15.58	17.17	9.18	18.02	9.01	13.82	17.25

Solution

The correlation can be computed by Eq. (19.2) and it comes out to 0.625. Spreadsheets, such as Microsoft Excel, contain a built-in function to compute correlation.

Figure 19.1 shows two cases of association between two variables. In the first case (left panel), r is about 0.9 which means strong correlation. In the second case (right panel), the variables do not have any significant linear dependence and r is nearly zero. A scatter plot of the data, such as Fig. 19.1, is a powerful way to explore the association between two variables.

Figure 19.2 shows six cases where the variables have very different associations, yet the correlation coefficient is about 0.6 in all the cases. These cases are shown here to emphasize that whenever correlation is computed, data should always be plotted for visual impression.

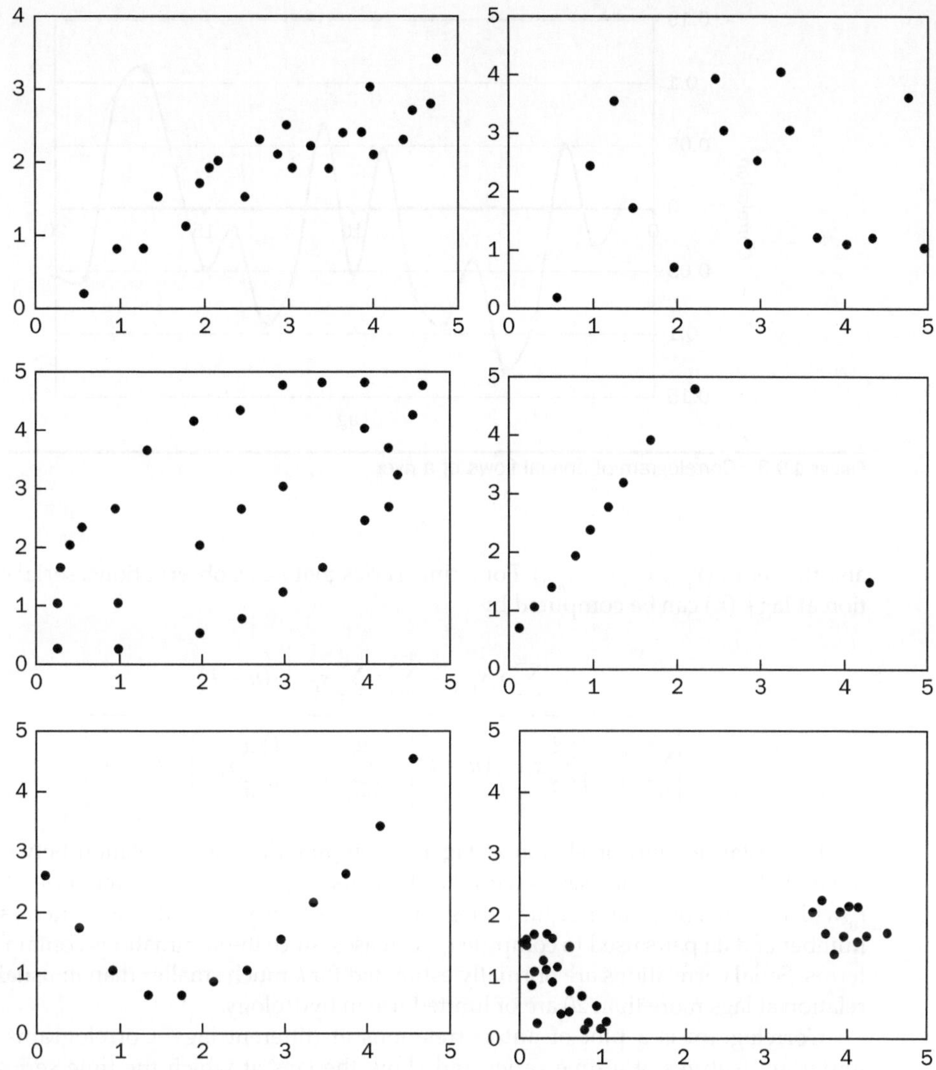

FIGURE 19.2 Various cases of association between two variables. Correlation coefficient is about 0.6 in each case.

If the scatter plot shows that a nonlinear relationship is likely between two variables, one may decide to transform data by, for example, logarithmic, square root, power, and so on, operations. In many cases, transformed series show higher linear correlation.

19.1.1 Serial Correlation or Autocorrelation

The autocorrelation (also called serial correlation) among the data in a series is the linear correlation between the values in the same time series, separated by a specified lag. Consider a time series where observations are taken at equally spaced time. Serial correlation at lag k (an integer) is defined as the correlation between the series $(x_1, x_2, \ldots x_n)$

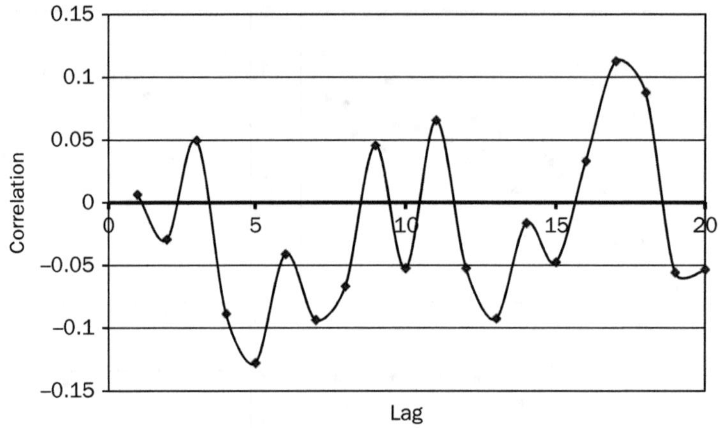

FIGURE 19.3 Correlogram of annual flows of a river.

and the series ($x_{1+k}, x_{2+k}, \ldots x_{n+k}$). For a time series that has n observations, serial correlation at lag k (r_k) can be computed by

$$r_k = \frac{\left[\sum_{i=1}^{n-k} x_i x_{i+k} - \sum_{i=1}^{n-k} x_i \sum_{i=1}^{n-k} x_{i+k}\right] \Big/ (n-k)}{\left[\sum_{i=1}^{n-k} x_i^2 - \left(\sum_{i=1}^{n-k} x_i\right)^2 \Big/ (n-k)\right]^{0.5} \left[\sum_{i=1}^{n-k} x_{i+k}^2 - \left(\sum_{i=1}^{n-k} x_{i+k}\right)^2 \Big/ (n-k)\right]^{0.5}} \tag{19.3}$$

For instance, autocorrelation at lag 1 is computed as the correlation between elements 1, 2, ..., $(n-1)$ of a series with the elements 2, 3, 4, ... n of the same series. Clearly, r_0 which is the correlation of the series with itself is unity. Note that as k increases, the number of data pairs used to compute r_k decreases, since the summations contain $(n-k)$ terms. Serial correlations are normally estimated for k much smaller than n; usually correlation at lags more than 20 are of limited use in hydrology.

Correlogram is a plot of autocorrelations at different lags. Correlograms reveal important features of a time series and show the lags at which the time series has a repetitive nature. Correlogram of a cyclic series, such as monthly temperature, is also cyclic.

Figure 19.3 shows the correlogram of the annual flows of a river up to 20 lags. It can be noted from the plot that this series has low autocorrelation.

19.1.2 Cross-Correlation

A high correlation between two variables is not always due to a cause-and-effect relation. For instance, monthly flows of two nearby rivers may have high correlation because the processes deriving flows are the same. But this high correlation does not imply that changes in the flow of the first river will cause similar changes in the flow of the other stream. Note that random variables that are independent are necessarily uncorrelated, but vice versa is not always true. Dependence in correlated variables is not always physical or cause-and-effect dependence.

Spurious correlation is the apparent correlation between two variables that do not have a cause-and-effect relation. Such correlation can lead to wrong conclusions if no attempt is made to find the physical reasons behind correlation. For example, flow data of two rivers in faraway regions (e.g., one in United States and the other in Australia) may show good correlation but this would be spurious.

19.1.3 Inferences on Correlation Coefficients

Uncorrelated variables X and Y have $r_{x,y} = 0$; if the variables are correlated, $r_{x,y} \neq 0$. Even though two variables do not have a cause-and-effect relation and the data is not showing any correlation, the correlation coefficient may not be zero. This nonzero correlation coefficient could be due to chance. Statistical tests are conducted to conclude if the deviation of the correlation coefficient from zero is due to chance or not. To conduct this statistical test, we need to make a null hypothesis that $r_{x,y}$ is zero or it has some nonzero value. In other words, H_o: $r_{x,y} = 0$ or H_o: $r_{x,y} = r^*$ where r^* is known. Next, an appropriate alternate hypothesis is also formulated.

Assume that X and Y are independent variables whose correlation is 0. Compute the test statistic

$$t = r_{xy}\sqrt{\frac{(n-2)}{(1-r^2)}} \tag{19.4}$$

which follows a t-distribution with $(n - 2)$ degrees of freedom. The null hypothesis H_o: $r_{x,y} = 0$ is rejected if $|t| > t_{1-\alpha/2,n-2}$, where α is the significance level.

19.1.4 Kendall's Correlation Coefficient

Kendall's correlation coefficient τ is an effective measure of correlation between two variables (Jain, 2011). Kendall's test is a rank-based test that is resistant to the effect of extreme values. Thus this test is well-suited for dependent variables which show highly skewed variation around the relationship. Dependent variables, such as river flow and sediment concentration, are some examples from hydrology where this kind of relationship is expected.

The steps to conduct Kendall's test for correlation are given here. The null hypothesis (H_o) is that the distribution of dependent variable y does not change with independent variable x:

1. Let there be n pairs of data (x_1, y_1), (x_2, y_2), … (x_n, y_n). Arrange the data as per the increasing value of x, such that $x_1 \leq x_2 \ldots \leq x_n$. Of course, y_i corresponds to x_i.
2. Compute statistic S

$$S = \sum_{k=1}^{n-1}\sum_{j=k+1}^{n} \text{sign}(y_j - y_k) \tag{19.5}$$

where $\text{sign}(\theta) = 1 \text{ if } \theta > 0$
$= 0 \text{ if } \theta = 0$
$= -1 \text{ if } \theta < 0.$

3. For $n > 10$, the test is conducted by using a normal approximation. The standardized test statistics Z is

$$Z = \begin{cases} \dfrac{S-1}{\sqrt{\mathrm{Var}(S)}} & S > 0 \\ 0 & S = 0 \\ \dfrac{S+1}{\sqrt{\mathrm{Var}(S)}} & S < 0 \end{cases} \tag{19.6}$$

Furthermore, Var $(S) = n(n-1)(2n+5)/18$.

4. Choose the significance level α. The null hypothesis is rejected if $|Z| > Z_{(1-\alpha/2)}$. Determine $Z_{(1-\alpha/2)}$ which is the value of the standard normal variate with a probability of exceedance of $\alpha/2$. If some of the x and/or y values are tied, a modified formula is used to compute Var (S). In case the number of samples is less than 10, tables for the S statistic are used.

Example 19.2

The correlation between the precipitation and runoff data of Example 19.1 (18 pairs of data were used) is 0.625. Test the null hypothesis H_o: runoff does not change as a function of rainfall.

Solution

Here for this data, 18 pairs of values are available which are arranged in ascending order. Using Eq. (19.5), the S statistic is found to be 71.

Hence $\mathrm{Var}(S) = 18\times(18-1)\times(2\times18+5)/18 = 697$.

Now, applying Eq. (19.6)

$$Z = (71-1)/\sqrt{697} = 2.65$$

Since $|z| > 1.96$, the null hypothesis is rejected at 5% significance level. It can be stated that runoff is a function of rainfall.

19.2 Regression

In the study of hydrological systems, the interest often lies in quantifying the relationship between two or more variables that have a cause-and-effect relationship. For example, rainfall over a catchment causes runoff. Regression is commonly applied to express such cause-and-effect relations mathematically. A relation is established to express the dependent variable Y (the variable whose values are to be estimated) as a function of one or more independent variable(s) X_i. Once established, this relationship can be used (1) to estimate or predict values of Y by using known values of independent variable(s); (2) at times, the observed data series of Y may have some gaps. These can be filled-in by using the corresponding data of independent variables; and (3) to validate the data series of variable Y.

The regression relation most suitable to describe the given relation depends on the physics of the process and the range of the data under consideration. The type of regression equations that are commonly applied in hydrology are simple linear regression, nonlinear regression, and multiple regression.

Regression techniques can be classified in two categories: linear or nonlinear.

Linear Regression A linear regression equation expresses the dependent variable as a linear function of independent variable(s)

$$Y = a + b_1X_1 + b_2X_1 + \dots + b_nX_n + \varepsilon \tag{19.7}$$

where, Y is the dependent or regressed variable, X is the independent or regressor variable, a and b_i = regression coefficients, and ε = the error term.

Nonlinear Regression In nonlinear regression, the dependent and independent variable(s) are related through a nonlinear relationship

$$Y = a \times X_1^{b1} \times X_2^{b2} \dots \times X_n^{bn} \tag{19.8}$$

By taking log of both sides. this nonlinear equation can be converted to a linear equation whose coefficients can be estimated by following the same procedure as for linear regression.

Depending on how many independent variables are involved, regression analysis can be divided in two categories (1) simple regression—where the dependent variable is regressed on a single independent variable; and (2) multiple regression—where the dependent variable is regressed on multiple independent variable.

The regression equation that is the best in a given situation is determined based on the variables whose data are available and the physics of the underlying process. It may also depend on the range of the data which is of interest since a relation that is nonlinear may be quite well represented by a linear equation within a (small) range of the variables.

19.3 Simple Linear Regression

The simple linear regression model most widely used in hydrology assumes that a linear relationship exists between two variables and is given by the following general form

$$y_i = a + bx_i + \varepsilon_i \quad i = 1, 2, \dots, n \tag{19.9}$$

where, y_i is the ith value of the dependent or regressed variable; x_i is the ith value of the independent or regressor or predictor variable; and a, b = the regression coefficients. The regression line given by Eq. (19.9) crosses the y axis at a point a (the intercept), has a slope b, and ϵ_i is the random error or residual term for the ith data. The need for residuals arises because all the actual (or observed) values of variable Y will not match the values estimated by the regression equation.

The variables involved in regression should be carefully chosen, following a logic based on physical principles. A scatter plot of y versus x should be drawn to confirm the linear dependence structure. Sometimes, the regression relation improves if the variable x is transformed by taking log or using a power transformation. A family of transformations, called ladder of powers, is defined by $z = x^p$ where p can be −2, −1, −0.5, −0.33, 0, 0.33, 0.5, 1, 2; p can be selected by plotting z and the dependent variable. The regression line is established such that it produces unbiased results and the variance of the residual σ^2_ε is minimum.

19.3.1 Estimation of Regression Parameters

The regression coefficients (a and b) can be estimated by minimizing the sum of squares of deviations of y_i from the values estimated by regression. For a data x_i, the value of y computed by the regression equation $\hat{y}_i$ is

$$\hat{y}_i = a + bx_i \tag{19.10}$$

The difference between the observed value of the dependent variable and that computed by the regression equation or the residual error is

$$e_i = y_i - \hat{y}_i \tag{19.11}$$

Residual error indicates how well the regression line represents the observed behavior of the data. If each data point falls on the regression line, the error e_i would be zero which is seldom the case. The sum of squares of errors is

$$S_{se} = \sum_{i=1}^{n} e_i^2 = \sum_{i=1}^{n} (y_i - \hat{y})^2 \tag{19.12}$$

The values of the regression coefficients a and b are determined by minimizing the sum of square of errors S_{se}. These can be computed by

$$b = \frac{\sum_{i=1}^{n} (x_i - \overline{x})(y_i - \overline{y})}{\sum_{i=1}^{n} (x_i - \overline{x})^2} = \frac{S_{XY}}{S_{XX}} \tag{19.13}$$

and

$$a = \overline{y} - b\overline{x} \tag{19.14}$$

S_{XX} is called sum of squares x and S_{XY} is called sum of cross products. Quantity $S_{YY} = \sum_{i=1}^{n} (y_i - \overline{y})^2$ is called sum of squares y.

Since a and b are computed by minimizing Σe^2, these estimates are called least squares estimates. After establishing regression relation, it is necessary to check how good the relationship is.

19.3.2 Goodness of Regression

A measure of the goodness of regression is how much of the variability of the dependent variable is explained by the regression relation. Figure 19.4 shows the key statistics of the variables involved in simple linear regression. Corresponding to the value of independent variable x_i, the observed value of dependent variable is y_i and that estimated by the regression equation is y_i.

In regression, the variance of the dependent variable can be partitioned in two components (Jain, 2011)

$$\text{Total variance} = \text{explained variance} + \text{unexplained variance} \tag{19.15}$$

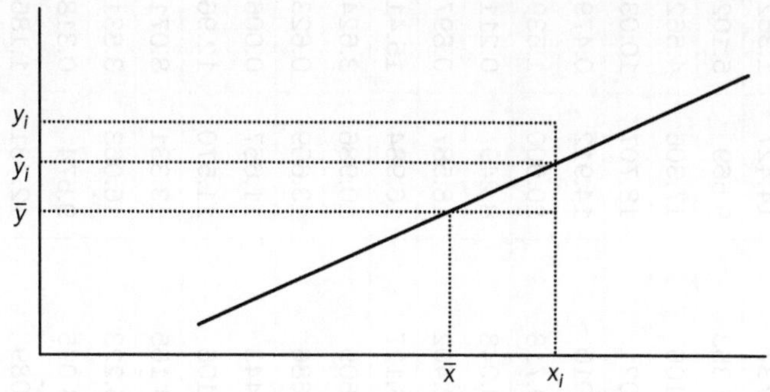

FIGURE 19.4 Graphical representation of regression related variables.

Obviously, the larger is the variance explained by regression (or variance due to regression), the smaller will be the unexplained variance (variance about regression). Since the regression equation is generally unable to explain the total variance of the hydrological variable being modeled, the series generated by the regression equation is smoother compared to the original series.

The coefficient of determination (COD) represents the fraction of variance that the regression equation is able to explain. COD is computed by

$$R^2 = 1 - S_{se}/S_{yy} \tag{19.16}$$

One can note from Eq. (19.16) that for a perfect relation, S_{se} would be zero and R^2 would be 1. Clearly, the closer COD is to unity, the better is the regression relation. However, a high COD alone is not a sufficient indicator of goodness of regression relation.

The residual error is an important tool to judge the regression model. The variation of the residuals about the regression line should also be plotted and investigated. Ideally, the residuals should have zero mean and zero autocorrelation. If the residuals are not randomly scattered about the regression line and are correlated, then it is likely that the variables might have a nonlinear relation and the unexplained variance is large.

Two other indicators of the goodness of regression are summarized here

$$\text{Mean square error } (mse)\text{:} \quad s^2 = S_{se}/(n - 2) \tag{19.17}$$

$$\text{Standard error of regression} \quad s = (mse)^{0.5} \tag{19.18}$$

Smaller values of *mse* and the standard error of regression are preferred.

Example 19.3

The precipitation and runoff for a catchment for a particular month are given in Table 19.1. (1) Develop the rainfall-runoff relationship in the form: $y = a + bx$, where y is runoff and x is precipitation. (2) What percent of the variation in runoff is explained by the developed regression equation?

Solution

In any regression exercise, it is always helpful to first plot the data and examine the variation in the dependent and independent variables. Figure 19.5 shows a scatter plot of the precipitation and runoff data and reveals that there is not much scatter around the line of best fit.

SN	Year	Precip (x)	Runoff (y)	$(x - \bar{x})$	$(y - \bar{y})$	$S_{XY} = (x - \bar{x}) \times (y - \bar{y})$	$S_{XX} = (x - \bar{x})^2$	$S_{YY} = (y - \bar{y})^2$	$\hat{y}$	$S_{se} = (y - \hat{y})^2$
1	1980	42.39	13.26	0.566	−0.858	−0.486	0.320	0.737	14.427	1.362
2	1981	33.48	7.31	−8.344	−6.808	56.808	69.620	46.353	9.569	5.102
3	1982	47.67	15.17	5.846	1.052	6.148	34.177	1.106	17.306	4.562
4	1983	50.24	15.54	8.416	1.422	11.965	70.831	2.021	18.707	10.032
5	1984	43.28	14.22	1.456	0.102	0.148	2.120	0.010	14.912	0.479
6	1985	52.61	21.24	10.786	7.122	76.815	116.340	50.718	20.000	1.539
7	1986	31.06	7.79	−10.764	−6.328	68.117	115.861	40.048	8.249	0.211
8	1987	50.02	17.64	8.196	3.522	28.864	67.176	12.402	18.587	0.897
9	1988	47.08	20.91	5.256	6.792	35.698	27.627	46.127	16.984	15.411
10	1989	36.08	12.89	−5.744	−1.228	7.055	32.992	1.509	10.986	3.624
11	1990	40.89	12.82	−0.934	−1.298	1.212	0.872	1.686	13.609	0.623
12	1991	37.31	11.58	−4.514	−2.538	11.458	20.375	6.443	11.657	0.006
13	1992	37.15	15.17	−4.674	1.052	−4.915	21.845	1.106	11.570	12.961
14	1993	40.38	10.49	−1.444	−3.628	5.239	2.085	13.165	13.331	8.071
15	1994	45.39	18.02	3.566	3.902	13.914	12.717	15.223	16.063	3.831
16	1995	33.49	9.01	−8.334	−5.108	42.572	69.454	26.095	9.574	0.318
17	1996	39.28	13.82	−2.544	−0.298	0.759	6.471	0.089	12.731	1.185
18	1997	45.03	17.25	3.206	3.132	10.040	10.279	9.807	15.866	1.914
	Sum	752.830	254.13	0.0	0.0	371.412	681.165	274.645	254.130	72.129
	Average	41.824	14.118	0.0	0.0	20.634	37.842	15.258	14.118	4.007

TABLE 19.1 Precipitation Runoff Data and Calculations

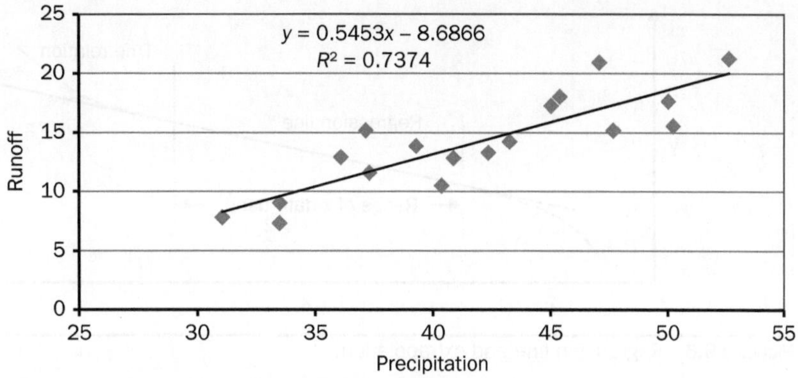

FIGURE 19.5 Plot of the precipitation and runoff data.

1. The values of various variables required to calculate a and b are computed in Table 19.1.

Here, $x = 752.83/18 = 41.82$, $\bar{y} = 254.13/18 = 14.12$. The regression coefficients are

$$b = S_{xy}/S_{xx} = 371.412/681.16 = 0.5453$$

and $a = \bar{y} - b\bar{x} = 14.12 - 0.5453 \times 41.82 = -8.68$

Hence the regression equation is: $y = -8.68 + 0.5453\,x$.

2. The percent of variation in y that is accounted for by the regression is obtained by multiplying the coefficient of determination (R^2) by 100. We have computed S_{se} in Table 19.1.

Coefficient of determination $R^2 = 1 - S_{se}/S_{yy} = 1 - 72.13/274.645 = 0.7374$.

Thus the regression equation is able to explain nearly 74% of variation in y. The remaining 26% variation is due to unexplained causes.

The coefficient of correlation (r) = square root of R^2

$$= \sqrt{0.7374} = 0.859.$$

Coefficient of correlation can also be computed as $r = \dfrac{S_{xy}}{\sqrt{S_{xx} \times S_{yy}}} = \dfrac{371.412}{\sqrt{681.165 \times 274.645}} = 0.859$

19.3.3 Extrapolation

At times, regression equation is employed to estimate the values of the dependent variable beyond the range of values of the independent variable that were used to calculate regression coefficients. Caution needs to be exercised here. As the point of interest moves away from the mean of the independent variable, the confidence intervals becomes wider. Furthermore, the relation between the dependent and independent variables may be nearly linear only over the range of data that was used in establishing the regression relation (Jain and Singh, 2003); it may be nonlinear over the full range of the variables (see Fig. 19.6).

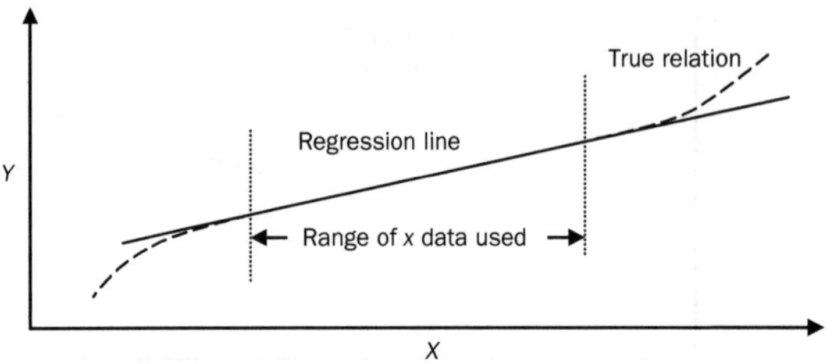

Figure 19.6 Regression line and extrapolation.

Therefore extrapolation should be attempted only after ascertaining that the relation between dependent and independent variables is linear over the range on interest.

19.3.4 Nonlinear Regression

In many cases, the relationship among the variables may be nonlinear and a linear model may not be appropriate. In some situations, the logarithmic transformation makes the data conform more closely to the linear regression assumptions. Assume that the data follow a generic relation

$$Q_i = aP_i^b T_i^c Y_i^d \varepsilon \tag{19.19}$$

which is not a linear relation. However, it can be linearized by using a logarithmic transformation

$$\ln(Q_i) = \ln(a) + b \times \ln(P_i) + c \times \ln(T_i) + d \times \ln(Y_i) + \varepsilon_T \tag{19.20}$$

or

$$Q_{Ti} = a_T + b \times P_{Ti} + c \times T_{Ti} + d \times Y_{Ti} + \varepsilon_T \tag{19.21}$$

where subscript T denotes transformed variable. If regression is performed by using Eq. (19.20), the error is minimized by using the logarithms rather than the original values. Note that (1) the regression coefficients are estimated such that $\sum(Q_{Ti} - \hat{Q}_{Ti})^2$ is a minimum and not the sum of squares of deviations in the original domain, and (2) the error term in the transformed equation is additive, whereas it is multiplicative in the original equation.

Rainfall-runoff relation is highly nonlinear in arid regions in general due to large variation in rainfall abstractions. In humid regions, the abstraction is comparatively less and the rainfall-runoff relation is comparatively more linear.

19.4 Multiple Linear Regression

In real-life problems, the dependent variable is frequently a function of more than one independent variables. For example, monthly water yield for a large catchment may be better modeled as a function of rainfall in the current and previous month(s) rather than the rainfall in the current month only. When the dependent variable is linearly related with a number of independent variables, this is known as multiple linear regression (MLR). If the association between the dependent and independent variables is nonlinear, sometimes a good linear relationship is obtained after transforming the variables. In such cases, the technique of multiple regression can be applied on the transformed data.

A general multiple linear regression equation would look like

$$Q_i = a + b_1P_i + b_2R_{i-1} + \ldots + b_mV_{i-m+1} + \varepsilon \tag{19.22}$$

where a and b_i = the regression coefficients, and ε = the error term. This model is linear with parameter vector b_j. Consider that n observations are available on the dependent variable and concurrently on each of the p independent variables. This will give n equations containing p unknown parameters, written as

$$\begin{aligned} y_1 &= b_1\,x_{1,1} + b_2\,x_{1,2} + \ldots + b_p\,x_{1,p} \\ y_2 &= b_1\,x_{2,1} + b_2\,x_{2,2} + \ldots + b_p\,x_{2,p} \\ &\ldots\ldots\ldots \quad \ldots\ldots \quad \ldots\ldots \\ y_n &- b_1\,x_{n,1} + b_2\,x_{n,2} + \ldots + b_p\,x_{n,p} \end{aligned} \tag{19.23}$$

where y_i is the ith observation of the dependent variable, $x_{i,1}, x_{i,2}, \ldots x_{i,p}$ are the ith observation on the independent variables, and $b_1, b_2, \ldots b_p$ are the unknown parameters.

In matrix notation, Eq. (19.23) can be written as

$$\underset{n \times 1}{\underline{Y}} = \underset{n \times p}{\underline{X}} \; \underset{p \times 1}{\underline{B}} \tag{19.24}$$

In establishing multiple regression relation, the dependent variable is defined by the problem and the independent variables are selected by the following steps: (1) a cause-and-effect analysis of the physical phenomenon is carried out to reveal the independent variables that determine the dependent variable, and (2) sufficiently long observed data series of dependent and independent variables should be present so that the parameters of the regression equation can be correctly estimated.

19.4.1 Estimation of Multiple Regression Coefficients

Parameters of a multiple linear regression equation are obtained by solving n equations for the p unknowns. Hence n must be greater than or equal to p; for the sake of good estimations, n should be much larger than p. Similar to simple linear regression, the regression coefficients are obtained by minimizing the sum of squares of errors. In matrix notation, the n MLR equations can be written in a compact form as

$$Y = Xb + e \tag{19.25}$$

where $Y = (n \times 1)$ column vector containing the dependent variable, $X = (n \times p)$ matrix of independent variables, $b = (p \times 1)$ column vector containing the regression coefficients, and $e = (n \times 1)$ column vector of residuals.

Now, the values of regression parameters are obtained by minimizing the residual sum of squares e^Te. Hence

$$e^Te = (Y - Xb)^T(Y - Xb) \tag{19.26}$$

Source	Sum of Squares	Degrees of Freedom
Total	$S_Y = Y^TY$	n
Mean	$n\bar{Y}^2$	1
Regression	$b^TX^TY - n\bar{Y}^2$	$p - 1$
Residual	$Y^TY - b^TX^TY$	$n - p$

TABLE 19.2 Analysis of Variance Table

is differentiated with respect to b. Equating the resulting expression to zero, this gives

$$X^TXb = X^TY \tag{19.27}$$

These equations are called the normal equations. Multiplying both sides with $(X^TX)^{-1}$ gives an explicit expression for b

$$\underset{(p\times 1)}{b} = \underset{(p\times n)(n\times p)}{(X^TX)^{-1}}\ \underset{(p\times n)}{X^T}\ \underset{(n\times 1)}{Y} \tag{19.28}$$

Table 19.2 summarizes the sum of squares quantities.

Example 19.4

Table 19.3 contains rainfall data for the months of July and August and discharge data for the August month for a catchment. Estimate the parameters of linear regression (LR) and of MLR. Also, check if it is advantageous to use MLR in comparison to LR for this problem?

Solution

Using the given data, a linear regression equation of the following form was established between observed rainfall and discharge data for the August month

$Q_A = a + b\,R_A$ where Q_A = discharge for August and R_A = rainfall for August.

The parameters were estimated as $a = 358.88$ and $b = 0.433$. Hence, the simple linear regression model is

$$Q_A = 358.88 + 0.433\,R_A$$

When this equation is used to estimate Q_A, the COD (R^2) was = 0.503.

In the case of MLR, the independent variables are the observed rainfall values for the months July and August and dependent variable is the observed discharge data for August. MLR model of the following form was established

$$Q_A = a + b_1\,R_J + b_2\,R_A$$

where R_J = the rainfall for July month. After computing the coefficients, the following MLR equation was obtained

$$Q_A = -2925.83 + 0.367\,R_J + 0.531\,R_A$$

Coefficient of determination $R^2 = 1 - 8.64 \times 10^6/5.2 \times 10^7 = 0.833$.

YEAR	RF-JUL (MCM)	RF-AUG (MCM)	Obs Q Aug (MCM)	$(Q_{obs} - Q_{obs,m})^2$	Comp. Q by Lin Reg (Q_L)	$(Qob - Q_L)^2$	Comp. Q by MLR (Q_M)	$(Qob - Q_M)^2$
1990	5130.04	14561.05	5994.94	3641293	6661.21	443917.4	6689.1	481869.2
1991	7781.13	6546.24	2657.92	2041486	3192.23	285492.1	3405.1	558333.4
1992	3002.36	12586.63	4895.52	654147	5806.64	830150.0	4860.0	1259.3
1993	8572.75	7532.13	5125.02	1078063	3618.94	2268264.5	4219.1	820671.2
1994	5242.03	5799.34	2532.37	2416000	2868.95	113287.4	2077.0	207331.4
1995	6311.05	9522.80	3274.52	659676	4480.55	1454510.1	4446.6	1373872.4
1996	6040.00	7285.46	4163.01	5820	3512.18	423580.8	3159.0	1008022.3
1997	1597.33	6922.49	2046.69	4161713	3355.08	1711877.9	1336.5	504351.9
1998	8561.71	6889.43	4190.08	10684	3340.77	721329.6	3873.7	100071.8
1999	7153.31	12566.82	6107.45	4083351	5798.07	95720.3	6372.2	70105.1
2000	5623.67	10263.08	5145.44	1120884	4800.96	118668.0	4587.6	311150.5
2001	4233.30	7108.91	2300.77	3189609	3435.77	1288208.4	2402.5	10345.1
2002	12076.88	10472.23	7994.09	15267488	4891.48	9626148.6	7066.0	861393.1
2003	6843.64	8068.47	3695.11	153360	3851.08	24328.1	3869.7	30465.3
2004	7819.49	9330.16	4870.40	614152	4397.17	223947.7	4897.7	744.3
2005	9403.82	7424.92	3943.46	20525	3572.54	137577.3	4467.0	274142.5
2006	7040.85	8306.55	3801.73	81222	3954.13	23227.0	4068.4	71133.5
2007	7380.56	9987.30	5895.90	3273123	4681.60	1474529.8	5085.7	656480.9
2008	8620.28	4283.79	1501.45	6683656	2212.99	506303.3	2511.4	1020067.0
2009	9113.46	5071.52	2670.74	2005007	2553.94	13641.6	3110.7	193556.8
2010	1296.93	11168.68	3192.95	798828	5192.92	3999885.7	3481.4	83188.8
2011	7393.84	7785.62	3908.33	31824	3728.66	32281.1	3921.3	167.4
Sum	146238.41	189483.61	89907.88	51991910	89907.88	2.58E+07	89907.9	8.64E+06
Average	6647.20	8612.89	4086.72	3641293	4086.72			

TABLE 19.3 Data and Computations for Multiple Linear Regression Example

The discharge for the August month was computed by LR and MLR models and the sum of squares of errors were computed for each case. The values were 2.58×10^7 for LR and 8.64×10^6 for MLR. Based on comparison of these values along with R^2 for the two cases, it can be concluded that MLR has given much improved estimates of discharge compared to LR.

19.4.2 Stepwise Regression

It is a process to arrive at the "best" regression equation that contains only relevant variables from among a number of independent variables. Here, the regression equation is built in "steps" by following certain strategies. This topic has been discussed in detail in Section 13.1.5. This is a good approach to rationally establish a regression relation.

19.4.3 Regression After Transforming Data

In many cases, transformation of variables significantly improves the regression relationship. Some variables may be related in a nonlinear manner but such relationship can be transformed to linear form and then the technique of linear regression can be applied. For example, let the variables X and Y be nonlinearly related as

$$Y = \alpha X^{\beta} \tag{19.29}$$

This relationship can be linearized by logarithmic transformation of the equation

$$\ln Y = \ln \alpha + \beta \ln X \tag{19.30}$$

or

$$s = a + b \times r \tag{19.31}$$

where $s = \ln Y$, $a = \ln \alpha$, $b = \beta$, and $r = \ln X$. Now, the regression technique can be applied to estimate parameters a and b and thereby α and β. Two points are important here: (1) parameters a and b are estimated by minimizing $\Sigma(s - s_{reg})^2$ and not by minimizing $\Sigma(Y - Y_{reg})^2$. Here s_{reg} and Y_{reg} are the value of s and Y estimated by the regression equation. (2) The error term is additive in the log-transformed equation ($s = a + br + c$), which means that it is multiplicative in the original equation

$$Y = \alpha X^{\beta} \varepsilon \tag{19.32}$$

19.4.4 Comments on Multiple Regression

Multiple regression is helpful when data of the dependent variable and several independent variables are available and the user wants to develop a model that has all independent variables that are important significantly. A relevant question is: Which independent variables should be used as predictors? A factor to be considered is that the predictor (independent) variables should not be highly correlated. If the correlation among independent variables in the regression equation is high, it may not be possible or advisable to provide a physical meaning to regressions coefficients; the sign and magnitudes of coefficients may be different than what is expected.

All variables contained in a regression equation should meaningfully and significantly contribute to the model unless there is a strong and valid reason to include a particular variable. A common practice is to set up regression equations by trying different combinations of independent variables and select the equation that best models the data. All the predictor variables in a regression equation should have physical significance. If there are variables that are highly correlated and make a significant contribution to the regression equation, then those that are easiest to obtain should be retained. A regression equation should have only those variables that are necessary.

If the regression model is able to predict the dependent variable by using those data of independent variables that were not used to set up the model, then the model is successful. Sometimes, a small part of data is not used to estimate the coefficients and is set aside to check the model.

Filling in Missing Data An important application of regression models is to use them to fill in missing data in the series. For example, missing runoff data can be easily filled-in by using rainfall data by using regression. Typically, this technique is applied to time series of monthly or larger time intervals.

19.5 Hypothesis Testing

In the analysis of hydrological data, typical questions that are asked are: Do the river discharge data at the given gauging site follow normal distribution? Has the prescribed water quality standard been violated at a particular location? And so on. Sometimes, the question can be answered based on knowledge and experience but sometimes it is necessary to conduct statistical tests to arrive at a decision that is objective and correct. These statistical tests constitute hypothesis testing.

Two broad categories of hypothesis tests are: parametric and nonparametric. In parametric tests, it is assumed that the distribution of the data is known and is specified. Most commonly, the data is assumed to follow the normal distribution. Test results will be wrong if the data does not follow the assumed distribution. Two examples of parametric tests are the t-test and the *F*-test. In the nonparametric or distribution free tests, it is not necessary to know which distribution the data follows; the data need not follow a particular distribution. The Kendall's Tau test and the Kruskal-Wallis test are nonparametric tests.

An important part for a statistical test is to state the hypotheses, which are the statements that inform what is the required answer. Hypotheses can be of two types: null and alternate. The null hypothesis is the hypothesis that is being tested—it is the hypothesis of no difference or no change. For example, the mean of two series is not different. The alternate hypothesis is the statement of change that is expected. For example, to test whether there is a step-change in the mean of a series or not, the null hypothesis could be that there is no change in the mean of the series. In this case, the alternative hypothesis would be that the mean of the series has suddenly changed.

To begin the test, it is assumed that the null hypothesis is true. Next, it is checked whether this hypothesis is supported by the data. The null hypothesis is what we assume to be true, unless proven otherwise. The null hypothesis is rejected if the data do not support it.

Statistical tests can be one sided or two sided. In one-sided tests, a large departure of the test statistic in only one direction would lead to the rejection of the null hypothesis. In one-sided tests, the null hypothesis, for example, could be "the spillway design flood exceeds 1000 cumec." In a two-sided test, larger or smaller departures from the null hypothesis would cause its rejection.

The null hypothesis can be written as

$$H_0: \mu_1 = \mu_0 \tag{19.33}$$

This statement says that the mean μ_1 of a dataset is similar to the mean μ_0 of another dataset.

After finalizing the null hypothesis, the alternative hypothesis is to be decided. The alternate hypothesis expresses some expected deviation from the null hypothesis. The null and the alternate hypotheses are mutually exclusive and all inclusive. The possible forms of alternate hypothesis are

$$H_1: \mu_1 > \mu_0 \tag{19.34a}$$

$$H_1: \mu_1 \neq \mu_0 \tag{19.34b}$$

$$H_1: \mu_1 < \mu_0 \tag{19.34c}$$

Equations (19.34a), (19.34b), and (19.34c) state that the mean of the population from which the sample was drawn is greater than, not equal to, or less than the specified population mean, respectively. Some examples of the null hypothesis are: (a) the discharge data at two river gauging stations follow different probability distributions, and (b) rainfall and discharge data for the catchment are related to each other.

If the null hypothesis is not confirmed by the data, it will be rejected. Hence, the alternate hypothesis should be general in nature. Rejection of the null hypothesis indicates that the true situation is somewhere in a wide arena of possible situations that are suggested by the alternate hypothesis. In hypothesis testing, the result is stated as either rejection of the null hypothesis or failure to reject the null hypothesis.

19.5.1 Test Statistic

To compare between the null and alternative hypotheses, a test statistic is selected whose significance is evaluated, based on the available evidence. The test of hypothesis basically checks whether the test statistic behaves in an expected manner (ignoring the possibility that this has happened only due to chance) if the null hypothesis was true. The test statistic is a numerical value that is calculated from the sample data being tested. It is a means employed to compare the null and alternative hypotheses. A good test statistic should clearly highlight the difference between null and alternate hypotheses. Slope of trend line is an example of a test statistic—it can be used to test for trend in the data. If the null hypothesis is that "there is no trend in the data," then for this hypothesis to be true, slope of trend line should have a value close to zero. If the alternative hypothesis is that "there is a significant trend in the data," then for this hypothesis to be true, slope of trend line should be very different from zero. Hence to carry out a statistical test, it is necessary to compare the computed test statistic with its expected value under the null hypothesis. The significance level of a test formally expresses this concept.

	Hypothesis Is Correct	Hypothesis Is Incorrect
Hypothesis is accepted	Correct decision. This outcome has a probability $1 - \alpha$	Type II error with a probability β
Hypothesis is rejected	Type I error with a probability of α	Correct decision with probability $1 - \beta$

TABLE 19.4 Hypothesis Testing: Possible Outcomes and Their Probabilities

After the hypothesis has been designed, the test statistic is computed from the sample data. Decision to accept or reject the null hypothesis is based on this test statistic. If the test statistic value falls in the rejection region, the null hypothesis is rejected. Note that the null hypothesis may be correct or incorrect. Further, it might be accepted or rejected. This gives rise to four possible combinations as shown in Table 19.4.

If a right hypothesis has been accepted or a wrong hypothesis has been rejected, it is a correct decision. Two types of errors can take place in a test result. If a null hypothesis, which is correct, is rejected, a type I error is committed. Significance level is the probability of making a type I error and this level α denotes the maximum allowable type I error. This level is specified before carrying out the test. If an incorrect hypothesis is accepted, it will lead to a type II error whose probability is β (it is also termed as power of the test).

The probability that a type II error will be made is not known; this error increases with increase in significance level α. The null hypothesis is formulated with the intention that it will be rejected because this eliminates the possibility of a type II error. A test with a low type II error probability is called powerful by statisticians. The power of the test is the probability of correctly detecting a trend when one is present. In general, more powerful tests are to be preferred.

19.5.2 Significance Level

The significance level measures whether a test statistic significantly differs from values that would normally be found under the null hypothesis. It is the probability of wrongly rejecting the null hypothesis. In hydrology, the significance levels of 0.05 (1 in 20) or 0.01 (1 in 100) are commonly adopted (Jain, 2011). A 5% significance level means that there is a 1 in 20 chance that the conclusion of the test is wrong. Note that when a two-sided test is being conducted at significance level α, the table of standard statistic must be seen against $\alpha/2$.

Figure 19.7 shows the rejection region for upper-tailed and lower-tailed t-tests. The shaded area in Fig. 19.7 is the area of rejection or the critical region. For a two-sided test (see Fig. 19.8) in which the test statistic under the null hypothesis is normal, each shaded area near the tails contains half of the rejection area. For the inequality-type alternative hypothesis mentioned in Eq. (19.34b), the null hypothesis is rejected if the test statistic is too high or too low. The choice of significance level depends upon the risks of damage or harm if a wrong decision has been taken.

The steps in hypothesis testing are (1) identify the question that is to be answered; (2) formulate the null and the alternative hypotheses; (3) choose the significance level α

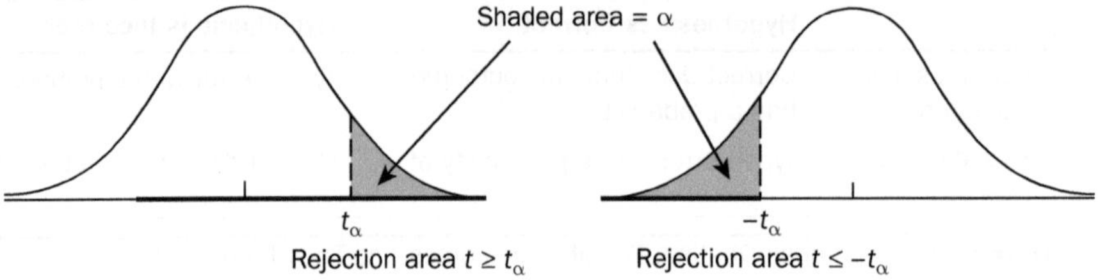

FIGURE 19.7 Rejection regions for t-tests: (a) for upper-tailed tests, (b) lower-tailed tests.

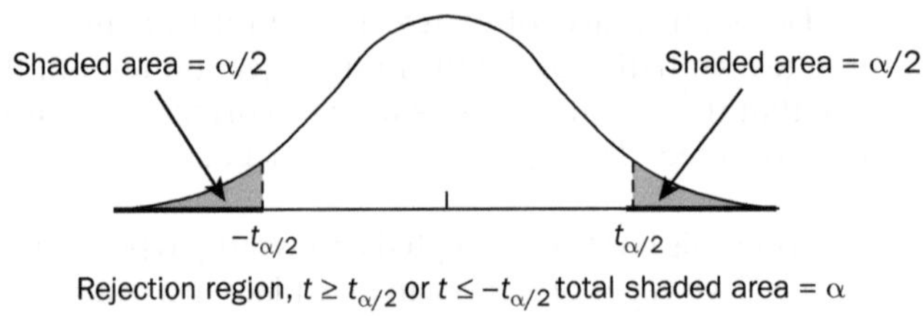

FIGURE 19.8 Rejection regions for t-test for two-tailed test.

and determine the rejection region; (4) collect and process the required sample data and compute the test statistic; and (5) decide whether H_0 should be accepted or cannot be rejected.

19.6 Probability Functions for Hypothesis Testing

Probability functions commonly used for hypothesis testing are the normal, student's t, chi-square, and *F* distributions.

19.6.1 Normal Distribution

Normal distribution, also known as the Gaussian distribution, is a bell-shaped distribution (see Fig. 19.9) that has two parameters: mean (μ) and standard deviation (σ). These are the population parameters and their estimates for sample data are called as the sample mean ($\bar{X}$) and standard deviation (s), respectively. If the random variable X follows a normal distribution, the standardized variates or normal reduced variates z can be defined as

$$z = \frac{x - \mu}{\sigma} \tag{19.35}$$

where the mean and standard deviation of z are 0 and 1, respectively. After transformation, the probability distribution function can be expressed as a function of z in the tables of standard normal distribution, which are widely available (see Table 19.5). Table 19.5 can be used to find the value of z corresponding to a certain probability. For example, if the probability is 0.05 (on the left tail), the value of z is -1.64.

Normal distribution is symmetric about the ordinate; the value of z equal to 1.64 (note positive value here) corresponds to 5% area under the curve $f(z)$, but on the right tail.

Normal distribution has been discussed in detail in Sec. 18.4.1.

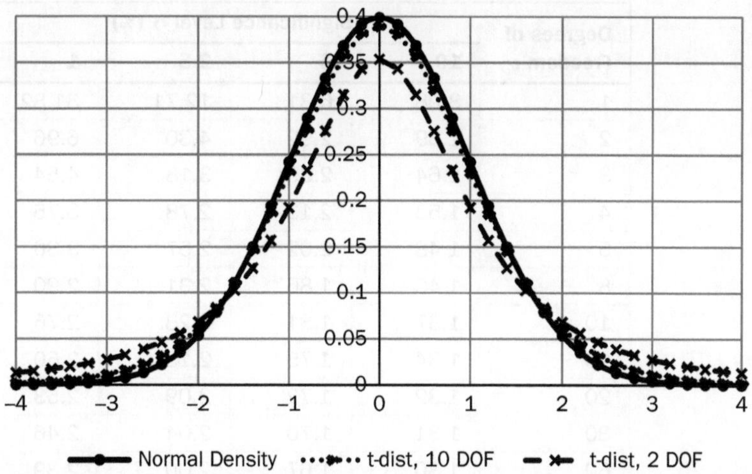

Figure 19.9 Density function of normal distribution (firm line) and t-distribution for 2 DOF (dashed line) and 10 DOF (dotted line). As DOF increases, t-distribution approaches normal distribution.

19.6.2 Student's t-Distribution

It is the distribution of the random variable which is the "best" that we can determine if the standard deviation of the data is not known. The density curves of student's t-distribution are symmetric and bell-shaped like the normal distribution with peak at 0. This distribution has only one parameter ν, which is known as the degrees of freedom (DOF). DOF is the number of independent observations that are used to compute the value of a parameter. In general, DOF for an estimate is equal to the number of data values minus the number of parameters to be estimated. For example, if 40 values of data are used to estimate two parameters, DOF will be $40 - 2 = 38$. The table of t-distribution lists the value of t-statistics for various levels of significance and DOF.

Parameter DOF controls the spread of the t-distribution. Figure 19.9 shows the density of the t-distribution for two values of ν. The density curve of normal distribution is also shown for comparison. It can be seen in Fig. 19.9 as well as Table 19.5 that as ν increases, the t-distribution approaches the normal distribution.

Table 19.5 gives the values of t-distribution for the desired DOF and significance level. In this table, the value of the t-statistics can be read from the row corresponding to DOF ν and the column corresponding to the significance level α. For example, for $\nu = 20$ the t value is 1.72 for 5% significance level. This means that for 20 DOF, 95% of the area of the distribution curve lies to the left of value 1.72. Noting that the t-distribution is symmetric about origin, 5% of the area below the curve in the left tail is to the left of t value of -1.72 for $\nu = 20$. When the interest is in 95% of the area but with 2.5% in each tail, the critical t value for $\alpha = 2.5$ is 2.09 for $\nu = 20$.

We will now discuss some applications of t-distribution.

19.6.3 Test Concerning Population Mean

This test is helpful in situations where the mean from a sample has been calculated and one would like to see if this is equal to the population mean μ_0. A different use of the test may arise when, for example, discharge in a river is estimated from a model and we want to check if this is a bias in the computed values.

Degrees of Freedom ν	Significance Level α (%)			
	10	5	2.5	1
1	3.08	6.31	12.71	31.82
2	1.89	2.92	4.30	6.96
3	1.64	2.35	3.18	4.54
4	1.53	2.13	2.78	3.75
5	1.48	2.02	2.57	3.36
8	1.40	1.86	2.31	2.90
10	1.37	1.81	2.23	2.76
15	1.34	1.75	2.13	2.60
20	1.32	1.72	2.09	2.53
30	1.31	1.70	2.04	2.46
60	1.30	1.67	2.00	2.39
120	1.29	1.66	1.98	2.36
∞ (normal)	1.282	1.645	1.96	2.32

Table 19.5 Values of t-Distribution for Selected Degrees of Freedom and Significance Level

The null hypothesis H_0 is: $\mu = \mu_0$ where μ is the population mean and the alternate hypothesis is $\mu \neq \mu_0$. The sample under test has n values and mean is $\bar{x}$. We assume that the data are independent, identically distributed, and follow a normal distribution (this assumption can be relaxed if n >30). The test statistic is computed as

$$t = (\bar{x} - \mu)/(s/\sqrt{n}) \tag{19.36}$$

The null hypothesis is rejected if $|t| > t_{(1-\alpha/2,n-1)}$ where $t_{(1-\alpha/2,n-1)}$ is the $(1 - \alpha/2)$ quantile of the student's t-distribution with $(n - 1)$ degrees of freedom.

Example 19.5

For a river gauging site, 21 values of annual runoff data are available. It is assumed that the data follows a normal distribution. For the data, the mean x_m is 15 mm and the standard deviation s_d is 5 mm. At 5% significance level, will it be right to say that the data is drawn from a population whose mean is 17 mm?

Solution

The null hypothesis is H_0: $\mu = 17$ mm against the alternate hypothesis H_1: $\mu \neq 17$ mm.

The test statistic is

$$t = (x_m - \mu)/(s_d/\sqrt{n}) = (15.0 - 17.0)/(5.0/\sqrt{21}) = -1.83$$

This is a two-tailed test. Hence, the null hypothesis will be rejected if the test statistic is either too high or too low. Therefore, to make a decision, we require a value of t for $\alpha/2$ and $\nu = n - 1$. From Table 19.5, $t_{\alpha/2,n-1} = t_{2.5,20} = 2.09$. Since $|t| = 1.83$, which is less than the critical value of 2.09, the statistic does not fall in the region of rejection. Hence, the null hypothesis is accepted.

19.6.4 *F*-Distribution

Suppose that one has two samples for which the sample variances have been calculated. It is desired to test the null hypothesis that the sample variances are estimates of the same parameter σ^2. We can define a variable F, which is the ratio of the two sample variances, as

$$F = \frac{s_1^2}{s_2^2} \tag{19.37}$$

The F-distribution is a well-known distribution which has two parameters. Tables of the F-distribution can be found in many textbooks, for example, Davis (2015); an abridged table is given in Table 19.6. Tables give values of the cumulative distribution for a range of degrees of freedom m and n and several levels of significance. Note that to calculate F, one must put the largest variance in the numerator and the smallest in the denominator. This is because the F-test has been set up as a one-sided test in which the alternative to the null hypothesis is the alternative hypothesis that s_1 is larger than s_2.

As an example, if $m = 10$ and $n = 20$, there is a 5% chance that F will be greater than 2.35. It is noted that the F-distribution table is not symmetric. For $m = 20$ and $n = 10$, the critical value for 5% area in the right tail is 2.77. Thus the values of m and n must not be switched.

19.6.5 Tests Concerning Variances of Two Populations

A hydrologist may be asked to answer: Are the variances of two populations the same or not? The F-distribution is employed for this purpose and the test is known as the F-test.

1. Null hypothesis: H_0: $\sigma_1^2 = \sigma_2^2$
2. Test statistic is computed as: $F = s_1^2/s_2^2$
3. Decision: depending upon the alternative hypothesis, reject H_0 if t-statistic falls in the rejection region as shown in Table 19. 7.

19.7 Trend Analysis

Water resources projects are designed and operated assuming that the data are stationary, which means that the statistical properties of data do not vary with time. If this assumption is incorrect, then existing procedures for hydrological analysis of data for designing levees, dams, reservoirs, and so on, will have to be revised. If the stationarity assumption is no longer valid, the systems may be over- or under-designed and they will be either unnecessarily expensive or will not be able to serve their intended purpose. Identification of trends in hydrologic data is helpful in modeling and predicting future behavior of the data. Trends can be spatial and temporal. Here, we will only discuss temporal trends.

A temporal trend is the general increase or decrease in observed values of a variable over time. It describes the long smooth movement of the variable, ignoring short-term changes. Trend analysis is performed to determine the significance of a trend (if present) and to estimate its magnitude.

Many parametric and nonparametric methods are applied for trend detection. In parametric tests, it is necessary to assume an underlying distribution for the data (the

	Degree of Freedom	Numerator													
		1	2	3	4	5	6	7	8	9	10	15	20	25	Infinity
	1	161.45	199.50	215.71	224.58	230.16	233.99	236.77	238.88	240.54	241.88	245.95	248.01	249.26	250.10
	2	18.51	19.00	19.16	19.25	19.30	19.33	19.35	19.37	19.39	19.40	19.43	19.45	19.46	19.46
Denominator	3	10.13	9.55	9.28	9.12	9.01	8.94	8.89	8.85	8.81	8.79	8.70	8.66	8.63	8.62
	4	7.71	6.94	6.59	6.39	6.26	6.16	6.09	6.04	6.00	5.96	5.86	5.80	5.77	5.75
	5	6.61	5.79	5.41	5.19	5.05	4.95	4.88	4.82	4.77	4.74	4.62	4.56	4.52	4.50
	6	5.99	5.14	4.76	4.53	4.39	4.28	4.21	4.15	4.10	4.06	3.94	3.87	3.83	3.81
	7	5.59	4.74	4.35	4.12	3.97	3.87	3.79	3.73	3.68	3.64	3.51	3.44	3.40	3.38
	8	5.32	4.46	4.07	3.84	3.69	3.58	3.50	3.44	3.39	3.35	3.22	3.15	3.11	3.08
	9	5.12	4.26	3.86	3.63	3.48	3.37	3.29	3.23	3.18	3.14	3.01	2.94	2.89	2.86
	10	4.97	4.10	3.71	3.48	3.33	3.22	3.14	3.07	3.02	2.98	2.85	2.77	2.73	2.70
	15	4.54	3.68	3.29	3.06	2.90	2.79	2.71	2.64	2.59	2.54	2.40	2.33	2.28	2.25
	20	4.35	3.49	3.10	2.87	2.71	2.60	2.51	2.45	2.39	2.35	2.20	2.12	2.07	2.04
	25	4.24	3.39	2.99	2.76	2.60	2.49	2.41	2.34	2.28	2.24	2.09	2.01	1.98	1.92
	30	4.17	3.32	2.92	2.69	2.53	2.42	2.33	2.27	2.21	2.17	2.01	1.93	1.88	1.84
	40	4.09	3.23	2.84	2.61	2.45	2.34	2.25	2.18	2.12	2.08	1.91	1.84	1.78	1.74
	50	4.03	3.18	2.79	2.56	2.40	2.29	2.20	2.13	2.07	2.03	1.87	1.78	1.73	1.69
	60	4.00	3.15	2.76	2.53	2.37	2.25	2.17	2.10	2.04	1.99	1.84	1.75	1.69	1.65
	70	3.98	3.13	2.74	2.50	2.35	2.23	2.14	2.07	2.02	1.97	1.81	1.72	1.66	1.62
	80	3.96	3.11	2.72	2.49	2.33	2.21	2.13	2.06	2.00	1.95	1.79	1.70	1.64	1.60
	90	3.95	3.10	2.71	2.47	2.32	2.20	2.11	2.04	1.99	1.94	1.78	1.69	1.63	1.59
	100	3.94	3.09	2.70	2.46	2.31	2.19	2.10	2.03	1.98	1.93	1.77	1.68	1.62	1.57
	Infinity	3.85	3.00	2.61	2.38	2.22	2.11	2.02	1.95	1.89	1.84	1.68	1.58	1.52	1.47

Table 19.6 Critical Values of *F* for the 0.05 Significance Level

Alternative Hypothesis	Rejection Region
$H_1: \sigma_1^2 > \sigma_2^2$	$F \geq F_{\alpha, m-1, n-1}$
$H_1: \sigma_1^2 < \sigma_2^2$	$F \leq F_{1-\alpha, m-1, n-1}$
$H_1: \sigma_1^2 \neq \sigma_2^2$	$F \geq F_{\alpha/2, m-1, n-1}$ or $F \leq F_{1-\alpha/2, m-1, n-1}$

Table 19.7 Rejection Region for *F*-test

normal distribution is most commonly adopted); data are also assumed to be independent. Note that these assumptions are not strictly true for many hydrological series; such series rarely follow a normal distribution and often have a temporal dependence, particularly if the time interval is short (e.g., a day). It may be helpful to transform such data so that the transformed data (nearly) follows the normal distribution (this may not be always feasible) and restrict analysis to series having long time interval. Nonparametric and distribution-free methods make fewer assumptions about the distribution of data. When these methods are used, it is not necessary to assume that the data follows a particular distribution. However, many of these methods assume that the data are independent.

19.7.1 Magnitude of Trend

There are two ways to determine the magnitude of trend in a time series: (a) by using regression analysis (parametric test), and (b) by using Sen's (1968) estimator (nonparametric method). In both these methods, it is assumed that there is a linear trend in the time series. The regression method has been described earlier in this chapter. In this method, a linear equation, $y = mt + c$, where c is the intercept and m is the slope or trend, is fitted. The trend or the slope is the rate of change (rise/fall) in the variable.

Sen's estimator is widely used for determining the magnitude of trend in hydrometeorological time series (Jain et al., 2013). In this method, slopes (T_i) of all data pairs are calculated by

$$T_i = \frac{x_j - x_k}{j - k} \quad \text{for } i = 1, 2, \ldots, N \tag{19.38}$$

where x_j and x_k are data values at time j and k ($j > k$), respectively. The median of these N values of T_i is Sen's estimator of slope which is calculated as

$$\beta = \begin{cases} T_{\frac{N+1}{2}} & N \text{ is odd} \\ \frac{1}{2}\left(T_{\frac{N}{2}} + T_{\frac{N+2}{2}}\right) & N \text{ is even} \end{cases} \tag{19.39}$$

A positive value of β indicates an upward (increasing) trend and a negative value indicates a downward (decreasing) trend in the time series.

Example 19.6

Annual rainfall at a station for 7 years is (in mm): 905, 830, 672, 667, 734, 866, and 799. Estimate the slope by using the Sen's estimator.

	1	2	3	4	5	6	7
	X1	X2	X3	X4	X5	X6	X7
	905	830	672	667	734	866	799
$j = 2$		−75.0	−116.5	−79.3	−42.8	−7.8	−17.7
$j = 3$			−158.0	−81.5	−32.0	9.0	−6.2
$j = 4$				−5.0	31.0	64.7	31.8
$j = 5$					67.0	99.5	44.0
$j = 6$						132.0	32.5
$j = 7$							−67.0
		-1	-2	-3	0	3	0

TABLE 19.8 Computation of Sen's Slope

Solution

The data are entered in Table 19.8 and are labeled *X*1 … *X*7. Now *T* values are computed as per Eq. (19.37). For example, in column 2, for $j = 2$ and $k = 1$, slope $T_i = (830 - 905)/(2 - 1) = -75$; in column 3, for $j = 3$ and $k = 1$, $T_i = (672 - 905)/(3 - 1) = -116.5$; for $j = 3$, $k = 2$, $T_i = (672 - 830)/(3 - 2) = -158$; and so on.

Now all the T_i are tabulated as shown below in Table 19.9 and then arranged in increasing order. Since we have 21 (odd) values, the Sen's slope β is the $(21 + 1)/2 =$ 11th value which is −6.2.

19.7.2 Significance of Trend: Mann-Kendall Test

In trend analysis for hydrologic variables such as temperature and precipitation, apart from the magnitude of trend, its statistical significant also needs to be checked. The nonparametric Mann-Kendall (MK) test can be employed for this purpose. The MK test checks the null hypothesis of no trend versus the alternative hypothesis of the existence of increasing or decreasing trend. An advantage of the MK test is that it has only a few assumptions. Thus, the trend may be linear or nonlinear, and the test makes no assumptions regarding the underlying statistical distribution. A drawback is that MK test is not robust when the data have high autocorrelation. If the series has high positive autocorrelation, it is quite likely that a positive trend is detected whereas there may actually be no trend. The case is reverse for negatively autocorrelated series where the trend may not be detected. This effect depends on the sample size as well as on the magnitude of the trend to be identified. Prewhitening has been used to detect trend in time series in the presence of autocorrelation. Prewhitening is an operation that processes a time series so that it behaves statistically like white noise. White noise is a discrete signal whose samples are regarded as a sequence of serially uncorrelated random variables with zero mean and finite variance.

i	Slope T_i	Slope T_i in Increasing Order
1	−75.0	−158.0
2	−116.5	−116.5
3	−79.3	−81.5
4	−42.8	−79.3
5	−7.8	−75.0
6	−17.7	−67.0
7	−158.0	−42.8
8	−81.5	−32.0
9	−32.0	−17.7
10	9.0	−7.8
11	−6.2	−6.2
12	−5.0	−5.0
13	31.0	9.0
14	64.7	31.0
15	31.8	31.8
16	67.0	32.5
17	99.5	44.0
18	44.0	64.7
19	132.0	67.0
20	32.5	99.5
21	−67.0	132.0
β	−6.2	mm/year

TABLE 19.9 T_i Tabulated and Arranged in Increasing Order

For prewhitening, the data series is first tested for presence of serial correlation. If the lag-1 auto-correlation (r_1) is insignificant at 95% confidence level, then the MK test is applied to the original data series ($x_1, x_2, \ldots, x_n$), otherwise if it is significant, the MK test is applied to "prewhitened" series obtained as ($x_2 - r_1x_1, x_3 - r_1x_2, \ldots, x_n - r_1x_{n-1}$).

The statistics (S) is computed as

$$S = \sum_{i=1}^{N-1} \sum_{j=i+1}^{N} \operatorname{sgn}(x_j - x_i) \tag{19.40}$$

where N is number of data points. Assuming $(x_j - x_i) = \theta$, the value of sgn(θ) is computed as follows

$$\text{sgn}(\theta) = \begin{cases} 1 & \text{if } \theta > 0 \\ 0 & \text{if } \theta = 0 \\ -1 & \text{if } \theta < 0 \end{cases} \tag{19.41}$$

This statistic represents the number of positive differences minus the number of negative differences for all the differences considered. For large samples ($N > 10$), the test is conducted using a normal distribution (Helsel and Hirsch, 1992) with the mean and the variance as follows

$$E[S] = 0$$

$$\text{Var}(S) = \frac{N(N-1)(2N+5) - \sum_{k=1}^{n} t_k(t_k - 1)(2t_k + 5)}{18} \tag{19.42}$$

where n is the number of tied (zero difference between compared values) groups, and t_k is the number of data points in the kth tied group. The standard normal deviate (Z-statistics) is then computed as (Hirsch et al., 1993)

$$Z = \begin{cases} \dfrac{S-1}{\sqrt{\text{Var}(S)}} & \text{if } S > 0 \\ 0 & \text{if } S = 0 \\ \dfrac{S+1}{\sqrt{\text{Var}(S)}} & \text{if } S < 0 \end{cases} \tag{19.43}$$

If the computed value of $|Z| > z_{\alpha/2}$, the null hypothesis (H_o) is rejected at the significance level α in a two-sided test.

According to Bayazit and Onoz (2007), no prewhitening of the data series needs to be carried out if the sample size is large ($n \geq 50$) and slope of trend is high (>0.01).

Example 19.7

For the data of Example 19.6, find if the trend is statistically significant or not.

Solution

We first compute the statistic S as shown in Table 19.10 below. The values in the rows marked 1, 2, ... are computed by using Eq. (19.40). Column sums are given in the last row. Summation of all these values give $S = -3$.

$$\text{Since } t_k = 0,\ \text{Var}(S) = \frac{N(N-1)(2N+5) - \sum_{k=1}^{n} t_k(t_k - 1)(2t_k + 5)}{18}$$

$$= \frac{7(7-1)(2 \times 7 + 5)}{18} = 43.3$$

$$Z = \frac{(-3+1)}{\sqrt{43.3}} = -0.30$$

For 5% significance level (α), $z_{\alpha/2} = 2.44$ (from table of t-distribution). Since $|Z| < z_{\alpha/2}$, the null hypothesis of no significant trend is rejected.

j	1	2	3	4	5	6	7
	X1	X2	X3	X4	X5	X6	X7
i	905	830	672	667	734	866	799
1		−1	−1	−1	−1	−1	−1
2			−1	−1	−1	1	−1
3				−1	1	1	1
4					1	1	1
5						1	1
6							−1
		−1	−2	−3	0	3	0

TABLE 19.10 Computation of Statistic S for Trend Analysis

19.8 Summary

This chapter has described statistical tools (correlation, regression, hypothesis testing, and trend analysis) that are very commonly used in hydrological analysis and design. In fact, besides hydrology, these are of immense use in many other fields.

19.9 Questions

19.1 Spurious correlation arises because

I. The data series are not homogenous

II. There is no cause and effect in data

III. The data do not follow normal distribution

IV. Data has random errors

19.2 Autocorrelation is zero for all lags for a

I. Stochastic process

II. For a probabilistic series

III. For a deterministic series

IV. For a pure stochastic series

19.3 Variables in which of the following pair are not likely to be correlated?

I. Catchment rainfall and river flow

II. River water temperature and dissolved oxygen

III. River water temperature and water level in a nearby well

IV. Rainfall and humidity.

19.4 Which of the following time series is likely to have weak autocorrelation?

I. Monthly discharge of a river

II. Daily water temperature of a river

III. Daily rainfall at Bhopal

IV. Daily water levels in a piezometer

19.5 List names of five pairs of hydrological variables which are likely to be correlated.

19.6 List names of five pairs of hydrological variables which are not likely to be correlated.

19.7 What is stepwise regression and what are its advantages?

19.8 Annual precipitation and runoff data (in cm) for a catchment for 6 years (1985–1990) are given in Table 19.11 below. Use linear regression to estimate runoff for 2 years, 1983 and 1984, whose precipitation values are 42.4 and 33.5 cm.

Year	Precipitation (x)	Runoff (y)
1985	47.67	15.17
1986	50.24	15.5
1987	43.28	14.22
1988	52.6	21.2
1989	31.06	7.7
1990	38.65	12.84

Table 19.11 Annual Precipitation and Runoff Data in a Catchment for 6 Years

19.9 For the regression relation derived in Question 19.8, compute the coefficient of determination.

19.10 Compute the correlation between the precipitation and runoff data of Question 19.8 and test the null hypothesis H_o: the distribution of runoff does not change as a function of precipitation.

19.11 Table 19.12 presents 18 years of annual rainfall and runoff data of a basin. Apply regression analysis and compute parameters.

Year	Rainfall (mm)	Runoff (mm)	Year	Rainfall (mm)	Runoff (mm)	Year	Rainfall (mm)	Runoff (mm)
1981	1130	592	1987	1670	872	1993	1650	1240
1982	1280	832	1988	1540	816	1994	1510	1030
1983	1270	768	1989	990	456	1995	1600	1340
1984	1040	488	1990	1190	780	1996	1300	870
1985	1080	472	1991	1520	1090	1997	1490	1060
1986	1150	656	1992	1370	960	1998	1280	990

Table 19.12 Rainfall and Runoff Data (in mm) for the Period 1981–1998

19.10 References

Bayazit, M., and B. Onoz. To prewhiten or not to prewhiten in trend analysis? *Hydrological Sciences Journal*, 52:611–624, 2007.

Davis, J. C. *Statistics and Data Analysis in Geology*. New York: John Wiley & Sons, 2015.

Haan, C. T. *Statistical Methods in Hydrology*. Ames, IA: Iowa State Press, 2002.

Helsel, D. R., and R. M. Hirsch. *Statistical Methods in Water Resources*. New York: Elsevier, 1992.

Hirsch, R. M., D. R. Helsel, T. A. Cohn, and E. J. Gilroy. Statistical analysis of hydrologic data. In *Handbook of Hydrology*, edited by D. R. Maidment. New York: McGraw-Hill Inc., 1993.

Jain, S. K., and V. P. Singh. *Water Resources Systems Planning and Management*. Amsterdam: Elsevier, 2003.

Jain, S. K. Hydrological Measurements and Analysis of Data. Web-based course under National Program on Technology Enhanced Learning (NPTEL). http://nptel.ac.in/courses/105107129/, 2011.

Jain, S. K., V. Kumar, and M. Saharia. Analysis of rainfall and temperature trends in North-East India. *International Journal of Climatology*, 33:968–978, 2013.

Sen, P. K. Estimates of the regression coefficient based on Kendall's tau. *Journal of the American Statistical Association*, 63:1379–1389, 1968.

CHAPTER 20

Hydrologic Modeling

Models enable us to study complex problems and synthesize different kinds of information. In this chapter, we will discuss about the mathematical models of hydrologic systems. Many hydrologic models are actually the conceptual representation of the processes that transform rainfall into movement of water on the surface and below the surface. Examples include the Tank Model and the Sacramento Soil Moisture Accounting (SAC-SMA) model. Models typically have two types of parameters. Physical parameters represent properties of the watershed that can be measured, for example, area and other geometric properties of the watershed, hydraulic conductivity of soil, and topographic slope. Process parameters represent properties of hydrologic processes in the watershed that cannot be directly measured, for example, the "effective" depth of soil, coefficient of friction, representative hydraulic conductivity, and the coefficient controlling the rate of recharge to groundwater.

A catchment or hydrologic model represents the physical, chemical, biological features of a catchment and, depending on the objective of modeling, simulates or duplicates the hydro-bio-geo-chemical processes taking place in a catchment. Emphasis has, however, been on hydrological or hydrochemical processes. Hydrologic models serve many useful purposes: determination of water yield of a catchment, simulation of the response of a catchment to inputs, extension of observed records, and flood forecasting.

In hydrology, models are applied for many different tasks. A model helps in decision-making, particularly where understanding of the system is incomplete, data are missing, or one has to choose from among numerous options (e.g., how much water to release from storage). Usually, it is not possible to work with a real-life system to understand the consequences of a decision. Models are very handy in such situations. The utility of a model is in its ability to extract the actionable information from the data. Finally, models are the best tool to answer: What if?

20.1 History of Hydrologic Modeling

Hydrological modeling has a long history. It goes back to the nineteenth century when the need arose to design and construct culverts, diversions, canals, urban water supply, drainage systems, and dams. Engineers of those times were doing manual calculations by using log-tables and slide rules. The origin of mathematical modeling in hydrology dates to around 1850 when the rational method was developed by Mulvany. Singh and Woolhiser (2002) presented state-of-the-art watershed modeling. Jain and Singh (2017) have also briefly discussed the history of hydrologic modeling. Singh (2018) has discussed progress and future direction in hydrologic modeling. Todini (2007) noted that the rational method of Mulvany is a clear exposition of the concept of time of concentration and its relation to maximum runoff. Subsequently, up until the 1950s, this field witnessed the development of numerous concepts and models of the hydrologic cycle and its components, such as interception, infiltration, evaporation, transpiration, overland and channel flow, and groundwater flow.

Starting from laws of physics, Green and Ampt developed infiltration theory in 1911. This theory is still in use for computing infiltration and is part of many watershed models. Sherman (1932) introduced the unit hydrograph (UH) concept which relates the direct surface runoff from a catchment to rainfall excess. The UH concept is based on the principles of proportionality and superposition and assumes that the catchment response is a linear function of excess rainfall. In fact, most of the nonlinearity of the rainfall-runoff process is taken care of, while determining the excess rainfall. The theory of UH has been discussed in great detail in Chapter 14. Subsequently, the concept of Instantaneous Unit Hydrograph (IUH), which is the impulse response of a linear system, was developed. An IUH is the UH when the unit duration of rainfall excess is infinitesimally small. Around the same time when the UH was developed, Horton (1933) developed a theory of infiltration. This helped in, among the other things, estimation of rainfall excess. About a decade later, Horton (1945) developed the concept of streamflow generation and formulated a set of empirical geomorphological laws, known as Horton's laws. In 1948, Thornthwaite and Penman made significant advancements in the modeling of evapotranspiration.

In 1956, the Soil Conservation Service (SCS) of the U.S. Department of Agriculture [now known as the Natural Resources Conservation Service (NRCS)] developed a procedure for computing the storm runoff. In this method, abstractions from rainfall are computed as a function of land use, hydrologic condition, soil type, and antecedent soil moisture condition. All these are specified by a parameter known as the curve number (CN). This method is popularly known as the Soil Conservation Service-Curve Number (SCS-CN) method. Although it was developed to model daily runoff by using data of rainfall, soil, and land use/cover, the method is extensively used to compute infiltration as well as runoff hydrograph for continuous hydrologic simulation. SCS-CN method is used either as a standalone model or as a part of many hydrologic models, such as the SWAT model (Neitsch, 2002). The SCS-CN method has been discussed in Section 13.2.

Theis (1935) investigated the subsurface component of the hydrologic cycle. He combined Darcy's law with the continuity equation to derive the relation between rate and duration of discharge of a well and the fall in the piezometric surface. This basic work of Theis paved the way for further developments in quantitative groundwater hydrology.

In 1955, Lighthill and Whitham developed the kinematic wave theory for flow routing in long rivers. This work became a mainstay in watershed runoff modeling (Singh, 1996). Around the same time, a major milestone in rainfall-runoff modeling was the work by Nash (1957) which led to the development of instantaneous unit hydrograph (IUH) theory by employing linear systems theory. Nash (1957) conceptualized a catchment as a cascade of N linear reservoirs, each with a residence time of K to develop the IUH. Dooge (1959) followed it and developed the generalized IUH theory.

With the advent of computers, hydrologists began to integrate models of individual components of the hydrologic cycle to develop integrated model of a watershed. One of the first comprehensive attempts toward watershed modeling was the Stanford Watershed Model-SWM (now called HSPF and BASINS) which was developed by Crawford and Linsley (1966). Subsequently, many catchment models were developed by hydrologists and were applied to a range of hydrologic problems. Some examples were the model developed by Dawdy and O'Donnell (1965), NWS River Forecast System (Burnash, 1975), and the SSARR model (Rockwood, 1982). In the United States, the Hydrologic Engineering Center (HEC) of the U.S. Army Corps of Engineers developed a series of models for various hydrologic problems. Some popular models in this series were HEC-1 (now called HEC-HMS), HEC-2 (now called HEC-RAS), and HEC-5 (now called HEC-ResSim). In the decade of the 1960s, many conceptual catchment models were developed in which the various catchment processes were represented through a series of tanks, for example, the Tank Models developed by Sugawara (1967). Around this time, attempts were initiated to develop physically based distributed models of catchment hydrology. The frontrunner and the most popular among these models was the SHE model which was jointly developed by three European organizations. Most (semi)-distributed catchment models in use these days also have geographical information system (GIS) capabilities. Currently, there are more than 100 watershed models in the world. A number of models have been discussed in Singh (1995), and Singh and Frevert (2002a, 2002b, 2006). The publication *Handbook of Applied Hydrology* (Singh, 2017) contains many articles relevant on hydrologic modeling.

20.2 Types of Hydrologic Models

Hydrologic models can be broadly divided in two categories: physical models and mathematical models. A physical model is constructed by "brick and mortar" and is a scaled version of the actual system. These days physical models are not common in hydrology. A mathematical model is a collection of mathematical equations (including ordinary or partial differential equations) and logical statements that describe the behavior of the hydrologic system. In a typical application, the initial and boundary conditions are prescribed and input data are provided to the model that produces output. The assumptions, structure, and simplifications in a hydrologic model depend on the purpose for which the model has been built.

Catchment models have been classified in many different ways, based on catchment representation, concept of deterministic or stochastic principles, application of linear or nonlinear principles, and so on. One way to classify the models from the standpoint of a practicing engineer can be the nature of problem and the data used/available for modeling; the simplest case being the estimation of peak discharge coming out of a catchment. Following this line of thought, hydrologic models can be classified as (1) models to estimate peak flow from a catchment; (2) models to estimate the volume of flow; (3) models to compute the full hydrograph; and (4) integrated catchment models.

Another scheme of classification of hydrologic models is shown in Fig. 20.2. According to this scheme, two broad groups of models are optimizing and nonoptimizing. Nonoptimizing models are classified as either deterministic or statistical models. Deterministic models can be further split into empirical, conceptual, and physically based distributed models. Statistical models can be subdivided into three categories: regression models, probabilistic models, and stochastic models types, each of these types have two classes, be spatially lumped/or spatially distributed, and linear or nonlinear models. Optimizing models have two groups, statistical decision theory and mathematical programming. In this book, nonoptimizing models will be discussed in detail.

As shown in Fig. 20.1, a model simulates the behavior of the real system (e.g., a catchment) with known inputs (e.g., rainfall, temperature) to produce output(s) (e.g., the hydrograph at the outlet, groundwater levels). In a model, the natural system may be described in different levels of details depending on the problems that the model is aimed to address, how the modeler perceives and conceptualizes the system, data availability, and required accuracy of the results.

Singh (1995) classified hydrologic models based on (1) process description, (2) time scale, (3) space scale, (4) techniques of solution, (5) land use, and (6) model use.

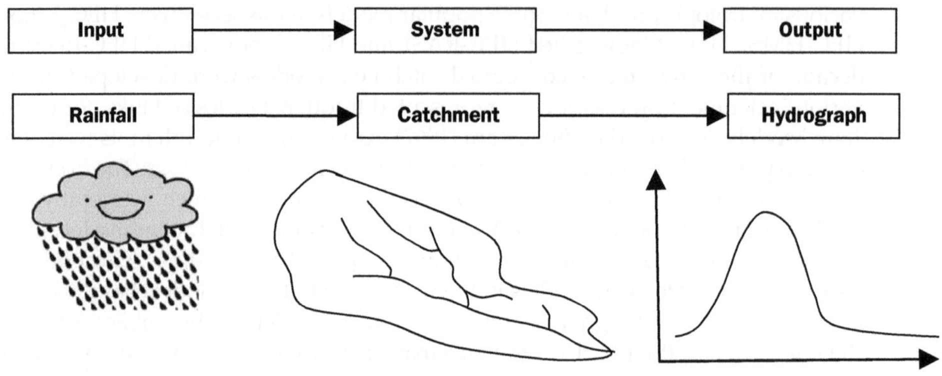

Figure 20.1 Representation of input, system, and output for a mathematical model.

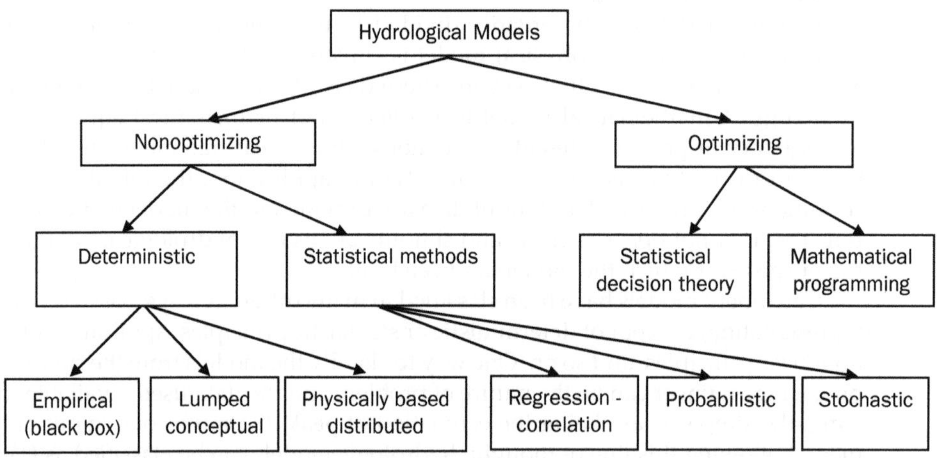

Figure 20.2 A classification of hydrologic models.

Models related with flood analysis can be categorized into several groups: rainfall-runoff models, flood routing models, and flood frequency analysis models. The first group may be again classified into three subclasses: (1) event-based precipitation runoff models and (2) continuous precipitation-runoff models. Subgroups of flow routing models are: (1) steady flow routing models, (2) unsteady-flow flood routing models, and (3) reservoir regulation models. ASCE (1996) has provided classification of flood analysis models. Table 20.1 suggests which type of model would be best suited, depending upon the data availability and the problem to be addressed.

Three main groups of hydrologic models are: black-box models, conceptual models, and physically based models and we will now discuss the underlying concepts of these model types.

20.2.1 Black-Box Models

Black-box models utilize observed input and output data to determine how the system behaves or models are calibrated using the observed data. The basic premise is: "let the data speak for itself." For example, consider the system shown in Fig. 20.3 whose

Available Data	Problem to be Addressed	Suitable Model(s)
Discharge	What is the design discharge for a culvert or bridge? What is the spillway design discharge? What is the surface water yield of the basin?	Rational method Statistical (frequency analysis) model Statistical model, flow duration curve
Long series of rainfall and short series of discharge	What is the long-term flow series at the given site? What is the water yield at the catchment outlet? What is the spillway design discharge? Flood forecasting	Statistical or conceptual model Statistical model, flow duration curve Frequency analysis Statistical or conceptual model
Time series of rainfall, discharge, and ET	Reconstruct flow series Flood forecasting Drought management	Statistical or conceptual model Statistical or conceptual model Statistical or conceptual model
Time series of rainfall, climate data, and discharge; catchment DEM, soil, LU/LC maps	What will be the impact of land use/landcover change on catchment response? What will be the impact of climate change on catchment response?	Physically based distributed catchment models Physically based distributed catchment models

TABLE 20.1 Showing Suitable Classes of Models Depending upon Data and Problem

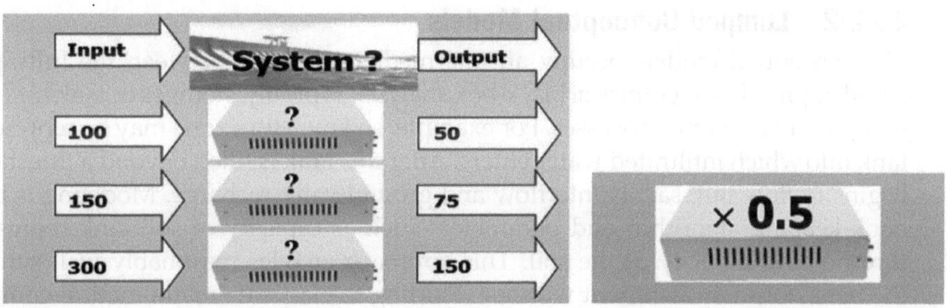

FIGURE 20.3 Determining properties of a black-box system by analysis of input and output.

properties are to be determined by observing input and output. Three pairs of input and output are available: input 100 → output 50, input 150 → output 75, and input 300 → output 150. Simple analysis will show that the system is multiplying the input by 0.5 to produce output.

In real life, things are not so simple. The system response may change with time (Fig. 20.4) or more than one input may be controlling the output.

Performance of black-box models when applied to the data within the range of data used for calibration is typically very good, because an implicit understanding of the physical system is captured by the mathematical relationship. However, application of these models outside the range of data used for calibration should be with caution since the assumptions of model may no longer be valid. Many black-box models assume that

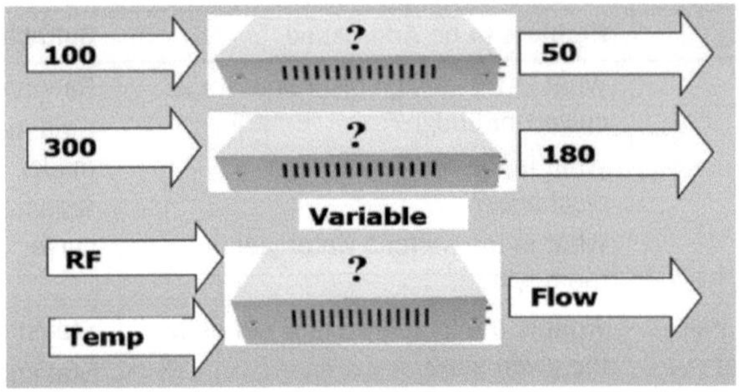

Figure 20.4 Response of a real-life system may change with time or more than one input may be controlling the output.

the hydrologic systems respond linearly but the real-world systems are nonlinear and thus the linearity assumption is not strictly valid. Probably the best-known black-box model in hydrology is the UH model. It is noted that the UH model cannot be employed for some practical problems, for example, to predict the impact of land-use change on surface runoff.

All black-box models may not be applicable to all catchments because the model may not be modeling all the hydrological processes. Hence, black-box models are unable to easily incorporate the relative importance of these processes.

In many cases, black-box models are incorporated in a larger model; for example, the UH is often used to compute runoff response of a watershed in conceptual rainfall-runoff models. Black-box models have been discussed in greater detail in Chap. 13.

20.2.2 Lumped Conceptual Models

The conceptual models occupy an intermediate position between the fully physically based approach and empirical black-box analysis. Typically, storages or tanks are employed to represent different processes. For example, soil moisture zone may be represented by a tank into which infiltrated water enters. After this tank is filled beyond a threshold, water begins to flow out, say as interflow and groundwater recharge. Modeling is basically a book-keeping algorithm and parameters such as capacity of soil tank represent moisture holding capacity of the soil. This approach enables reasonably well simulation of the behavior of a catchment without resorting to elaborate mathematical computations. At the same time, it does not require elaborate data of catchment properties and hydrologic variables. Conceptual models have modest computational requirements which was a helpful feature in earlier times when powerful computers were not widely available.

Lumped conceptual models consist of a small number of components, each representing an element of the hydrologic system. Typically, each model component consists of a nonlinear tank/reservoir in which the relationship between outflow (Q) and storage (S) is given by

$$Q_i = f(K, S^n) \tag{20.1}$$

where K and n are coefficients that are calibrated by using the observed historical data. The model operation is essentially a sequence of mathematical and logical operation to determine the movement of moisture in various storages and keeping track of water balance at each time step. Nonlinear behavior and resemblance with the real system in determining excess rainfall, soil moisture storage and movement, and surface/subsurface runoff are partially addressed by providing thresholds of different storages. Calibration of lumped conceptual models is in some ways a curve-fitting exercise in which selected parameters are systematically varied to match the observed behavior with that computed by the model.

An example of the lumped conceptual models from the 1960s is the Tank Model developed by Sugawara (1967). This simple model has been used and proved to be helpful in many studies. The HBV model was developed at the Swedish Meteorological and Hydrological Institute. Bergstrom (1976) has described this model which is similar to the Tank Model in some ways. HBV model has also been applied in numerous studies. It is also being used operationally for flood forecasting of floods and inflows to reservoirs of several hydropower systems in Sweden and Norway. Numerous such models have been developed subsequently. Some of these became popular and some were short-lived.

Parameters of black-box and conceptual models do not have any physical meaning and cannot be obtained by measuring catchment characteristics. Hence these models must be calibrated for each study area by the use of hydrometeorological data series of sufficient length.

20.2.3 Fully Distributed, Physically Based Models

Over the last 60 plus years, model developments in hydrology have followed a path from black-box (rational, regression) models to gray box (lumped, conceptual) models and then to physically based, distributed (white box) models. This progress has been driven by the needs of the hydrologic community at various times to (1) address the increasingly complex water assessment and management problems due to rising demands and diminishing water resources; (2) provide better forecasts for protection from water-related disasters; (3) better understand the impacts of land-use and land-cover changes on hydrologic response; (4) tackle increasing pollution of water bodies/resources; and (5) address problems arising due to impacts of climate change. Progress in modeling has been supported by three developments: (1) increasing coverage, frequency, and reliability of terrain and hydrometeorological data being collected by better instruments and space-based platforms; (2) improved understanding of the physics, chemistry, biology, and geology of hydrological processes; and (3) developments in information technology that have provided powerful and affordable computer hardware and software.

Physics-based distributed models apply principles of physics to model the hydrologic behavior of a watershed. In these models, typically the watershed is subdivided in small elemental areas which may be square (most common) grids or subcatchments/response units. Movements of mass, momentum, and energy in the watershed are determined by solving the governing partial differential equations. These are the Saint-Venant equations for surface flow, the Richards equation for moisture movement in unsaturated zone, and the Boussinesq equation for groundwater flow or saturated zone. Governing equations are solved by numerical methods with given boundary conditions and inputs. By definition, physically based models are spatially distributed, since the underlying governing equations generally involve one or more space coordinates (Jain and Singh, 2017). Due to this feature, distributed models give values of the hydrologic variables, for example, river flow, groundwater level, and actual ET, at any location in a watershed. Also note that such models are costly to develop, apply, and computational time and data requirements are large. Hydrologic literature contains many publications, such as Beven (2000, 2001), Vieux (2016), and Jain and Singh (2017) which provide excellent overview of the topic.

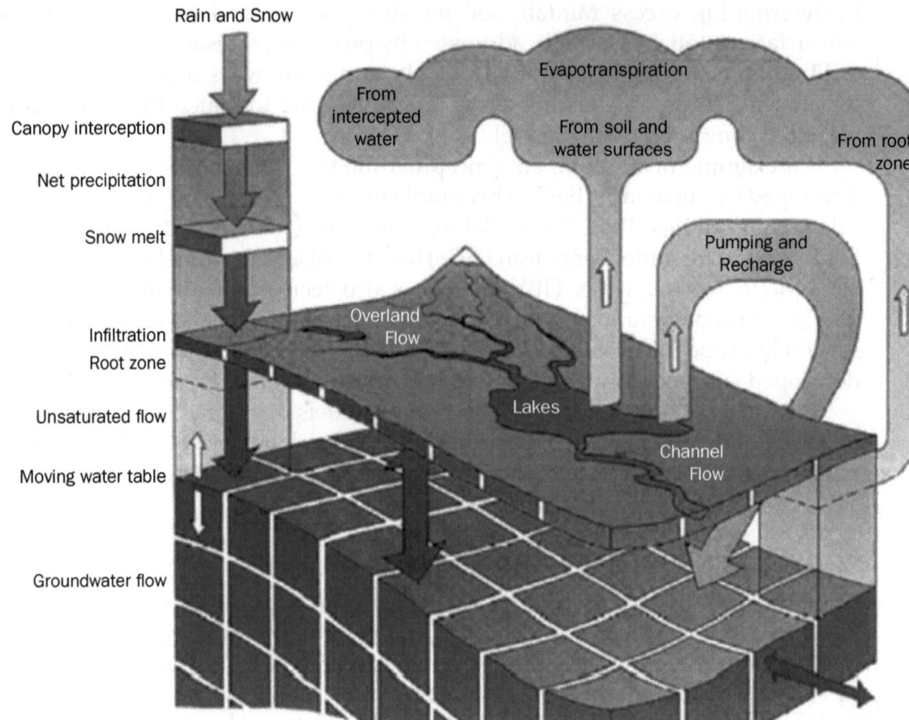

FIGURE 20.5 Schematic diagram of a catchment and a quasi-three-dimensional physically based distributed model: the SHE model. (Source: http://mikebydhi.com.)

Development of physics-based distributed models and their application began in the 1970s–1980s. However, difficulties were faced in applying these models, because computer and data requirements were quite high, compared more so to the situation in the 1980s. There were additional issues, such as numerical stability and ensuring mass balance at each time step. These difficulties were gradually resolved and several physically based distributed models were developed. Prominent among these models were the Système Hydrologique Européen (SHE; Fig. 20.5), a grid-based model developed by a consortium of three European institutes: the Danish Hydraulic Institute (Denmark), Institute of Hydrology (UK), and SOGREAH (France) (Abbott et al., 1986); the Institute of Hydrology Distributed Model (IHDM) (Beven et al., 1987) and ANSWERS (Beasley et al., 1980). Initially, these models were tested by using the data of small catchments having good database. These applications were aimed to remove errors in the code and improve the models. The SHE model was tested on watersheds in a range of climates and with size varying from tens of hectares to nearly 1000 km^2 (Jain et al., 1992). Later, development of the SHE model was vigorously taken up by DHI and these days it is a widely used commercial package under the name MikeSHE (http://mikebydhi.com).

These days, the Soil and Water Assessment Tool (SWAT) model has become highly popular. It is a semi-distributed model in which the catchment is subdivided in homogenous areas, called Hydrologic Response Units, or HRUs. SWAT code and model documentation are freely available online at https://swat.tamu.edu/.

20.2.4 Advantages and Limitations of Physically Based Distributed Models

The concept of physically based distributed models has many advantages.

1. Physically based models, in principle, can be applied to data of any catchment since the laws of physics are valid everywhere.
2. Parameters of physically based models have physical interpretation and can, in principle, be estimated from field/laboratory measurements. Thus such models can be applied to catchments with even short hydrometeorological records. Further, such models are the best tools to predict catchment response in presence of changes in land use, climate inputs, etc.
3. In principle, physically based models should not require calibration. However, in practice these models need some calibration, since modeling eventually involves some lumping, approximations, and simplifications. Hence limited calibration for some parameters is required which can be carried out with short hydrometeorological records.
4. Physically based models can beneficially use almost all available relevant information from various parts of study area (e.g., topography, slopes, river channel cross-sections and properties, soil and geology data, vegetation properties and maps, information on extreme events) and may improve with time as new insights into hydrological processes are developed.

Currently, there is growing concern about the impacts of anthropogenic activities, such as changes in land use/land cover and climate change on hydrologic response, including the movement of contaminants originating from diffuse sources (agricultural areas) and point sources (municipal and industrial areas). Many catchment changes are due to growing population and increasing water demands and it is helpful to understand how the response of a catchment will be impacted due to changes. Physically based models can be of immense value in such studies.

Parameters of black-box and conceptual models lack physical basis and therefore cannot be specified for the altered catchment conditions. Further, spatially lumped models cannot account for spatial variations and spatial changes in land use/land cover and hydrometeorological inputs. Model parameters calibrated by using the past data will not be valid if the catchment state changes substantially. In contrast, parameters of physically based distributed models have a physical basis and can, in principle, be measured in field. Thus these parameters can be specified for the future changed catchment state. This makes physically based models eminently suitable for predicting the hydrologic response of a catchment in the changed conditions.

Limitations Inputs to these models typically include large spatial data and the outputs may also be spatially distributed. Thus large data storage capacity is needed, although this is not a serious issue now. Computer memory and time required to run a physically based distributed catchment model at finer spatial and temporal discretization are rather high. Computers having a fast processing speed are required because complex calculations are to be performed at each time step at a number of (grid) points or sub-basins/HRUs. Short time steps are often required to ensure the stability of numerical schemes, more so during periods of sharp changes in rainfall, and so on. Of course, some of the computer-related issues are not critical these days due to a steep decline in hardware costs. But these days, finer spatial resolution are being used in many studies. This means that even with faster processing speed, computational time is not reduced.

Calibration of physically based models is more difficult than other models. The number of parameters of these models is very large and the values of parameters typically vary with location. Hence automatic calibration routines are not available (or advisable) for most such models. Many hydrologists argue that such models should not require any calibration in the first place but, in practice, some calibration is required and is carried out because catchment properties are not uniform even at finer discretization, such as over a grid of 50 m × 50 m. To properly carry out manual calibration, the modeler should be familiar with the study area and the model and should know the physically acceptable range of parameters. Furthermore, new users of such models require more rigorous training and a longer training period.

Distributed models face problems in parameterizing subgrid scale processes (Beven, 2001). For example, we know that soil properties have large spatial variabilities but they are often assumed to be constant over a grid, since sufficient information to realistically describe the spatial variations is lacking. Beven (2001) states that the problem of nonlinearity is at the heart of many of the problems being faced in the application of the distributed modeling concepts in hydrology.

Regarding the development and application of physically based distributed hydrologic modeling, the following points highlighted by researchers (Beven, 2000, and others) need to be considered:

- How can the catchment description in physically based models be improved without increasing the number of parameters?
- Even a physically based model requires some calibration depending upon the spatial scale. Can this be overcome? If yes, how?
- Most distributed models are calibrated by using the discharge data at the outlet which essentially reduces them to a lumped rainfall-runoff model. What's the best way to carry out multivariable calibration?
- A distributed model has a large number of parameters and erroneous values of some parameters may be compensated for by others. How to ensure that the calibration has identified the correct parameter set?
- Given the uncertainties in the modeling process, is it worthwhile to use a complex model, particularly in data deficient situations?
- Since field measurements require huge resources (instruments, manpower, logistics), how should the field investigations be planned and conducted to better estimate the parameters of a distributed model?
- What is the best way to present uncertainties to the users and utilize these to arrive at better inferences and decisions?

Wider application of distributed models is imperative to solve the emerging hydrological problems. To that end, one of the prerequisites is hydrometeorological, pedospheric, and geological databases at fine resolutions in space and time.

20.2.5 Statistical Models

While deterministic models rely on physical laws and/or their simplifications, in statistical models, generally a relationship is established between various hydrological variables using observed data. Statistical models that are used in hydrology apply the laws of probability, statistical theory, and so on. These have been discussed in other chapters.

20.3 Model Calibration and Validation

Once a particular model has been chosen to solve a given problem, it is necessary to determine the right values of its parameters. Results of a model will be useful only if the assumptions are valid and parameters are correct. After a suitable model has been selected, the next task is to find the values of model parameters such that the model correctly simulates the catchment behavior with the right parameters.

Two possible approaches to determine model parameters are measurement and calibration. In general, it is not possible to obtain the parameters of hydrologic models by field measurements. In fact, in most cases, the parameters do not correspond to a measurable feature of a catchment. In addition, a typical catchment may extend to a few hundred square kilometers. For example, a double ring infiltrometer, which is used to measure infiltration and, indirectly, soil hydraulic conductivity, covers an area less than about 1 square meter. But a grid or HRU in a distributed model may be several hundred square kilometers and there is no technique to satisfactorily relate the measured value of soil properties with the effective value for the model grid. If measured values of a parameter at locations in the catchment are available, a practical approach is to use these values as the starting point and calibrate to find the effective value for modeling. Model calibration can be performed manually or using an automated procedure. In either case, an objective function or a measure of goodness of fit of the model results is required.

20.3.1 Objective Function and Measures of Model Performance

An objective function is a numerical measure of the closeness between the observed (measured) output of interest and that computed by the model. The calibration algorithm tries to find the model parameters that optimize (minimize or maximize) the objective function. A commonly used objective function is the weighted least squares (WLS) function

$$F(\theta) = \sum_{i=1}^{n} w_i . [Q_{\text{oi}} - Q_{\text{ci}}(\theta)]^2 \tag{20.2}$$

where Q_{oi} = the observed (measured) output of interest at time i; $Q_{\text{ci}}(\theta)$ = the model computed output at time i; θ = the vector of model parameters; w_i = the weight at time or step i; and n = the number of data points to be compared. Weights w_i specify the importance given to fitting a particular hydrograph feature (e.g., peak). If the weights are equal to 1.0, the WLS function turns to the simple least squares (SLS) function. The minimum possible value of $F(\theta)$ is zero, which is obtained when the model perfectly reproduces all observed values. However, in practice a zero value is not attained as the model results seldom, if at all, exactly match the observed values. The calibration algorithm finds the set of parameters θ which minimizes function $F(\theta)$. Note that most commonly observed discharge at the catchment/subcatchment outlet is compared with the model computed discharge. But this could be any other variable, such as groundwater level or evapotranspiration.

In modeling, a common measure of the goodness of fit is the sum of squares of errors, also called error variance. Taking the squares of residuals, both positive (over-simulation) as well as negative (under-simulation) contributions get added up for all the time steps. The sum of squares of errors, σ_ε^2, is computed as

$$\sigma_\varepsilon^2 = \frac{1}{n-1} \sum_{i=1}^{n} (Q_{\text{oi}} - Q_{\text{ci}})^2 \tag{20.3}$$

It is also helpful to compute the serial correlation of residuals which should ideally be independent. A large serial correlation value for the residuals implies that some information is left in the residuals which the model could not extract. In rainfall-runoff modeling of rivers with high seasonality, the variance of residuals is usually high during periods of high flows.

The model performance is typically evaluated by comparing simulated and observed variable of interest (commonly discharge) in terms of mean, standard deviation, maximum daily discharge, and the total discharge using commonly used indices. Many statistical indices are applied by hydrologists to evaluate the model performance (Singh, 2015): root mean square error (RMSE), RMSE-standard deviation ratio (RSR), coefficient of determination (COD or R^2), Nash and Sutcliffe efficiency (NSE), and percent bias (PBias). These can be computed as described next.

RMSE is a popular performance measure which is computed as

$$\mathrm{RMSE} = \sqrt{\frac{1}{n}\sum_{i=1}^{n}(Q_{\mathrm{oi}} - Q_{\mathrm{ci}})^2} \tag{20.4}$$

RMSE indicates error in the units of the variable. An RMSE of 0 indicates a perfect match; a value less than 0.5 SD (standard deviation) is considered low and acceptable.

Another useful index is the RMSE-SD ratio (RSR) which is RMSE scaled by SD. It can also be used to compare results of different applications. It is computed by

$$\mathrm{RSR} = \frac{\mathrm{RMSE}}{\mathrm{SD}} = \frac{\sqrt{\sum_{i=1}^{n}(Q_{\mathrm{oi}} - Q_{\mathrm{ci}})^2}}{\sqrt{\sum_{i=1}^{n}(Q_{\mathrm{oi}} - Q_{o,\mathrm{mean}})^2}} \tag{20.5}$$

RSR varies from zero (perfect model simulation) to a large positive value; a low RSR indicates better model simulation.

Coefficient of Determination (COD or R^2) Willmott (1981) and Legates and McCabe (1999) defined COD as

$$R^2 = \left(\frac{\sum_{i=1}^{n}(Q_{\mathrm{oi}} - Q_{o,\mathrm{mean}})(Q_{ci} - Q_{c,\mathrm{mean}})}{\sqrt{\sum_{i=1}^{n}(Q_{\mathrm{oi}} - Q_{o,\mathrm{mean}})^2}\sqrt{\sum_{i=1}^{n}(Q_{\mathrm{ci}} - Q_{c,\mathrm{mean}})^2}}\right)^2 \tag{20.6}$$

COD lies in the range 0–1; higher values imply better simulation. Models with systematic over- or under-prediction may also give good R^2. Hence, COD should not be used alone. The coefficient of regression line (b) multiplied by R^2 or bR^2 is a better measure of model performance.

Nash-Sutcliffe Efficiency (NSE) Nash and Sutcliffe (1970) developed this popular index as

$$\mathrm{NSE} = 1 - \left[\frac{\sum_{i=1}^{n}(Q_{\mathrm{oi}} - Q_{\mathrm{ci}})^2}{\sum_{i=1}^{n}(Q_{\mathrm{oi}} - Q_{o,\mathrm{mean}})^2}\right] \tag{20.7}$$

NSE lies in the range $-\infty$ to 1. NSE is 1 for a perfect fit when all $Q_{oi} = Q_{ci}$; it has the value of 0 when numerator and denominator are equal which shows that the model is basically a "no-knowledge" model whose prediction is the mean of observed data for all time steps. Negative values of NSE are obtained when the model performs worse than this "no-knowledge" model. It is easy to get comparatively high values of NSE if the data has high variance. This index is not sensitive to systematic over- or under-prediction by the model. Logarithmic NSE is calculated with logs of observed and predicted values. By doing this, peaks are flattened and the influence of low values increases. Logarithmic NSE is more sensitive to systematic over- or under-prediction.

Percent Bias This measures the average tendency of simulated data to be larger or smaller than observed values. Its best value is 0 and it is computed by

$$\text{PBIAS} = \frac{\sum_{i=1}^{n}(Q_{oi} - Q_{ci}) * 100}{\sum_{i=1}^{n} Q_{oi}} \tag{20.8}$$

While using a hydrologic model with monthly data, Moriasi et al. (2007) suggested ranges of RSR, NSE, and PBIAS to label model performance from very good to unsatisfactory (see Table 20.2).

Example 20.1

A catchment model was used to simulation discharge at the outlet of a catchment. Monthly values of observed and computed hydrographs (average in cumecs) are given in Table 20.3. Compute the following performance indices: RMSE, RSR, COD, NSR, and PBIAS.

Performance Rating	RSR	NSE	PBIAS (%) Streamflow	PBIAS (%) Sediment
Very good	$0.00 \le \text{RSR} \le 0.50$	$0.75 < \text{NSE} \le 1.00$	$\text{PBIAS} \le \pm 10$	$\text{PBIAS} < \pm 15$
Good	$0.50 < \text{RSR} \le 0.60$	$0.65 < \text{NSE} \le 0.75$	$\pm 10 \le \text{PBIAS} \le \pm 15$	$\pm 15 \le \text{PBIAS} < \pm 30$
Satisfactory	$0.60 < \text{RSR} \le 0.70$	$0.50 < \text{NSE} \le 0.65$	$\pm 15 \le \text{PBIAS} \le \pm 25$	$\pm 30 \le \text{PBIAS} < \pm 55$
Unsatisfactory	$\text{RSR} > 0.70$	$\text{NSE} \le 0.50$	$\text{PBIAS} \ge \pm 25$	$\text{PBIAS} \ge \pm 55$

TABLE 20.2 Guidelines Suggested by Moriasi et al. (2007) to Determine Model Performance Based on RSR, NSE, and PBIAS

Month	Jan	Feb	Mar	Apr	May	Jun	Jul	Aug	Sep	Oct	Nov	Dec
Observed	157.86	150.22	172.05	242.24	450.14	919.86	1952.17	2387.91	1429.04	542.60	274.51	193.74
Computed	146.39	141.35	176.01	284.10	277.29	482.89	1304.78	1779.18	1336.13	644.76	313.06	185.13

TABLE 20.3 Observed and Simulated Discharge for Example Problem

Solution

We first calculate the following quantities:

$\Sigma(Q_{oi}-Q_{ci})$	$\Sigma(Q_{oi}-Q_{ci})^2$	$\Sigma(Q_{oi}-Q_{om})$	$\Sigma(Q_{oi}-Q_{om})^2$	$\Sigma(Q_{ci}-Q_{cm})$	$\Sigma(Q_{ci}-Q_{cm})^2$	$\Sigma(Q_{oi}-Q_{om})(Q_{ci}-Q_{cm})$
1801.27	1033100.75	0.00	6587200.11	0.00	3497675.58	4661077.73

By using the above quantities and relevant equations, the various indices are calculated as follows:

RMSE = 84.701

RSR = 0.396

COD = 0.943

NSE = 0.843

PBIAS = 20.302

20.3.2 Manual Calibration

To manually calibrate a model, we first select the variable that the model is expected to simulate, for example, discharge hydrograph at the catchment outlet in case of rainfall-runoff modeling. Next, we select the initial values of model parameters. A trial-and-error procedure is adopted in which the model is run with the requisite inputs and selected parameters and the simulated response is obtained. The goodness of fit is determined by using one or more of the indices described earlier. In addition, a visual comparison between the time series of observed and computed values of the variable (Fig. 20.6) is very helpful. Depending upon the mismatch between observed and simulated responses, the model parameters are systematically adjusted such that improved values of the goodness of fit indices are obtained. It is advisable to change one parameter at a time and observe the impact on catchment response. With modeling experience and familiarities with the model, it is not difficult to obtain good model calibration by following the manual approach.

With user-friendly software, graphical plots of the results and visualizations are easy. Note that in manual calibration, different persons may end up with different parameter sets for the same catchment. Manual calibration of a model with a large number of parameters can be tedious and frustrating.

20.3.3 Automatic Calibration

There are several reasons behind the development of computer-based methods for automatic calibration of catchment models: (1) the need to quickly calibrate models with a large number of parameters; (2) to overcome the shortage of people with requisite expertise of a particular model; and (3) to make model calibration more objective. Although there are many general as well as model-specific tools, it is not advisable to completely depend on these tools. One should always manually check the results to make sure that the parameter values make hydrologic sense.

A general automatic calibration task consists of five elements: (1) objective function, (2) calibration/validation data, (3) optimization method, (4) stopping criteria, and (5) model validation. We will now discuss optimization methods for model calibration.

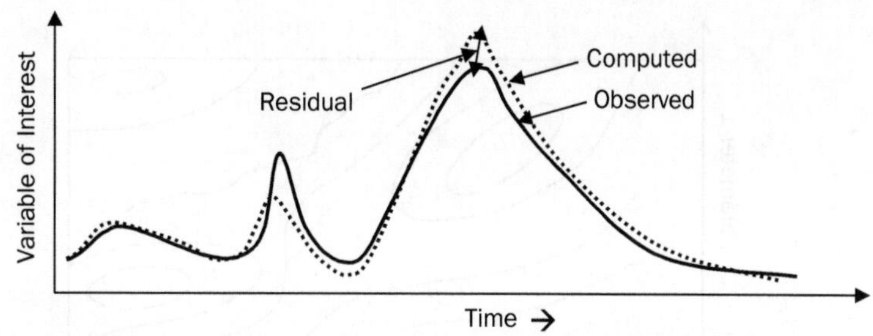

FIGURE 20.6 Comparing observed and simulated variable of interest.

20.4 Optimization Methods for Calibration of Hydrologic Models

When we have more than two parameters, the objective function forms a surface in the parameter space, called the response surface. An optimization algorithm searches the response surface (subject to constraints on the feasible values of parameters) and finds the optimum value of the objective function. When the number of parameters is more, the objective function forms a complex shape. Two broad categories of optimization techniques are local search and global search. The following discussion assumes that we are interested in finding the minimum of objective function.

20.4.1 Local Search Methods

Local search methods are employed to find the minimum of unimodal functions—the functions that have only one extreme value. For unimodal functions, any algorithm that continuously proceeds downhill (a direction of improving function value) will eventually reach at the minimum function value. Imagine that a visually impaired person is standing somewhere on a mountain and desires to reach the lowest point of the valley. An intelligent person will (1) examine the topography and determine the best direction in which to move; (2) take a few steps in that direction; and (3) repeat the steps 1 and 2 until satisfied that (s)he has reached the bottommost point and no further improvement is feasible.

In the above algorithm, three decisions are required at each step: (1) the best direction to move; (2) how far to move in that direction; and (3) how to identify that the person is at the lowest. Local search strategies differ the way in which these decisions are made. Accordingly, local search methods are classified as "direct search" and "gradient" methods.

A direct search method uses only the information on the function value in the above decision process. Starting at the initial point, the algorithm selects a direction to move and step size. Reaching at the new point, the function value is evaluated. Based on the differences in function values between the initial and new point(s), the best (trial) direction to move to improve the function and the step size are predicted. A step is taken in the trial direction, and function value at the new point is evaluated. If the function value is lower than the previous point, it replaces the old value and the procedure is repeated. If the function value at new point is worse than the previous point, the step size is reduced and another trial is made. The search is terminated when the strategy is unable to find a direction in which improvement is possible.

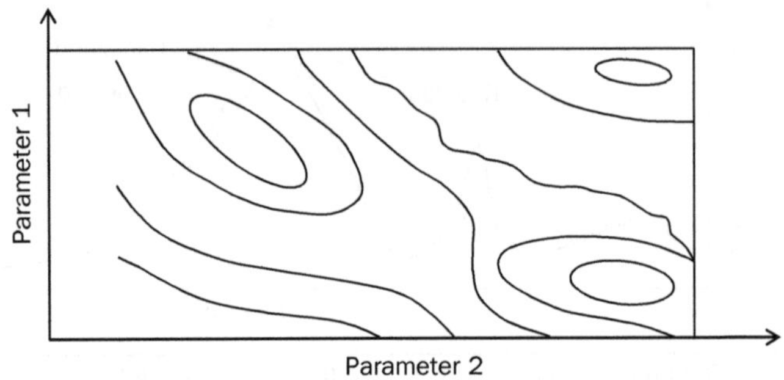

FIGURE 20.7 A complex response surface in a two-parameter space showing multiple local optima.

Some direct methods make use of the "gradient" of the objective function for this purpose, since it is known that the value of a function changes fastest by moving in the direction of the gradient. Since a gradient is a local property which changes with location, it is necessary to evaluate the gradient at many points. Thus these methods are iterative. A simple gradient search algorithm is

$$x^{(k+1)} = x^{(k)} - \alpha^{(k)} \nabla f\{x^{(k)}\} \tag{20.9}$$

where $x^{(k)}$ is the solution at kth step, $\alpha^{(k)}$ is the step-length parameter, and $\nabla f(x)$ is the gradient. At the optimum point, the gradient value will be very close to zero.

Local search methods are not good for calibration of hydrologic models, because response surfaces of such models may have multiple peaks. In these situations, the algorithm may terminate at a local optimum and may not be able to find the global optimum.

20.4.2 Global Search Methods

Consider the response surface of a model shown in Fig. 20.7 which has multiple optima. It is difficult to find the global optimum in such cases, since the optimization algorithm may be trapped by one of the local optima. Most calibration problems involve more than two parameters. It is difficult to visualize the response surface if there are more than three parameters. Here, different starting values of parameters may give different optimal parameter values. The problem of multiple local optima can make hill-climbing optimization particularly difficult. One of the peaks shown in Fig. 20.7 will be the global optimum.

Multistart approaches run the local search algorithm repeatedly from different starting parameter-sets in the feasible parameter space. The SCE-UA (shuffled complex evolution–University of Arizona) global optimization method developed by Duan et al. (1992) is a powerful multistart algorithm popularly used to calibrate rainfall-runoff models. It combines the strengths of the simplex procedure of local search with the concepts of controlled random search competitive evolution and complex shuffling. This combination gives the SCE-UA global optimization method good global convergence properties over a broad range of problems encountered in hydrology.

In view of the above, a successful optimization procedure must be globally based and possess global convergence properties. The algorithm must be able to come out of a local optima; it must not be compulsory for the algorithm to require explicit analytical expressions for the objective function. Algorithm must be robust keeping in view parameter interaction and (possible) nonconvexity of the objective function. Since catchment models typically involve a large number of parameters, the algorithm must work well in a high dimensional space. The SCE-UA (Duan et al., 1992) global optimization method is claimed to be an effective and efficient method for calibration of catchment models.

Parameter optimization is essentially a mathematical exercise, not based on hydrologic principles. Therefore it is quite possible that some of the parameter values obtained after automatic calibration make little sense from a hydrologic viewpoint. Constraints may partially address this problem but the user also needs to ensure that the chosen parameter set is consistent with the catchment properties. A detailed discussion on optimization techniques is available in many textbooks, such as Rao (2009) and Ravindran et al. (2006).

20.5 Model Validation

After calibrating a model, it is necessary to examine the suitability of calibrated parameters. This is done by running the model for the input data that were not used for model calibration. If the model has been calibrated properly, it will perform well on the validation data. If not, it is necessary to go back to calibration and improve it till the results of model validation are acceptable. Such a scheme is called a split-sample test: part of the data is used to calibrate the model and the remaining part for validation. If the results of validation are not acceptable, either the data contains some error or the assumptions behind the model do not hold good for the study catchment and another model may be chosen.

20.5.1 Calibration Data

If good calibration data is available, many difficulties can be reduced during model calibration. However, it is tough to answer: What constitutes "good" calibration data? How much data are required for acceptable calibration? What kind of data will give the best calibration?

A common practice is to use about 60%–70% data for calibration and the remaining for validation. Sorooshian et al. (1983) had stated that if the calibration procedure was properly designed, only 1 year of daily data was sufficient to obtain conceptually realistic estimates for the SMA-NWSRFS (soil moisture accounting-National Weather Service River Forecast System) model. According to them, the use of longer data series for calibration serves only marginally to improve the parameter estimates. A guideline is that the length of a dataset should be at least 20 times the number of parameters to be estimated. Gupta and Sorooshian (1985) showed that the standard error (*j*) of the estimate of parameter (*j*) decreases with sample size n approximately according to

$$\sigma(\text{j}) \alpha \frac{1}{\sqrt{n}} \tag{20.10}$$

Since changes in $1/\sqrt{n}$ with respect to n become smaller as n increases beyond, say 1000, this suggests that about 3 years of calibration data may be sufficient, provided the data are of the *right* kind.

Quality of Data Regarding the quality of the data for model calibration, this is dependent on the information content (about parameter calibration) and the noise (errors) in the data. Obviously, the information content of the data should be as high as possible and the noise the least.

Informativeness For example, if the data for a dry year have been used for calibration, certain processes that generate overland and channel flow may not get activated and the parameters that control these processes may not be properly calibrated. However, if the data selected are from a year of high rainfall when the watershed mostly remains saturated, the model parameters that control subsurface and low flow components may not be properly calibrated.

The best scenario is a dataset that contains considerable hydrologic variability; if during the calibration period, the hydrologic regime frequently changes from dry to wet modes, all the important model parameters will be used to produce output, resulting in better calibration.

Data Errors Instrumental and measurement errors degrade the quality of data, thereby resulting in parameters with poor reliability. Streamflow data are often the most reliable but still the error margin is about 10% or more. It is desirable that the data used for model calibration be carefully screened and the errors be removed.

20.5.2 Equifinality of Parameters

The concept of "equifinality of parameters" states that for a given model structure and catchment, there may be more than one set of parameters which may simulate the hydrologic response of the catchment nearly equally well (Beven, 2001). In calibration of a model, even if the "optimum" parameters have been found, there may be some other sets of parameters that might simulate the catchment response equally well. This concept recognizes the approximations and simplifications of both model structure and the data used for calibration.

20.5.3 Sensitivity Analysis

Small changes in values of some parameters cause large changes in model results. Such parameters are termed "sensitive parameters." Sensitivity analysis is carried out to identify such parameters before starting calibration and is helpful. Another objective of sensitivity analysis is to determine the "region of indifference" for the parameters—the region around the best value of parameters in which the function value varies slowly. The size and shape of this indifference region also give an idea of the amount by which the parameter can be in error without significantly impacting the results.

Consider that we have a two-parameter model. The model is run with calibration data and a goodness of fit measure, such as RMSE, is calculated. A number of runs of the model are taken by systematically increasing the parameter values by a predetermined increment, one at a time, and RMSE values are computed. In this two-parameter case, the feasible range of values for each parameter can be divided in 10 increments, and the model can be run for each combination (total 100 runs of the model) of parameter values. When these RMSE values are plotted, we may obtain a graph similar to Fig. 20.8. Note that as the discretization becomes finer, the number of model runs will rise exponentially. Information contained in Fig. 20.8 helps in understanding the interaction of model parameters, identifying the shape of response surface, region of indifference, and so on.

It is easy to visualize that as the number of parameters increases, the shape of response surface may also become more and more complex, and it is difficult to visualize its shape when more than three parameters are involved. The form of response surface is seldom smooth (e.g., as shown in Fig. 20.8) and this makes it tough to find the global optimum.

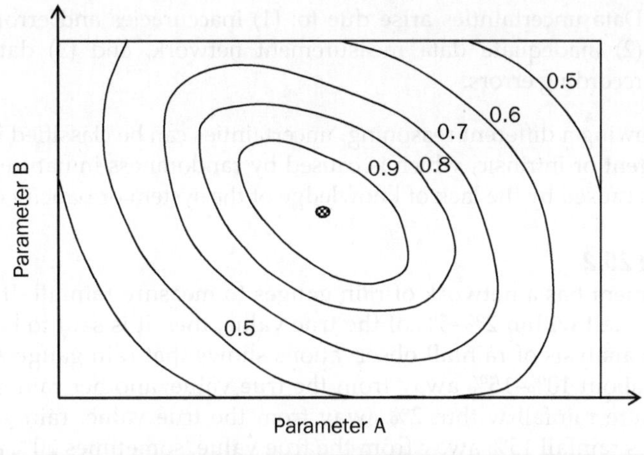

FIGURE 20.8 Response surface for a two-parameter case. The performance index is represented as contours. Small hatched circle refers to optimal solution.

Another problem encountered in calibration is parameter insensitivity which arises when changes in values of a parameter cause little changes in the model output in a given range of parameter values. This may happen, if the corresponding model component is not activated during a run or the parameter physically itself does not control a major process.

20.6 Uncertainty in Model Calibration

All model results have uncertainty associated with them, which arises due to a combination of factors: errors in inputs, errors in model and parameters, and errors in the initial and boundary conditions. A good discussion of these sources of uncertainty may be found in Singh et al. (2007). The sources of uncertainty in hydrologic modeling can be grouped into four categories:

1. Natural uncertainties which are associated with inherent spatial and temporal variations in hydrologic processes and/or parameters controlling them, such as spatial variability in catchment properties, variability in climatic inputs, occurrence of rainfall, hydrologic extremes, and so on.
2. Model structure uncertainty which is caused when the hydrologic model does not precisely represent the process behavior or when the assumptions in modeling are not valid. For example, the Manning's formula is used to compute discharge in a channel which is an empirical formula.
3. Model parameter uncertainties which reflect the possible errors in determining the values of parameters to be used in the model, for example, uncertainty in determining value of Manning's roughness factor n in Manning's formula

$$Q = \frac{1}{n} A R^{2/3} S^{1/2} \tag{20.11}$$

4. Data uncertainties arise due to: (1) inaccuracies and errors in measurement, (2) inadequate data measurement network, and (3) data observation and recording errors.

Following a different reasoning, uncertainties can be classified into two categories: (1) inherent or intrinsic, which is caused by randomness in nature; and (2) epistemic, which is caused by the lack of knowledge of the system or paucity of data.

Example 20.2

A catchment has a network of rain gauges to measure rainfall. If a rain gauge measures rainfall within 2%–5% of the true value, then it is said to be accurate and precise. An analysis of rainfall observations shows that rain gauge A always measures rainfall about 10%–15% away from the true value; another rain gauge B was found to measure rainfall within 2% away from the true value; rain gauge C sometimes measures rainfall 15% away from the true value, sometimes 20% away from the true value, and sometimes very close to the true value. Another rain gauge, called D, sometimes measures rainfall 5% higher, sometimes 5% lower, but always within 5% away from the true value. What can be said about the measurements of these rain gauges? This example has been drawn from Singh et al. (2007).

Solution

Rain gauge A has the repeatability quality. Hence its measurements are precise but inaccurate. Rain gauge B is accurate and precise, because it has repeatability quality and its observations are close to the true values. Rain gauge C is inaccurate and imprecise because its measurements are neither repetitive nor accurate. Rain gauge D is imprecise but accurate because its observations are close to the true values but this characteristic is not repeatable. The characteristics of rain gauges are shown in Fig. 20.9.

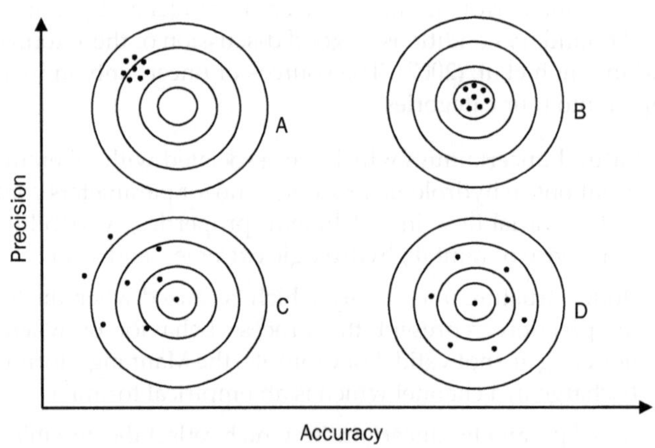

FIGURE 20.9 Measurement of rainfall by four rain gauges. The innermost circle indicates the true value. Gauge A is precise, inaccurate; gauge B is precise, accurate; gauge C is imprecise, inaccurate; gauge D is imprecise, accurate.

20.7 Summary

The field of hydrologic modeling is witnessing exciting developments. While using a model, it is necessary to ensure that the model works right for the right reasons. Students of hydrology are to be encouraged to critically evaluate models and identify the gaps in process understanding and representation in order to improve modeling. As better spatial and time series data are becoming widely available, hydrologic models are finding increasing use in water resources planning and management and the trend is likely to accelerate.

20.8 Questions

20.1 What are the ways in which hydrological models are classified? Discuss briefly?

20.2 What are the main advantages and limitations of physically based distributed models?

20.3 Describe any three measures of model performances?

20.4 A catchment model was used to simulate discharge at the terminal point. Monthly values of observed and computed hydrographs (average in cumecs) are given in Table 20.4. Compute the following performance indices: RMSE, COD, NSR, and PBIAS.

Month	Jan	Feb	Mar	Apr	May	Jun	Jul	Aug	Sep	Oct	Nov	Dec
Observed	20	60	180	210	160	110	68	40	24	14	12	9
Computed	18	52	165	192	179	135	74	46	31	16	10	9

TABLE 20.4 Monthly Values of Observed and Computed Hydrographs

20.9 References

Abbott, M. B., J. C. Bathurst, J. A. Cunge, P. E. O'Connell, and J. Rasmussen. An introduction to the European Hydrologic System-Systeme Hydrologique Europeen, SHE, 1: History and philosophy of a physically-based, distributed modeling system. *Journal of Hydrology*, 87:45–59, 1986.

ASCE. *Handbook of Hydrology, ASCE Manual and Rep. on Engineering Practice No. 28*. New York: ASCE, 1996.

Beasley, D. B., L. F. Huggins, and E. J. Monke. ANSWERS: A model for watershed planning. *Transactions of ASAE*, 23:938–944, 1980.

Bergstrom, S. Development and application of a conceptual runoff model for Scandinavian countries. *SMHI Rep. No. 7*. Norrkoping, Sweden, 1976.

Beven, K. On the future of distributed modelling in hydrology. *Hydrology Process*, 14:3183–3184, 2000.

Beven, K. J. *Rainfall-Runoff Modelling: The Primer*. New York: Wiley, 2001.

Beven, K. J., A. Calver, and E. Morris. The Institute of Hydrology distributed model. *Institute of Hydrology Rep. No. 98*. Wallingford, UK, 1987.

Burnash, R. J. C. Chapter 10: The NWS river forecast system catchment modeling. In *Computer Models of Watershed Hydrology*, edited by V. P. Singh. Littleton, CO: Water Resources Publications, 1975.

Crawford, H. H., and R. K. Linsley. *Digital Simulation in Hydrology: Stanford Watershed Model IV*. Technical Report No. 39, Department of Civil and Environmental Engineering, Stanford University, Stanford, 1966.

Dawdy, D. R., and T. O'Donnell. Mathematical Models of Catchment Behavior, *Am. Soc. Civil Engineers Proc.*, 91(HY4):123–137, 1965.

Dooge, J. C. I. A general theory of the unit hydrograph. *Journal of Geophysical Research*, 64(2):241–256, 1959.

Duan, Q., S. Sorooshian, and V. K. Gupta. Effective and efficient global optimization for conceptual rainfall-runoff models. *Water Resource Research*, 28(4):1015–1031, 1992.

Gupta, V. K., and S. Sorooshian. The relationship between data and the precision of parameter estimates of hydrologic models. *Journal of Hydrology*, 81(1–2):57–77, 1985.

Horton, R. E. The role of infiltration in the hydrologic cycle. *Transactions of the American Geophysics Union*, 145:446–460, 1933.

Horton, R. E. Erosional development of streams and their drainage basins: Hydrophysical approach to quantitative geomorphology. *Bulletin of the Geological Society of America*, 56:275–370, 1945.

Jain, S. K., B. Storm, J. C. Bathurst, J. C. Refsgaard, and R. D. Singh. Application of the SHE to catchments in India—Part 2: Field experiments and simulation studies with the SHE on the Kolar subcatchment of the Narmada river. *Journal of Hydrology*, 140:25–47, 1992.

Jain S. K., and V. P. Singh. Hydrological cycles, models and applications to forecasting. In *Handbook of Hydrometeorological Ensemble Forecasting*, edited by Q. Duan, F. Pappenberger, J. Thielen, A. Wood, H. Cloke, J. Schaake. Berlin, Heidelberg: Springer, 2017.

Legates, D. R., and G. J. McCabe. Evaluating the use of goodness-of-fit measures in hydrologic and hydroclimatic model validation. *Water Resources Research*, 35(1): 233–241, 1999.

Moriasi, D. N., J. G. Arnold, M. W. Van Liew, R. L. Bingner, R. D. Harmel, and T. L. Veith. Model evaluation guidelines for systematic quantification of accuracy in watershed simulations. *Transactions of the ASABE*, 50(3):885–900, 2007.

Nash, J. E. The form of the instantaneous unit hydrograph. *Hydrological Sciences Bulletin*, 3:114–121, 1957.

Nash, J. E., and J. V. Sutcliffe River flow forecasting through conceptual models. 1: Discussion of principles. *Journal of Hydrology*, 10(3):282–290, 1970.

Neitsch, S. L., J. C. Arnold, J. R. Kiniry, J. R.Williams, and K. W. King. Soil and Water Assessment Tool: Theoretical Documentation. Version 2000. Texas Water Resources Institute, College Station, Texas, 2002.

Rao, S. S. *Engineering Optimization—Theory and Practice*. Hoboken, NJ: John Wiley and Sons, 2009.

Ravindran, A., K. M. Ragsdell, and G. V. Reklaitis. *Engineering Optimization—Methods and Applications*. New York: John Wiley and Sons, 2006.

Rockwood, D. M. Theory and practice of the SSARR model as related to analyzing and forecasting the response of hydrologic systems. In *Applied Modeling in Catchment Hydrology*, edited by V. P. Singh, 87–106. Littleton, CO: Water Resources Publications, 1982.

Sherman, L. K. Stream flow from rainfall by the unit graph method. *Engineering News-Record*, 108:501–505, 1932.

Singh, V. P. Watershed hydrology. In *Computer Models of Watershed Hydrology*, edited by V. P. Singh. Fort Collins, CO: Water Resources Publications, 1995.

Singh, V. P. *Kinematic Wave Modeling in Water Resources: Surface Water Hydrology*. New York: Wiley, 1996.

Singh, V. P. *Entropy Theory in Hydrologic Science and Engineering*. New York: McGraw-Hill Education, 2015.

Singh, V. P. Hydrologic modeling: Progress and future directions. *Geoscience Letters*, 5(15):1–18, 2018. doi.org/10.1186/s40562-018-0113-z.

Singh, V. P., and D. K. Frevert, D.K. *Mathematical Models of Large Watershed Hydrology*. Highland Ranch, CO: Water Resources Publications, 2002a.

Singh, V. P., and D. K. Frevert. *Mathematical Models of Small Watershed Hydrology and Applications*. Highland Ranch, CO: Water Resources Publications, 2002b.

Singh, V. P., and D. K. Frevert. *Watershed Models*. Boca Raton, FL: CRC Press, 2006.

Singh, V. P., and D. K. Woolhiser. Mathematical modeling of watershed hydrology. *Journal of Hydrologic Engineering*, 7(4):270–292, 2002.

Singh, V. P., S. K. Jain, and A. Tyagi. *Risk and Reliability Analysis: A Handbook for Civil and Environmental Engineers*. New York: ASCE Press, 2007.

Singh, V. P. (Editor). *Handbook of Applied Hydrology*. McGraw Hill, New York, 2017.

Sorooshian, S., V. K. Gupta, and J. L. Fulton. Evaluation of maximum likelihood parameter estimation techniques for conceptual rainfall runoff models—Influence of calibration data variability and length on model credibility. *Water Resource Research*, 19:251–259, 1983.

Sugawara, M. The flood forecasting by a series storage type model. International Symposium on Floods and their Computation. *International Association of Hydrologic Sciences*, 1–6, 1967.

Theis, C. V., The relation between the lowering of the piezometric surface and the rate and duration of discharge of a well using groundwater storage, *Am. Geophys. Union Trans.*, 16:519-524, 1935.

Todini, E. Hydrological catchment modelling: Past, present and future. *Hydrology and Earth System Sciences*, 11: 468–482, 2007. doi:10.5194/hess-11-468-2007.

Vieux B. E. Introduction to physics-based distributed hydrology. In *Distributed Hydrologic Modeling Using GIS*. Water Science and Technology Library Vol. 74. Dordrecht: Springer, 2016.

Willmott, C. J. On the validation of models. *Physical Geography*, 2:184–194, 1981.

CHAPTER 21

Design Storm and Design Flood Estimation

Design rainfall constitutes the basis for computing design flood which, in turn, is fundamental to water resources project design. Different methods are available for computing design rainfall as well as design flood. Regardless of the computation method, there are uncertainties associated with computed design rainfall and design flood which cannot be entirely eliminated. This chapter presents simple methods of computing design rainfall. For large projects, such as large dams and nuclear power plants, design rainfall is obtained by computing probable maximum precipitation, which is also discussed in the chapter.

21.1 Design Flood Estimation

Estimation of design flood is fundamental to water resources project planning, design, and operation. The type of flood adopted for the design of a project or its components depends upon the importance of the structure and the consequences if it fails. It is neither possible nor desirable to provide for the complete safety of any structure against the maximum possible flood likely to be generated by its catchment. For example, if a small road culvert fails, it will cause temporary disruption of traffic but rarely any property damage and/or loss of life. If, however, a major dam fails, it may cause considerable loss of life and property in the downstream areas. Obviously, the type, importance, and socioeconomic profile of the area likely to be impacted dictate the criteria for choosing the design flood of a structure. Before discussing procedures for computing design floods for hydraulic structures, some important definitions are given here based on WMO METEOTERM (http://www.wmo.int/pages/prog/lsp/meteoterm_wmo_en.html), CWC (2001), and others.

Design Storm (1) Rainstorm, either observed or synthetic, which is chosen as the basis for the design of a hydraulic structure. (2) Rainfall amount and distribution adopted for a given drainage area, used in determining the design flood.

Design Flood Flood hydrograph or the instantaneous peak discharge adopted for the design of a hydraulic structure or river control after accounting for political, social, economic, and hydrological factors. It is the maximum flood against which the project is protected; its selection involves choosing safety criteria and estimating the flood magnitude that meets the criteria. According to CWC (2001), the value of design flood should increase with increasing consequences of the failure of the structure.

Probable Maximum Precipitation (PMP) In a severe storm, all the moisture available in the storm region does not usually precipitate; typically, rainstorm efficiency is less than 25%. This leads to the concept of PMP, defined as "theoretically the greatest depth of precipitation for a given duration that is physically possible over a size storm area at a particular geographical location at a certain time of year" (WMO, 1986). Estimates of PMP for different durations over an area are required to calculate the maximum flood that can be realistically generated in a catchment, or the probable maximum flood (PMF). Note that precipitation exceeding the calculated PMP can occur but the cost of protection from the maximum possible flood (if such a flood existed) would be too high to justify the project and hence a lower level of protection is generally accepted. The terms PMP and probable maximum flood (PMF) cover this lower level of protection.

Standard Project Storm (SPS) This is defined as that rainstorm which is reasonably capable of occurring in the region of the study basin.

Spillway Design Flood This is a flood that a spillway must be able to pass to provide the desired degree of protection for the dam.

Construction Flood This is a flood for which reasonable precautions will be taken to avoid flooding of construction sites and thereby to prevent damage to a project during its construction.

Probable Maximum Flood (PMF) This is the largest flood that may be expected at a site, taking into account all pertinent factors of location, meteorology, hydrology, and terrain. It essentially has a very long (almost infinite return period) and can be selected as the design flood to prevent a major disaster.

Standard Project Flood (SPF) This is a flood resulting from the most severe combination of meteorological and hydrological conditions that are considered reasonable characteristics of the geographical region involved, excluding extremely rare combinations. It has a long but unspecified return period and may be selected as a design flood for structures of great importance.

Frequency-Based Flood (FBF) This is a flood determined by employing frequency analysis of flood flows or rainfall data by performing one of the following: (1) frequency analysis of rainfall data to estimate a frequency-based design storm, which is then converted to design flood using a rainfall-runoff model; (2) frequency analysis of flood flows available at the site to directly estimate the design flood; or (3) regional frequency analysis to estimate the design flood.

The criteria used to select the design flood for various hydraulic structures vary from country to country. Table 21.1 gives a summary of the guidelines adopted by CWC (2001) to select design floods in India.

21.2 Approaches to Design Flood Estimation

Depending upon the size of a water resources development project and the consequences if it fails, three types of design floods are computed: PMF, SPF, and FBF.

Dambreak studies may be conducted for important projects to decide whether PMF needs to be used as design flood or not.

S. N.	Structure	Recommended Design Flood
1.	Spillways for major and medium projects with storage more than 60 mm³	(a) PMF determined by unit hydrograph and PMP. (b) If (a) is not applicable or possible, flood frequency analysis with return period T = 1000 years.
2.	Permanent barrage and minor dams with capacity less than 60 mm³	(a) SPF determined by unit hydrograph and SPS, which is usually the largest recorded storm in the region. (b) Flood with a return period of 100 years. (a) or (b), whichever gives higher value.
3.	Pickup weirs	Flood with a return period of 100 or 50 years depending upon the importance of the project.
4.	Aqueducts	
	(a) Waterway	Flood with T = 50 years.
	(b) Foundations and free board	Flood with T = 100 years.
5.	Projects with scanty or inadequate data	Empirical formulae.

TABLE 21.1 Guidelines for Selecting Design Floods in India

Two approaches are generally adopted for design flood estimation: (1) hydrologic modeling—determine the design storm for the project and then compute the design flood that will be generated by this storm, and (2) statistical approach. The steps involved in the hydrologic modeling approach are as follows:

1. Select major storms from the historical data.
2. Determine the duration of the design storm. This duration should typically be approximately equal to the time of concentration of the watershed.
3. Determine the PMP or SPS; it depends on the size of the project.
4. Estimate the storm depth for the selected frequency and duration.
5. Estimate the abstractions, and determine the effective rainfall, its intensity and its time distribution.
6. Select an appropriate rainfall-runoff model, such as the unit hydrograph (UH) method for the watershed.
7. Determine the design hydrograph by using the rainfall-runoff model or by convoluting the UH with the time distribution of the effective rainfall.

The UH method is simple, versatile, and popular, and is widely used for the estimation of floods for design of hydraulic structures, drainage systems, small bridges, culverts, parking lots, and so on. The methods to develop UH have been discussed in Chap. 14. The design flood hydrograph is then determined by convoluting the UH with the design storm. At times, it is helpful to divide a large watershed into a number of small watersheds (subwatersheds) to account for the spatial variability of rainfall and watershed characteristics. The number of subwatersheds depends upon the watershed size, rainfall pattern, vegetation, soil, and topographic characteristics.

The statistical approach is based on frequency analysis of flood data which can be station-based or regional. Also, it is important to compute the degree of risk associated with the estimated design flood. This then involves cost-benefit analysis, hence integrating hydrologic design and economic design. These days, environmental and ecological considerations take a center stage in any project design. Also, social and political considerations are important and cannot be ignored in design.

21.3 Design Storm

Design storms (rainfall) typically used for estimating design floods are the probable maximum storms and the standard project storms; the design storm used for other hydraulic structures, such as bridges and culverts, is generally of some specific small return period. To fix the size of spillway of a dam, the common practice is to route the design flood impinging at the Full Reservoir Level (FRL) and derive the best combination of the Maximum Water Level (MWL) and spillway discharge capacity (length and crest height). It is essential to take into account the volume associated with SPF/PMF for arriving at the surcharge storage capacity. After the major portion of the surcharge storage has been filled up in regulating high flow, if another flood wave brings in additional large volume into the reservoir, it may render the situation critical. Hence determination of the critical rainfall duration is important.

The estimation of design storm begins with rainstorm analysis. Abbi (1972) defined a rainstorm as a spatial distribution of heavy rainfall, yielding an average depth of precipitation that equals or exceeds a certain threshold value over a region in association with some meteorological phenomena, namely, low pressure areas, depression, or cyclonic storms, etc. The design storm is a storm with high intensity and magnitude of rainfall, occurring over the catchment of the project. It has three elements: the amount of rainfall, its areal coverage and distribution, and duration of the storm and distribution of rain with time. The duration of design storm depends on the size of the drainage basin, duration of past large storms in the catchment, and the type of structure. The relationship between these three elements—depth, area, and duration is—of vital importance in estimation of design flood.

The data required for design storm studies include daily and hourly records of stations in and around the study area, storm dew point and the maximum dew point temperature. Extremely high floods are caused by extreme storms, which have very low probability of occurrence. Extreme rainfall events are few and hence the designer needs to consider the storms that have been observed over the watershed as well as those that could have taken place.

21.3.1 Design Storm Selection

Sound knowledge of the hydrometeorology of large storms in the project catchment derived from the study of past major storms is necessary to select the design storm. A set of storms that are representative of the project catchment are selected. To decide the appropriate rainfall depth for design purposes, either of the two types of analyses is carried out: basin-centered or storm-centered.

For preliminary selection of design storm, relevant records, such as storm rainfall data, storm tracks, cyclones, low pressure depressions, history of storms, and flood data are studied. Then, an appropriate threshold value of rainfall depth is selected to screen storms. The threshold values vary with the watershed size—a higher value for smaller watersheds and a lower value for larger watersheds. If the total rainfall depth of two storms on 2 days equals or exceeds the 2-day threshold value, then both storms are included in the analysis. Studies have shown that in the most severe 3-day rainstorms in India, 50% of rainfall is received on the heaviest of the 3 days, 30% on the second heaviest day, and only 20% on the third day. Design flood of a project is not related with the mean annual rainfall in the project catchment. A catchment with low mean annual rainfall could have high design flood and vice versa.

By analyzing daily or multi-hourly rainfall data, details of past storms, storm tracks, and high-flow data, a database of potential high-flow storms is prepared. For each storm, the average rainfall is computed by using data of rain gauges located in and near the catchment. This average rainfall is compared with a pre-set threshold and all the storms whose average daily depth is equal or more than the threshold are chosen for further consideration. Next, isohyetal maps are prepared for all the shortlisted storms. It is common to prepare maps showing total storm depth of 1 day, 2 days, and 3 days; this depends on size of the catchment and the duration of storms in the region. Where the movement of storms is over larger, elongated catchment areas, isohyetal maps may have to be prepared for longer duration. For smaller catchments, such maps may be prepared for multi-hourly durations, for example, 3-hour, 6-hour, and 12-hour.

21.3.2 Design Storm Duration

A rainfall storm event denotes a period of high rainfall that is preceded and followed by periods (of more than a day) of insignificant rainfall. Storm duration is the time elapsed from the beginning of the rainfall of significant intensity to the end of the event. It is commonly reported in hours but it may be (less commonly) in terms of minutes for small catchments. The rainstorm depth, duration, and intensity vary widely, depending on the geographic location of the catchment, climate, orography, and time of the year. Typical rainfall intensities may be up to 50 mm/h, but these can be much larger in extreme cases.

The amount of moisture precipitated over an area by a moving storm depends on the storm duration. Regarding the critical rainstorm duration, the important parameters are the shape of the catchment, its size, the time base of UH, and the direction of storm movement (whether upstream, or downstream, or across). In general, the larger the duration of the storm, the better would be the estimate of the flood hydrograph. If the storm is moving in the upstream direction, the duration of flood hydrograph would be larger. As far as the estimation of flood peak is concerned, it would not help to go for a very large duration. This would only increase the amount of computations.

If the UH approach is being used, a value equivalent to the base period of UH adjusted to the next nearest value in multiples of 12 hours may be adopted. If the time base of UH is between 24 hours to less than 48 hours, a 2-day design storm should be considered. For basins having the UH base of more than 48 hours, the design storm duration of 3 days should be appropriate. Generally, the storm durations of more than 72 hours are not considered in the calculations of design floods.

21.3.3 Depth-Duration Analysis

WMO (1969) describes the procedure for Depth-Duration (DD) and Depth-Area-Duration (DAD) analysis for storms. The DD analysis is performed to determine rainfall depth over the watershed for different durations. To carry the DD analysis, the boundary of the catchment up to project site is used and the storm depths over the catchment are estimated. Here, data of both in situ and transposed storms are used and the average rainfall depths for different durations are computed. These depths constitute the DD data.

Data of in situ as well as transposed storms are included in the analysis. The average watershed rainfall depth can be computed by using any appropriate method. This average depth corresponds to a specific duration. The value of the areal average depth corresponding to different durations are tabulated and plotted on a graph that shows the relation between rainfall depth and duration.

The rainfall depth increases with storm duration. This can be mathematically described as

$$P = at^n \tag{21.1}$$

where P is the storm rainfall depth, t is the storm duration, and a is a coefficient. Further, n is a positive number whose value is less than 1; commonly, n varies from 0.2 to 0.5. If the values of a and n are known for a region, Eq. (21.1) can be used to estimate the depth of rainfall as a function of storm duration for that region.

If the data of a number of recording rain gauges is provided at short intervals, say hourly, the DD analysis can be performed for rainfall at shorter intervals (1-hour, 3-hour, 6-hour, 12-hour, etc.). Alternatively, observed rainfall data at nonrecording stations can be distributed in multihourly values by using the data of nearby and representative recording rain gauge stations, and then DD analysis can be performed on it.

21.4 Depth-Area-Duration Analysis

Generally, as the catchment area increases, the spatially averaged rainfall depth for the catchment falls. Data on the depth of rainfall for the storms is commonly presented as tables or graphs showing maximum average depth of storm rainfall over different areas and is called the Depth-Area-Duration (DAD) data. Thus the DAD analysis provides the relationship between storm rainfall depth, corresponding area, and duration. The DAD analysis is carried out to determine the maximum rainfall depth over areas of various sizes in different times, for example, maximum depth over 500 sq. km in one day. Two methods are used to construct the DAD relationship: (1) the mass-curve method and (2) the incremental isohyetal method. In each of these methods, the time division of rainfall is treated differently. In the mass-curve method, the mass curves of rainfall for individual stations are constructed, then an average mass curve is prepared for the entire catchment, and finally the isohyetal map for the total storm rainfall is prepared. When rainfall data are available at multi-hourly intervals, mass curves can be constructed by using such data.

In the incremental isohyetal method, first isohyetal maps are prepared for the maximum first-day, second-day, etc. These maps are studied and the map for the day that yielded the maximum depth over various areas is used to calculate the maximum one-day depth area. This step requires examination of maps prepared for various days. Next, the isohyetal maps for the maximum 2-day rainfall depths are prepared for various areas. Likewise, the DAD values for the maximum 3-day rainfall depths etc. are obtained.

A mathematical expression for the depth-area-duration relation can be given as

$$P = \frac{at}{(A + bt)^m} \tag{21.2}$$

where P is the rainfall depth (cm); t is the time duration (h); A is the catchment area (km^2); and a, b, and m are the parameters determined from regression analysis.

Example 21.1

Determine the 1-d maximum rainfall depth for catchment areas of 250 and 5000 km^2 located in the United States. For the study area: $a = 4.31 \times 10^3$, $b = 2.77 \times 10^2$, and $m = 0.803$.

Solution

Given: $t = 1$ day $= 24$ h and $A = 250$ and 5000 km^2, the maximum rainfall depth for $A = 250$ km^2 is computed as

$$P = \frac{at}{(A + bt)^m} = \frac{4.31 \times 10^3 \times 24}{(250 + 277 \times 24)^{0.803}} = 85.54 \text{ cm}$$

Similarly, for $A = 5000$ km^2, P can be computed as 56.17 cm.

The storm-centered DAD analysis for 1-day storm using the isohyetal method for a project is presented in Table 21.2.

It may be seen from the isohyetal maps of various storm durations that (1) in general, the areal spread for a given duration decreases as the storm duration decreases; and (2) the average rainfall corresponding to the given storm duration over the area surrounding the storm center decreases as one moves farther from the storm center.

Since the network of precipitation stations is sparse in many countries, the probability that the highest point precipitation has been recorded is low. Hence, while studying rainstorms covering large area in such regions, it is fine to assume that the highest observed precipitation at a station represents the average rainfall depth over a sizeable area and not the maximum point rainfall. When plotting rainfall depth area curves, conventionally, the logarithmic scale is used for depicting the area and the linear scale for the rainfall depths.

The rainfall depth for a small area is reasonably represented by the rainfall depth recorded at a station but the ratio of mean areal rainfall to the point rainfall decreases as the areas increases. This ratio is much more sensitive for high intensity short duration rainfalls. Hershfield and Wilson (1960) reported the values of the ratio of average rainfall over the whole area to that over a 25 km^2 as shown in Table 21.3. Numerous depth-area relationships have been developed. Leclerc and Schaake (1972) presented a

Isohyetal Value (Point)	Area Planimetered in cm^2	Area in km^2	Net Area in km^2	Average Depth (mm)	Volume in mm × km^2		Average Rainfall Depth
					Increment	Cumulative	
488	—	—	—	—	—	—	488.0
450	13.4	1,340	1,340	469	628,460	628,460	469.0
400	38.1	3,810	2,470	425	1,049,750	1,678,210	440.5
300	91.2	9,120	5,310	350	1,858,500	3,536,710	387.8
200	177.1	17,710	8,590	250	2,147,500	5,684,210	321.0

TABLE 21.2 Depth-Area-Duration Analysis Computation for One-Day Storm (August 30, 1982)

Time (h)	Area (km²)				
	250	500	1250	2500	12500
6	0.85	0.80	0.75	0.65	0.46
12	0.89	0.85	0.79	0.72	0.51
24	0.92	0.88	0.82	0.76	0.58
48	0.93	0.90	0.84	0.80	0.63

TABLE 21.3 Ratio of Mean Rainfall to That Over 25 km²

relation between the ratio of areal precipitation depth P_m to point precipitation depth (P) as

$$K = \frac{P_m}{P} = 1 - \exp\left(-1.1t^{0.25}\right) + \exp\left(-1.1t^{0.25} - 0.01A\right) \tag{21.3}$$

where t is the duration in hours and A is the area in square miles.

For further details on the DAD analysis, one may refer to WMO (1969). IMD (1972) and CWC (2001) have provided detailed procedure for constructing DAD curves.

21.4.1 Rainfall Intensity-Duration–Frequency Analysis

Rainfall frequency is the time elapsed on an average between two rainfall events of the same depth and duration. The actual time elapsed may vary by a wide margin. Events with larger rainfall depths are not very frequent, but events with smaller depths are frequent.

It is helpful to have a long historical data series of observations to determine the statistical properties of the annual maximum rainfall. When such long data series are not available, designers resort to extrapolations by using statistical tools to estimate rainfall depths corresponding to the desired return period. However, the extrapolated values have a certain amount of risk and uncertainty. A good design is expected to provide security against any danger; it is necessary that the risk is contained in the acceptable limits. Hence extrapolation also takes into account the meteorological factors that are also reasonably maximized.

The relationship between intensity (I), duration (D), and frequency (F) for a given project catchment can be obtained by the analysis of short-interval data.

The data pertaining to the amount of rainfalls of different durations, such as 10 min, 20 minutes, 30 minutes, 60 minutes, 120 minutes, and so on, are processed to determine the rainfall depths for various frequencies. Plots of average rainfall intensity vs duration for a particular frequency give the intensity-duration curves.

In many problems related to the design of water resources projects, it is required to know rainfall intensities for different durations and return periods. The relationship between rainfall intensity i (cm/h), duration D (hour), and return period T (years) can be expressed by the following general form

$$i = \frac{kT^n}{(b+t)^m} \tag{21.4}$$

where k, b, m, and n are empirical coefficients for a given catchment that are estimated using measured data of the same catchment or region or experience. Typical values of these constants Values of these coefficients for a few places in India, based on studies by

Place	k	n	b	m
Bhopal	6.93	0.189	0.50	0.878
Chandigarh	5.82	0.160	0.40	0.750
Nagpur	11.45	0.156	1.25	1.032
Raipur	4.68	0.139	0.15	0.928

TABLE 21.4 Representative Values of Constants in Eq. (21.4)

Indian workers, are provided in Table 21.4. Note that for the values presented in Table 21.5, the units are: rainfall intensity, i, measured in cm/h; duration, t, in hours; and return period, T, in years. Similar to the depth-duration-frequency curves, IDF curves are plotted with intensity on the ordinate (vertical or y-axis) and duration on the abscissa (horizontal or x-axis); frequency (or return period) is plotted as the third variable.

Bell (1969) reported an empirical relationship between high intensity short duration rainfall depth and return period as

$$\frac{P_T^t}{P_{10}^t} = 0.21 \ln T + 0.52, \quad 2 \le T \le 100 \text{ years} \tag{21.5}$$

in which P_T^t is the depth of a t-minute T-year return period rainfall, and P_{10}^t is the depth of a t-minute 10-year return period rainfall. He also expressed the ratio of P_T^t to P_T^{60} as a function n of t as

$$\frac{P_T^t}{P_T^{60}} = 0.54t^{0.25} - 0.50, \quad 5 \le t \le 120 \text{ min} \tag{21.6}$$

Equations (21.5) and (21.6) can be combined to yield

$$P_T^t = (0.21 \ln T + 0.52)(0.54t^{0.25} - 0.50)P_{10}^{60}, \quad 2 \le T \le 100 \text{ years}, \ 5 \le t \le 120 \text{ min} \tag{21.7}$$

When records are insufficient for determining 60 minutes, 10-year rainfall, the following relationships can be applied

$$P_2^{60} = 6.69 \times 10^{-3} MN^{0.33}, \ 0 < M \le 2.0, \ 1 < N \le 80 \tag{21.8}$$

$$P_2^{60} = 8.27 \times 10^{-3} M^{0.67} N^{0.33}, \ 2.0 < M \le 4.5, \ 1 < N \le 80 \tag{21.9}$$

where P_2^{60} is the 60-minute, 2-year return period rainfall depth in mm, M is the mean maximum annual daily rainfall in mm, and N is the mean annual number of thunderstorm days. For other durations and frequencies, the 60 minutes, 2-year rainfall can be employed to determine values as

$$P_T^t = (0.35 \ln T + 0.76)(0.54t^{0.25} - 0.50)P_2^{60}, \ 2 \le T \le 100 \text{ years}, \ 5 \le t \le 120 \text{ min} \tag{21.10}$$

Example 21.2

Determine 1-h and 6-h maximum rainfall depths and intensities for return periods of 2, 10, 20, 50, and 100 years for a project. The values of required coefficients are given as $k = 5.82$, $n = 0.160$, $b = 0.40$, and $m = 0.75$.

Solution

Here, $t = 1$ h and 6 h. The rainfall intensity for $t = 1$ h and $T = 2$ years is computed from Eq. (21.2) as

$$i_o = \frac{kT^n}{(b+t)^m} = \frac{5.82 \times 2^{0.160}}{(0.40+1)^{0.75}} = 5.05 \text{ cm/h}$$

The corresponding maximum rainfall depth is: $5.05 \times 1 = 5.05$ cm. Results for other cases are given in Table 21.5.

IDF curves can be drawn in the same way as the depth-duration-frequency (DDF) curves are drawn, but the difference is that the IDF curves use the data of maximum rainfall intensity and the DDF curves use the data of the maximum rainfall depth. Figure 21.1 shows typical DDF curves. Note that the maximum rainfall depth increases as the duration and return periods increase. If the data of maximum rainfall depth and its duration is available, the maximum rainfall intensity can be determined by dividing the maximum rainfall depth by the corresponding duration to yield the data that is required to plot IDF curves.

		Return Period (Years)				
Rainfall	**Duration (h)**	**2**	**10**	**20**	**50**	**100**
Intensity (cm/h)	1	5.05	6.54	7.30	8.46	9.45
	6	1.62	2.09	2.34	2.70	3.02
Depth (cm)	1	5.05	6.54	7.30	8.46	9.45
	6	9.72	12.54	14.04	16.20	18.12

TABLE 21.5 Computation of Rainfall Depths and Intensities for Various Return Periods

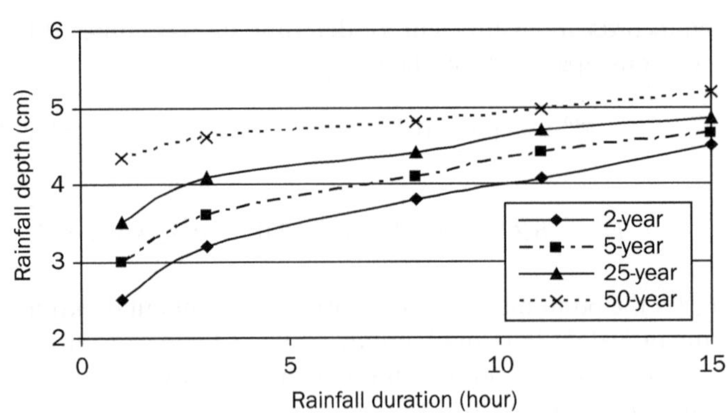

FIGURE 21.1 Typical variation of the maximum rainfall depth of various durations with return period.

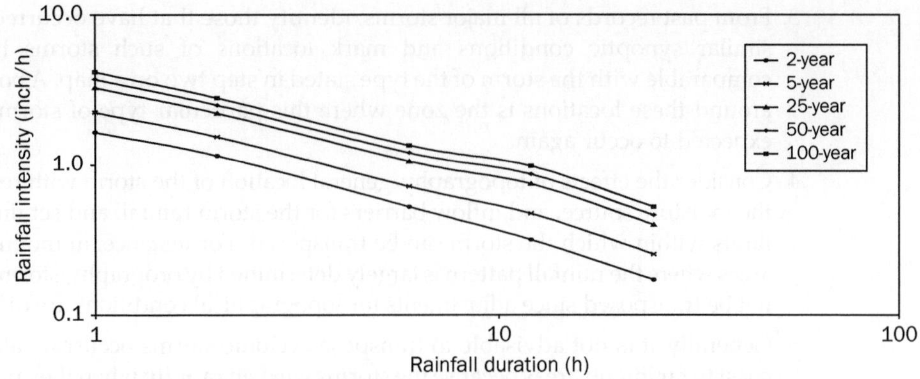

FIGURE 21.2 Rainfall intensity-duration-frequency plots for Baton Rouge Airport.

For the construction of IDF curves, the maximum rainfall intensities of various durations can be calculated from the data of maximum rainfall depth and corresponding duration. Frequency analysis is performed by using, for example, Weibull's plotting position formula. An appropriate distribution, such as the lognormal distribution, is fitted to these curves:

$$y = a \ln(x) + b \tag{21.11}$$

where y is the maximum rainfall intensity (mm/h), x is the return period (year), and a and b are the distribution parameters.

As an example, the IDF curves at Baton Rouge Airport are plotted in Fig. 21.2. The figure shows that as the rainfall duration increases, the rainfall intensity decreases and vice versa. Further, the intensity increases with increase in return period.

21.5 Storm Transposition

Storm transposition is the transfer of storm data from the place of its occurrence to another place (typically the catchment of the project under investigation) with the meteorological homogenous region. Storm transposition enhances the storm database for a project as the designer can additionally consider the storms that have occurred in the nearby meteorologically homogeneous region. A region may have areas that have not experienced severe storms of the magnitude observed in adjacent areas and hence transposition of the severe storms is performed to supplement the database.

Two aspects are involved in storm transposition. One is to define the meteorological homogeneous region in which the project catchment lies. The second concerns how much change in orientation of a storm is permitted when it is transposed to the catchment of the project under consideration. The aerial limits of transposition of a particular storm may be determined as follows:

1. Identify the area where the heaviest rain in the storm has fallen by inspecting an isohyetal map prepared from the daily rainfall data of the area.
2. Identify the causes of the storm rainfall in terms of synoptic situations associated with the particular storm by the study of weather maps.

3. From past records of all major storms, identify those that have occurred under similar synoptic conditions and mark locations of such storms that are comparable with the storm of the type stated in step two on a map. A boundary around these locations is the zone where this particular type of storm can be expected to occur again.
4. Consider the effects of topography, general location of the storm with respect to the moisture source, and inflow barriers for the storm rainfall and set final areal limits within which the storm can be transposed. For instance, in mountainous areas where the rainfall pattern is largely determined by orography, storm should not be transposed since adjustments for topographical conditions are difficult.
5. Generally, it is not advisable to transpose cyclonic storms occurring along the coast to far inland areas because the storms weaken rapidly when they penetrate inland areas.
6. Transposition should not be attempted, if it involves crossing a barrier or change in elevation more than 800 m as it may affect the storm characteristics.
7. In general, no storm should be transposed beyond 2° latitude or longitude from the place of its occurrence.

The orientation of the storm rainfall pattern in the basin is important in determining the maximum runoff. In attempting to get the maximum runoff effect, the orientation should not be adjusted by more than 20° as suggested in the *Manual for Estimation of PMP* (WMO, 1986).

The probability of occurrence of a storm over all the places of its transposition areas is not necessarily the same because the synoptic features associated with a particular storm are likely to occur with varying frequencies over different places. Each storm must, therefore, be considered individually to decide its transposition limits. The recommended practices for limits of transposition of storms are as follows:

a. Where the project catchment is intercepted by a dam, the transposed depths for upper and lower catchments shall be evaluated for two conditions of transposition when the storm is centered over the upper catchment and over the lower catchment. The severity of flood potential shall be examined after routing the flood through reservoirs and channels.
b. Where the catchment is elongated, the effect of moving storm should be accounted for, while transposing the isohyetal pattern of 1st day, 2nd day and 3rd day, and so on, instead of transposing 1-day, 2-day, and 3-day isohyetals.
c. Where the project catchment is big and much elongated, the probability of occurrence of another storm with almost similar weather conditions in the downstream part of the catchment and consequent synchronization of flood should be studied.

- As per WMO (1986), transposition to catchments larger than 50,000 km^2 may result in unreasonably high PMP estimates.
- The shape and orientation of the catchment vis-à-vis the shape and storm axis of an isohyetal pattern and its extent may make the transposed depths of various durations unrealistic. Hence it is essential to transpose a number of severe storms of varied intensities, durations, and isohyetal patterns for obtaining PMP for very large catchments.

Alexander (1963) discussed the probability of transposition which takes account of the area of the homogenous region (A_h), specific catchment area (A), mean depths (P_m), and duration of rainfall. The probability that the center of a storm would fall within a specific catchment can be given as $p_c = A/A_h$. This storm has a specific depth. If there is a record of N storms which are arranged in ascending order then the probability of a storm of rank r occurring is $p_r = r/N$. It may be noted that the depth of storm rainfall varies with the area, the ranking should be related to the area. Furthermore, observations show that the maximum observed rainfall is greater with denser rainfall network. The probability of occurrence of a transposable storm (p) over the given catchment in any specified year is $p = p_c \times p_r$. This product shows that transposition leads to greater floods than observed. For example, if $p_c = 1/25$ and $p_r = 1/40$, then $p = 1/1000$, indicating a probability level 25 times the record length.

For a given duration, the ranking of storms for decreasing mean depths can be computed for the size of the area covered, for example, 50, 100, 150, 200, ..., km². Plotting mean depths against the logarithm of rank, Alexander (1963) obtained a family of curves which were linear and tended to lower values of mean depth occurring approximately concurrently (e.g., at P_1, r_1). The rainfall depth for a specified area can be stated as

$$P_m - P_1 = a \log\left(\frac{r_1}{r}\right) \tag{21.12}$$

in which a is a function of area. Equation (21.12) can be recast as

$$P_m - P_1 = b \log\left(\frac{A_1}{A}\right) \tag{21.13}$$

since r is a function of A and noting that P_m decreases with decreasing log A, meaning P_m can be plotted against log A. In Eq. (21.13) b is a function of r.

Equation (21.13) shows that log rainfall intensity is related to log (duration + c). These three relationships can be combined to yield

$$P_m - P_1 = c \log\left(\frac{A_1}{A}\right) \log\left(\frac{r_1}{r}\right) \log\left(\frac{t_1}{t}\right) \tag{21.14}$$

in which c is a constant. Hershfield (1962) noted that the depth of heavy rainfall increases with the logarithm of return period T of rainfall of given depth and duration, which is a characteristic of several probability distributions.

21.6 Maximization of Selected Storms

The storm maximization is carried out to answer by how much the rainfall in a given storm could increase due to physically feasible rise in the meteorological factors that have generated the storm. The procedure described here is based on CWC (2001). The relevant meteorological factors are divided into two types: (1) the "mechanical efficiency" of the storm in changing atmospheric water vapor and droplets to rain, and (2) the moisture content of the air mass involved in the storm.

The CWC (2001) practice is to maximize the storm for moisture content only. Since the moisture adjustments are performed for the major storms in record, a reasonable assumption is that these historical storms were efficient mechanically.

The moisture adjustment for an observed storm is carried out to determine the rain that would fall when the moisture in that storm was the highest available to the storm in its transposed position over the project basin, assuming that the mechanical efficiency remains unchanged. The ratio between the adjusted storm rainfall to the observed storm rainfall is assumed to be the same as the ratio between the maximum moisture charge over the basin to the moisture charge of the observed storm. The moisture adjustment factor (MAF) that is applied to the observed rainfall from a particular storm is defined as

$$\text{MAF} = \frac{\text{Maximum precipitable water over the basin}}{\text{Storm precipitable water}} \tag{21.15}$$

To determine the MAF for a storm under consideration for transposition to the project catchment, first the amount of moisture available in the storm where it had occurred is determined. A representative dew point for the storm is computed by averaging the highest dew point data of stations that lie in the air flow path for the area of storm. Then, the depth of the storm precipitable water corresponding to the representative dew point is computed. This depth is adjusted for barriers to moisture flow, if any, between the sea (moisture source) and the storm area.

The next step is to compute the maximum moisture that may be available to the storm after it is transposed to the project catchment. Long-duration (30–50 years) data of surface dew points at all stations in the region are examined and all high values for each station are plotted against date. A smooth seasonal envelope is drawn. Monthly values at mid-month are read from, adjusted to 1000-hPa elevation, and plotted on monthly maps. The maximum dew point for the catchment is obtained for the month in which the storm has occurred and the MAF is determined.

21.7 Estimation of Probable Maximum Precipitation

Two approaches are mainly followed for estimation of PMP: physical and statistical. In the physical approach, data of observed major rainstorms in the project basin or nearby areas are analyzed. The severest storm patterns are transposed to the project catchment to compute the maximum rainfall depths for different durations. These rainfall depths are adjusted for storm maximization. The procedure for storm maximization and computation of MAF has been described in the previous section. The maximum DAD data for all storms are multiplied by the MAF. An envelope of adjusted depth-duration data of different storms is drawn to determine PMP. The PMP estimation using the physical method, as followed in India, is described in detail by IMD (1972) and CWC (2001).

In the statistical approach, PMP for a particular project is determined by the frequency analysis of the annual maximum rainfall series by using, e.g., Gumbel or the log-normal distribution method (this may vary with country). Hershfield (1962, 1965) developed a statistical method to estimate PMP for small areas by using the modified form of Chow's general frequency equation

$$X_e = \bar{X} + k_m \sigma \qquad (21.16)$$

$$k_m = (X_1 - \bar{X}_{n-1})/\sigma_{n-1} \qquad (21.17)$$

where X_e is the PMP rainfall for a given station for a specific duration; k_m is the frequency factor; and X_1, $\bar{X}_n$, and σ are, respectively, the highest rainfall, mean rainfall, and standard deviation for a series of n annual maximum rainfall values of a given duration. Further, $\bar{X}_{n-1}$ and σ_{n-1} are, respectively, the mean and standard deviation for this series after dropping the highest value from the series.

21.8 An Illustrative Example for Design Flood Estimation

To illustrate the methodology of design flood estimation, data of a project located in Central India is selected where the UH approach was used to determine the design flood. Note that procedure described here may not be applicable to other countries and thus the national guidelines need to be followed for design flood estimation.

To apply the UH approach, the design storm for the project area has to be developed. In India, the PMP values for the study area are provided by the India Meteorological Department (IMD). The time distribution of 24-hours, 48-hours, and 72-hours storm rainfall are also supplied by IMD along with PMP, and are given in Table 21.6.

Temporal distribution of PMP was also provided as shown in Table 21.7.

As per the IMD guidelines, the 1-day PMP value is increased by 15% to convert it into any 24-hour value.

Critical Sequencing of the PMP The critical sequencing of a PMP hyetograph is performed as per the recommended guidelines given by CWC (2001). The detailed step-by-step calculations of the sequencing are discussed here.
Step 1: Since the 1-day, 2-day, and 3-day PMP depths are 341 mm, 641 mm, and 824 mm respectively, the incremental rainfall depths on the 1st, 2nd, and 3rd day are 341 mm, 300 mm, and 183 mm, respectively (Table 21.8). These are arranged in such a way that the largest depth is preceded by the 2nd largest and succeeded by the 3rd largest. Thus the rainfall depths on the 1st, 2nd, and 3rd day are 300 mm, 341 mm, and 183 mm, respectively. Note that no clock hour corrections have been performed here.

Duration in Days	Duration in Hours	PMP Values (cm)
1-Day	24	34.1
2-Day	48	64.1
3-Day	72	82.4

TABLE 21.6 The PMP Values for the Existing Project

	Temporal Distribution of Percentage of Design Storm		
Duration (hours)	**24-h storm**	**48-h storm**	**72-h storm**
3	30	23	26
6	46	30	32
9	60	36	38
12	71	42	44
15	80	48	49
18	88	54	54
21	95	59	59
24	100	65	63
27		70	67
30		75	71
33		80	75
36		84	79
39		88	82
42		92	85
45		96	88
48		100	90
51			92
54			94
57			96
60			97
63			98
66			99
69			100
72			100

Table 21.7 Time Distribution for the PMP of the Project

Duration	PMP in mm	Increment Depths in mm	
1-Day	341.0	1st Day	**341.0**
2-Day	641.0	2nd Day	**300.0**
3-Day	824.0	3rd Day	**183.0**

Table 21.8 Incremental 1-, 2-, and 3-Day PMP Depths for the Project

Step 2: The rainfall depths are arranged in 2 bells per day. The storm depths for 12-hour blocks are determined by using the 12-hour to 24-hour ratio of 71% (as per IMD). The calculations are shown in Table 21.9.

Step 3: Using the distribution percentages given above, cumulative depths and incremental depths for the 12-hours blocks can be obtained as shown in the Table 21.10.

1st Day Depth = 300 mm		2nd Day Depth = 341 mm		3rd Day Depth = 183 mm	
1st 12 h	2nd 12 h	1st 12 h	2nd 12 h	1st 12 h	2nd 12 h
0.71	0.29	0.71	0.29	0.71	0.29
213 mm	87 mm	242.11 mm	98.89 mm	129.93 mm	53.07 mm

TABLE 21.9 Incremental 12-hour Storm Distribution Ratios for 1-, 2-, and 3-Day PMP Depths

			Cumulative Rainfall in Each Bell (mm)					
Time (hours)	Fractional distri[b] of 12 h PMP	12 h distr[b] conver-ted to %	1st 12 h of day 1 (bell-1)	2nd 12 h of day 1 (bell-2)	1st 12 h of day 2 (bell-3)	2nd 12 h of day 2 (bell-4)	1st 12 h of day 3 (bell-5)	2nd 12 h of day 3 (bell-6)
1	0.11	0.155	33.0	13.5	37.5	15.3	20.1	8.2
2	0.21	0.296	63.0	25.7	71.6	29.2	38.4	15.7
3	0.3	0.423	90.0	36.8	102.3	41.8	54.9	22.4
4	0.37	0.521	111.0	45.3	126.2	51.5	67.7	27.7
5	0.41	0.577	123.0	50.2	139.8	57.1	75.0	30.6
6	0.46	0.648	138.0	56.4	156.9	64.1	84.2	34.4
7	0.51	0.718	153.0	62.5	173.9	71.0	93.3	38.1
8	0.55	0.775	165.0	67.4	187.6	76.6	100.7	41.1
9	0.6	0.845	180.0	73.5	204.6	83.6	109.8	44.8
10	0.64	0.901	192.0	78.4	218.2	89.1	117.1	47.8
11	0.67	0.944	201.0	82.1	228.5	93.3	122.6	50.1
12	0.71	1.000	213.0	87.0	242.1	98.9	129.9	53.1

	Incremental Rainfall Depth in Each Bell (mm)					
Time (hours)	1st 12 h of day 1	2nd 12 h of day 1	3rd 12 h of day 2	4th 12 h of day 2	5th 12 h of day 3	6th 12 h of day 3
1	33.0	13.5	37.5	15.3	20.1	8.2
2	30.0	12.3	34.1	13.9	18.3	7.5
3	27.0	11.0	30.7	12.5	16.5	6.7
4	21.0	8.6	23.9	9.7	12.8	5.2
5	12.0	4.9	13.6	5.6	7.3	3.0
6	15.0	6.1	17.1	7.0	9.2	3.7
7	15.0	6.1	17.1	7.0	9.2	3.7
8	12.0	4.9	13.6	5.6	7.3	3.0
9	15.0	6.1	17.1	7.0	9.1	3.7
10	12.0	4.9	13.6	5.6	7.3	3.0
11	9.0	3.7	10.2	4.2	5.5	2.2
12	12.0	4.9	13.6	5.6	7.3	3.0

TABLE 21.10 Cumulative and Incremental Depths for 12-hour Blocks for a Three-Day Period

Time (hours)	UH ordinate	Critical Sequence of Rainfall Increments in Each Bell (mm)					
		1st bell	2nd bell	3rd bell	4th bell	5th bell	6th bell
1	159.2	12.0	4.9	13.6	5.6	7.3	3.0
2	163.0	12.0	4.9	13.6	5.6	7.3	3.0
3	165.4	15.0	6.1	17.1	7.0	9.2	3.7
4	167.2	15.0	6.1	17.1	7.0	9.2	3.7
5	168.1	27.0	11.0	30.7	12.5	16.5	6.7
6	**168.3**	**33.0**	13.5	37.5	15.3	20.1	8.2
7	163.2	30.0	12.3	34.1	13.9	18.3	7.5
8	158.3	21.0	8.6	23.9	9.7	12.8	5.2
9	134.6	15.0	6.1	17.1	7.0	9.1	3.7
10	115.9	12.0	4.9	13.6	5.6	7.3	3.0
11	102.9	12.0	4.9	13.6	5.6	7.3	3.0
12	92.0	9.0	3.7	10.2	4.2	5.5	2.2
Sum		213.0	87.0	242.1	98.9	129.9	53.1

TABLE 21.11 Critical Sequence of Rainfall at Unit Duration

Step 4: For this purpose, the largest incremental depth is placed against the peak of UH, then the next largest incremental depth against the next largest UH ordinate, and so on until all incremental rainfall depths are arranged. Next, the sequence is reversed to get the critical sequencing as shown in Tables 21.11 and 21.12, respectively.

Step 5: Now, the first 12-hour block is interchanged with the next 12-hour block to create the most critical situation as shown in Table 21.13.

It is to be noted here that the maximum depth for any duration within this sequence should not exceed the design storm depth for that duration. Specifically, the maximum depth in any 24-hour sequence shall be checked to ensure that this is not more than the design storm depth for 24 hours. It means that the sum of rainfall depths for any successive 24-, 48-, and 72- hour blocks should not be greater than the 24-, 48-, and 72- hour PMP values, respectively. This is checked in Tables 21.13 and 21.14. As can be seen from the results the calculated values are well within the limits. The same procedure is also repeated for other spells.

The rainfall loss rate for this project is 3.1 mm/h. This was used to calculate the excess rainfall hyetograph. The excess rainfall hyetograph values are given in Table 21.15.

Now this excess rainfall PMP hyetograph can be convoluted with the UH for the catchment to compute a surface runoff hydrograph. A design flood hydrograph is obtained by adding baseflow to the surface runoff.

Time (hours)	Critical Sequence of Rainfall Increments after Changing of 1st and 2nd Spells					
	1st bell	2nd bell	3rd bell	4th bell	5th bell	6th bell
1	9.0	3.7	10.2	4.2	5.5	2.2
2	12.0	4.9	13.6	5.6	7.3	3.0
3	12.0	4.9	13.6	5.6	7.3	3.0
4	15.0	6.1	17.1	7.0	9.1	3.7
5	21.0	8.6	23.9	9.7	12.8	5.2
6	30.0	12.3	34.1	13.9	18.3	7.5
7	33.0	13.5	37.5	15.3	20.1	8.2
8	27.0	11.0	30.7	12.5	16.5	6.7
9	15.0	6.1	17.1	7.0	9.2	3.7
10	15.0	6.1	17.1	7.0	9.2	3.7
11	12.0	4.9	13.6	5.6	7.3	3.0
12	12.0	4.9	13.6	5.6	7.3	3.0
Sum	213.0	87.0	242.1	99.0	129.9	53.1

Table 21.12 Reverse Critical Sequence

Interchanging of 1st and 2nd 12-h Block (Bells)						
Time	Critical sequence of rainfall increments after interchanging of 1st & 2nd spells					
	1st bell	2nd bell	3rd bell	4th bell	5th bell	6th bell
1	3.7	9.0	4.2	10.2	5.5	2.2
2	4.9	12.0	5.6	13.6	7.3	3.0
3	4.9	12.0	5.6	13.6	7.3	3.0
4	6.1	15.0	7.0	17.1	9.1	3.7
5	8.6	21.0	9.7	23.9	12.8	5.2
6	12.3	30.0	13.9	34.1	18.3	7.5
7	13.5	33.0	15.3	37.5	20.1	8.2
8	11.0	27.0	12.5	30.7	16.5	6.7
9	6.1	15.0	7.0	17.1	9.2	3.7
10	6.1	15.0	7.0	17.1	9.2	3.7
11	4.9	12.0	5.6	13.6	7.3	3.0
12	4.9	12.0	5.6	13.6	7.3	3.0
Sum	87.00	213.00	98.89	242.11	129.93	53.07

Table 21.13 Check for 24-Hour Blocks

Sum of 12-h Blocks	1st and 2nd Bell	2nd and 3rd Bell	3rd and 4th Bell	4th and 5th Bell	5th and 6th Bell	6th and 1st Bell
24-h PMP	300	311.89	341	**372.04**	183	140.07
Sum of 12-h blocks	**1 to 4**	**2 to 5**	**3 to 6**	**4 to 1**	**5 to 2**	**6 to 3**
48-h PMP	641	**683.93**	524	512.11	483	451.96
72-h PMP				**824.0**		

Table 21.14 Check for Successive 24-, 48-, and 72-Hour PMP Blocks of Table 21.10. The bold values are the highest values in the row.

Time (hours)	Excess Rainfall Depth (mm)	Time (hours)	Excess Rainfall Depth (mm)
1	0.6	25	7.1
2	1.8	26	10.5
3	1.8	27	10.5
4	3.0	28	14.0
5	5.5	29	20.8
6	9.2	30	31.0
7	10.4	31	34.4
8	7.9	32	27.6
9	3.0	33	14.0
10	3.0	34	14.0
11	1.8	35	10.5
12	1.8	36	10.5
13	5.9	37	1.1
14	8.9	38	2.5
15	8.9	39	2.5
16	11.9	40	3.9
17	17.9	41	6.6
18	26.9	42	10.8
19	29.9	43	12.2
20	23.9	44	9.4
21	11.9	45	3.9
22	11.9	46	3.9
23	8.9	47	2.5
24	8.9	48	2.5

Table 21.15 48-Hour PMP Excess Rainfall Hyetograph

21.9 Summary

This chapter has discussed design storm analysis for computation of design flood. It is noted here that the exact procedure to be followed for design storm analysis and design flood computations will vary from country to country and applicable national standards should be followed.

21.10 Questions

21.1 What is design storm? What factors govern the duration of design storm?

21.2 Determine the 2-d maximum rainfall depth for a catchment whose area is 1000 km^2. Assume $a = 4.3 \times 10^3$, $b = 2.8 \times 10^2$, and $m = 0.8$.

21.3 Determine the 1-d maximum rainfall depth for a catchment area of 500 km^2. Assume $a = 4.1 \times 10^3$, $b = 2.7 \times 10^2$, and $m = 0.8$.

21.4 Determine 2-h maximum rainfall depths and corresponding intensities for return periods of 2, 10, 50, and 100 years that are likely to occur at a project. Assume $k = 5.8$, $n = 0.160$, $b = 0.45$, and $m = 0.75$.

21.5 Determine the maximum 2-h and 3-h rainfall for DAD analysis for the records in Table 21.16:

Time	Hourly Rainfall (cm)
8:00 AM	0
9:00	0.5
10:00	0.4
11:00	0.6
12:00	0.0
1:00 PM	0.2
2:00	0.6
3:00	0.1
4:00	0.8
5:00	0.4
6:00	0.1
7:00	0.7
8:00	0.0

TABLE 21.16 Rainfall Data for Question 21.5

21.6 Construct a depth-area duration curve for a storm that occurred from 06:00 on June 3 to about 05:00 on June 6, 1981, in the Middle Brazos-Bosque watershed (HUC 120602) (1894 km^2 in area) as shown in Fig. 21.3. The watershed is located southwest of Dallas between Waco and Stephenville. Its climate is such that for Waco, the annual mean precipitation = 924 mm and average temperature is 19.2°C and for Stephenville, the annual mean precipitation = 802 mm and the average temperature is 17.6°C. Select time intervals as 2, 4, 6, 8, 10, 12, and 24 h. The data is given in Table 21.17.

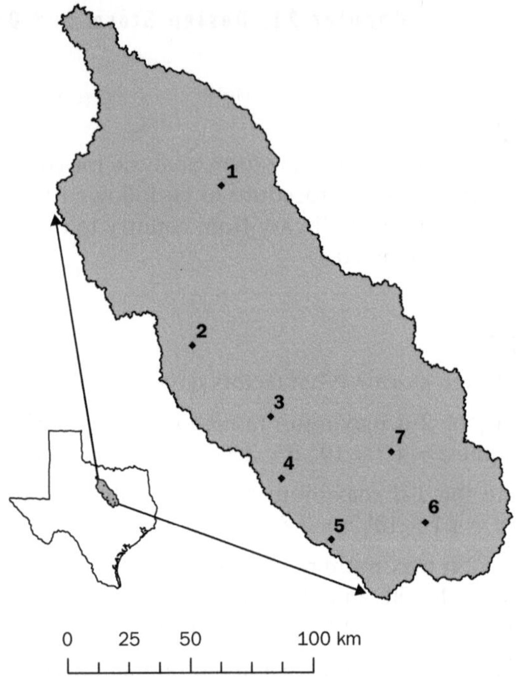

No	Station ID	Name
1	**415957**	MINERAL WELLS 1 SSW
2	**418623**	STEPHENVILLE 1 N
3	**414476**	IREDELL
4	**412086**	CRANFILLS GAP
5	**412024**	CORYELL CITY
6	**419419**	WACO REGIONAL AIRPORT
7	**419715**	WHITNEY DAM

FIGURE 21.3 Middle Brazos-Bosque watershed (HUC 120602).

Station ID				**415957**	**418623**	**414476**	**412086**	**412024**	**419419**	**419715**
Y	**M**	**D**	**T**	**1**	**2**	**3**	**4**	**5**	**6**	**7**
1981	6	3	6	0	0	0	0	0	0	0
1981	6	3	7	0	0	0	0	0	0	0
1981	6	3	8	0	0	0	0	0	0	0
1981	6	3	9	0	0	0	0	0	0	0
1981	6	3	10	10.16	10.16	10.16	10.16	10.16	0	10.16
1981	6	3	11	2.54	2.54	2.54	2.54	2.54	1.02	2.54
1981	6	3	12	7.62	7.62	7.62	7.62	7.62	0.76	7.62
1981	6	3	13	0	0	0	0	0	16.76	0
1981	6	3	14	5.08	5.08	5.08	5.08	5.08	2.03	5.08
1981	6	3	15	35.56	35.56	35.56	35.56	35.56	27.43	35.56
1981	6	3	16	7.62	7.62	7.62	7.62	7.62	11.18	7.62

TABLE 21.17 Precipitation Data for the Storm of June 3–5, 1981 (Unit = mm)

Station ID				415957	418623	414476	412086	412024	419419	419715
Y	**M**	**D**	**T**	**1**	**2**	**3**	**4**	**5**	**6**	**7**
1981	6	3	17	0	0	0	0	0	0	0
1981	6	3	18	0	0	0	0	0	0	0
1981	6	3	19	0	0	0	0	0	0	0
1981	6	3	20	0	0	0	0	0	0	0
1981	6	3	21	0	0	0	0	0	0	0
1981	6	3	22	0	0	0	0	0	0	0
1981	6	3	23	0	0	0	0	0	0	0
1981	6	4	0	0	0	0	0	0	0	0
1981	6	4	1	0	0	0	0	0	0	0
1981	6	4	2	0	0	0	0	0	0	0
1981	6	4	3	0	0	0	0	0	1.52	0
1981	6	4	4	5.08	5.08	5.08	5.08	5.08	1.02	5.08
1981	6	4	5	2.54	2.54	2.54	2.54	2.54	0	2.54
1981	6	4	6	5.08	5.08	5.08	5.08	5.08	0.51	5.08
1981	6	4	7	5.08	5.08	5.08	5.08	5.08	4.83	5.08
1981	6	4	8	0	0	0	0	0	0	0
1981	6	4	9	0	0	0	0	0	0.25	0
1981	6	4	10	0	0	0	0	0	1.27	0
1981	6	4	11	0	0	0	0	0	4.06	0
1981	6	4	12	2.54	2.54	2.54	2.54	2.54	2.54	2.54
1981	6	4	13	5.08	5.08	5.08	5.08	5.08	5.59	5.08
1981	6	4	14	5.08	5.08	5.08	5.08	5.08	1.52	5.08
1981	6	4	15	2.54	2.54	2.54	2.54	2.54	3.3	2.54
1981	6	4	16	7.62	7.62	7.62	7.62	7.62	4.32	7.62
1981	6	4	17	0	0	0	0	0	1.02	0
1981	6	4	18	0	0	0	0	0	0	0
1981	6	4	19	0	0	0	0	0	0	0
1981	6	4	20	0	0	0	0	0	0	0
1981	6	4	21	0	0	0	0	0	0	0
1981	6	4	22	0	0	0	0	0	0	0
1981	6	4	23	0	0	0	0	0	0	0
1981	6	5	0	0	0	0	0	0	1.02	0
1981	6	5	1	0	0	0	0	0	0.51	0
1981	6	5	2	2.54	2.54	2.54	2.54	2.54	0.76	2.54
1981	6	5	3	2.54	2.54	2.54	2.54	2.54	0	2.54
1981	6	5	4	0	0	0	0	0	1.27	0
1981	6	5	5	0	0	0	0	0	0	0

Table 21.17 Precipitation Data for the Storm of June 3–5, 1981 (Unit = mm) (*Continued*)

21.7 Hourly rainfall values for a storm that occurred from 00:00 on August 1 to about 23:00 on August 2, 1989 are given in Table 21.18 for several stations on Lower Brazos River watershed (HUC 120701) (2047 km^2) between College Station and Brenham, as shown in Fig. 21.4. The climate of the watershed is such that for College Station, the annual mean precipitation = 1018 mm and average temperature is 20.5°C, and for Brenham, the annual mean precipitation = 1145 mm and average temperature is 19.8°C. Determine the maximum depth area curves for 2-, 6-, 12- , and 24-hour durations.

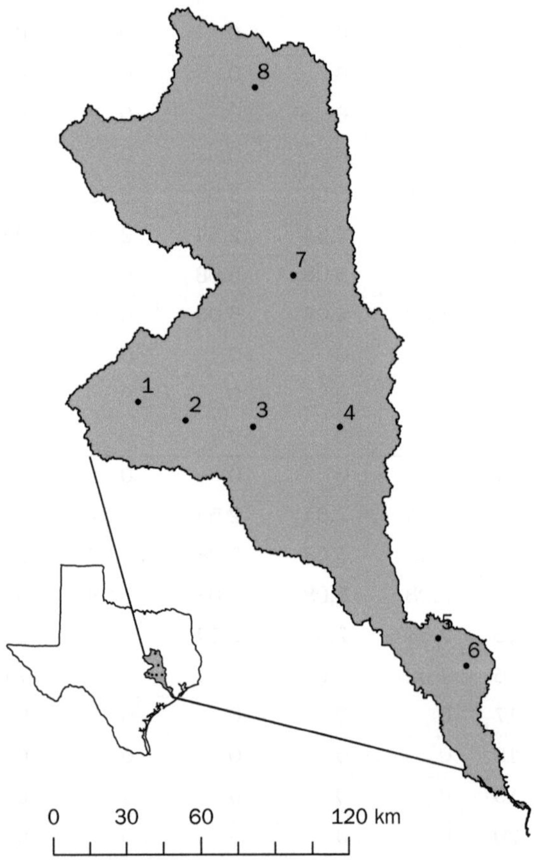

No	Station ID	Name
1	**415193**	LEXINGTON
2	**412462**	DIMEBOX
3	**418446**	SOMERVILLE DAM
4	**419491**	WASHINGTON STATE PARK
5	**417594**	RICHMOND
6	**418996**	THOMSONS 3
7	**419665**	WHEELOCK
8	**413771**	GROESBECK 2

FIGURE 21.4 Lower Brazos River watershed (HUC 120701).

Station ID				415193	412462	418446	419491	417594	418996	419665	413771
Y	M	D	T	1	2	3	4	5	6	7	8
1989	8	1	0	0	0	0	0	0	0	0	0
1989	8	1	1	0	0	0	0	0	0	0	0
1989	8	1	2	0	0	0	0	0	0	0	0
1989	8	1	3	0	0	0	0	0	0	0	0
1989	8	1	4	0	0	0	0	0	0	0	0
1989	8	1	5	0	0	0	0	0	0.76	0	0
1989	8	1	6	0	0	0	0	0.76	1.52	0	0
1989	8	1	7	0	0	0	0	2.03	6.1	0	0
1989	8	1	8	0	0	0	0	2.79	4.83	0	0
1989	8	1	9	0	0	0	0	7.87	9.14	0	0
1989	8	1	10	0	0	0	7.62	23.37	27.94	0	0
1989	8	1	11	0	0	5.08	5.08	23.62	22.61	0	0
1989	8	1	12	0	0	2.54	7.62	16	19.3	2.54	0
1989	8	1	13	2.54	0	7.62	5.08	37.59	46.23	0	0
1989	8	1	14	2.54	1.27	12.7	2.54	19.81	45.47	0	0
1989	8	1	15	2.54	3.56	17.78	5.08	13.21	40.89	2.54	0
1989	8	1	16	7.62	6.1	17.78	2.54	27.94	3.3	0	0
1989	8	1	17	2.54	11.68	5.08	2.54	1.02	1.78	7.62	0
1989	8	1	18	0	11.43	5.08	0	1.52	0	33.02	0
1989	8	1	19	2.54	3.05	0	0	0	0.51	30.48	2.54
1989	8	1	20	5.08	0.25	0	2.54	0	0	5.08	2.54
1989	8	1	21	10.16	1.02	2.54	0	0	0	5.08	0
1989	8	1	22	0	1.27	0	0	0	1.02	2.54	2.54
1989	8	1	23	0	2.79	0	0	0	0.51	0	2.54
1989	8	2	0	0	6.1	0	0	1.02	0.25	2.54	0
1989	8	2	1	2.54	0.76	0	0	0	0	0	0
1989	8	2	2	2.54	1.52	0	0	0	0	0	0
1989	8	2	3	0	1.02	0	0	0	0	2.54	0
1989	8	2	4	0	0	0	0	0	0	7.62	2.54
1989	8	2	5	0	0.51	0	0	0	0	0	7.62
1989	8	2	6	0	0.51	0	0	0	0	2.54	5.08
1989	8	2	7	0	0	0	0	0	0	0	0
1989	8	2	8	0	0	0	0	0	0	0	2.54
1989	8	2	9	0	0	0	0	0	0	0	0
1989	8	2	10	0	0	0	0	0	0	2.54	0

TABLE 21.18 Precipitation Data for the Storm of August 1–2, 1989 (Unit = mm)

Station ID				415193	412462	418446	419491	417594	418996	419665	413771
Y	M	D	T	1	2	3	4	5	6	7	8
1989	8	2	11	0	0	0	0	0	0	0	0
1989	8	2	12	0	0	0	0	0	0	0	0
1989	8	2	13	0	0	0	0	0	0	0	0
1989	8	2	14	0	0	5.08	0	0	0	0	0
1989	8	2	15	0	0	2.54	0	0	0	0	2.54
1989	8	2	16	0	0	0	0	0	0	0	0
1989	8	2	17	0	0	0	0	0	0	2.54	0
1989	8	2	18	0	0	0	0	0	0	0	0
1989	8	2	19	0	0	0	0	0	0	0	0
1989	8	2	20	0	0	0	0	0	0	0	0
1989	8	2	21	0	0	0	0	0	0	0	0
1989	8	2	22	0	0	0	0	0	0	0	0
1989	8	2	23	0	0	0	0	0	0	0	0

TABLE 21.18 Precipitation Data for the Storm of August 1–2, 1989 (Unit = mm) *(Continued)*

21.8 Construct rainfall-frequency curves for Austin, Dallas, Houston, and San Antonio, Texas. Data can be obtained from the National Weather Service website.

21.11 References

Abbi, S. D. S. *Hydrometeorological Analysis*. Manual of Hydrometeorology, Part I, pp. 45–69. India Meteorology Department, New Delhi, 1972.

Alexander, G. N. Using the probability of storm transposition for estimating the frequency of rare floods. *Journal of Hydrology*, 1:46–57, 1963.

Bell, F. C. Generalized rainfall duration frequency relationships. *Journal of Hydraulics Division, ASCE*, 95(1):311–327, 1969.

CWC. *Manual for Estimation of Design Flood.* New Delhi: Central Water Commission, 2001.

Hershfield, D. M. Extreme rainfall relationships. *Proceedings of Hydraulics Division, ASCE*, 88(HY6):73–92, 1962.

Hershfield, D. M. Method for estimating probable maximum rainfall. *Journal of the American Waterworks Association*, 57:965–972, 1965.

Hershfield, D. M., and W. T. Wilson. A comparison of extreme rainfall depths from tropical and nontropical storms. *Journal of Geophysical Research*, 65:969–982, 1960.

IMD. *Manual of Hydrometeorology, Part I.* New Delhi: India Meteorological Department, 1972.

Leclerc, G., and J. C. Schaake. *Derivation of Hydrologic Frequency Curves.* Massachusetts Institute of Technology Technical Report, 142, 1972.

WMO. Manual for Depth–Area–Duration Analysis of Storm Precipitation. WMO No. 237. Geneva: World Meteorological Organization, 1969.

WMO. Manual for Estimation of Probable Maximum Precipitation. WMO No. 332. Geneva: World Meteorological Organization, 1986.

CHAPTER 22

Climate Change and Its Impact on Water Resources

Global warming and climate change have been receiving much attention in recent years. It is now widely accepted that climate change is real and is not just a natural aberration. This chapter therefore discusses rudimentary aspects of what is causing the change, how the change can be simulated, and more importantly what the impacts on water resources of the change are. Also discussed is what can be done to mitigate the impacts of climate change.

22.1 Introduction

The climate of a place represents the long-term (more than 30 years) average of weather. Thus it may be defined as "average weather" or the statistical description (mainly mean and variability) of relevant weather variables over an extended time period. According to the Intergovernmental Panel on Climate Change (IPCC), climate change refers to a change in the state of the climate that can be identified (e.g., by using statistical tests) by changes in the mean and/or the variability of its properties, and that persists for an extended period, typically decades or longer. Earth's climate has been changing in the past also. But the concerns about climate change at present are due to the high rate and extent of current and predicted changes that are adversely impacting the society. Climate change is a major global environmental problem and an issue of great social concern.

22.2 Causes of Climate Change

Climate change may occur due to forcings that are external to Earth or due to internal forcings. Regarding the external forcings, Earth orbits around the Sun and rotates on its axis; both the revolution and rotation rates are changing. The natural cycles in Earth's movement are collectively known as the "Milankovitch cycles," after the Serbian astronomer who calculated their magnitude. These cycles change very slowly and cause changes in climate over long-term. Three Milankovitch cycles influence the seasonality and distribution of solar energy on Earth. First, the Earth's axis is tilted with respect to the plane of orbit and this angle changes between 22.1° and 24.5° in about 41,000 years. This tilt changes the solar radiation received on Earth in space and time, although the total radiation received remains the same. Second, the orientation of the Earth's axis of rotation relative to fixed stars is gradually shifting. This is called axial precession, which takes about 26,000 years for one cycle. When this axis of rotation points toward the Sun during perihelion, one hemisphere will observe a greater difference between seasons but the other will have lesser difference. Third, eccentricity is the shape of the Earth's orbit around the Sun, which ranges between more and less elliptical (0% to 5% ellipticity) in about 100,000 years. Thus the distance between Earth and Sun changes, resulting in variation in the amount of radiation received by Earth in different seasons.

The forces of climate change that operate within the climate system are called internal forcing mechanisms. Huge amounts of gases, aerosols, and ash enter in the atmosphere during volcanic eruptions. These impact climate, as the solar radiation reaching Earth is reduced. Combustion of fossil fuels to generate electricity, heating, and transport account for about 70% of total emissions of greenhouse gases (GHGs), which are the main cause of global warming these days. In addition, the movement of tectonic plates causes uplift of land mass and formation of mountains, and impacts atmospheric circulation and hydrologic cycle. This also has some impact on climate.

22.2.1 Greenhouse Effect

Shortwave radiation from the Sun reaches the Earth's surface after passing through the atmosphere. Gases present in the atmosphere reflect a part of this radiation back into space, some part is absorbed by the atmosphere, and the remainder reaches the Earth's surface which either absorbs it or reflects. As depicted in Fig. 22.1, the Earth's surface emits longwave radiation upwards. Some gases present in the atmosphere absorb a part of this longwave radiation and some part of this longwave radiation is reradiated back to Earth, causing warming. Thus these gases change the heat balance of Earth by retaining the longwave radiation that would have otherwise gone to space. This effect is called the greenhouse effect. The gases that cause the greenhouse effect are called GHGs. The principal GHGs present in the atmosphere include carbon dioxide (CO_2), nitrous oxide (N_2O), methane (CH_4), water vapor, chlorofluorocarbons (CFCs), and ozone (O_3).

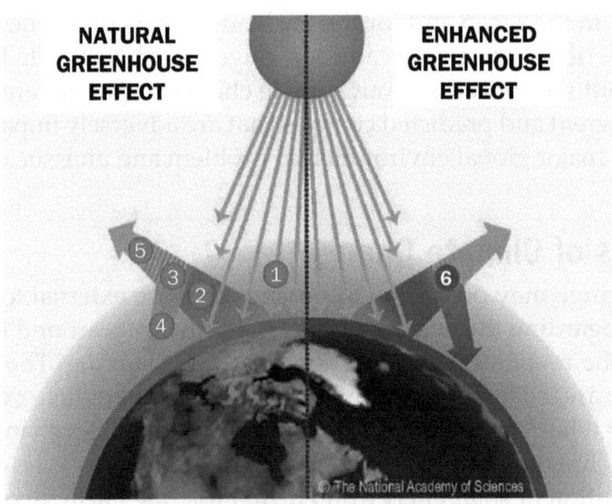

Figure 22.1 Illustration of the greenhouse effect. Some of the sunlight striking Earth is (1) absorbed and converted to infrared radiation (heat). The surface (2) emits infrared radiation to the atmosphere, where some of it (3) is absorbed by GHGs and (4) reemitted toward the surface; some of the infrared radiation is not trapped by GHGs and (5) escapes into space. Human activities that emit additional GHGs to the atmosphere (6) increase the amount of infrared radiation that gets absorbed before escaping to space, thus enhancing the greenhouse effect and amplifying the warming of Earth. (Source: National Academy of Sciences, 2012.)

Obviously, GHGs have an important role in controlling the temperature of Earth and keeping it sufficiently warm so that life can survive. But excess of GHGs is harmful. As the concentration of GHGs rises, it would lead to more warming, which is likely to have a major impact on the Earth's climate. Figure 22.2 shows the progressive rise in concentration of CO_2 in the atmosphere in past 70 odd years. Global atmospheric concentrations of CO_2, methane, and nitrous oxide have increased from 280 ppm to 406 ppm, from 722 ppb to 1855 ppb, and from 270 ppb to 330 ppb, respectively, between the preindustrial period (1750) and 2017. Water vapor is the most abundant GHG in the atmosphere. In addition, the excess presence of CFCs damages the protective ozone layer, which prevents the harmful shortwave rays from reaching Earth.

Global warming caused by the increased atmospheric concentration of GHGs due to anthropogenic activities has emerged as one of the most crucial environmental issues faced by humanity. Concerns over global climatic changes triggered by growing atmospheric concentrations of GHGs have increased in recent years. This concern arises because our daily activities are leading to changes in the composition of Earth's atmosphere that can significantly alter the planet's heat and radiation balance. Scientific understanding of climate and its likely response to increased concentrations of GHGs have greatly advanced in recent times. However, the global climate system is so vast and complex that it is difficult to understand it fully and accurately. Consequently, outcomes of analyses contain uncertainties.

22.2.2 Evidence of Climate Change

The Fifth Assessment Report of the Intergovernmental Panel on Climate Change (IPCC, 2014b) has produced many evidences which clearly show that global warming

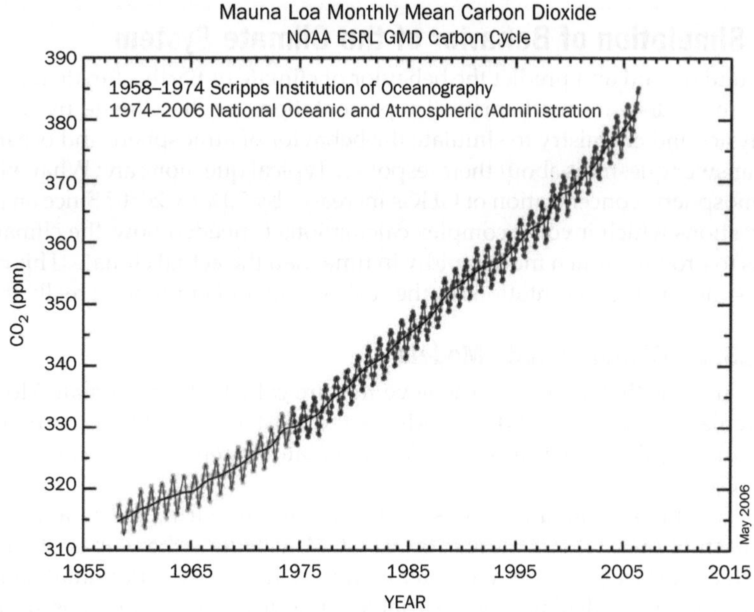

FIGURE 22.2 Changes in concentration of CO_2 in the atmosphere in recent times. (Source: NOAA, 2018.)

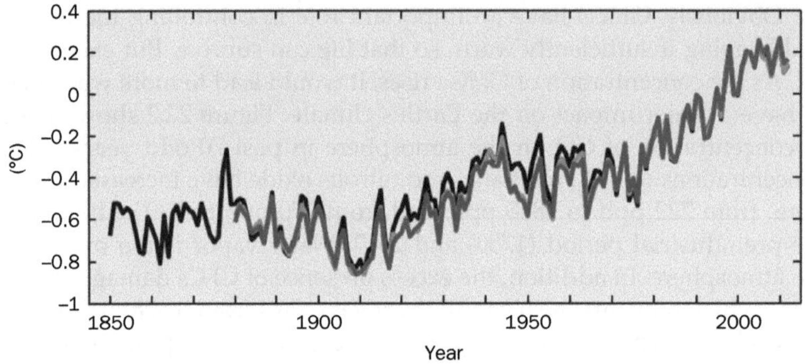

FIGURE 22.3 Globally averaged combined land and ocean surface temperature anomaly. (Source: IPCC, 2014b.)

is indeed happening. Measured temperature data dating back to 1850 are available at many places on Earth. Record-high average global surface temperatures have been observed in recent times. Each of the last three decades has been successively warmer at the Earth's surface than any preceding decade since 1850. IPCC (2014b) notes that the period from 1983 to 2012 was likely the warmest 30-year period of the last 1400 years in the Northern Hemisphere. Figure. 22.3 depicts that the globally averaged combined land and ocean surface temperature data as calculated by a linear trend show a warming of 0.85°C (0.65–1.06°C) over the period 1880–2012, when multiple independently produced data sets are used.

22.3 Simulation of Behavior of the Climate System

To understand and predict the behavior of climate of Earth, climate models are needed. These models are mathematical representations of the climate by using the laws of physics and chemistry to simulate the behavior of atmosphere and oceans with the aim to answer questions about their response. Typical questions are: What will happen if the atmospheric concentration of GHGs increases by 50% by 2050? Since an answer to these questions which involve complex calculations is needed now, the climate in the model has to progress much more rapidly in time than the actual climate. This can be achieved by simplified representations of the real system and employing really fast computers.

22.3.1 Global Climate Models

The models that are used to achieve this are called Global Climate Models or General Circulation Models (GCMs). GCMs are the most advanced tools currently available for simulating the response of the global climate system to increasing concentrations of GHGs.

To simulate climatic processes, GCMs divide Earth, ocean, and atmosphere into a mesh of (nearly) rectangular grids. GCMs calculate the values of climatic variables such as surface pressure, wind, temperature, humidity, and rainfall at each grid point over time. To calculate these variables, the governing equations need to be solved at each time step at the grid points. This involves an enormous number of calculations. Hence very powerful computers are necessary to run GCMs. In order to limit the computational time to reasonable limits, climate models are set up with large grid spacing and longer time steps. Actually, among other things, the length of a time step depends on the grid size. In fact, the computational time for one simulation run of 50 years, for example, may be several months. Due to immense requirements of human expertise, computer resources, and other infrastructure, there are limited centers in the world where GCMs are run.

22.3.2 Coupled Model Systems

Earlier models of climate system were able to consider the properties of the atmosphere in three dimensions and had simplified representations of land surface processes. These models were called atmospheric general circulation models. At the same time, oceanographers were busy with the simulation of ocean behavior by using their properties in three dimensions and developed ocean general circulation models. It is known that there is a considerable amount of interaction between atmosphere, land, and oceans and exchange of mass, energy, and momentum between them. Hence to simulate the behavior of a climate system over long periods of time, it is essential to use models that integrate the behavior of climatic, land surface, and the oceanic processes. Coupling mechanisms in such models consider exchanges of heat, moisture, and momentum among the three subsystems. These models are known as coupled GCMs. There are numerous challenges in coupling atmospheric, land, and oceanic processes. For example, oceans have large thermal capacity compared to atmosphere and therefore they can supply large amounts of heat to the atmosphere or extract heat from it. Oceans also absorb GHGs. To correctly model these exchanges and storages, a 3-D model is necessary, which can also simulate oceanic currents which transport mass and energy.

The climate of Earth has many short-term events. For example, the El-Nino Southern Oscillation (ENSO) represents shorter-term events that have a significant impact on global climate. Likewise, thunderstorms and cyclones are also such events that may not be important on global scales but are important at local and regional scales. Further, it is necessary to couple other components of climate system, especially the cryosphere and the biosphere, to realistically simulate climate on decadal and longer timescales. The models must also be able to correctly simulate such events.

GCMs depict the climate using a three-dimensional grid over the globe, a typical grid size in horizontal is between 200 and 400 km. About 10–20 vertical layers are used to model the atmosphere and about 30 layers to model the oceans (Fig. 22.4). Although this resolution may appear to be acceptable, considering the total number of computational nodes it generates, these resolutions are coarse relative to the scale at which data are required in most impact assessments.

Large computer resources and time are required to solve the governing equations. Usually, GCM model runs are taken for times spanning several decades and the time step size could be several minutes. A number of organizations have developed GCMs. These differ in terms of model structure, solution scheme, details of climate representation, data requirement, and grid size adopted. The components considered in the models also vary, for example, modeling of clouds, aerosols, and so on. Results of these simulations are available as time series of climatic variables. For example, hydrologists may be interested in the time series of temperature at a location for the period 2025–2075.

GCMs are very complex models and simpler models can provide globally or regionally averaged estimates of the climate response. But only GCMs can provide geographically and physically consistent estimates of regional climate which are

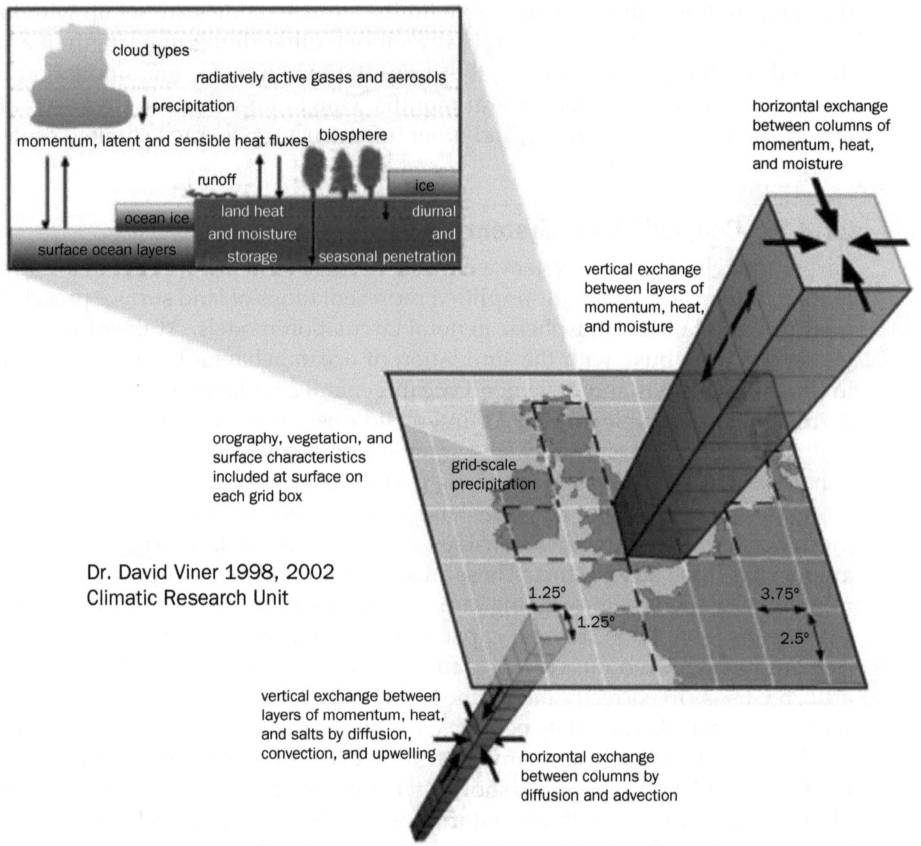

FIGURE 22.4 Discretization scheme used in a GCM. (Source: IPCC, 2013.)

required in impact evaluation and analysis (http://www.ipcc-data.org/guidelines/pages/gcm_guide.html). Many atmospheric processes, such as cloudbursts and thunderstorms, occur at scales that are much smaller than the grid size of GCMs. These processes are important from a hydrological point of view and should not be ignored. However, it is not possible to adopt a small grid size and a time step size to model these processes, since the computational and data storage requirements will drastically increase. An alternative is to represent the combined impact of all small-scale processes occurring in a grid cell by simplified equations whose parameters are adjusted accordingly. This approach is called parameterization. This a source of uncertainty in simulations of future climate by use of GCM. Other sources are related with the simulation of various feedback mechanisms in GCMs, such as water vapors, aerosols, clouds, ocean circulation, radiation, and ice and snow albedo. For this reason, different GCMs may respond quite differently to the same forcing because certain processes and feedbacks may be modeled differently. For example, some GCMs can simulate the Indian summer monsoon rainfall quite well but many GCMs can't do it so well.

22.4 Methodology to Study Impacts of Climate Change on Water Resources

To identify the actions that are needed to overcome the adverse impacts of climate change on water resources, it is necessary to first find out what are the likely impacts. To identify the likely impacts, particularly on water resources, the following methodology can be adopted:

1. Select a GCM that closely simulates the relevant climatic variables for the region of interest. Also, choose a hydrologic model to simulate response of the catchment of interest.
2. Using the likely scenario(s) of future GHG emission, predict the future climate at the place of interest by employing the selected GCM.
3. Downscale (see Sec. 22.5) the relevant GCM variables as per the requirement of the chosen hydrologic model and for the chosen climate change scenario (see Sec. 22.4.1) and area(s) of interest.
4. Calibrate and validate the chosen hydrologic model and apply it to simulate the response of the study of the catchment area under future climatic conditions.
5. Use outputs of the hydrological models as inputs to water management models to update or improve river basin planning, flood management, reservoir operation policy, and so on.

22.4.1 Scenarios and Their Purpose

Future GHG emissions are the result of a very complex system and depend on divergent driving forces, such as demographic and socioeconomic developments, and technological change (IPCC, 2000), where future evolution is highly uncertain. Scenarios are alternative descriptions of what the future might be. These are needed to analyze how the driving forces may influence the future emission of GHGs and aerosols. Thus scenarios assist in assessment of impacts of climate change, adaptation, and mitigation. Since the future is highly uncertain, it is difficult to say which emission path will finally take place.

Based on a detailed analysis of driving forces and emissions in the scenario literature, alternative modeling approaches, and wide participation and feedback, the Special Report on Emissions Scenarios (SRES) team (IPCC, 2000) came out with four narrative storylines. These are termed as A1, A2, B1, and B2 (Fig. 22.5) and describe the relationships between the forces driving GHGs and aerosol emissions and their evolution during the twenty-first century.

Each storyline represents different demographic, technological, economic, social, and environmental development expected in future. The scenarios cover a wide range of main driving forces causing GHG emissions. Each of the four scenarios represents a specific interpretation of a storyline in quantitative terms. A "family" is composed of all the scenarios that are based on the same storyline.

An A1 family refers to a world that will see very rapid economic growth, low population growth, and rapid implementation of new and more efficient technologies. An A2

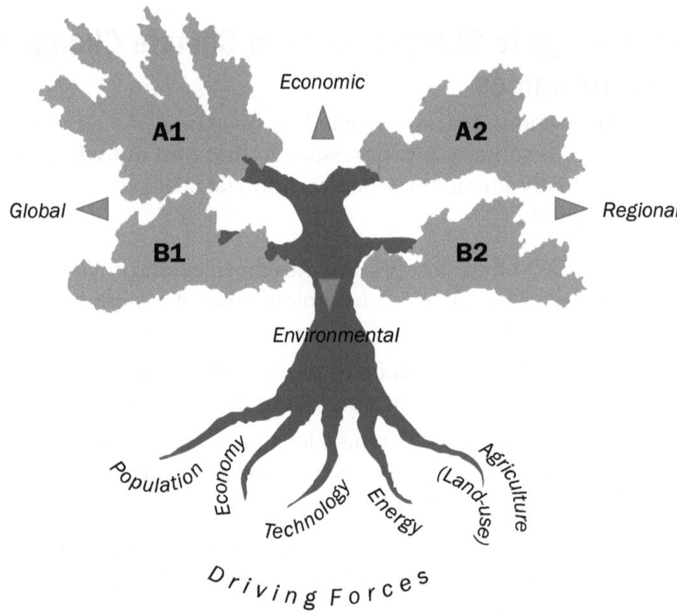

FIGURE 22.5 Four families of climate change scenarios and driving forces, as per IPCC (2000).

scenario family, on the other hand, describes a very heterogeneous world; the focus will be on self-reliance and preservation of local identities. According to B1 family, the world would be convergent and will have low population growth (same as in the case of A1 family), but there will be fast changes in economic structures. There will be reduction in material intensity and there will be more use of clean and resource-efficient technologies (IPCC 2000). A B2 family corresponds to a world in which it gives more importance to local solutions to achieve social, economic, and environmental sustainability. The world of a B2 family has moderate population growth, intermediate levels of economic development, and less fast and diverse technological changes as compared to the B1 and A1 families.

Figure 22.6 shows the main characteristics of all four storylines. Altogether, 40 SRES scenarios have been developed by six modeling teams, and all are equally valid with no assigned probabilities of occurrence. The set of scenarios consists of six scenario groups drawn from the four families: one group each in A2, B1, B2, and three groups within the A1 family, characterizing alternative developments of energy technologies: A1FI (fossil fuel intensive), A1B (balanced), and A1T (predominantly nonfossil fuel). Within each family and group of scenarios, some share "harmonized" assumptions on global population, gross world product, and final energy. These are marked "HS" or harmonized scenarios. "OS" denotes scenarios that explore uncertainties in driving forces beyond those of harmonized scenarios. The number of scenarios developed in each category is also shown. For each of the six scenario groups, an illustrative scenario (always harmonized) is given. Four illustrative marker scenarios, one for each scenario family, were used in draft form in the 1998 SRES open process and are included in revised form in the *IPCC Special Report Emissions Scenarios* (2000). Two additional illustrative scenarios for the groups A1FI and A1T are also provided and complete a set of six that illustrates all scenario groups.

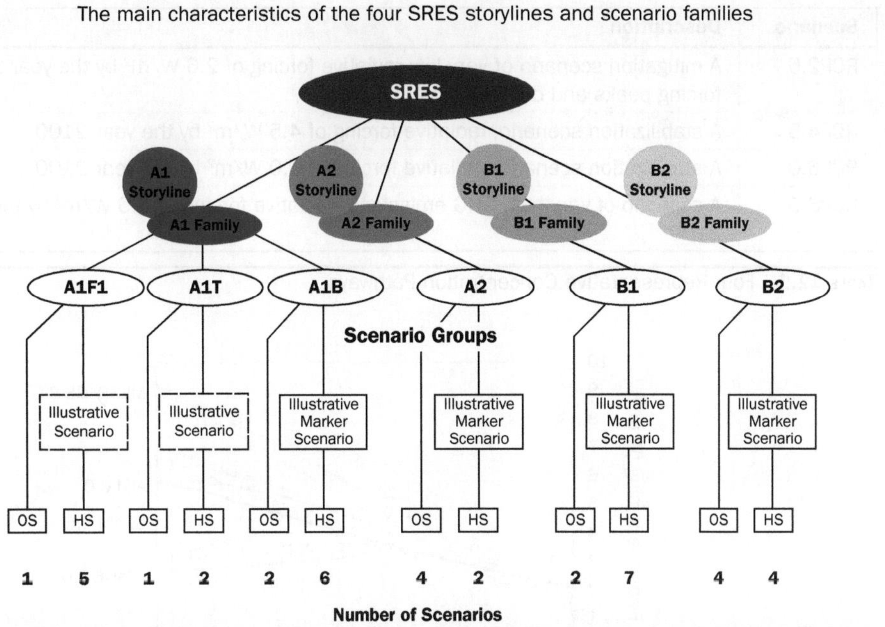

FIGURE 22.6 Schematic illustration of SRES scenarios. Four qualitative storylines yield four sets of scenarios called "families": A1, A2, B1, and B2. (Source: IPCC Special Report Emissions Scenarios 2000.)

22.4.2 Representative Concentration Pathways

Radiative forcing (watt/m^2) is the difference between the solar energy absorbed by Earth and that radiated back to space. If radiative forcing is positive, Earth is receiving more energy than it is radiating back, which will cause warming. Negative radiative forcing means Earth is receiving less energy than it is radiating back, which will cause cooling. Typically, radiative forcing is assessed at the top of the atmosphere. By far, the dominant factor in causing warming is positive forcing from CO_2 followed by methane.

Representative Concentration Pathways (RCPs) are the latest generation of scenarios that are input to climate models (van Vuuren et al., 2011). This term RCP was adopted to emphasize the reason behind the use of these pathways. RCPs are representative— these are one of several possible scenarios that have identical radiative forcing and emission characteristics. RCPs are pathways—their main purpose is to provide time-dependent projections of concentrations of GHG in the atmosphere. A pathway emphasizes that not only a specific long-term concentration or radiative forcing outcome is of interest but the trajectory leading to that concentration is also important. The magnitude of changes in GHG concentrations, land use/land cover, and so on, depends on population growth, changes in technology, economic growth, innovation, urbanization, and so on.

Each RCP was developed by a different independent modeling group; these cannot be directly compared. The main difference between the RCPs and previously developed scenarios is that RCPs are not associated with any fixed assumptions concerning population growth, technology, and economic development. Hence, the same level of radiative forcing can be associated with different socioeconomic futures. This enables

Scenario	Description
RCP2.6	A mitigation scenario of very low radiative forcing of 2.6 W/m² by the year 2100, forcing peaks and declines.
RCP4.5	A stabilization scenario, radiative forcing of 4.5 W/m² by the year 2100
RCP6.0	A stabilization scenario, radiative forcing of 6.0 W/m² by the year 2100
RCP8.5	A scenario of very high GHG emissions, radiative forcing of 8.5 W/m² by the year 2100

TABLE 22.1 Four Representative Concentration Pathways

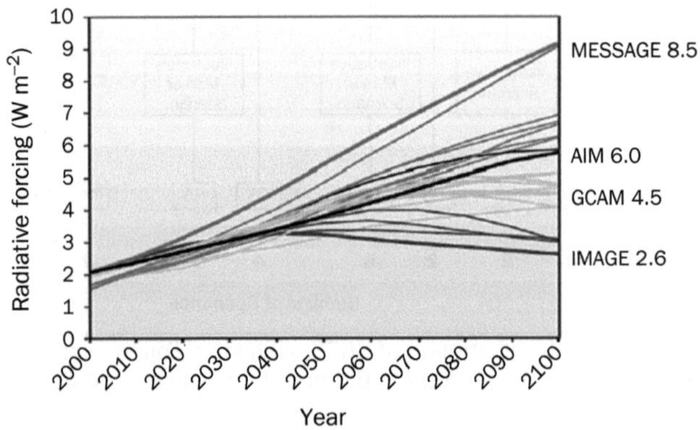

FIGURE 22.7 Changes in radiative forcing relative to preindustrial conditions for four RCPs.

scientists to test various combinations of climate policies and technological, social, and economic positions. For example, at global level, higher population or higher energy consumption could be compensated by higher use of renewable energy.

Four RCPs have been developed and adopted by IPCC for the Fifth Assessment Report (AR5). These are named after approximately total radiative forcing in the year 2100 relative to 1750. Details are shown in Table 22.1.

Each RCP has spatially distributed data sets of land use change and emissions of air pollutants for various sectors. RCPs also specify annual concentrations of GHG and anthropogenic emissions up to the year 2100. For the four RCPs, approximate combined CO_2-equivalent concentrations by the year 2100 set by IPCC are 475 ppm (RCP2.6), 630 ppm (RCP4.5), 800 ppm (RCP6.0), and 1313 ppm (RCP 8.5). Figure 22.7 (drawn from IPCC) shows changes in radiative forcing relative to preindustrial conditions. Bold lines show four RCPs; thin lines show individual scenarios from nearly 30 candidate RCPs and provide information on all major factors that affect radiative forcing.

22.5 Downscaling Climate Data

In climate change impact studies, the time step could vary from very short interval of 10 minutes (for studies on urban floods) to a year (for large river basins). Similarly, the space scales could vary from a few square kilometers (for urban watersheds) to thousands of square kilometers (for large river basins). Prediction of climate change impacts

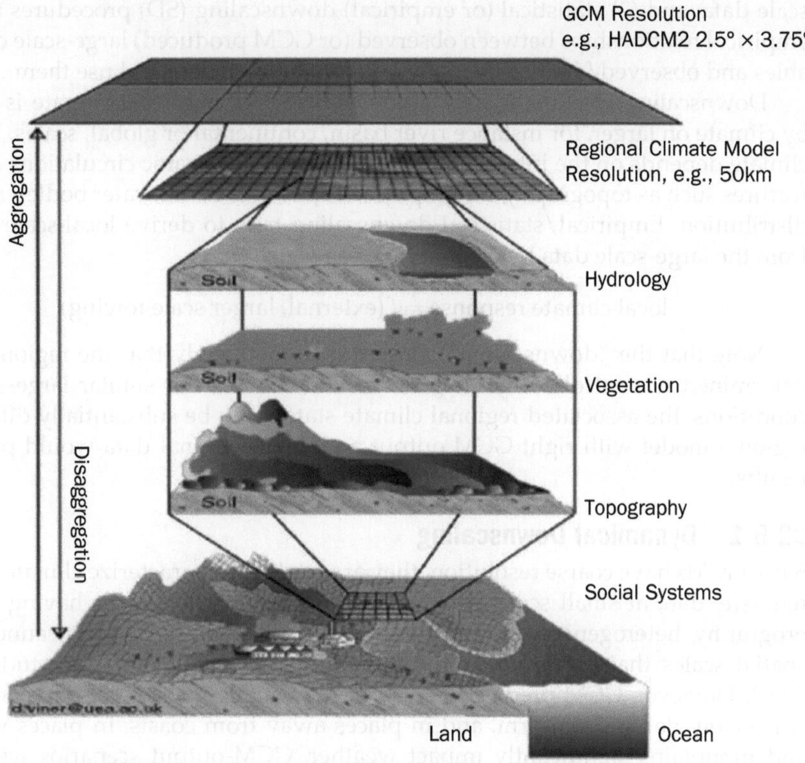

FIGURE 22.8 A schematic illustrating the general approach and need for downscaling. (Source: http://www.ciesin.org/documents/Downscaling_CLEARED_000.pdf.)

at local, regional, and national scales is required to plan adaptation and mitigation. In the water sector, such predictions involve reassessment of water availability, strengthening or planning additional infrastructure, review of design standards, revising reservoir operation policies, etc.

Among the tools available to compute the climatic variables needed for future predictions, GCMs are the best. However, so far, GCMs are unable to reproduce the climatic variables at temporal and spatial scales of relevance to hydrological studies. As noted earlier, GCMs produce outputs at a resolution that is too coarse for climate change impact studies, particularly in water sector. Hydrologic models that are applied to study the impact of climate change may need information at a fine resolution, say less than 10 km. Hence a method is needed to obtain data at the small scale by using the large-scale data.

Downscaling aims to estimate required variables at the small scale (often station level) based on large-scale variables produced by GCMs. Hence, it is a technique to overcome the scale mismatch between what is available (coarse-resolution output) and what is needed (regional or local/catchment scale output) for climate change impact assessment using hydrological models. Figure 22.8 depicts the general approach and need for downscaling.

In impact assessment studies, two broad groups of downscaling techniques are employed: (1) dynamical downscaling (DD) techniques that involve modeling of regional climate by use of large-scale GCM output and extraction of regional/local scale data, and (2) statistical (or empirical) downscaling (SD) procedures that develop empirical relationships between observed (or GCM produced) large-scale climatic variables and observed/analyzed small-scale (or stations) data and use them.

Downscaling of climatic data assumes that regional/local climate is conditioned by climate on larger, for instance river basin, continental or global, scales. The regional climate depends on the interaction of atmospheric or oceanic circulations and regional features such as topography, orography, land use and cover, water bodies, and land-sea distribution. Empirical/statistical downscaling tries to derive local-scale information from the large-scale data by using a function f such that

$$\text{local climate response} = f\,(\text{external, larger scale forcing}) \qquad (22.1)$$

Note that the "downscaling" concept does not imply that the regional climate is determined or controlled by large-scale state climate. For similar large-scale climate conditions, the associated regional climate states may be substantially different. But a regional model with right GCM output and right regional data would provide good results.

22.5.1 Dynamical Downscaling

Since GCMs have coarse resolution, they are unable to characterize climate and provide requisite data at small scale in many regions, more so in places having pronounced orography, heterogeneous land cover, or coasts. Therefore, "GCMs cannot access the spatial scales that are required for climate impact and adaptation studies" (WMO, 2002). However, GCM predictions may be reasonably good where the land surface is reasonably flat and uniform, and in places away from coasts. In places where coasts and mountains significantly impact weather, GCM-output scenarios will be unable to capture the local details needed for impact assessment. Also, at coarse resolutions, short-time extreme events, such as cyclones or heavy rainfall, are either not captured or may be incorrectly modeled.

Dynamical downscaling (DD) uses regional climate models (RCMs) to simulate finer-scale physical processes consistent with large-scale weather simulated by a GCM. In DD, a regional climate model (RCM) is run by using GCM output as initial and boundary conditions over the region of interest. A typical GCM has horizontal resolution of hundreds of kilometers. In contrast, regional models can resolve small size features; they have a more accurate representation of many surface features, such as mountains and coastlines.

An RCM has a high resolution (typically 50 km or less) and covers a limited area of the globe. It is a comprehensive mathematical model in which all the important climate system processes are represented. RCMs are models for limited areas. At the boundaries of study area, time series data of large-scale fields, such as temperature, water vapor, wind, and pressure, are provided which may have been obtained using measuring instruments or a GCM. RCM predictions of medium or long term are needed to generate robust climate statistics, for example, distributions of daily rainfall or intraseasonal variability.

Nested modeling shown in Fig. 22.9 is an approach to overcome the limitations due to coarse spatial resolution of GCMs. In this approach, the model of small area at fine resolution is linked with a model of large area at coarse resolution to provide a detailed representation of local conditions. Results of the large-area model (GCM) provide initial and boundary conditions and input variables to derive the high-resolution

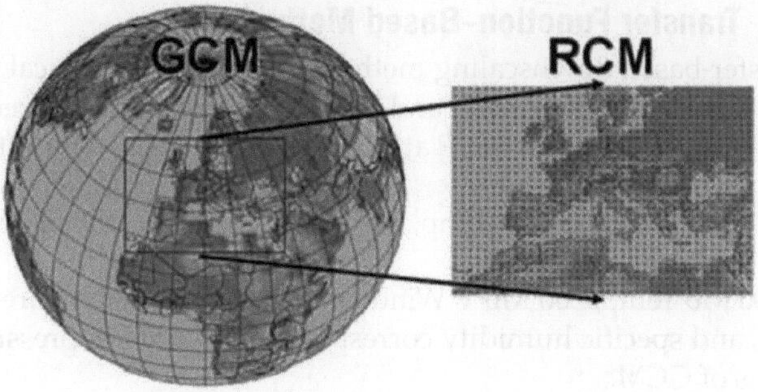

Figure 22.9 Regional Climate Model nesting approach. (Source: Ethiopian Panel on Climate Change, 2015.)

local model or RCM. Since the resolution of RCM is much finer, depiction of topography and orography is more detailed and realistic, and parameterization is better. Due to this, RCM can be used for more detailed and small-scale modeling of regional climatology. This approach is valuable for mountains and coastal areas where complex topography cannot be correctly represented in coarse-resolution GCMs. Many RCMs have been developed by modeling teams working in different countries around the world.

Regarding advantages, RCMs simulate current climate more realistically and predict climate change with more detail. RCMs also allow realistic representation of small islands and peninsulas. In a global model, due to small size, their climate would be that of the surrounding ocean. Finer resolution RCMs are able to simulate these realistically. RCMs are much better at simulating extremes of weather. A limitation of RCMs is that their predictions depend on the correctness of the global model whose data are used. Errors in the GCM predictions will be reflected in the RCM predictions.

22.5.2 Statistical (or Empirical) Downscaling

Statistical (or empirical) downscaling (SD) methodologies use statistical techniques for downscaling. According to the mathematical function used to describe the predictor-predictand relationship, the computational technique used, or the selection of the predictor variables, the approaches can be classified into three categories: transfer function–based methods, stochastic weather generators, and weather typing. The past observations used in SD should cover a wide range; to the extent feasible, the expected future variations in the predictors should be covered in the data used to set up the model. Three implicit assumptions in statistical downscaling are as follows:

1. The predictors are variables of relevance and are realistically modeled by the GCM at the scale considered.
2. The transfer function is valid also under altered climatic conditions. This assumption cannot be verified in advance.
3. The predictors employed fully represent the climate change signal.
4. The local-scale surface meteorological variables are a function of synoptic forcing.

22.5.3 Transfer Function-Based Methods

The transfer-based downscaling methods rely on the empirical statistical relationships between large-scale predictors and local-scale variables or predictands. For example, predictand daily precipitation P at Roorkee (T_{Rke}) can be computed as

$$T_{Rke} = f(200 \text{ Mb Temp}, 200 \text{ Mb V Wind}, 200 \text{ Mb SpecHum}) \tag{22.2}$$

where 200 Mb Temp, 200 Mb V Wind, and 200 Mb SpecHum are the temperature, vertical wind, and specific humidity corresponding to 200 Mb pressure. These are obtained as outputs of GCM.

A simple solution is to develop a multiple regression model by relating the values of GCM outputs at grid points to required hydrometeorological variables. An alternative to linear regression is to use nonlinear regression. In general, the main advantage of the regression technique is simplicity, computationally less demand, and intuitive appeal. However, this method can be applied only when a good predictor-predictand relationships could be found. This may not be possible in all conditions. Two emerging nonlinear approaches are the artificial neural networks (ANN) and the Support Vector Machines (SVM).

22.5.4 Weather Generators

A weather generator produces a synthetic time series of weather data for a location by using the statistical properties of observed weather at that place. Stochastic weather data is generated in two steps: (1) modeling daily precipitation, and (2) modeling the other variables of interest, such as daily maximum and minimum temperature, humidity, radiation, and wind speed. These generated data show the conditional probability of precipitation. Seasonality and cross-correlations among variables may be addressed by using different models/parameters for different seasons/months to take care of seasonal variability.

22.5.5 Weather Typing

In this approach, the atmospheric circulation pattern is classified into several classes. Weather types are simulated by stochastic models and linkages between rainfall occurrence and weather types are set by conditional probabilities. Next, rainfall and other variables are generated by using weather types. This approach can link climate on the large scale with weather at local scale. Long data series can be generated based on short historical data.

22.5.6 Predictors and Predictands in Statistical Downscaling

In SD in hydrology, most commonly, predictands are precipitation and temperature, as well as monthly, seasonal, or annual statistics of climatic variables. While selecting predictor variables, all important variables and those that may influence the behavior of predictands should be included. One may consider that a predictor may not be significant in the downscaling model under present condition, but its influence under the future climate may be important. For example, the downscaled scenario of future precipitation may significantly change if humidity is included as a predictor.

SD methods have several advantages over DD methods. SD methods can be easily adapted to specific study. These are inexpensive to apply and have modest computing requirements. Several SD models have been developed and applied to relate local weather variables and GCM outputs.

22.6 Climate Change: Adaptation and Mitigation

According to IPCC, "adaptation to climate change refers to the adjustment in natural or human systems in response to actual or expected climatic stimuli or their effects, which moderate, harm, or exploit beneficial opportunities." Adaptation activities have been categorized into several different types. Anticipatory or proactive adaptation is initiated before impacts of CC are observed. Autonomous or spontaneous adaptation is not a conscious response to climatic stimuli; it is triggered by changes in natural system and by market or welfare changes in the society. Planned adaptation is the result of a deliberate policy decision, based on an awareness that conditions have changed or are about to change and that action is required to return to, maintain, or achieve a desired state. Adaptive capacity is the potential to adjust so that the negative impacts are minimized and benefits from changes are maximized.

Adaptation is basically an understanding of how individuals, groups, and natural systems can prepare for and respond to changes in climate. Adaptation is crucial to reduce vulnerability to climate change. A successful adaptation can reduce vulnerability by building on and strengthening existing coping strategies.

IPCC defines mitigation as "An anthropogenic intervention to reduce the sources or enhance the sinks of greenhouse gases (GHG)" (IPCC, 2014a). Regarding CC, mitigation is any action taken to either permanently eliminate or reduce the risk and hazards of CC to human life and property in long-term. Since atmospheric processes have teleconnections, mitigation of CC is a responsibility of all. Note that mitigation tackles the causes of CC and adaptation takes care of the effects. Obviously, if mitigation is taken up vigorously, the adverse impacts to which the society will be required to adjust will be lesser and the society will be exposed to less risks. If mitigation is less, CC will be greater and more adaptation will be needed. However, if the adaptation is more, the impacts felt with given level of climate change will be softer. Hence, mitigation measures should commence at the earliest or the reduction in emission of GHGs should urgently start. Climate mitigation and adaptation are not alternatives to each other, they are combined set of actions in a comprehensive strategy to fight CC.

22.6.1 Adaptation and Mitigation Strategies

All adaptation activities are not uniformly good. If adaptation is improperly carried out, it may increase vulnerability rather than reducing. Such actions are called maladaptation. While implementing adaptation, checks should be carried out to identify and delete maladaptive practices. It is also necessary that the exposure of various groups to risk is carefully managed. For example, risk to farmers can be reduced or transferred through crop insurance or diversified livelihoods, such as integrated agriculture–animal husbandry or agriculture–aquaculture systems. If such provisions are made, farmers can shift their focus if there is short-term fall in the availability of water to grow crops (drought). Safety nets will be required where the possibilities of diversification are limited.

Key messages regarding adaptation are as follows:

1. The world needs to adapt to climate change in water management without delay.
2. Water has a central role in a large number of sectors, e.g., agriculture and energy generation, which require water in certain quantity and of certain quality. Hence, proper water management can enhance or restrict adaptation in related sectors.
3. Implementing integrated water resources management supports adaptation.
4. Water resources are under pressure due to many reasons. Adaptation policy needs to take a holistic view and consider climate change as one of the many stresses.
5. Earth has 263 transboundary lake and river basins. Hence, transboundary cooperation will be of high value in adaptation programs.

IPCC has listed seven sectors of mitigation: energy supply, transportation and its infrastructure, residential and commercial buildings, industry, agriculture, forestry, and waste management. Many water-related activities, such as water supply for drinking, irrigation, and wastewater treatment, consume large amount of energy. Energy saving here can reduce emission of GHGs. Besides, cultivation of bio-energy crops and afforestation will require water and have the potential to reduce streamflow and deplete groundwater resources. Thus it is important to factor-in the influence of mitigation measures on water resources and their management at the planning stage.

Reduction of GHG emissions through saving in energy and switching to energy sources that do not produce GHG is at the core of most mitigation ideas. Energy consumption can be reduced by increasing fuel efficiency of vehicles, changes in lifestyle, and more efficient production processes. Large-scale deployment of renewable energy plants that produce hydropower, solar power, tidal and ocean power, geothermal power, and wind power can help in reduction of GHG emission. Nuclear power projects can also be helpful, although this technology has become controversial due to safety issues. Economic instruments such as carbon credits and green taxes are aimed to promote use of cleaner technology. Geoengineering techniques range from carbon sequestration (the process of removal of carbon dioxide from atmosphere) projects to solar radiation management schemes such as the creation of stratospheric sulfur aerosols.

Besides reduction in emission of GHGs, increase in carbon sequestration is also a mitigation measure and can be implemented through management of forests, agricultural areas, and wetlands.

22.6.2 Land Use Management

Many land use management activities may form a part of mitigation plans and will impact water resources development and management. Practices for soil conservation, such as reduced tillage, afforestation, planting perennial crops, and crop management, also help in water conservation. But if not done carefully, these may result in some adverse effects on water resources, such as more nutrients or pesticides flowing in surface water and groundwater.

Land use planning involves disciplines of ecology, economics, and social sciences. Its aim is to meet current and future needs of society for the utilization and conservation of land and natural resources. This is a specialized task. Guidance manuals providing frameworks for development of low-emission land use plans are available, for example, Barber et al (2015).

22.6.3 Afforestation and Reforestation

Plants are sinks of carbon since they take up carbon dioxide to carry out photosynthesis. Therefore, an increase in forest area is a way to mitigate climate change, which yields many other benefits. It also helps in improving environment. But it should noted that the role of forests in the hydrologic cycle is highly context dependent and there are several myths concerning forests. Depending upon the species and growth stage, in general, forests consume more water by evapotranspiration as compared to many crops or grass. Therefore, after extensive afforestation, water consumption from the lands is likely to increase and significantly impact hydrologic cycle. Newly planted trees use more water than the mature trees. Trees in dry climates may have root systems to extract water during prolonged dry seasons. Hence extensive afforestation or reforestation in the dry regions can significantly impact river flows and groundwater.

Afforestation and reforestation may also have beneficial hydrological effects. After afforestation in wet areas, the amount of direct runoff initially decreases rapidly, then gradually becomes constant, and baseflow increases slowly as the age of trees increases (Metz et al. 2001). Further, reforestation and afforestation help in reducing intensity of small floods and increase infiltration of water but do not have significant impact on flows when rainfall is intense. A substantial rise in water consumption in forests will reduce availability for other sectors, including ecosystems. This will not be an issue in the places where surplus water is left after meeting all demands.

Afforestation and reforestation activities are commonly placed high in the climate change mitigation agenda. However, recent research suggests that addition of new forest in mid-latitudes decreased the albedo (reflectivity of Earth's surface) and increased local and global temperature. In view of concerns, there is a need for a holistic approach to afforestation and reforestation by considering all aspects including carbon sequestration, biodiversity, soil conservation, and livelihood issues in a coordinated way.

22.7 Impacts of Climate Change on Water Resources

Many hydrological processes are impacted by temperature changes in different ways. A warmer climate may intensify the hydrological cycle and lead to higher rates of evaporation, increase in liquid precipitation, more frequent occurrences of extreme events, and changes in spatial and temporal patterns of precipitation. Climate change may affect the spatial and temporal distribution of runoff, soil moisture, groundwater storage, and so on. Global warming will also have impacts on water quality as many chemical and biological processes depend on temperature of water and air.

22.7.1 Impacts on Surface Water and Groundwater

Many rivers across the world carry substantial contribution from the snow and glacier melt. Changes in the form of precipitation (liquid or solid), temperature pattern, humidity, and other climatic variables will change the amount and distribution of runoff in rivers. The flow regime in the Himalayan rivers, for instance, is expected to alter due to climate change because higher fraction of precipitation will fall as rain in warmer climate and seasonality of river will change due to early beginning of snow and ice melt. Changes in runoff regime of mountainous rivers would directly impact drinking water, irrigation, energy generation, and other uses of water. Faster glacier melt under warm climate would lead to retreat of glaciers. In case the glaciers retreat significantly, summer flows in some river systems will increase for a few decades, followed by a reduction in flow.

Rural and urban water supply and agriculture in many countries critically depend on groundwater. Increased rainfall intensity would lead to higher runoff and reduced groundwater recharge. Increased frequency and severity of floods may affect groundwater quality in alluvial aquifers. Warmer climate is likely to increase evaporation. Rising sea water levels may lead to increased saline water intrusion into coastal and island aquifers. Large depletion in groundwater reserves and quality of their waters would adversely impact water and food security in vast swath of areas.

Frequency of extreme events (floods and droughts) may increase in many nations during the twenty-first century. Changes in the amount and pattern of rainfall would affect demand for water. More severe flooding and droughts and higher flood water levels would cause substantial damages to key economic sectors: agriculture, infrastructure, and housing. The high-magnitude floods are also likely to bring more sediments which may settle in reservoirs. Although floods and droughts affect all sections of the society, the rural and urban poor are the first and hardest hit and need support.

The projected climate changes will alter the relative magnitudes of various components of water balance in various river basins of the world. All water users including farmers will have to adapt in view of these changes. It would be prudent to incorporate possible effects of climate change in the design and management of water resources systems. Similarly, the managers of water resources systems will also need to modify practices accordingly. Although the uncertainties are high at present, the possibility of changes in such extreme events is alarming. National standards for hydro-project designs as well as for flood and drought control schemes will have to be revised keeping in view the changes in frequency and severity of extreme events.

22.7.2 Adaptation Needs in the Water Sector

Climate change will influence the Earth's ecosystem mainly through water. Although CC will impact most nations, the consequences are expected to be more severe in those developing countries whose economies are chiefly dependent on agriculture, infrastructure is weak, and natural resources are stressed due to increasing demands for freshwater, energy, and food. Island nations may face inundation due to rising sea water levels. Hence, adaptation in the water sector with a long-term view is an urgent necessity that requires policy shifts and significant investments.

For long-term resilience, a necessary requirement is to create adequate water infrastructure and stronger institutions. The following adaptation actions may be taken up in the water sector to check the adverse impacts of climate change:

1. Strengthen hydrometeorological data observation network and databases. Hydrological knowledge and information should be shared for better analysis and decisions.
2. Compute river-basin-wise water availability and demands in current and future scenarios.
3. Generate reliable downscaling of relevant hydrometeorological variables from GCM results to regional and watershed levels.
4. Assess the impact of climate and other changes on surface water and groundwater resources.
5. Assess the impact of climate change on rainfall intensity-duration-frequency relationships and on frequency and magnitude of floods.
6. Assess the impact on magnitude-duration-frequency of droughts (agricultural, meteorological, and hydrological) and develop adaptation plans.
7. Assess the impact of these changes on soil erosion, sediment loads, and water quality and initiate suitable management actions.
8. Review hydrological planning, design, and operating standards in view of changed scenarios and non-stationarities in data.
9. Develop adequate infrastructure to cope up with enhanced variability in the water sector including floods and droughts.
10. Implement mainstream adaptation within the overall development context and strengthen water governance.

22.8 Summary

Variability is an inherent feature of the climate system but the recently observed and projected changes are worrying. Analysis based on observed and reconstructed data by IPCC and various groups, including the heat content of oceans, shows that Earth is receiving positive external radiative forcing. As a result of dedicated efforts worldwide, scientific understanding of the causes and consequences of climate change has greatly improved in the past few years. It may also be added that not everyone has the similar views about CC. A group of people believe and climate change is either not occurring, or is not real, or is not caused by humans.

Like any other scientific field, research on climate change has some fundamental gaps. Schiermeier (2010) has listed four areas—regional climate forecasts, precipitation forecasts, aerosols, and palaeoclimate data—that require greater open discussion. Also termed as "holes" in present-day understanding, this may mean that future climate changes could be worse than currently projected.

This chapter has provided basic understanding of the complex subject of climate change. As new discoveries are being made frequently, interested readers are advised to update themselves regularly.

22.9 Questions

22.1 Check the definitions of climate change provided by the IPCC and the United Nations Framework Convention on Climate Change (UNFCC). What are the similarities and differences in the two definitions?

22.2 Collect data on mass balance or spatial extent of a few glaciers from the area of your interest. Analyze these to find the changes in recent times and try to ascertain the causes.

22.3 Pick up a GCM from among the models used in CMIP5—Coupled Model Intercomparison Project Phase 5 (https://cmip.llnl.gov/cmip5/) and study it.

22.4 What do you understand by representative concentration pathways?

22.5 List key impacts of climate change on water resources and briefly describe them.

22.6 What are the adaptation options in the water sector to check the adverse impacts of climate change?

22.10 References

Barber, J., P. Stephen, D. Saah, and P. C. Pham. Guidance on low emission land use planning. Published by USAID (http://ledsgp.org/wp-content/uploads/2015/08/Guidance-on-low-emission-land-use-planning2015.pdf), 2015.

Ethiopian Panel on Climate Change. *First Assessment Report, Working Group I Physical Science Basis*. Technical Report. Addis Ababa: Ethiopian Academy of Sciences, 2015.

IPCC. What is a GCM? *IPCC Data Distribution Cen*tre. 2013. http://www.ipcc-data.org/guidelines/pages/gcm_guide.html._

IPCC. Annex II: Glossary [Mach, K.J., S. Planton and C. von Stechow (eds.)]. In *Climate Change 2014: Synthesis Report. Contribution of Working Groups I, II and III to the Fifth Assessment Report of the Intergovernmental Panel on Climate Change*, edited by R. K. Pachauri and L. A. Meyer, 117–130. Geneva: Intergovernmental Panel for Climate Change, 2014a.

IPCC. *Climate Change 2014 Synthesis Report: Summary for Policymakers*. Geneva: Intergovernmental Panel for Climate Change, 2014b.

IPCC, SRES, N. Nakićenović, and R. Swart (eds.). *Special Report on Emissions Scenarios: A Special Report of Working Group III of the Intergovernmental Panel on Climate Change*. Cambridge: Cambridge University Press, 2000.

Metz, B., O. Davidson, R. Swart, and J. Pan (eds.). *Climate Change 2001: Mitigation: Contribution of Working Group III of Intergovernmental Panel on Climate Change*. Cambridge: Cambridge University Press.

National Academy of Sciences. *America's Climate Choices: Figure 3*. 2012. https://nas-sites.org/americasclimatechoices/files/2012/10/Figure-3.png.

NOAA. NOAA in situ carbon dioxide measurements. *Earth System Research Laboratory, Global Monitoring Division*. 2018. https://www.esrl.noaa.gov/gmd/obop/mlo/programs/esrl/co2/co2.html.

Schiermeier, Q. The real holes in climate science. *Nature*, 463:284–287, 2010.

van Vuuren, D. P., J. Edmonds, M. Kainuma, K. Riahi, A. Thomson, K. Hibbard, G. C. Hurtt, et al. The representative concentration pathways: An overview. *Climatic Change*, 109:5, 2011. doi:10.1007/s10584-011-0148-z.

WMO. *Atmospheric RCMs: A Multipurpose Tool? Report of the Joint WGNE/WGCM ad hoc Panel on Regional Climate Modelling*. Geneva: World Meteorological Agency, 2002.

Index

A

B

C

F

G

H

I

K

L

M

N

O

P

9 781259 64197

Printed by Libri Plureos GmbH in Hamburg,
Germany